T0331766

Nonparametric Estimation under Shape Constraints

This book treats the latest developments in the theory of order-restricted inference, with special attention to nonparametric methods and algorithmic aspects. Among the topics treated are current status and interval censoring models, competing risk models and deconvolution. Methods of order-restricted inference are used in computing maximum likelihood estimators and developing distribution theory for inverse problems of this type.

The authors have been active in developing these tools and present the state of the art and open problems in the field. The earlier chapters provide an introduction to the subject, while the later chapters are written with graduate students and researchers in mathematical statistics in mind. Each chapter ends with a set of exercises of varying difficulty. The theory is illustrated with the analysis of real-life data, which are mostly medical in nature.

PIET GROENEBOOM is Professor Emeritus of Statistics at Delft University of Technology, the Netherlands.

GEURT JONGBLOED is Professor of Statistics at Delft University of Technology, the Netherlands.

CAMBRIDGE SERIES IN STATISTICAL AND PROBABILISTIC MATHEMATICS

This series of high-quality upper-division textbooks and expository monographs covers all aspects of stochastic applicable mathematics. The topics range from pure and applied statistics to probability theory, operations research, optimization, and mathematical programming. The books contain clear presentations of new developments in the field and also of the state of the art in classical methods. While emphasizing rigorous treatment of theoretical methods, the books also contain applications and discussions of new techniques made possible by advances in computational practice.

A complete list of books in the series can be found at www.cambridge.org/statistics.
Recent titles include the following:

13. *Exercises in Probability*, by Loïc Chaumont and Marc Yor
14. *Statistical Analysis of Stochastic Processes in Time*, by J. K. Lindsey
15. *Measure Theory and Filtering*, by Lakhdar Aggoun and Robert Elliott
16. *Essentials of Statistical Inference*, by G. A. Young and R. L. Smith
17. *Elements of Distribution Theory*, by Thomas A. Severini
18. *Statistical Mechanics of Disordered Systems,* by Anton Bovier
19. *The Coordinate-Free Approach to Linear Models*, by Michael J. Wichura
20. *Random Graph Dynamics*, by Rick Durrett
21. *Networks*, by Peter Whittle
22. *Saddlepoint Approximations with Applications*, by Ronald W. Butler
23. *Applied Asymptotics*, by A. R. Brazzale, A. C. Davison and N. Reid
24. *Random Networks for Communication*, by Massimo Franceschetti and Ronald Meester
25. *Design of Comparative Experiments*, by R. A. Bailey
26. *Symmetry Studies*, by Marlos A. G. Viana
27. *Model Selection and Model Averaging*, by Gerda Claeskens and Nils Lid Hjort
28. *Bayesian Nonparametrics*, edited by Nils Lid Hjort et al.
29. *From Finite Sample to Asymptotic Methods in Statistics*, by Pranab K. Sen, Julio M. Singer and Antonio C. Pedrosa de Lima
30. *Brownian Motion*, by Peter Mörters and Yuval Peres
31. *Probability (Fourth Edition)*, by Rick Durrett
32. *Analysis of Multivariate and High-Dimensional Data*, by Inge Koch
33. *Stochastic Processes*, by Richard F. Bass
34. *Regression for Categorical Data*, by Gerhard Tutz
35. *Exercises in Probability (Second Edition)*, by Loïc Chaumont and Marc Yor
36. *Statistical Principles for the Design of Experiments*, by R. Mead, S. G. Gilmour and A. Mead
37. *Quantum Stochastics*, by Mou-Hsiung Chang

Nonparametric Estimation under Shape Constraints

Estimators, Algorithms and Asymptotics

Piet Groeneboom
Delft University of Technology

Geurt Jongbloed
Delft University of Technology

Shaftesbury Road, Cambridge CB2 8EA, United Kingdom

One Liberty Plaza, 20th Floor, New York, NY 10006, USA

477 Williamstown Road, Port Melbourne, VIC 3207, Australia

314–321, 3rd Floor, Plot 3, Splendor Forum, Jasola District Centre, New Delhi – 110025, India

103 Penang Road, #05–06/07, Visioncrest Commercial, Singapore 238467

Cambridge University Press is part of Cambridge University Press & Assessment,
a department of the University of Cambridge.

We share the University's mission to contribute to society through the pursuit of
education, learning and research at the highest international levels of excellence.

www.cambridge.org
Information on this title: www.cambridge.org/9780521864015

First published 2014

A catalogue record for this publication is available from the British Library

Library of Congress Cataloging-in-Publication data
Groeneboom, P.
Nonparametric estimation under shape constraints : estimators, algorithms and asymptotics / Piet Groeneboom,
Technische Universiteit Delft, The Netherlands, Geurt Jongbloed, Technische Universiteit Delft, The Netherlands.
pages cm. – (Cambridge series in statistical and probabilistic mathematics ; 38)
Includes bibliographical references and index.
ISBN 978-0-521-86401-5 (hbk.)
1. Nonparametric statistics. 2. Multivariate analysis. 3. Mathematical statistics. 4. Estimation theory.
I. Jongbloed, Geurt, 1968– II. Title.
QA278.8.G76 2014
519.5′4–dc23 2014022148

ISBN 978-0-521-86401-5 Hardback

Additional resources for this publication at http://statistics.tudelft.nl/CUPbook

Contents

Preface and Acknowledgments

Research on nonparametric estimation under shape constraints started in the 1950s. Papers such as Ayer et al., 1955, and Van Eeden, 1956, appeared on estimation of functions under the restriction of monotonicity or unimodality, more generally called isotonic estimation. An isotonic estimator is an estimator that is computed under an order restriction, where the order can be a partial order. The order restriction can also be imposed on the derivative of the estimator, so an estimator of a convex function (in dimension one or higher), which is itself also convex, is also called an isotonic estimator.

A summary of the early work was given in the well-known book by Barlow et al., 1972, on isotonic regression. Originally, the focus was on defining and constructing estimators satisfying these order constraints. As an example, in Grenander, 1956, it is shown that the (nonparametric) maximum likelihood estimator (MLE) of a monotone decreasing density can be constructed as the left-continuous slope of the least concave majorant of the empirical distribution function. Developing asymptotic distribution theory for these isotonic estimators turned out to be rather difficult. Nonnormal limit distributions appear and rates of convergence are slower than the square root of the sample size. This behavior is now commonly classified as belonging to the area of nonstandard asymptotics. In the case of the mentioned Grenander MLE, the rate of convergence of this estimator (evaluated at a fixed point, under some local assumptions) is the cube root of the sample size. Moreover, the nonnormal asymptotic distribution of the estimator is (after rescaling) the so-called Chernoff distribution, which is (up to a factor 2) the distribution of the derivative of the greatest convex minorant of two-sided Brownian motion with parabolic drift, evaluated at zero.

Research on isotonic regression received new impetus in the 1990s when it became clear that it was the right setting for studying (nonparametric) MLEs of the distribution function in inverse problems. Examples of such problems include interval censoring models such as the current status model, deconvolution problems and the classical Wicksell corpuscle problem. Current status data or interval censored data are quite common in medical research, but are also relevant for econometric models such as the binary choice model. In the context of the current status model, the same (Chernoff) limit distribution appears for the MLE of the (by definition) monotone distribution function as for the Grenander estimator of a monotone density. Whether this Chernoff distribution also gives the (pointwise) limit behavior of the MLE of the distribution function in the more general interval censoring problem or a large class of deconvolution problems is still an open question. Very specific conjectures for the convergence to this limit distribution have been formulated, though.

As mentioned, the Chernoff distribution is the distribution of a functional of Brownian motion. Also, other functionals of Brownian motion appear in the limit theory for nonparametric estimators, for example, in the situation of estimating a convex function and its derivative at a fixed point. The local limit of a nonparametric least squares estimator of a smooth convex regression function can be characterized as the second derivative of an "invelope" of integrated Brownian motion plus the 4th power of the time variable, instead of Brownian motion with parabolic drift, which figured in the limit distribution of monotone estimators. The estimators have pointwise rates of convergence $n^{2/5}$ for the convex function and $n^{1/5}$ for its derivative, rather than the $n^{1/3}$ occurring in the situation of estimating a monotone function.

It is the purpose of this book to introduce the subject of shape-restricted statistical inference, to present the current state of the theory and also to describe still open problems. The subjects covered include those discussed in part 2 of the book by Groeneboom and Wellner, 1992. That book is still available in a Kindle edition, but a lot of theory has been developed since 1992. As an example, in applying the maximum likelihood theory in inverse (often medical) problems, the maximum likelihood estimator will usually be a piecewise constant jump function, so estimation of a hazard or density function is only possible after some kind of smoothing. Theory about this, and theory about the smoothed maximum likelihood estimator (SMLE) and the maximum smoothed likelihood estimator (MSLE), has only recently been developed and is discussed in the present book.

Another direction of considerable progress has been the analysis of so-called smooth functionals in inverse problems. Although the local rate of convergence of the MLE in the inverse problems is usually slower than $n^{1/2}$, there are often smooth functionals of moment type that can be estimated at rate $n^{1/2}$ with normal limit distributions. This theory, which is far from complete, is treated in this book. It depends on properties of solutions of certain integral equations, which in many situations do not have explicit solutions.

Also, theory has been developed for testing problems in models with shape constraints. Examples are the two- and k-sample tests based on interval censored data. In contrast with testing theory in the presence of right censored data, the theory of these problems has had a very slow start. The main challenge is to construct test statistics that have distributions that do not depend on the observation time distributions in the samples. Such tests are presented in this book and compared with tests not having this property. The techniques used here are different from both the theory used for the local limits and the smooth functional theory. Moreover, various forms of the bootstrap play a very important role in determining the critical values for these tests.

We also discuss confidence intervals for distribution functions, densities and hazards, constructed by bootstrap procedures. These are compared with intervals, based on plug-in estimators for the variance of the asymptotic distribution and intervals based on likelihood ratio tests, using the MLE (which only exist for distribution functions and not for densities or hazards). The confidence intervals also are computed for some real-life data sets, such as the Bangkok cohort data, which were kindly provided to us by the researchers in the Bangkok Metropolitan Administration Injecting Drug Users cohort study and Michael Hudgens (see the acknowledgments). We also constructed confidence intervals for the hepatitis A data, provided to us by Niels Keiding.

Throughout the years, we have written computer programs related to the subject of this book. We will make relevant programs (some of these need some rewriting) public via the website http://statistics.tudelft.nl/CUPbook.

The book can be used as the textbook for an advanced undergraduate course. Working knowledge of basic probability theory and mathematical statistics is assumed. Chapters 1 through 7 are rather general and focus on modeling of data and the derivation of nonparametric estimators for functions within these models. Parts of Chapter 3 could be skipped for an undergraduate course. Chapter 7 focuses on algorithms that can be used to compute the shape-constrained estimators in particular models. The later chapters are more technical and can be used as ingredients of a graduate course. These chapters can certainly be used to define concrete research projects in the area. Every chapter concludes with a number of exercises of varying levels. Some of these are meant to fill in details of arguments given in the text. Others extend results obtained in the text or present additional results.

Apart from being used as a textbook, we hope the book will also be used by colleagues and inspire them to work in the field of shape-restricted statistical inference. We also hope that people involved in medical statistics, econometrics and other fields of application will use the book to learn about shape-restricted statistics and benefit from the progress made in this field.

Acknowledgments

We would like to thank Diana Gillooly for her support in the process of writing this book. Also we thank our co-authors of various papers in the area of shape-constrained estimation, including Lutz Dümbgen, Bert van Es, Ronald Geskus, Gerard Hooghiemstra, Vladimir Kulikov, Steven Lalley, Rik Lopuhaä, Marloes Maathuis, Frank van der Meulen, Stefanie Michael, Nico Temme, Jon Wellner and Birgit Witte.

We also wish to express our thanks to Niels Keiding for providing us with various data sets and Kachit Choopanya, Dwip Kitayaporn, Timothy Mastro, Philip Mock and Suphak Vanichseni for allowing us to use the data from the Bangkok Metropolitan Administration Injecting Drug Users cohort study.

Piet Groeneboom and Geurt Jongbloed
Delft, 2014

1

Introduction

To give a feeling of what this book is about, it is perhaps best to take a look at some real-life examples. Real-life examples have the disadvantage of giving rise to a lot of discussion on the interpretation of the data, as the authors have experienced when they started a lecture with a real-life example. This often distracted the audience from the main message of the lecture. But they have the advantage of "sticking in the mind," which might be more important than the temporary distraction they might cause. Therefore, the first four sections of this chapter are about real data. Section 1.1 is concerned with the estimation of the expected duration of ice (in days) at Lake Mendota in Wisconsin, assuming these expected durations decrease in time. In Section 1.2, a data set on time-till-onset of a nonlethal lung tumor for mice is studied. There are two groups of mice, one living in a conventional environment and the other in a germ-free environment. The main question then is whether the distribution of the time-till-onset of the tumor is affected by the choice of environment. The complication is that the times of onset are not precisely observed, but subject to censoring. The third example, in Section 1.3, concerns the estimation of a relatively complicated quantity, the transmission potential of a disease, also based on censored data on hepatitis A in Bulgaria. Section 1.4 introduces the Bangkok Metropolitan Administration injecting drug users cohort study, which is further analyzed in Chapter 12, using methods that were developed for competing risk models.

In Section 1.5, a particular shape constrained estimation problem is considered. It is argued that this problem (and many of the other problems to be considered in this book) can also be viewed from another perspective; for example, as inverse problem, mixture model, or censoring problem. As will be seen later in this book, these points of view immediately suggest methods one could use for estimating shape constrained functions and methods one could use to compute these. Finally, Section 1.6 gives an outline of the content of this book.

1.1 Is There a Warming-up of Lake Mendota?

Lake Mendota has been called the most studied lake in the United States. One of the reasons we start with this example is that it appealed very much to one of the authors when he first read the book Barlow et al., 1972. In that book it is also the first example. The authors study the number of days until freezing in the years $1854 + i$, $i = 1, \ldots, 111$, and state: "According to a simple, useful (if not completely realistic) model, the days till freezing X_i are observations on a normal distribution with unknown means μ_i, $i = 1, 2, \ldots, 111$, and a common variance σ^2." The maximum likelihood estimates of μ_i under the restriction

Table 1.1 *Number of Days that Lake Mendota Was Frozen during Winter Seasons, Starting with the Year 1855*

118	151	121	96	110	117	132	104	125	118	125	123	110	127
131	99	126	144	136	126	91	130	62	112	99	161	78	124
119	124	128	131	113	88	75	111	97	112	101	101	91	110
100	130	111	107	105	89	126	108	97	94	83	106	98	101
108	99	88	115	102	116	115	82	110	81	96	125	104	105
124	103	106	96	107	98	65	115	91	94	101	121	105	97
105	96	82	116	114	92	98	101	104	96	109	122	114	81
85	92	114	111	95	126	105	108	117	112	113	120	65	98
91	108	113	110	105	97	105	107	88	115	123	118	99	93
96	54	111	85	107	89	87	97	93	88	99	108	94	74
119	102	47	82	53	115	21	89	80	101	95	66	106	97
87	109	57											

Note: The order is in increasing years from left to right and (next) row-wise.

$\mu_1 \leq \cdots \leq \mu_{111}$ minimize (as a function of the μ_i):

$$\sum_{i=1}^{111} (X_i - \mu_i)^2 ,$$

subject to $\mu_1 \leq \cdots \leq \mu_{111}$. This is a so-called isotonic estimator: the maximum likelihood estimates of $(\mu_1, \ldots, \mu_{111})$ under the restriction that the μ_i are nondecreasing in i (time).

We choose to use the data on duration of ice in days and estimate by isotonic regression (not assuming normality) the nonparametric regression function on these for 157 seasons, that is, we minimize

$$\sum_{i=1}^{157} (Y_i - \nu_i)^2 ,$$

subject to $\nu_1 \geq \cdots \geq \nu_{157}$, where Y_i is the number of days the lake was frozen in season i. Note that we have 157 seasons instead of 111, since we have more data on seasons than in 1972. The data are obtained from http://www.aos.wisc.edu/~sco/lakes/Mendota-ice.html and given in Table 1.1, and start in the year 1855.

How can this isotonic regression estimate be computed? We consider the so-called cumulative sum (or cusum) diagram, consisting of the points

$$(0, 0), (1, Y_1), (2, Y_1 + Y_2), \ldots, \left(i, \sum_{j=1}^{i} Y_j\right), \ldots, \left(157, \sum_{j=1}^{157} Y_j\right).$$

For this set of points we compute the least concave majorant. The solution is the left continuous slope of the least concave majorant of the y-values in the diagram. To show a more clearly visible difference between the cusum diagram and its least concave majorant, we subtract the trend (line between endpoints) in Figure 1.1b.

The resulting estimate is shown in Figure 1.2a and its smoothed version is shown in Figure 1.2b. The hypothesis that there is indeed a warming-up is not tested in Barlow et al., 1972, but can be tested with the methods of the present book, either using the isotonic least

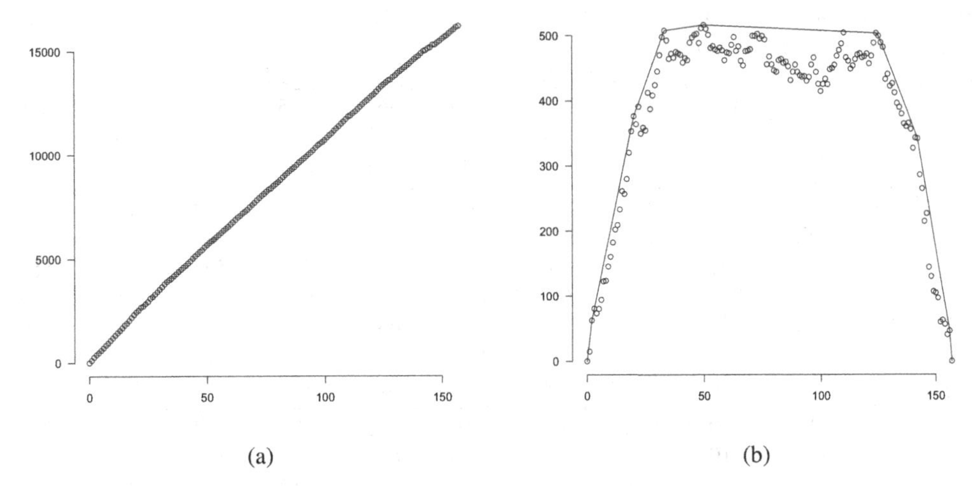

Figure 1.1 Cusum diagram for Lake Mendota data, without least concave majorant (a) and with least concave majorant (and minus the line connecting the two end points) (b).

squares (LS) estimate or the smoothed isotonic LS estimate. The smoothed isotonic LS estimate avoids the bad behavior of the ordinary isotonic LS estimate at the boundary, and will generally be consistent in situations where the LS estimate itself will be inconsistent, as will be discussed in this book.

1.2 Onset of Nonlethal Lung Tumor

For two groups of mice, the ages at death (in days) were measured. One group was kept in a germ-free environment and the other in a conventional environment. The distribution of interest is that of the age of onset of a lung tumor of type RFM. For mice, this type of tumor is nonlethal (according to Hoel and Walburg, 1972, from which this example is taken). At the

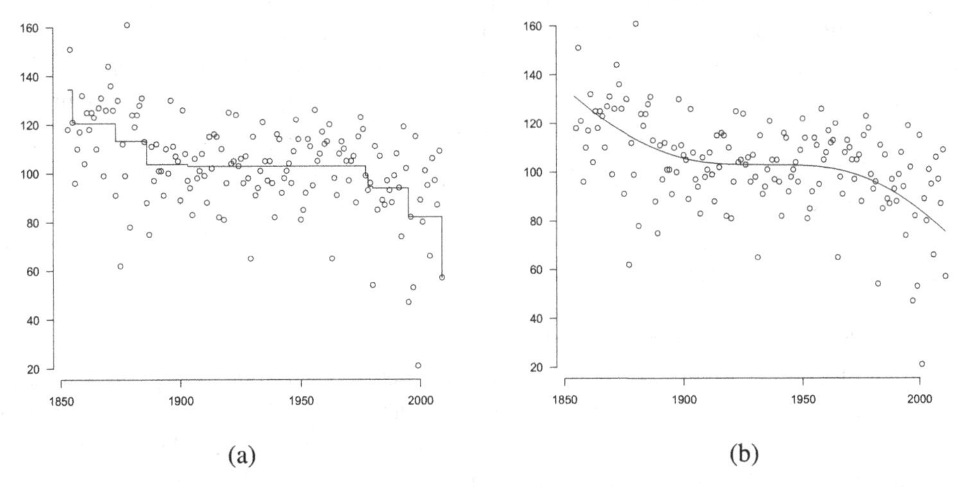

Figure 1.2 Isotonic estimators for the warming trend of Lake Mendota, without smoothing (a) and with smoothing (b).

Table 1.2 *Inspection Times (Ages at Death) of the Mice, with Indicator Whether Tumor was Found at Time of Inspection ($\Delta = 1$) or Not ($\Delta = 0$)*

CE & $\Delta = 1$	381	477	485	515	539	563	565	582	603	616
	624	650	651	656	659	672	679	698	702	709
	723	731	775	779	795	811	839			
CE & $\Delta = 0$	45	198	215	217	257	262	266	371	431	447
	454	459	475	479	484	500	502	503	505	508
	516	531	541	553	556	570	572	575	577	585
	588	594	600	601	608	614	616	632	632	638
	642	642	642	644	644	647	647	653	659	660
	662	663	667	667	673	673	677	689	693	718
	720	721	728	760	762	773	777	815	886	
GE & $\Delta = 0$	546	609	692	692	710	752	773	781	782	789
	808	810	814	842	846	851	871	873	876	888
	888	890	894	896	911	913	914	914	916	921
	921	926	936	945	1008					
GE & $\Delta = 1$	412	524	647	648	695	785	814	817	851	880
	913	942	986							

Note: The first group concerns mice living in a conventional environment (CE), the second mice living in a germ-free environment (GE).

time of death, it was checked whether the mouse did develop the lung tumor or not. The ages at death can therefore be viewed as "inspection times," whereas the event time of interest in this context is the time at which the tumor starts to grow. The data, taken from Hoel and Walburg, 1972, are given in Table 1.2.

Hoel and Walburg, 1972, first treat the lung tumors as a lethal disease, although they mention that this is incorrect, viewing the data as right censored. Then they calculate the Kaplan-Meier estimates for the distribution functions of the mortality due to lung cancer in the two groups under this assumption. The Kaplan-Meier estimates are shown in Figure 1.3. The figure suggests that the conventional group had a higher incidence or earlier occurrence of lung tumors than the germ-free group. They also applied the Breslow test for statistical significance, which was found to be significant an the 5% level. But they also note that in their opinion this is actually due to the incorrect assumption that the lung cancer is lethal for these mice and that the right estimator for the onset of the lung cancer is given by the maximum likelihood estimator (MLE) for current status data. The terminology "current status data" is still not used, but they propose an estimator that is actually just the MLE for current status data and refer for this to Ayer et al., 1955.

An exposition on the current status model is given in Section 2.3. The setting is as follows. We have a (unobservable) sample $X_1, X_2, \ldots, X_n$, drawn from a distribution with distribution function F, in this case representing the times of onset of the lung cancer. Instead of observing the X_is, one only observes for each i whether or not $X_i \le T_i$ for some random T_i (independent of the other T_js and all X_js), where in this case the T_i are the ages at death. More formally, instead of observing X_is, one observes

$$(T_i, \Delta_i) = (T_i, 1_{[X_i \le T_i]}).$$

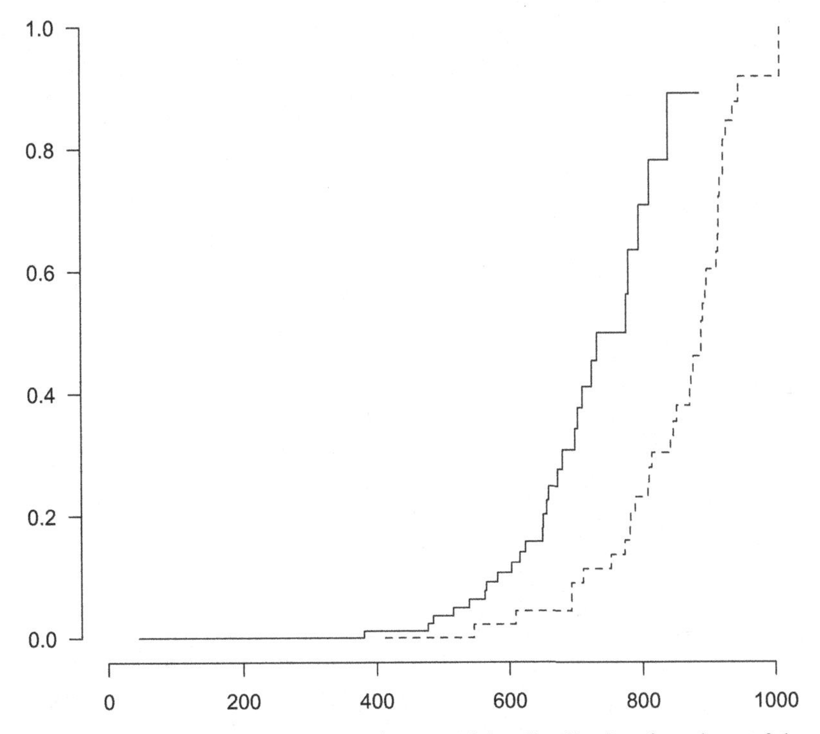

Figure 1.3 The Kaplan-Meier estimates of the distribution functions of the mortality due to lung cancer in the two groups for the data of Hoel and Walburg, under the assumption that the lung cancer is lethal. The solid curve is the estimate for the conventional group and the dashed curve the estimate for the mice in the germ-free environment.

One could say that the ith observation represents the current status of item X_i (onset of lung cancer) at time T_i.

The problem is to estimate the unknown distribution function F of the X_i, using the indirect information in the data. Denote the realized T_i by t_i and the associated realized values of the Δ_i by δ_i. For this problem the log likelihood function in F (conditional on the T_is) can be shown to be

$$\ell(F) = \sum_{i=1}^{n} \{\delta_i \log F(t_i) + (1 - \delta_i) \log(1 - F(t_i))\} . \tag{1.1}$$

The (nonparametric) MLE maximizes ℓ over the class of all distribution functions. Since distribution functions are by definition nondecreasing, computing the maximum likelihood estimator poses a shape restricted optimization problem in a natural way. As can be seen from (1.1), the value of ℓ only depends on the values that F takes at the observed time points t_i; about the values between these points we have no information. Hence one can choose to consider only distribution functions that are constant between successive observed time points t_i. The MLEs can then actually be computed by a procedure similar to the procedure in Section 1.1, since they can be characterized as the left-continuous slopes of the appropriate cusum diagrams.

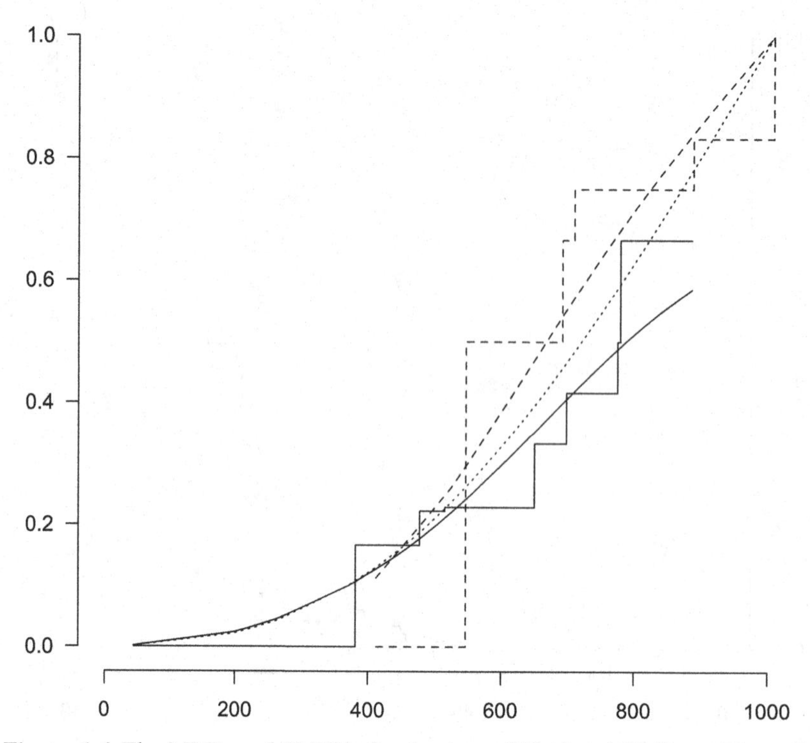

Figure 1.4 The MLEs and SMLEs for the data of Hoel and Walburg. The solid curves are the estimates of the distribution function of the time of start of the lung tumor of type RFM for the conventional group and the dashed curves the estimates for the mice in the germ-free environment. The dotted curve is the SMLE, based on the combined samples, with bandwidth $h = 1000n^{-1/5} \approx 370.1$.

The MLEs $\hat{F}_{ni}$ together with the smoothed maximum likelihood estimators (SMLEs) $\tilde{F}_{ni}$ for the two groups are shown in Figure 1.4. If $\hat{F}_{ni}$ is the MLE for group i, $i = 1, 2$, where, for example, 1 corresponds to the conventional group and 2 to the germ-free group, then the corresponding SMLEs are given by

$$\tilde{F}_{ni}(t) = \int \mathbb{K}\left(\frac{t-x}{h}\right) d\hat{F}_{ni}(x), \tag{1.2}$$

where $h > 0$ is a bandwidth, which is chosen to be $h = 1000n^{-1/5} \approx 370.1$ in this case and $\mathbb{K}$ is the integrated kernel

$$\mathbb{K}(x) = \int_{-\infty}^{x} K(u)\,du, \tag{1.3}$$

where K is a symmetric kernel with support $[-1, 1]$, for example the triweight kernel

$$K(u) = \frac{35}{32}\left(1 - u^2\right)^3 1_{[-1,1]}(u).$$

For values close to the boundary, we use in fact a boundary correction, explained later in the book. The MLEs and SMLEs for the two groups are shown in Figure 1.4.

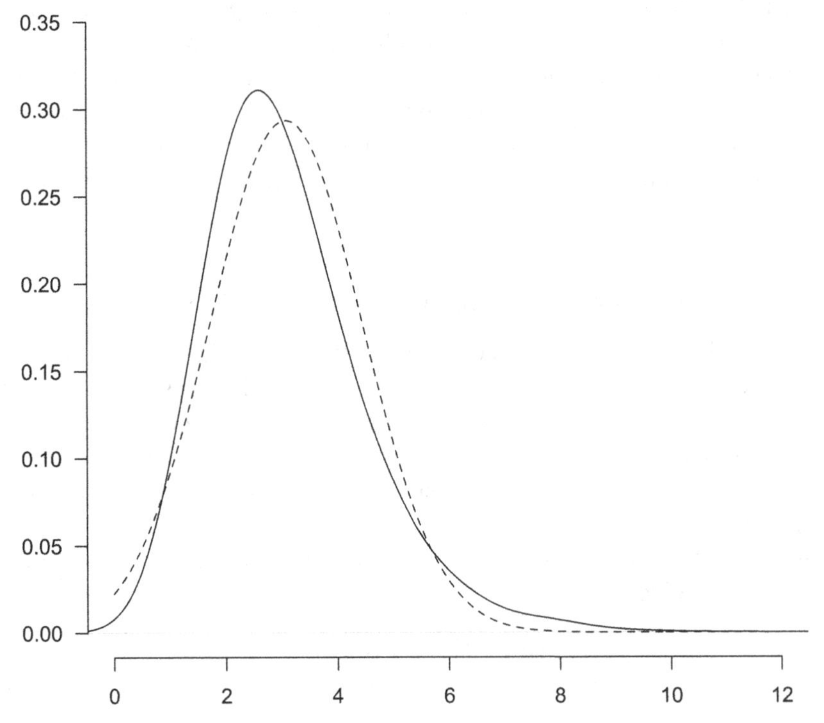

Figure 1.5 The kernel estimate of the density of the 10,000 bootstrap values of V_n^* from $\tilde{F}_n$ for the data of Hoel and Walburg, using a bandwidth $h = 0.5$. The dashed curve is the corresponding normal density, scaled with the mean and variance of the bootstrap values.

We can now test the hypothesis of equality of the two distributions with a likelihood ratio test, based on the test statistic, V_n, defined by

$$V_n = \sum_{i=1}^{n_1} \left\{ \Delta_i \log \frac{\hat{F}_{n1}(T_i)}{\hat{F}_n(T_i)} + (1 - \Delta_i) \log \frac{1 - \hat{F}_{n1}(T_i)}{1 - \hat{F}_n(T_i)} \right\}$$

$$+ \sum_{i=n_1+1}^{n_1+n_2} \left\{ \Delta_i \log \frac{\hat{F}_{n2}(T_i)}{\hat{F}_n(T_i)} + (1 - \Delta_i) \log \frac{1 - \hat{F}_{n2}(T_i)}{1 - \hat{F}_n(T_i)} \right\},$$

where $\hat{F}_n$ is the MLE, based on the combined samples.

In the present case we have: $n_1 = 96$, $n_2 = 48$ and $n = n_1 + n_2 = 144$. The test statistic has the value $V_n = 2.5580$. For the purpose of bootstrapping, the distribution function of the onset of the tumor, under the null hypothesis of no difference in the distribution in the two samples, was estimated by the SMLE $\tilde{F}_n$, based on the combined samples (the dotted curve in Figure 1.4). Next the values of the Δ_is were resampled, keeping the observation times T_i fixed, by letting the Δ_i^* be independent Bernoulli $(\tilde{F}_n(T_i))$ random variables, where $\tilde{F}_n$ was the SMLE, based on the combined samples, with the bandwidth $h = 1000n^{-1/5} \approx 370.1$.

This gave, for 10,000 bootstrap samples, a p-value of 0.6009. A picture of an estimate of the density of V_n^* for these 10,000 bootstrap samples, made in R, is shown in Figure 1.5.

Directly resampling from the MLE for the combined samples gave a p-value of 0.599, so a value very close to the values obtained by resampling from the smooth estimate, based on the SMLE. Further work on tests of this type can be found in Chapter 9. There we also discuss the possible validity or invalidity of bootstrap resampling from $\tilde{F}_n$ or $\hat{F}_n$ in this context. In any case, the analysis based on the current status model instead of the right-censoring model gives no indication of a difference in susceptibility to a lung tumor of type RFM in the two groups.

1.3 The Transmission Potential of a Disease

Keiding, 1991, analyzed demographical data on hepatitis A in Bulgaria. He notes that for the planning of vaccination programs it is important to estimate the transmission potential R_V, which measures the number of secondary cases one case could produce during the infectious period. Informally stated: the transmission potential is the expected number of other people one infects if one is infected. If this number is bigger than 1, there is the danger of an epidemic spread.

It was shown by Dietz and Schenzle, 1985, that for virus infections with a short infectious period this number is given by:

$$R_V = \frac{\int_0^\infty \exp\left\{-\int_0^a \mu(u)\,du\right\} \lambda(a)^2 V(a)\,da}{\int_0^\infty \exp\left\{-\int_0^a \mu(u)\,du\right\} \lambda(a)^2 \exp\left\{-\int_0^a \lambda(u)\,du\right\}\,da}, \tag{1.4}$$

where $\lambda(a)$ is the infection intensity, $V(x)$ the probability that an individual of age x has not yet been vaccinated and the mortality μ can usually be taken to be known from official vital statistics. Table 2 in Keiding, 1991, which is reproduced in Table 1.3, contains the prevalence data from Bulgaria on the presence of antibodies for hepatitis A, which can be used in estimating the infection intensity λ; $V(a)$ is a quantity one can manipulate. The ages 71, 84 and 85, for which there were no observations in Table 2 in Keiding, 1991, are omitted from our Table 1.3.

Just as in Section 1.2, the data available for estimating λ are current status data: if a person in the survey has antibodies, it is clear the he/she has been infected at a time preceding the check on antibodies, otherwise this person can still obtain antibodies in future or may never get the disease. In this case, the survey contained 850 people, and the MLE $\hat{F}_n$ of the distribution function of age at which people were infected is shown in Figure 1.6a, together with the corresponding SMLE $\tilde{F}_n$, given by

$$\tilde{F}_n(t) = \int \mathbb{K}\left(\frac{t-x}{h}\right)\,d\hat{F}_n(x), \tag{1.5}$$

where $\mathbb{K}$ is an integrated kernel, just as in (1.2) (see (1.3)).

The corresponding density estimate is defined by

$$\tilde{f}_n(t) = h^{-1}\int K\left(\frac{t-x}{h}\right)\,d\hat{F}_n(x), \tag{1.6}$$

where the bandwidth is usually larger than in estimating the distribution function (the typical orders are $n^{-1/7}$ and $n^{-1/5}$, respectively). An estimate of the hazard is given in Figure 1.6b, where the bandwidths in estimating F and f were 35 and 45, respectively. Note that by

Table 1.3 *Current Status Data on Hepatitis A in Bulgaria*

Age	Virus Positive	Total	Age	Virus Positive	Total
1	3	16	43	7	10
2	3	15	44	5	5
3	3	16	45	7	7
4	4	13	46	9	9
5	7	12	47	9	9
6	4	15	48	22	22
7	3	12	49	6	7
8	4	11	50	10	10
9	7	10	51	6	6
10	8	15	52	13	14
11	2	7	53	8	8
12	3	7	54	7	7
13	2	11	55	13	13
14	0	1	56	11	11
15	5	16	57	8	8
16	13	41	58	8	8
17	1	2	59	9	10
18	3	6	60	13	16
19	15	32	61	5	5
20	22	37	62	5	6
21	15	24	63	5	5
22	7	10	64	5	5
23	8	10	65	10	10
24	7	11	66	8	8
25	12	15	67	4	4
26	5	10	68	5	5
27	10	13	69	4	5
28	15	19	70	8	8
29	9	12	72	9	9
30	9	9	73	1	1
31	9	14	74	4	4
32	8	10	75	7	7
33	9	11	76	6	6
34	8	9	77	2	2
35	9	14	78	3	3
36	13	14	79	2	2
37	6	7	80	4	4
38	15	16	81	1	1
39	11	13	82	1	1
40	6	8	83	2	2
41	8	8	86	1	1
42	13	14			

choosing the bandwidths in this way, $\tilde{f}_n$ is no longer the derivative of $\tilde{F}_n$. If one wants to keep this relation, one has to take equal bandwidths for $\tilde{F}_n$ and $\tilde{f}_n$, as was done in Groeneboom's discussion in Keiding, 1991; the estimator of the hazard obtained in this way was not very different from our estimator in 1.6b, though. Bootstrap methods for determining the

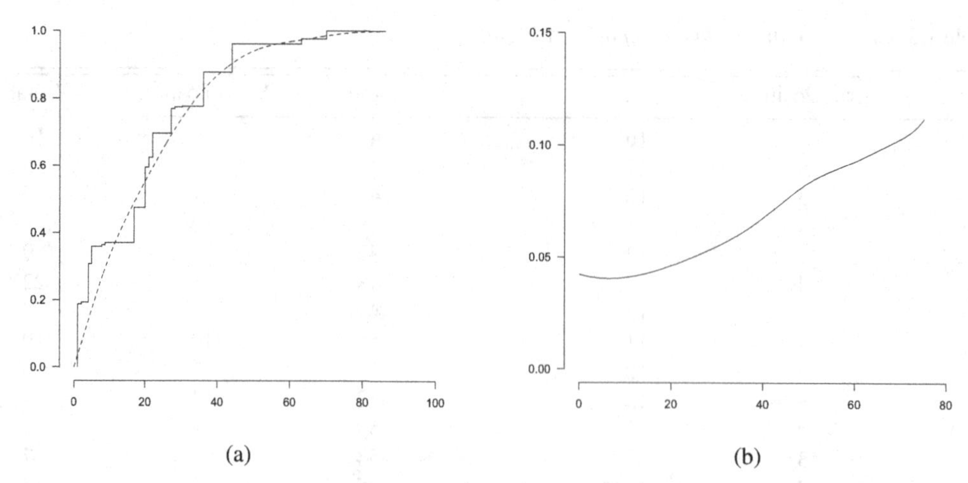

(a) (b)

Figure 1.6 MLE (step function) and SMLE (dashed) of the distribution function (a) and estimate of the hazard rate of the age of infection (b) based on the hepatitis A data.

bandwidths for this example can also be found on p. 400–401 of the discussion in Keiding, 1991, and in Groeneboom et al., 2010.

By methods of the present book one can derive distribution theory for estimates of the transmission potential (1.4). Smoothing methods are unavoidable; note, for example, that one cannot (sensibly) differentiate the MLE itself, since it is a step function, so one cannot estimate the infection intensity (hazard) α without applying some kind of smoothing. On the other hand, the transmission potential is a global functional, so one has to combine local and global methods for obtaining its distribution. The interplay between local and global methods is one of the themes of our book.

1.4 The Bangkok Cohort Study

The Bangkok Metropolitan Administration injecting drug users cohort study (Kitayaporn et al., 1998, and Vanichseni et al., 2001) was started in 1995 to assess (among other things) the feasibility of conducting a phase III HIV vaccine efficacy trial for injecting drug users in Bangkok. The data on a subset of 1,365 injecting drug users who were below 35 years of age in this study were analyzed by Maathuis and Hudgens, 2011, and Li and Fine, 2013. In this group, 392 were HIV positive, with 114 infected with subtype B, 237 infected with subtype E, 5 infected by another mixed subtype, and 36 infected with missing subtype. The subjects with other, mixed, or missing subtypes were grouped in a single category.

In Maathuis and Hudgens, 2011, the maximum likelihood estimator (MLE) for the subtype-specific cumulative incidence of HIV is computed, as well as a so-called naive estimator, based on analyzing one category such as the type B subjects, ignoring the data on the other types. There also confidence intervals are given, based on the naive estimators, using the likelihood ratio test method developed in Banerjee and Wellner, 2001, and Banerjee and Wellner, 2005.

Li and Fine, 2013, compute both the regular MLE and a smoothed version of the MLE (called the SMLE) and use theory developed in Groeneboom et al., 2010, for constructing

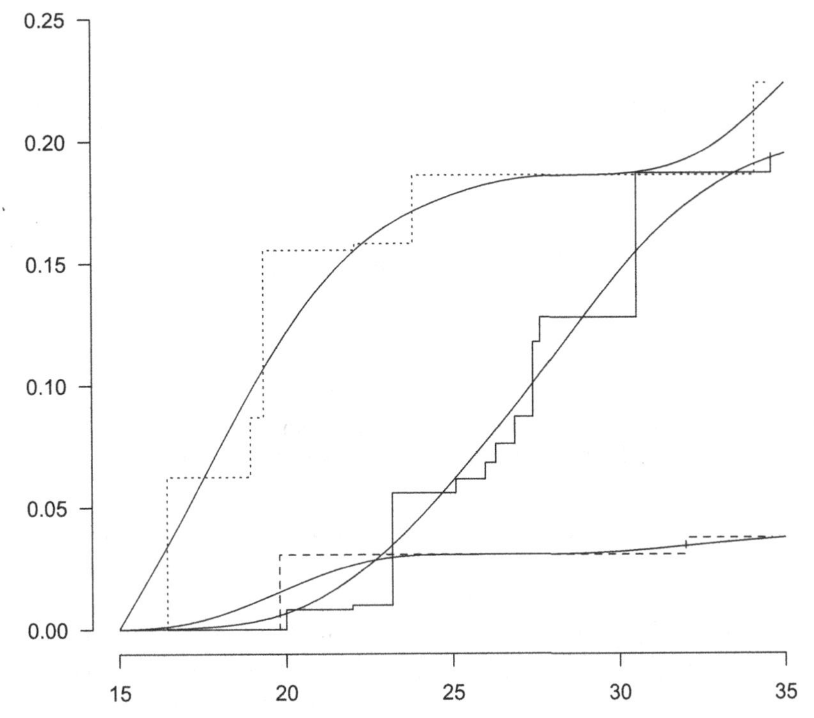

Figure 1.7 The MLE for the three categories in the Bangkok cohort study. The piecewise constant curves give the subdistribution functions, based on the MLE, for the cumulative incidence of HIV in the different categories; dotted, type E; solid, type B; dashed, other types. The smooth solid curves give the corresponding estimates, based on the SMLE.

confidence intervals. They also estimate the hazard and construct confidence intervals for the hazard, again using Groeneboom et al., 2010. The regular MLE cannot directly be used for this purpose because it corresponds to a discrete distribution, so that some kind of smoothing is needed to estimate the hazard and to construct the confidence intervals.

We also analyze these data in Section 12.1, using methods that are somewhat different from the methods in Maathuis and Hudgens, 2011, and Li and Fine, 2013. Our treatment is closest to the treatment in Li and Fine, 2013, though. The MLE and corresponding SMLE for the subtype-specific incidence of HIV in the three categories are shown in Figure 1.7. Bootstrap confidence intervals will be constructed in Section 12.1, both for the subdistribution functions and the corresponding hazards for the categories type B and type E.

1.5 Inverse Problems, Censoring, Mixture Models and Shape Constraints

Having seen four real-life examples illustrating the need and relevance of shape constrained statistical inference, we now consider some types of statistical problems where these naturally occur. We start with a description of the familiar classical parametric density estimation

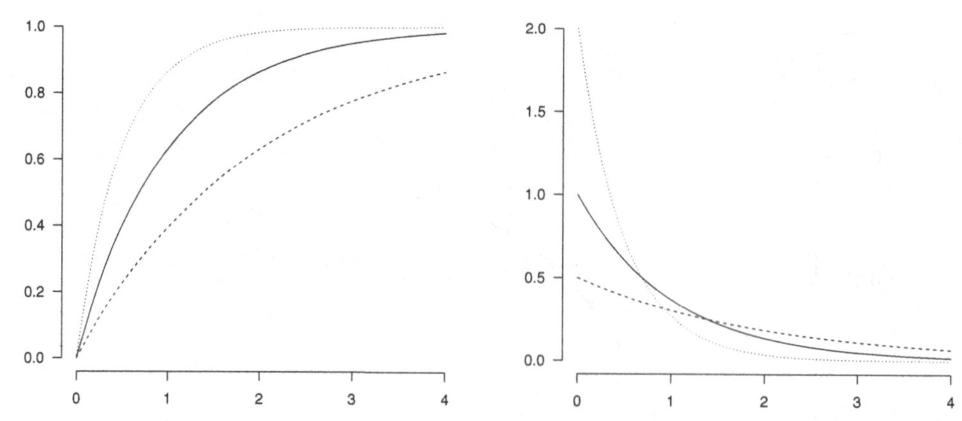

Figure 1.8 Exponential distribution functions F_θ and corresponding densities $f_\theta = F'_\theta$ for $\theta = 0.5$ (dashed); 1 (solid); and 2 (dotted).

problem. Given a sample of independent and identically distributed random variables

$$X_1, X_2, \ldots, X_n$$

from an unknown distribution function F_0, estimate this F_0 (or certain aspects of it) as well as possible. In order to do that, a statistical model $\mathcal{F}$ is assumed. This is a set of distribution functions to which the underlying F_0 is assumed to belong. Write this set as

$$\mathcal{F} = \{F_\theta : \theta \in \Theta\}.$$

The set Θ is referred to as parameter space, usually a subset of $\mathbb{R}^k$ for some small k. An example is the class of exponential densities on $[0, \infty)$, where

$$F_\theta(x) = 1 - \exp\{-\theta x\} \text{ on } [0, \infty) \text{ and } \theta \in \Theta = \mathbb{R}^+. \tag{1.7}$$

General methods such as maximum likelihood can be used to obtain an estimator for the parameter θ_0 corresponding to the underlying distribution function F_0. Looking at some of the distribution functions (and densities corresponding to) the family $\mathcal{F}$ (see Figure 1.8), it is clear that this family is rather rigid in the sense that all distribution functions it contains have the same shape.

Figure 1.9 shows two histograms of data sets that would result in the same estimate of the underlying distribution function if it is assumed to belong to $\mathcal{F}$. The histograms show that exponentiality is doubtful for both underlying distributions.

Nonparametric statistical methods aim to "let the data speak for themselves." A histogram is in fact a nonparametric density estimator. It has a high value when locally there are relatively many observations and a low value if observations are locally sparse. Also other nonparametric procedures, such as kernel density estimators, spline smoothers and wavelet methods, have been developed to estimate the distribution function (and density) in the context sketched here.

Where these nonparametric methods often use minimal (or no additional) assumptions, there are classes of models where a priori information is known on the shape of the distribution function of interest. For example, in survival analysis, where one can cannot directly observe the event time of interest. One only approximately knows when a certain event

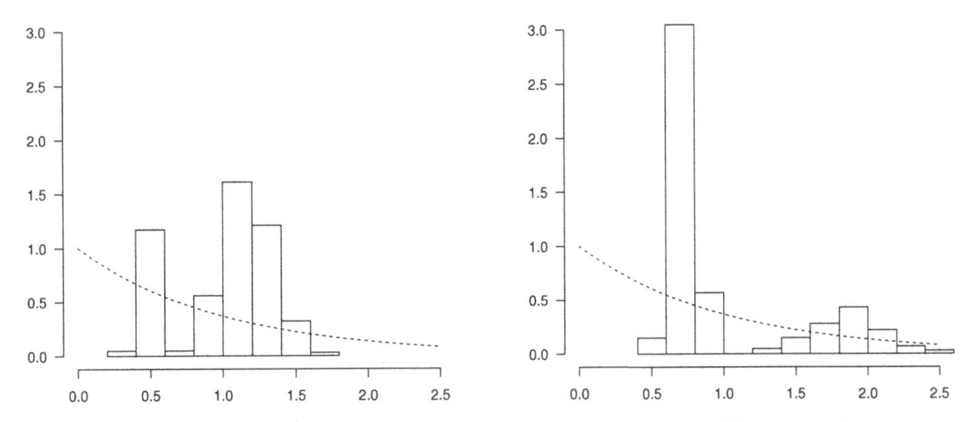

Figure 1.9 Histograms of two datasets with mean value one with the maximum likelihood fit under the assumption of exponentiality (dashed).

happened, leading to censored data. The current status problem as encountered in Section 1.2 and 1.3 is an example of such a model. The distribution of the actually observable (two-dimensional) data is related to the (one-dimensional) event time distribution of interest, but clearly not the same. In stereology, one is interested in the distribution of some aspects of three-dimensional objects, being able to observe only two-dimensional projections of these. The sampling strategy and nature of the visible projection then lead to observable data that are related to, but not equal to, the data one actually is interested in. These types of problems can be viewed as inverse problems (see Figure 1.10). An example relation

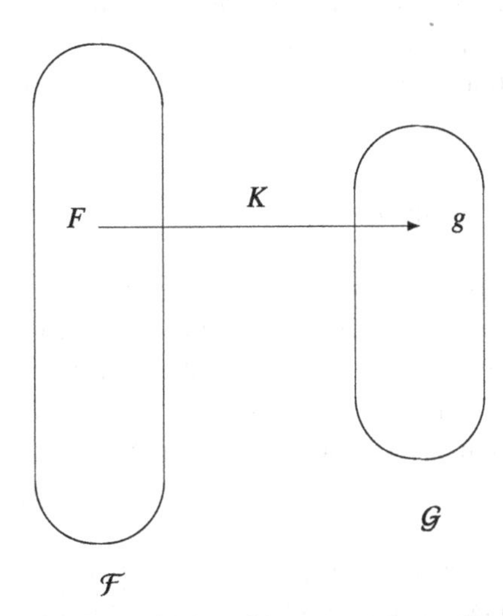

Figure 1.10 The random quantities of interest X have distribution function F contained in the set $\mathcal{F}$ of distribution functions. The observable random quantity, Z, has a probability density g that can be expressed in terms of distribution function F: $g = K(F)$. The inverse problem is: estimate F based on a sample from density $g = K(F)$. The class of possible densities $\mathcal{G}$ equals $K(\mathcal{F})$.

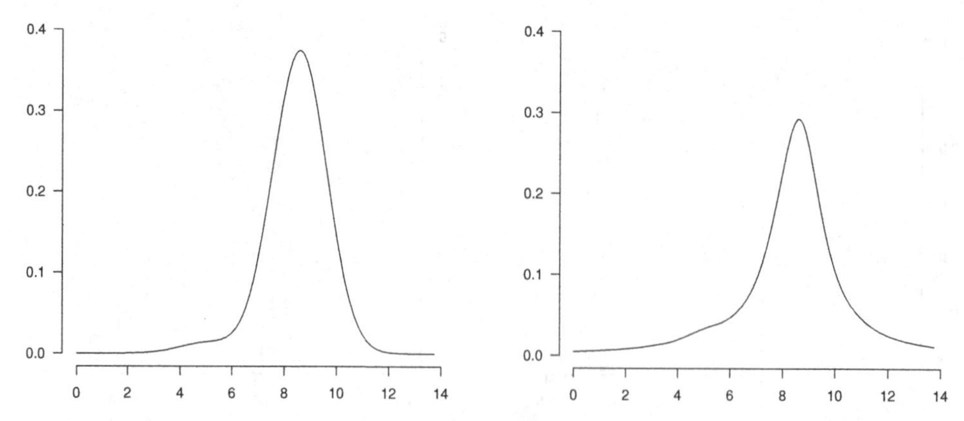

Figure 1.11 Two densities; one is a location mixture of standard normal densities, the other is not.

(to be encountered in Section 2.2) is given by

$$g(z) = \frac{1 - F(z)}{\int (1 - F(u))\, du} \quad \text{for } z \geq 0, \tag{1.8}$$

where F is a distribution function on $[0, \infty)$ with $0 < \int_z z\, dF(z) < \infty$. Whatever assumptions are imposed on the underlying distribution function F, the sampling density g will be bounded and monotonically decreasing on $[0, \infty)$ because all distribution functions F are increasing. This shows that not all distributions are possible as sampling distribution. All possible sampling densities share the same shape constraint of being monotonically decreasing and bounded.

Shape constraints can also be less obvious to identify visually. Mixture models are models where the sampling distribution is an unknown (scale or location) mixture of a basic distribution. In the deconvolution model, to be discussed in Section 2.4 and 4.6, the sampling density is given by

$$g(z) = \int k(z - x)\, dF(x) \tag{1.9}$$

where k is a known kernel function (density of the "noise" in the deconvolution setting) and F is a mixing distribution. Identifying whether a given density g is of this type is not straightforward (certainly not as straightforward as determining whether the density is bounded and decreasing). Nonetheless, the class of densities that can be expressed as location mixture of the kernel function k is drastically smaller than the class of all densities on $\mathbb{R}$, and it usually pays off to take advantage of this prior knowledge on the class of sampling densities. Figure 1.11 shows two densities, one of which is a (location) mixture of standard normal densities and the other of which is not. Which is which?

Another view of representations (1.8) and (1.9) is that one samples from density $g = g_F$ that is parameterized by an underlying "infinite dimensional" parameter, being the distribution function F. Shape constrained models can really be viewed as intermediate between fully nonparametric models (such as the class of all distribution functions) and finite dimensional parametric models.

1.6 Outline of the Book

In Chapter 2 we introduce basic examples related to estimating a monotone function. In those examples, it is argued that monotonicity assumptions are natural and defendable. From the computational point of view, the resulting monotone estimators can be elegantly determined by constructing a diagram of points along with a convex hull of these points, as already seen in Sections 1.1, 1.2 and 1.3.

Chapter 3 deals with typical pointwise asymptotic properties of estimators introduced in the basic examples. The "convex minorant" (or "concave majorant") representation of the estimators readily leads to consistency proofs. This representation can actually also be pushed to the limit by proper rescaling and localizing, leading to asymptotic distributions that can be expressed as functionals of convex minorants (or concave majorants) of processes such as Brownian bridge on [0, 1] or Brownian motion with a parabola added to it on the whole of $\mathbb{R}$. Heuristic as well as rigorous arguments are given for various examples from Chapter 2. Moreover, the most commonly encountered nonnormal limiting distributions (one of which is known as the Chernoff distribution) are discussed in two separate sections.

In Chapter 4 more examples of problems where shape constraints are present are discussed. These models are more complex than those considered in Chapter 2 in several ways. Most important is that natural nonparametric estimators cannot be constructed explicitly as for the examples in Chapter 2. Sometimes the shape constraints are also somewhat indirect, in the sense that it is hard to judge whether a particular sampling density satisfies the shape constraint or not (as the deconvolution problem mentioned in Section 1.5). Also, higher dimensional censoring problems lead to shape constrained models. In Chapter 5 some of these models are introduced and discussed.

In general statistical estimation problems, one aims at using the data as efficiently as possible. The concept of minimax risk can be used to quantify how hard an estimation problem intrinsically is. In Chapter 6 some methods are discussed that can be used to derive (in quite some examples sharp) lower bounds to a (local) minimax risk. It is desirable to have an estimator that attains the lower bound (at least as far as the rate of convergence is concerned). For some of the examples already seen, it will be shown that the derived estimators are rate optimal in a minimax sense.

As mentioned before, for some examples it is straightforward to compute shape constrained estimators. In other examples there is not an explicit method to construct these. Chapter 7 introduces various algorithms that can be used to compute nonparametric estimators under shape constraints. The methods are also illustrated using some of the previously introduced problems. The various perspectives from which shape constrained models can be viewed, as discussed in Section 1.5, can be used to determine a natural algorithm to use when computing a shape constrained estimator in such models.

In Chapter 8, estimation methods are introduced where shape constraints are joined with smoothness assumptions. The resulting estimators have better asymptotic properties than the nonsmoothed counterparts. Moreover, these smooth shape-constrained estimators turn out to be well suited for bootstrap analyses. Apart from estimation, testing hypotheses is also an important subject in shape constrained inference. Testing whether a shape constraint is valid is one example. Testing for equality of two functions within shape-constrained models is another. Examples of both types will be discussed in Chapter 9.

The final chapters are concerned with asymptotic theory for shape constrained estimators and test statistics. Some recent developments are described and problems posed. Chapter 10 deals with the asymptotic distribution of functionals of the underlying distribution that depend on the underlying distribution in a smooth sense; for example, moments of these distributions. Smooth functional theory is the way to derive asymptotic results in this setting. In Chapter 11, asymptotic results are derived for shape constrained functions evaluated at a fixed point. The optimality conditions for the estimators are important in these situations. In contrast to the approach described in Chapter 3, these optimality conditions are only implicit. Where Chapter 11 is only concerned with univariate problems, in Chapter 12 pointwise asymptotic results are derived in the context of some of the bivariate problems introduced in Chapter 5. Chapter 13 is concerned with the asymptotic distribution theory for test statistics that are based on global deviation measures.

Each chapter is concluded with problem section and a section with bibliographic remarks. Most of the exercises are of a level that should be within reach for graduate students in statistics. To produce the pictures, we used the R software environment (R Development Core Team, 2011).

Exercises

1.1 Consider the situation where $X_1, \ldots, X_n$ are independent and identically distributed exponential random variables with parameter θ; so each X_i has distribution function (1.7).

a) The maximum likelihood estimator (MLE) of θ is defined as the maximizer of the log likelihood function

$$\ell(\theta) = \sum_{i=1}^{n} \log f_\theta(X_i)$$

where $f_\theta = F'_\theta$. Show that this estimator is given by $1/\bar{X}_n$ in this situation. Here $\bar{X}_n = \frac{1}{n}\sum_{i=1}^{n} X_i$.

The method of maximum likelihood can be used also in models where the class of possible distributions of the data is much larger, say nonparametric. In the following also another method, which is not so common in the parametric context, will be used in nonparametric situations.

b) Denote the empirical distribution function of the X_is by $\mathbb{F}_n$ and define an estimator (least squares) as minimizer of

$$Q(\theta) = \frac{1}{2}\int_0^\infty f_\theta(x)^2\,dx - \int_0^\infty f_\theta(x)\,d\mathbb{F}_n(x).$$

Show that

$$Q(\theta) = \theta\left\{\frac{1}{4} - \frac{1}{n}\sum_{i=1}^{n} \exp\{-\theta X_i\}\right\}$$

and think of a method to minimize this function.

c) Also use the estimation methods of (a) and (b) to find estimators for the parameter θ in the class of uniform distributions:

$$\mathcal{F} = \{F_\theta : \theta > 0\} \text{ with } F_\theta(x) = \frac{x}{\theta} \text{ on } [0, \theta].$$

1.2 Consider the setting of Figure 1.10 with the transformation given in (1.8).

a) Determine the class of densities $\mathcal{G}$ when

$$\mathcal{F} = \{F_\theta : \theta > 0\}, \text{ with } F_\theta(x) = 1_{[\theta,\infty)}(x).$$

b) The same as under (a), but now with $\mathcal{F}$ as defined in (1.7).

c) Again as (a), but now with $\mathcal{F}$ the class of concave distribution functions on $[0, \infty)$.

1.3 Let k be a fixed probability density on $\mathbb{R}$ and $\mathcal{K} = \{k_\theta : \theta > 0\}$ with

$$k_\theta(x) = \frac{1}{\theta} k\left(\frac{x}{\theta}\right)$$

the scale family of densities generated by k. Let F be a distribution function on $(0, \infty)$. The scale mixture of k with mixing distribution F is then given by

$$g(x) = \int_{(0,\infty)} k_\theta(x)\, dF(\theta).$$

a) Identify the class scale mixtures of the standard uniform density, when F runs through the class of all distribution functions on $(0, \infty)$ with $\int_0^\infty y^{-1}\, dF(y) < \infty$.

b) The same as in (a), but now with F running through the class of (degenerate) point measures on y, for $y > 0$.

c) The same as in (a), but now with F running through the class of distributions with densities

$$f_c(y) = (c+1)y^c 1_{[0,1]}(y)$$

for $c > 0$.

1.4 For the deconvolution model, suppose the convolution kernel is the standard uniform kernel and the class of distribution functions $\mathcal{F}$ consists of the exponential distributions as given in (1.7). Construct the class of possible sampling densities $\mathcal{G}$.

Bibliographic Remarks

There are books dedicated to censoring problems. These include Andersen et al., 1993, Klein and Moeschberger, 2003, Sun, 2006, and Fleming and Harrington, 2011. Within other areas of mathematics, there is also a lot interest in (ill posed) inverse problems. Also within the statistics literature, there is interest in a functional analytic approach to these problems. See, e.g., Kaipio and Somersalo, 2005. There are two classical books on estimation of parameters (and also functions) under order restrictions (a special type of shape restriction), dealing with isotonic regression. The first is Barlow et al., 1972, and the second is Robertson et al., 1988. A more recent reference is Silvapulle and Sen, 2005. In Groeneboom and Wellner, 1992, nonparametric maximum likelihood estimators in interval censoring and deconvolution models are thoroughly studied. A book on mixture models is Lindsay, 1995.

2

Basic Estimation Problems with Monotonicity Constraints

The most basic shape constraint for a real valued function on $\mathbb{R}$ is monotonicity. In many situations the nature of the data imposes this constraint in a straightforward way. In other situations the mere fact that one wants to estimate a distribution function, which is by definition monotone, dictates the constraint. In this chapter we introduce various models where a monotonicity constraint is part of the model, and consider nonparametric estimation procedures that can be used to estimate the monotone function of interest. In Section 2.1 the monotone regression problem is introduced. Monotonicity of a regression function is a natural assumption in many applications. Section 2.2 is concerned with the problem of estimating a decreasing density on $[0, \infty)$. The method of sampling from a certain population can lead in a natural way to a monotonicity property of the sampling density. The interval censoring, case I, or current status model is widely applied in biostatistics. In this model it is a distribution function one wants to estimate. This model is studied in Section 2.3. In Section 2.4 deconvolution problems are introduced. In those models, one wants to estimate an unknown distribution function based on a sample from another distribution, which is the convolution of the distribution of interest and another (known) distribution. An interesting family of problems are generalized isotonic regression problems. These are discussed in Section 2.5. In fact, these problems can look quite unrelated, but turn out to be equivalent to isotonic regression problems as considered in Section 2.1. The last example, in Section 2.6, comes from reliability theory, where one is interested in estimating an increasing hazard rate based on a sample of survival data.

2.1 Monotone Regression

Consider the standard simple regression context, where one observes data $\{(x_i, y_i): 1 \le i \le n\}$. Here the x_is are considered fixed and increasing in order whereas y_i is a realization of the random variable

$$Y_i = r(x_i) + \epsilon_i. \tag{2.1}$$

The random variables $\epsilon_1, \epsilon_2, \ldots, \epsilon_n$ are i.i.d. with $E\epsilon_i = 0$. The problem then is to estimate the function r from the data. In the simple linear regression context, r is assumed to belong to a class of regression functions that can be parameterized linearly using two real parameters. In this section, we first give an example data set. Then we characterize the least squares estimator of the monotone regression function r.

Table 2.1 *Height Measurements of 42 School Girls of Different Age, the Corresponding Mean Heights with Standard Deviations*

Age i							$\bar{x}(i)$	st-dev(i)
12	1.62	1.51	1.55	1.54	1.71	1.63	1.59	0.07
13	1.67	1.61	1.83	1.54	1.53	1.55	1.62	0.11
14	1.61	1.60	1.56	1.57	1.68	1.51	1.59	0.06
15	1.70	1.56	1.66	1.68	1.75	1.50	1.64	0.09
16	1.58	1.75	1.56	1.65	1.67	1.67	1.65	0.06
17	1.84	1.66	1.71	1.82	1.79	1.88	1.68	0.08
18	1.68	1.64	1.65	1.73	1.65	1.83	1.70	0.07

Example 2.1 Consider the (generated small) data set given in Table 2.1, consisting of measurements of the heights of 42 school girls where the age ranges from 12 to 18. For each age the heights of six girls were measured. It is natural to assume that the expected height of school girls is increasing as a function of age. On the other hand, it is obvious that the natural estimates for these expected heights given by the mean values of the samples do not have to share this monotonicity property. The second last column in Table 2.1 shows that in this example the means are indeed not increasing with age.

The least squares estimator in model (2.1), minimizing the weighted sum of squared residuals over all nondecreasing functions r, is defined as

$$\hat{r} = \text{argmin}_{r \in \mathcal{M}} \frac{1}{2} \sum_{i=1}^{n} (y_i - r(x_i))^2 w_i \tag{2.2}$$

Here the weight vector w has length n and positive components. The function class $\mathcal{M}$ is defined by

$$\mathcal{M} = \{f : \mathbb{R} \to \mathbb{R} : f(u) \leq f(v) \text{ for all } u \leq v\}.$$

This least squares estimator is often called the *isotonic* regression of the vector y, indicating that the ordering of the components of y goes in the same direction as that of the components of x. If the ordering is reverse, the corresponding least squares estimator is often called the *antitonic* regression of the vector y. Note, however, that the antitonic regression is often again called isotonic regression.

Let us now solve minimization problem (2.2). The first observation is that the objective function only depends on the values of the function r at the observed x_is. Hence, as long as the monotonicity constraint is satisfied, changing r between two observed x_is does not change the value of the objective function. With that in mind, we solve (2.2) over all functions in $\mathcal{M}$ that are constant between successive x_is. Abusing notation slightly, we write $r = (r_1, r_2, \ldots, r_n) = (r(x_1), r(x_2), \ldots, r(x_n)) \in \mathbb{R}^n$.

Lemma 2.1 *Let $y \in \mathbb{R}^n$ and $w = (w_1, \ldots, w_n) \in (0, \infty)^n$ fixed. Then the vector $\hat{r} \in \mathbb{R}^n$ minimizes the function*

$$r \mapsto Q(r) = \frac{1}{2} \sum_{i=1}^{n} (r_i - y_i)^2 w_i$$

over the closed convex cone $C = \{r \in \mathbb{R}^n : r_1 \le r_2 \le \cdots \le r_n\}$ *if and only if*

$$\sum_{j=1}^{i} \hat{r}_j w_j \begin{cases} \le \sum_{j=1}^{i} y_j w_j & \text{for all } i = 1, 2, \ldots, n \\ = \sum_{j=1}^{i} y_j w_j & \text{if } \hat{r}_{i+1} > \hat{r}_i \text{ or } i = n \end{cases} \tag{2.3}$$

Remark Note that the (unique) $\hat{r}$ satisfying these (in)equalities can easily be constructed. Define the cumulative sum diagram consisting of the points $P_i = (\sum_{j=1}^{i} w_j, \sum_{j=1}^{i} y_j w_j)$, for $1 \le i \le n$, and $P_0 = (0, 0)$. Then construct the (greatest) convex minorantof these points, i.e., the maximal convex function lying entirely below the diagram of points. Then $\hat{r}_i$ is given by the left derivative of this convex minorant evaluated at the point P_i.

Proof Observe that Q is a strictly convex function on $\mathbb{R}^n$, having a unique minimizer $\hat{r}$ over C. We now prove that conditions (2.3) are necessary. Define for $i = 1, 2, \ldots, n$ the vectors $v^{(i)}$ by $v_j^{(i)} = 1_{\{1,2,\ldots,i\}}(j)$. Then for each i, $\hat{r} - \epsilon v^{(i)} \in C$ for all $\epsilon > 0$. Hence by convexity of Q we have for all i

$$0 \le \lim_{\epsilon \downarrow 0} \epsilon^{-1} \left(Q(\hat{r} - \epsilon v^{(i)}) - Q(\hat{r}) \right) = \sum_{j=1}^{i} (y_j - \hat{r}_j) w_j.$$

Note that if $\hat{r}_{i+1} > \hat{r}_i$ or $i = n$, it also holds that $\hat{r} + \epsilon v^{(i)} \in C$ for all $\epsilon > 0$ sufficiently small, implying that for all such i

$$0 \le \lim_{\epsilon \downarrow 0} \epsilon^{-1} \left(Q(\hat{r} + \epsilon v^{(i)}) - Q(\hat{r}) \right) = \sum_{j=1}^{i} (y_j - \hat{r}_j) w_j.$$

Together with the inequality in the opposite direction, the equality part of (2.3) follows. This furnishes necessity of (2.3).

Since (in)equalities (2.3) uniquely define the vector $\hat{r}$ (this follows from the construction as derivative of the convex minorant), existence and uniqueness of the least squares estimator yields sufficiency of (2.3). See Exercise 2.2 for a more direct argument to show sufficiency. □

For the data given in Table 2.1, Figure 2.1 shows the cumulative sum diagram based on the values $\bar{x}(i) - \bar{x}$, where $\bar{x}$ denotes the mean of the $\bar{x}(i)$s. This centering is purely for illustrative purposes; the solution of the minimization problem is obtained from Figure 2.1 by simply adding $\bar{x}$ to the derivative of the convex minorant obtained.

Example 2.2 At Lake Monona in Wisconsin, the duration (in days) of ice has been measured in the years 1855–2012. Figure 2.2a shows the scatter plot of these durations against the year. In view of possible consequences of the greenhouse effect, one could propose a model that the expected duration (in days) of ice in Lake Monona is a decreasing function of time, i.e.

$$EY_i = r_i \text{ for } 1855 \le i \le 2013.$$

The mean vector r can then be estimated by the (constant weight) antitonic (decreasing) regression of the observed durations y_i, $1855 \le i \le 2013$. Figure 2.2b shows the cumulative

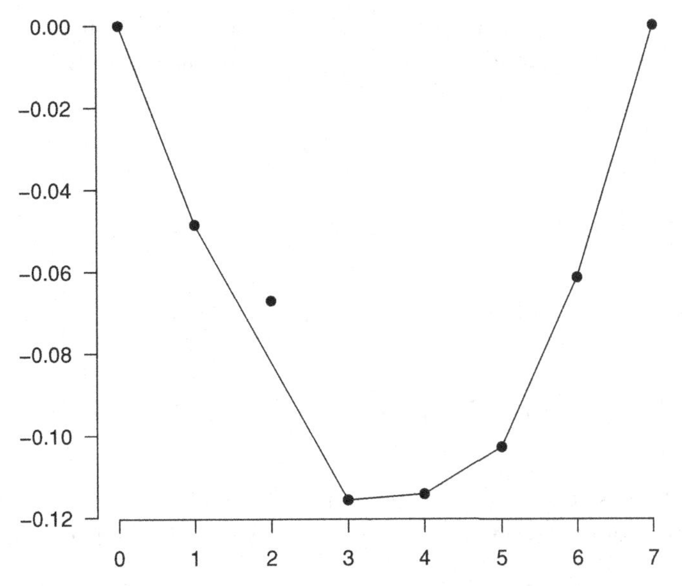

Figure 2.1 The cusum diagram based on the values $\bar{x}(i) - \bar{x}$ and its convex minorant. This gives $\hat{r} = (1.590, 1.605, 1.605, 1.640, 1.650, 1.680, 1.700)$ as solution to the isotonic regression problem.

sum diagram based on the points $y_i - \bar{y}$, $1855 \leq i \leq 2012$. A useful corollary of Lemma 2.1 that will be used later in the book is the following.

Corollary 2.1 *Let $\hat{r}$ be the isotonic regression of y with weights w as characterized in Lemma 2.1. Then, for every $r \in C$,*

$$\sum_{i=1}^{n} (\hat{r}_i - y_i) r_i w_i \geq 0.$$

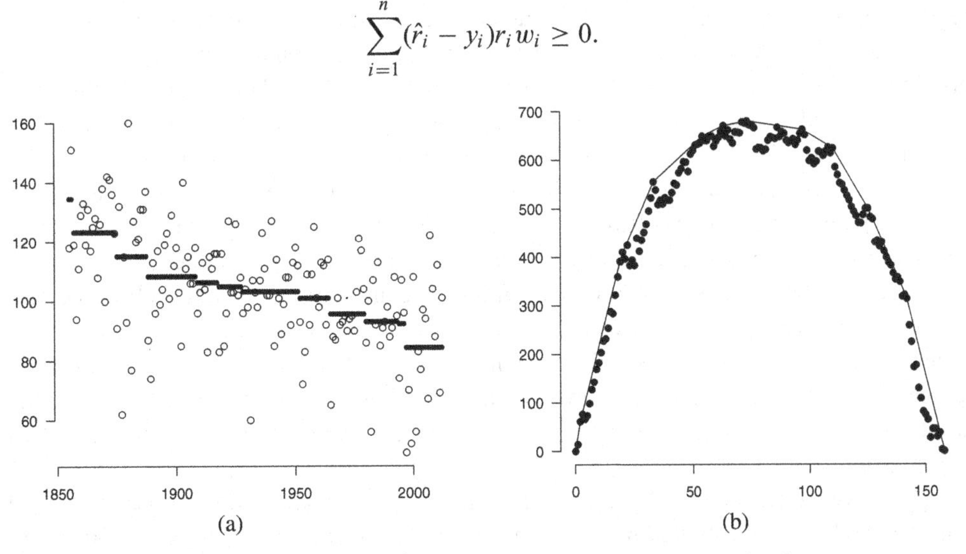

Figure 2.2 (a) Scatter plot of the duration of freezing (in days) of Lake Monona against the year. (b) The cusum diagram based on the centered (by mean subtraction) temperatures together with its least concave majorant. The resulting antitonic regression is added to the scatter plot.

Proof Define for $0 \le i \le n-1$ the vectors $\mu^{(i)} \in \mathbb{R}^n$ such that it contains i zeros followed by $n-i$ ones. Then, write for arbitrary $r \in C$ and the isotonic regression $\hat{r}$ at hand

$$r = \sum_{j=0}^{n-1} \alpha_j \mu^{(j)}.$$

Here $\alpha_j \ge 0$ for $1 \le j \le n-1$ and $\alpha_0 \in \mathbb{R}$. Then

$$\sum_{i=1}^{n} (\hat{r}_i - y_i) r_i w_i = \sum_{i=1}^{n} (\hat{r}_i - y_i) \sum_{j=0}^{n-1} \alpha_j \mu_i^{(j)} w_i = \sum_{j=0}^{n-1} \alpha_j \sum_{i=1}^{n} (\hat{r}_i - y_i) \mu_i^{(j)} w_i$$

$$= \alpha_0 \sum_{i=1}^{n} (\hat{r}_i - y_i) w_i + \sum_{j=1}^{n-1} \alpha_j \sum_{i=j+1}^{n} (\hat{r}_i - y_i) w_i \ge 0.$$

Indeed, the first term is zero by the equality part of (2.3) with $i = n$. For the sum at the right hand side we know for each term that $\alpha_j \ge 0$ and, also taking the inequality part of (2.3) into account,

$$\sum_{i=j+1}^{n} (\hat{r}_i - y_i) w_i \ge 0.$$

□

2.2 Monotone Density Estimation

Estimating a probability density based on an independent and identically distributed sample generated by it is a common problem in mathematical statistics and application fields. In the parametric context, the problem is reduced to estimating a finite dimensional parameter from the data. If only smoothness assumptions are imposed on the sampling density, often kernel or spline estimators are applied to estimate the density based on the given data. In this section we introduce the problem of estimating a density on $[0, \infty)$ based on a sample from its corresponding distribution, only using the assumption that the density is decreasing.

There are various contexts where this problem arises. One example is related to the inspection paradox described in Ross, 2010, Section 7.7. The aim is to estimate the interarrival distribution of a stationary renewal process, based on data from n independent realizations of the process. For each process the residual waiting time until the next event is observed, from time point zero. This observation can be viewed as uniform random fraction of a draw from the length-biased distribution associated with the interarrival distribution. In the following we derive the equation relating the interarrival distribution function F of interest with the sampling density. It will be seen that the sampling scheme imposes the sampling density to be decreasing.

Suppose we have a sample $Z_1, \ldots, Z_n$ from the length-biased distribution associated with an unknown distribution function F of interest. This means that the distribution function of Z_i is given by

$$\bar{F}(z) = P(Z_i \le z) = \frac{1}{m_F} \int_0^z x \, dF(x) \tag{2.4}$$

where $m_F = \int_0^\infty x\, dF(x)$ is assumed to be nonzero and finite. However, instead of observing the values of Z_i directly, we only observe the data $X_1, \ldots, X_n$ where X_i is a uniform random fraction of Z_i. More specifically, we observe

$$X_i = U_i Z_i,$$

where $U_1, \ldots, U_n$ is a random sample from the uniform distribution on $[0, 1]$, independent of the Z_is. Now the density of X_i can be seen to be

$$g(x) = \frac{1}{m_F}(1 - F(x)), \quad x \geq 0. \tag{2.5}$$

See also Exercise 2.4. Hence, by monotonicity of the initial distribution function F and the fact that $0 < m_F < \infty$, it follows that g is bounded and decreasing on $[0, \infty)$. Moreover, if no additional assumptions are imposed on F, any density of this type can be represented by (2.5).

We now introduce the (nonparametric) maximum likelihood estimator for the density g based on a realized sample $x_1, \ldots, x_n$ from that density. This estimator is also known as the Grenander estimator. Denoting the empirical distribution function of these data by $\mathbb{G}_n$, the log likelihood of a specific density g is given by

$$\ell(g) = \frac{1}{n}\sum_{i=1}^n \log g(x_i) = \int \log g(x)\, d\mathbb{G}_n(x).$$

Maximizing this function over all decreasing densities on $[0, \infty)$ boils down to maximizing ℓ over the set

$$\mathcal{G} = \left\{g\colon [0, \infty) \to [0, \infty)\colon g \text{ decreasing and } \int g(x)\, dx = 1\right\}.$$

Lemma 2.2 *The function $\hat{g}$ maximizing ℓ over the function class $\mathcal{G}$ is constant on intervals of the type $(x_{(i-1)}, x_{(i)}]$, where $x_{(0)} = 0$ by convention and $x_{(1)} < x_{(2)} < \cdots < x_{(n)}$ are the ordered observations $x_1, \ldots, x_n$. The value $\hat{g}(x_{(i)})$ is the left derivative of the least concave majorant $\hat{G}_n$ of (i.e., least concave function dominating) the empirical distribution function $\mathbb{G}_n$ of the data evaluated at the point $x_{(i)}$.*

Proof We first prove that the maximizer of ℓ over $\mathcal{G}$ has to belong to the subclass $\mathcal{G}_p$ of $\mathcal{G}$ consisting of functions that are constant on intervals of the type $(x_{(i-1)}, x_{(i)}]$. Let g be an arbitrary bounded decreasing density on $[0, \infty)$ with $\ell(g) > -\infty$ (i.e. $g(x_{(n)}) > 0$) and define the function $\tilde{g}$ by

$$\tilde{g}(x) = \begin{cases} g(x_{(i)}) & \text{for } x_{(i-1)} < x \leq x_{(i)} \\ 0 & \text{for } x > x_{(n)} \end{cases}$$

See Figure 2.3a. Because g is decreasing, $\tilde{g}(x) \leq g(x)$ for all x, implying that $\tilde{g}$ is a subdensity with $0 < a = \int \tilde{g}(x)\, dx \leq 1$. Moreover, $\ell(\tilde{g}) = \ell(g)$ since $\tilde{g}(x_{(i)}) = g(x_{(i)})$ for all $1 \leq i \leq n$. Hence, defining the density $\bar{g} = \tilde{g}/a \in \mathcal{G}_p$ (see Figure 2.3b), we see that $\ell(\bar{g}) = \ell(g) - \log a \geq \ell(g)$, implying that in maximizing ℓ over $\mathcal{G}$, attention may be restricted to $\mathcal{G}_p$.

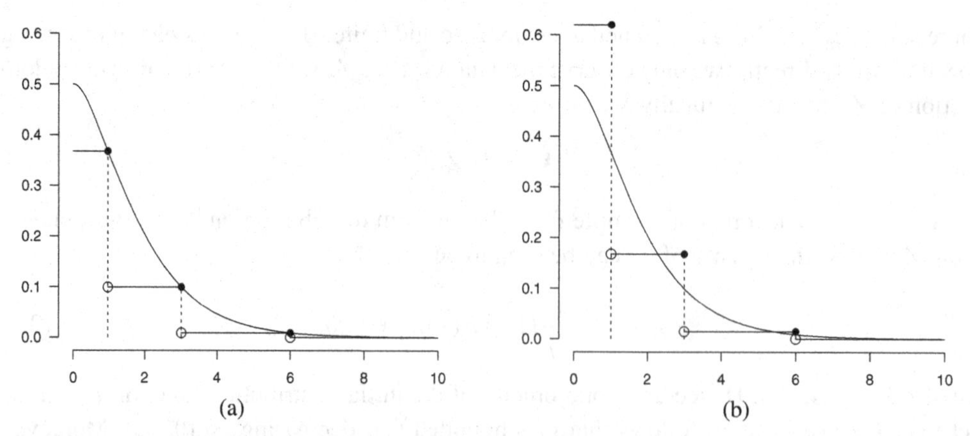

Figure 2.3 (a) An "arbitrary" decreasing density g on $[0, \infty)$ together with the piecewise constant version $\tilde{g}$ coinciding with g at the data points (in this example $1, 3$ and 6). (b) Rescaling the resulting subdensity to a density $\bar{g}$ yields piecewise constant density. Clearly, $\bar{g}(x) > g(x)$ at the data points.

Define the function $\hat{\ell}$ by

$$\hat{\ell}(g) = \int \log g(x)\, d\hat{G}_n(x).$$

We now prove that $\hat{g}_n$ as defined here indeed maximizes ℓ over $\mathcal{G}_p$ by showing that for each $g \in \mathcal{G}_p$,

$$\ell(g) \le \hat{\ell}(g) \le \hat{\ell}(\hat{g}_n) = \ell(\hat{g}_n). \tag{2.6}$$

Without loss of generality, we take g with $g(x_{(n)}) > 0$.

For the first inequality in (2.6), note that the (piecewise constant) logarithm of g can be represented as

$$\log g(x) = \sum_{i=1}^{n} \alpha_i 1_{[0,x_{(i)}]}(x),$$

with $\alpha_i \ge 0$ for $1 \le i \le n-1$ and $\alpha_n = \log g(x_{(n)}) \in \mathbb{R}$. This implies that

$$\hat{\ell}(g) = \sum_{i=1}^{n} \alpha_i \hat{G}_n(x_{(i)}) \ge \sum_{i=1}^{n} \alpha_i \mathbb{G}_n(x_{(i)}) = \ell(g),$$

where we use that $\hat{G}_n(x) \ge \mathbb{G}_n(x)$ for all x and $\hat{G}_n(x_{(n)}) = \mathbb{G}_n(x_{(n)})$.

The second inequality in (2.6) follows from Jensen's inequality:

$$\hat{\ell}(g) - \hat{\ell}(\hat{g}_n) = \int \log \frac{g(z)}{\hat{g}_n(z)}\, d\hat{G}_n(z) \le \log \int \frac{g(z)}{\hat{g}_n(z)}\, d\hat{G}_n(z) = \log 1 = 0.$$

Observing that $z \mapsto \log \hat{g}_n(z)$ is constant on intervals $(u, v]$, where u and v are successive vertices of $\hat{G}_n$ and the functions $\mathbb{G}_n$ and $\hat{G}_n$ have equal increments on these intervals, the equality in (2.6) easily follows. □

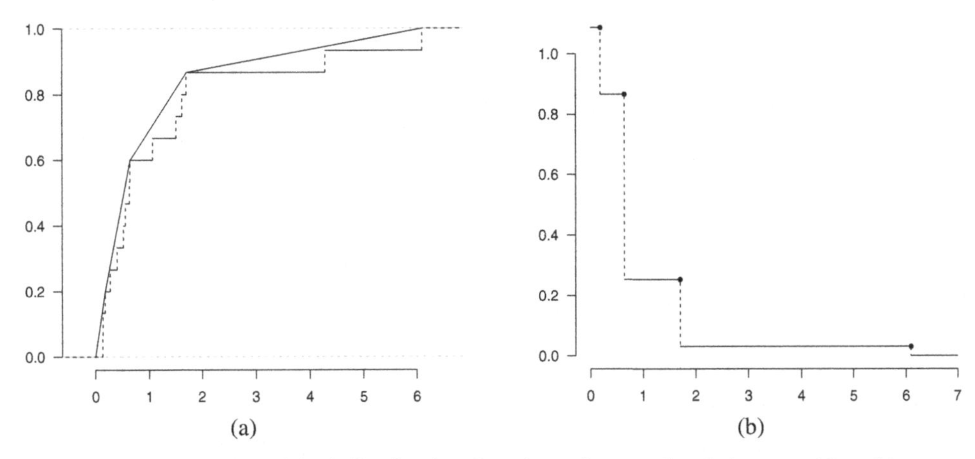

Figure 2.4 (a) Empirical distribution function of a sample of size $n = 15$ and its concave majorant. (b) The resulting Grenander estimate.

Figure 2.4 shows the empirical distribution function of a sample of size 15 generated from the standard exponential distribution together with its concave majorant. The right picture shows the resulting Grenander estimate as derivative of this concave majorant.

Example 2.3 In Slama et al., 2012, an interesting data set of current durations of pregnancy in France is studied. The aim is to estimate the distribution of the time it takes for a woman to become pregnant after having started unprotected sexual intercourse. For 867 women the current duration of unprotected intercourse was recorded and this is the basis of part of the research reported in Slama et al., 2012.

Given that the women in the study are currently trying to become pregnant, the actual recorded data (current duration) can be viewed as uniform random fraction of the true, total duration. In that sense, the model as given in (2.5) is not unreasonable. Figure 2.5a shows a part of the empirical distribution function of 618 recorded current durations, kindly provided to us by Niels Keiding, where the data are truncated at 36 months and are of a nature similar to the data in Slama et al., 2012. Based on the least concave majorant, Figure 2.5b is computed, showing the resulting MLE of the decreasing density of the observations together with its smoothed version, the smoothed maximum likelihood estimator (SMLE), defined by

$$\tilde{f}_{nh}(t) = -\int \mathbb{K}((t-x)/h)\, d\hat{f}_n(x), \qquad \mathbb{K}(x) = \int_x^\infty K(u)\, du, \tag{2.7}$$

where $\hat{f}_n$ is the Grenander estimator (the MLE) and K is a symmetric kernel, for which we took the triweight kernel

$$K(u) = \frac{35}{32}\left(1-u^2\right)^3 1_{[-1,1]}(u), \qquad u \in \mathbb{R}.$$

The bandwidth h was defined by

$$h = 36n^{-1/5} \approx 9.95645,$$

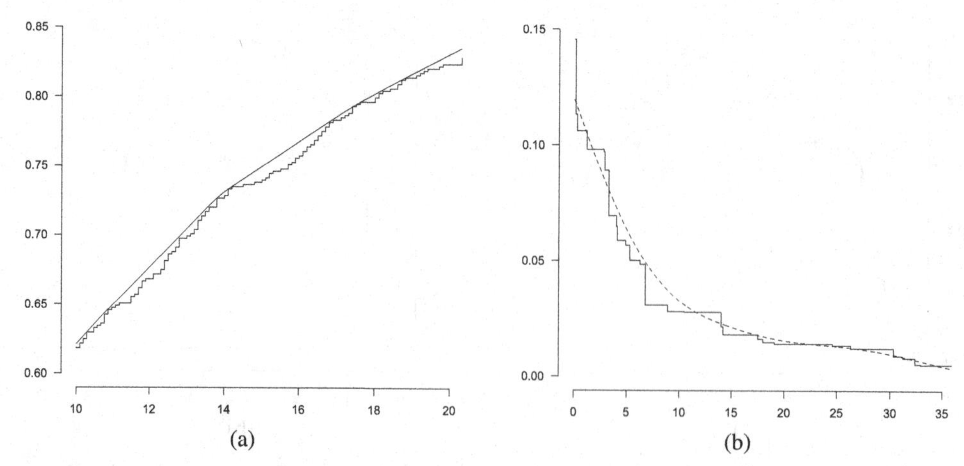

Figure 2.5 (a) The empirical distribution function and its least concave majorant for the values between 10 and 20 months of the 618 current durations $\le$ 36 months. (b) The resulting Grenander estimate (the MLE) of the observation density on the interval [0, 36], together with its smoothed version (dashed, the SMLE)

where $n = 618$. Near the boundary points 0 and 36 the boundary correction, defined in Section 9.2, was applied. The SMLE is asymptotically equivalent to the ordinary density estimator

$$\int K_h(t-x)\,d\mathbb{F}_n(x), \qquad K_h(u) = h^{-1}K(u/h), \tag{2.8}$$

(with the boundary correction corresponding to the boundary correction in 9.2), which, however, will in general not be monotone, so will not belong to the allowed class.

The survival function for the time until pregnancy or end of the period of unprotected intercourse is given by $g(x)/g(0)$, where g is the density of the observations, see Exercise 2.4. The 95% confidence for the survival function at the 99 equidistant points $0.36, 0.72, \dots, 35.64$, are constructed from 1,000 bootstrap samples $T_1^*, \dots, T_n^*$, also of size n, drawn from the original sample, and in these samples we computed

$$\tilde f_{nh}^*(t)/\tilde f_{nh}^*(0) - \tilde f_{nh}(t)/\tilde f_{nh}(0), \tag{2.9}$$

where $\tilde f_{nh}$ and $\tilde f_{nh}^*$ are the SMLEs in the original sample and the bootstrap sample, respectively. The chosen bandwidth was $36n^{-1/4} \approx 7.2203$, so (according to the method of undersmoothing, see Section 9.5) smaller than the bandwidth in Figure 2.5, which uses a rate for which the squared bias and variance are approximately in equilibrium. The 95% asymptotic confidence intervals are given by

$$\left[\tilde f_{nh}(t)/\tilde f_{nh}(0) - U_{0.975}^*,\, \tilde f_{nh}(t)/\tilde f_{nh}(0) - U_{0.025}^*\right],$$

where $U_{0.025}^*$ and $U_{0.975}^*$ are the 2.5% and 97.5% percentiles of the bootstrap values (2.9). The result is shown in part (a) of Figure 2.6 and should be compared with the confidence intervals

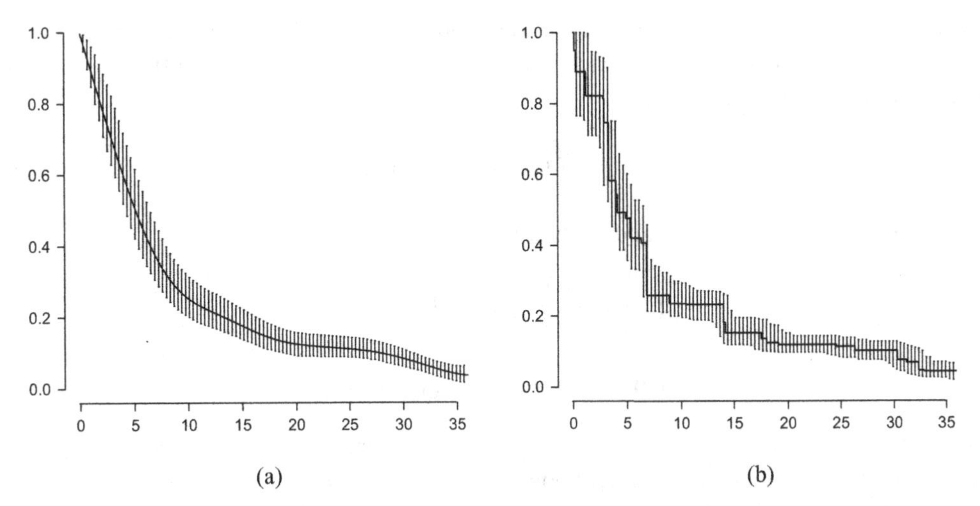

Figure 2.6 95% confidence intervals, based on the SMLE (part (a)) and MLE (part (b)), respectively, for the survival functions in Slama et al. (2012) at the points 0.36, 0.72, ..., 35.64. The chosen bandwidth for the SMLE was $36n^{-1/4} \approx 7.2203$ and the MLE was restricted to have the same value as the (consistent) SMLE at zero.

in part A of Figure 2, p. 1495 of Slama et al., 2012, based on a parametric (generalized gamma) model.

We have here the easiest, but also somewhat unusual, situation in which the isotonic estimator is asymptotically equivalent to an ordinary nonisotonic estimator. The more usual situation is that we only can find a so-called toy estimator, which is asymptotically equivalent to the MLE or SMLE, but still contains parameters that have to be estimated. We will see examples of the latter in the current status and, more generally, the interval censoring model.

In Slama et al., 2012, and Keiding et al., 2012, parametric models are also considered for analyzing these data. We compute the MLE as the slope of the smallest concave majorant of the data ≤ 36 months, where the x values are only the strictly different values, and where we use the number of values at a tie as the increase of the second coordinate of the cusum diagram. In this way we get 618 values ≤ 36, but only 248 strictly different ones. It is clear that the SMLE has a somewhat intermediate position with respect to the parametric models and the fully nonparametric MLE, considered in Slama et al., 2012, and Keiding et al., 2012. The fully nonparametric MLE is inconsistent at zero and can therefore not be used as an estimate of $g(0)$ and therefore also not as an estimate of the survival function $g(x)/g(0)$. This is in contrast with the SMLE, which is consistent at zero; see Section 9.2. This difficulty with the inconsistency of the MLE at zero for the present model is discussed in Keiding et al., 2012.

Remark The mentioned difficulties have very recently been resolved in Groeneboom and Jongbloed, 2014, see the discussion at the end of Section 9.5. We can perform likelihood ratio tests with MLEs that are restricted to have the value of a consistent estimator at zero. The 95% confidence intervals based on the MLEs are shown in part (b) of Figure 2.6. For a further discussion of these matters, see Groeneboom and Jongbloed, 2014.

As alternative nonparametric estimator, one could use a least squares density estimate. This estimate minimizes an L_2 distance to an initial nonmonotone density estimate f_n over the class of decreasing densities:

$$Q(g) = \int (g(x) - f_n(x))^2 \, dx, \tag{2.10}$$

where, defining $x_0 = 0$, the initial density estimate f_n is given by

$$f_n(x) = \begin{cases} \dfrac{w(x_i)}{n(x_i - x_{i-1})} & , \quad \text{for } x \in (x_{i-1}, x_i], \\ 0 & , \quad \text{elsewhere}, \end{cases} \tag{2.11}$$

and where $w(x_i)$ is the number of observations in the interval $(x_{i-1}, x_i]$. This least squares estimate is exactly the maximum likelihood estimate. (See Exercise 2.5.)

The model considered in this section is a so-called mixture model, in the sense that it is a scale mixture of uniform distributions:

$$g(x) = \int_{(0,\infty)} \frac{1}{t} 1_{[0,t]}(x) \, dQ(t).$$

2.3 Estimating a Distribution Function from Current Status Data

A problem that is often encountered in reliability theory is that of estimating a distribution function of the lifetime of a certain product. In medical science, conceptually the same question is relevant when the distribution of the survival time of a person with a certain disease is to be estimated. In business, a similar question arises when one is interested in the duration of a subscription to a certain newspaper. The whole of the methods and techniques within the field of statistics that is related to these questions is called survival analysis.

One problem that is typical in survival analysis is that of censoring. One would like to have a sample from the distribution of interest, but because of some mechanism, it is not possible to observe such a sample. To illustrate the current status model, we will use an interesting data set studied in Keiding et al., 1996.

In order to investigate the age-specific immunization intensity of rubella seroprevalence, a cross-sectional sample of 230 Austrian males older than 3 months was taken during the period 1–25 March 1988. Of interest is the distribution of the time to infection by rubella in the male population, whereas it is assumed that immunization, once achieved, is lifelong. For each person, the exact birthday was known and the current immunization status was tested at the Institute of Virology, Vienna. In particular, the exact time of immunization is not known. The data are represented graphically in Figure 2.7. Each individual in the study corresponds to a horizontal line in the picture. This line denotes the set of possible values of T for this individual. An individual with $\Delta = 1$ is denoted by a line starting at zero, ending at the censoring time T. An observation with $\Delta = 0$ starts at the observed censoring time T and continues to the right till the maximum age, which we (somewhat arbitrarily) put equal to 100 here.

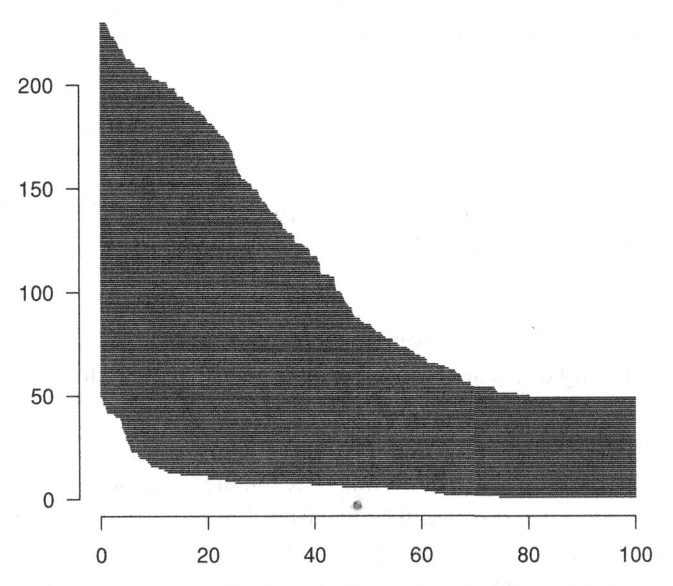

Figure 2.7 Rubella data studied in Keiding et al., 1996.

Imposing some assumptions, we use the following model for the data. Consider a sample $X_1, X_2, \ldots, X_n$ is drawn from a distribution with distribution function F. However, instead of observing the X_is, one only observes for each i whether or not $X_i \le T_i$ for some random T_i (independent of the other T_js and all X_js). More formally, instead of observing X_is, one observes

$$(T_i, \Delta_i) = (T_i, 1_{[X_i \le T_i]}). \tag{2.12}$$

One could say that the ith observation represents the current status of item i at time T_i.

Now the problem is to estimate the unknown distribution function F based on the data given in (2.12). Denote the ordered realized T_is by $t_1 < t_2 < \cdots < t_n$ and the associated realized values of the Δ_is by $\delta_1, \ldots, \delta_n$. For this problem the log likelihood function in F (conditional on the T_is) is given by

$$\ell(F) = \sum_{i=1}^{n} (\delta_i \log F(t_i) + (1 - \delta_i) \log(1 - F(t_i))) \tag{2.13}$$

(see also Exercise 2.9). The (nonparametric) maximum likelihood estimator maximizes ℓ over the class of *all* distribution functions. Since distribution functions are by definition nondecreasing, computing the maximum likelihood estimator poses a shape restricted optimization problem in a natural way. As can be seen from the structure of (2.13), the value of ℓ only depends on the values that F takes at the observed time points t_i; the values of F in between are not relevant as long as F is nondecreasing. Hence one can choose to consider only distribution functions that are constant between successive observed time points t_i. Lemma 2.3 shows that this estimator can also be characterized in terms of a convex minorant of a certain diagram of points.

Lemma 2.3 *Consider the cumulative sum diagram consisting of the points* $P_0 = (0, 0)$ *and*

$$P_i = \left(i, \sum_{j=1}^{i} \delta_j \right), \quad 1 \le i \le n,$$

recalling that the δ_i*s correspond to the* t_i*s, which are sorted. Then* $\hat{F}(t_i)$ *is given by the left derivative of the convex minorant of this diagram of points, evaluated at the point* i*. This maximizer is unique.*

Remark The *left* derivative of the convex minorant at P_i determines the value of $\hat{F}$ at t_i and hence (by right continuity of the step function) on $[t_i, t_{i+1})$, a region to the *right* of t_i.

Proof First note that for maximizing the likelihood in terms of F, we may without loss of generality assume that $\delta_1 = 1 - \delta_n = 1$. If this were not the case, the $F(t_i)$s corresponding to the first sequence of δ_is equal to zero can be fixed at value zero and the $F(t_i)$s corresponding to the last string of δ_is equal to one can be fixed at value one. Consequently, the terms $\log F(t_1)$ and $\log(1 - F(t_n))$ force the range constraints for distribution functions to be automatically satisfied by the MLE.

Next, note that the function ℓ in (2.13) is strictly concave in the vector $(F(t_1), \ldots, F(t_n))$, giving uniqueness of the maximizer if it exists. Hence, showing for the constructed vector $(\hat{F}(t_1), \ldots, \hat{F}(t_n))$ that for every vector $(F(t_1), \ldots, F(t_n))$ satisfying $F(t_1) \le \cdots \le F(t_n)$,

$$\lim_{\epsilon \downarrow 0} \epsilon^{-1} \left(\ell(\hat{F} + \epsilon(F - \hat{F})) - \ell(\hat{F}) \right) \le 0, \tag{2.14}$$

the result follows. In other words: starting at $\hat{F}$, no feasible direction of ascent can be found for ℓ.

Now, using that $\log(1 + u) \sim u$ for $u \to 0$, observe that

$$\lim_{\epsilon \downarrow 0} \epsilon^{-1} \left(\ell(\hat{F} + \epsilon(F - \hat{F})) - \ell(\hat{F}) \right) = \sum_{i=1}^{n} \frac{(F(t_i) - \hat{F}(t_i))(\delta_i - \hat{F}(t_i))}{\hat{F}(t_i)(1 - \hat{F}(t_i))} = I_1 - I_2.$$

Writing $0 = i_0 < i_1 < \cdots < i_k = n$ for the locations of the bend points of the convex minorant of the cumulative sum diagram and using that $\hat{F}(t_i) = \hat{F}(t_{i_j})$ for $i_{j-1} < i \le i_j$, the second term can be written as

$$I_2 = \sum_{j=1}^{k} \sum_{i=i_{j-1}+1}^{i_j} \frac{\delta_i - \hat{F}(t_i)}{1 - \hat{F}(t_i)} = \sum_{j=1}^{k} \frac{1}{1 - \hat{F}(t_{i_j})} \sum_{i=i_{j-1}+1}^{i_j} (\delta_i - \hat{F}(t_i)) = 0.$$

The last equality follows since the inner summation is zero, essentially because the increase of the cumulative sum diagram and the increase of its convex minorant are the same between consecutive points where these coincide. For I_1, again using that $\hat{F}(t_i) = \hat{F}(t_{i_j})$ for $i_{j-1} < i \le i_j$,

$$I_1 = \sum_{j=1}^{k} \frac{1}{\hat{F}(t_{i_j})(1 - \hat{F}(t_{i_j}))} \sum_{i=i_{j-1}+1}^{i_j} F(t_i)(\delta_i - \hat{F}(t_i)).$$

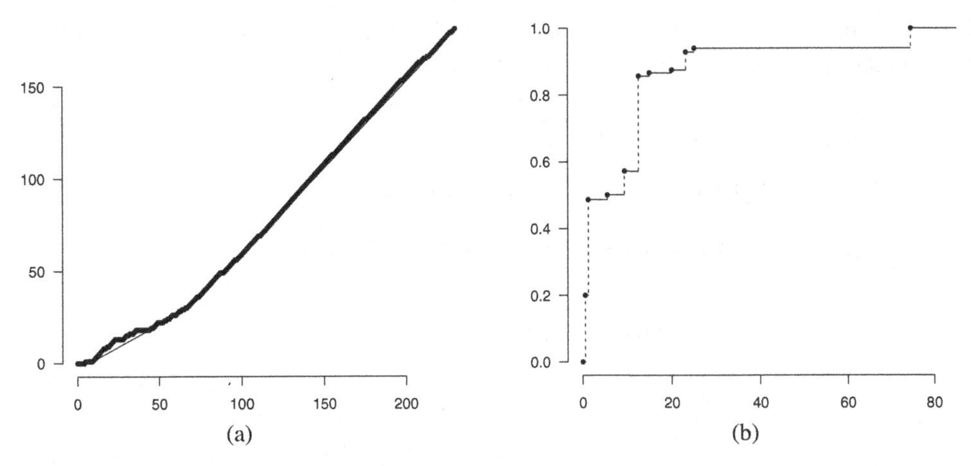

Figure 2.8 (a) The cumulative sum diagram based on the Rubella data and its convex minorant. (b) The associated MLE of the distribution function F.

Now, consider the inner sum, and for $i_{j-1} < i \le i_j$ represent $F(t_i) = \sum_{m=i_{j-1}+1}^{i} \alpha_m$ with $\alpha_m \ge 0$. Then

$$\sum_{i=i_{j-1}+1}^{i_j} F(t_i)(\delta_i - \hat{F}(t_i)) = \sum_{i=i_{j-1}+1}^{i_j} \left(\sum_{m=i_{j-1}+1}^{i} \alpha_m \right) (\delta_i - \hat{F}(t_i)) = \sum_{m=i_{j-1}+1}^{i_j} \alpha_m \sum_{i=m}^{i_j} (\delta_i - \hat{F}(t_i)) \le 0.$$

This last inequality follows from the interpretation of the convex minorant. (See also Exercise 2.11.) This proves that $I_1 \le 0$ and hence (2.14). □

Figure 2.8 shows the MLE based on the rubella data, obtained via the characterization in Lemma 2.3.

2.4 Deconvolution Problem with Jump Kernel

Consider a sample $X_1, \ldots, X_n$ generated by an unknown distribution function F. Instead of observing this sample directly, a sample $Z_1, \ldots, Z_n$ is observed where $Z_i = X_i + Y_i$ for each i. Here the Y_is are independent and identically distributed with (known) probability density k. Moreover, all Xs and Ys are independent. This means that the distribution of Z_i is given by the convolution of the k and F. In this case it has a density g, given by

$$g(z) = [k * dF](z) = \int_{-\infty}^{\infty} k(z-x)\, dF(x). \tag{2.15}$$

Since k is known, one can state the estimation problem as that of "deconvolving" G with k. In Section 4.6 this problem will be studied in more generality. Here we consider a specific class of deconvolution models, where X and Y are nonnegative and the kernel k can be represented as

$$k(x) = k(0)\left(1 - \int_0^x l(u)\, du\right), \quad x \ge 0 \tag{2.16}$$

where l is a bounded measurable function on $[0, \infty)$. Then, it can be shown that the (type one) resolvent p of k as solution of the integral equation

$$[p * k](x) = \int_0^x p(z)k(x - z)\,dz = x \tag{2.17}$$

is well defined. Using properties of convolutions, we get

$$\int_0^x p(x - z)g(z)\,dz = [p * g](x) = [p * k * dF](x)$$
$$= [id * dF](x) = \int_0^x (x - y)\,dF(y) = \int_0^x F(y)\,dy.$$

Hence,

$$F(x) = \frac{d}{dx}\int_0^x p(x - z)g(z)\,dz. \tag{2.18}$$

Example 2.4 *(Exponential deconvolution)* Suppose $k(x) = e^{-x}$ on $(0, \infty)$ and $F(0) = 0$. Then

$$p(x) = 1 + x \text{ on } [0, \infty) \tag{2.19}$$

and the explicit inverse relation boils down to

$$F(x) = g(x) + \int_0^x g(y)\,dy. \tag{2.20}$$

This relation can be used to estimate F based on a sample from g. Indeed, denote by G the distribution function corresponding to g and define the convex function

$$U(x) = \int_0^x F(y)\,dy = G(x) + \int_0^x G(y)\,dy.$$

This function can be estimated by its empirical counterpart

$$U_n(x) = \mathbb{G}_n(x) + \int_0^x \mathbb{G}_n(y)\,dy$$

where $\mathbb{G}_n$ is the empirical distribution function of the observed data. Note that U_n is an increasing function that is linear between successive observation points. At these points the function has jumps of size $1/n$ and after each jump the slope of the function is increased by $1/n$. Obviously, U_n is not differentiable. See Figure 2.9, where the empirical distribution function $\mathbb{G}_n$ based on a sample of size $n = 10$ is shown and the function U_n corresponding to it.

One could define an estimator for F directly based on U_n by, e.g., taking the derivative of the piecewise linear function connecting the points $(z_i, U_n(z_i-))$. However, this derivative will in general not be monotone because the linearly interpolated function U_n is not convex. Moreover, it will generally not satisfy the range constraints for distribution functions, that it takes values in $[0, 1]$. See Figure 2.10a. In the spirit of estimators previously met in this chapter, one can define an estimator $\hat{F}_n$ as the right derivative of the (greatest) convex minorant of the function U_n. Obviously, $\hat{F}_n(0) = 0$, $\lim_{x\to\infty} \hat{F}_n(x) = 1$, $\hat{F}_n$ is monotone and right continuous. Hence, it is a proper estimator of F. See Figure 2.10b.

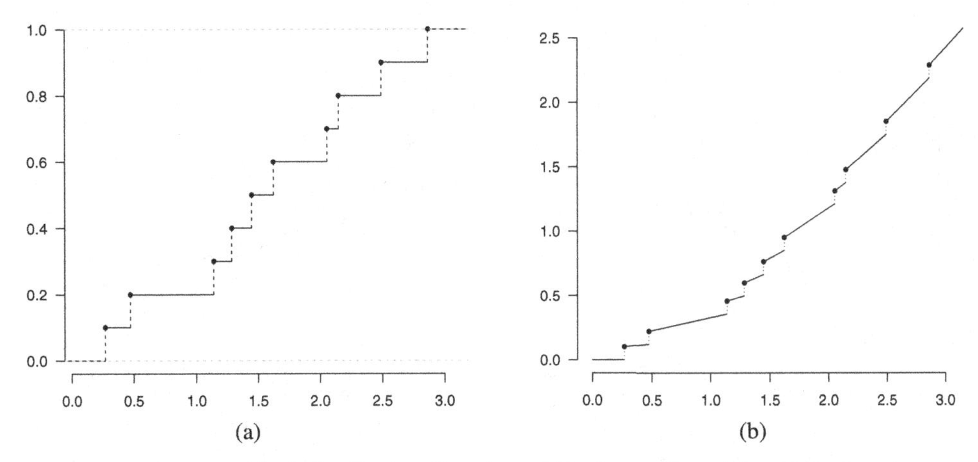

Figure 2.9 Empirical distribution function of 10 observations (a) and the function U_n based on this (b).

2.5 Generalized Isotonic Regression Problems

In Section 2.1 the constructive solution to the isotonic regression problem was derived. In particular, the least squares estimator for the monotone function can be obtained by taking the derivative of a convex minorant of a diagram of points. Interestingly, estimators other than least squares estimators for monotone functions can be computed in the same way. An example was already encountered in Section 2.3, where the maximum likelihood estimator in the current status model was seen to equal a particular least squares estimator. A whole class of estimation problems where this phenomenon occurs is the class of generalized isotonic regression problems. We first give an example.

Example 2.5 *(Poisson extremum problem)* Let $X_1, X_2, \ldots, X_n$ be n independent random variables such that X_i has the Poisson distribution with parameter λ_i and, moreover, the

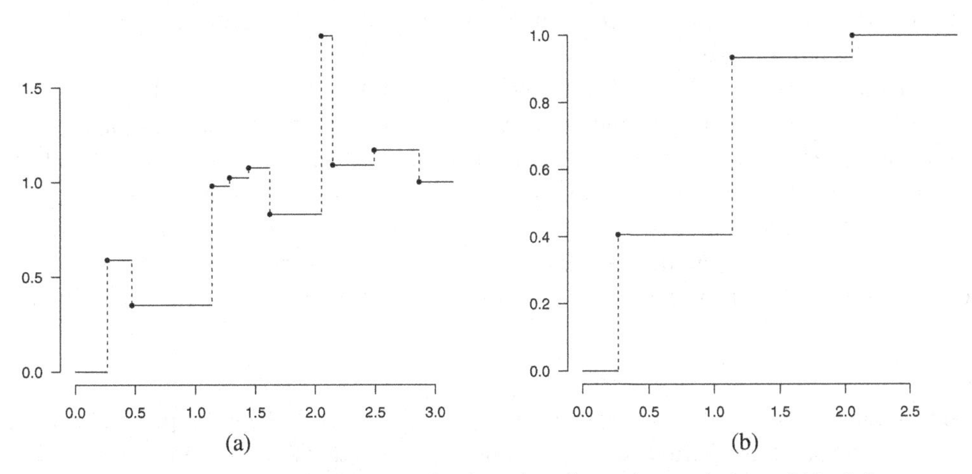

Figure 2.10 Naive estimator of F based on linear interpolation of (the left continuous version of) U_n in Figure 2.9 (a) and the estimator based on the convex minorant of U_n (b).

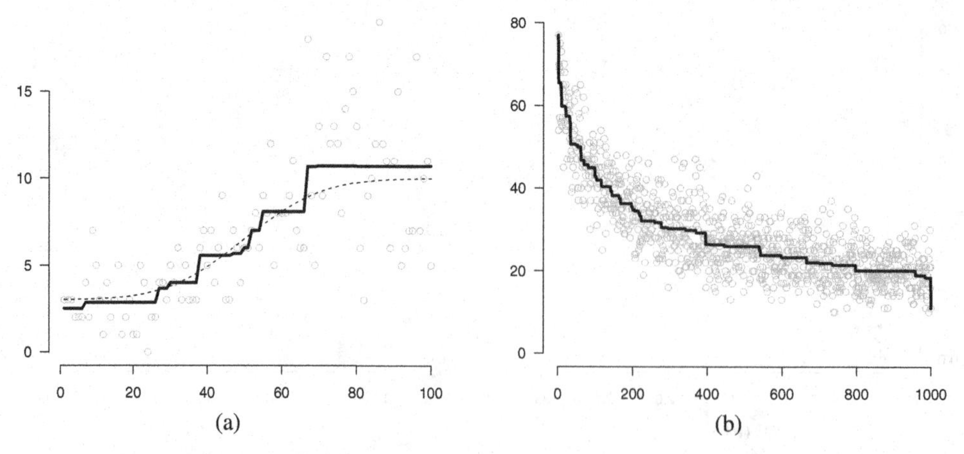

Figure 2.11 (a) Scatter plot of a data set of size $n = 100$ with underlying parameters λ_i (dashed curve) in the Poisson extremum problem. (b) A measured luminescence signal, together with the maximum likelihood estimates of the (decreasing) Poisson parameters.

parameters are increasing: for $i = 1, \ldots, n$

$$P(X_i = k) = \frac{\lambda_i^k}{k!} e^{-\lambda_i} \text{ for } k = 0, 1, 2, \ldots \text{ and } 0 \le \lambda_1 \le \lambda_2 \le \cdots \le \lambda_n.$$

Denote the observed data by $x_1, \ldots, x_n$ (an example data set is given in Figure 2.11). Estimating the parameter vector $(\lambda_1, \ldots, \lambda_n)$ via maximum likelihood entails maximizing the function

$$\ell(\lambda_1, \ldots, \lambda_n) = \sum_{i=1}^{n} (x_i \log \lambda_i - \lambda_i) \tag{2.21}$$

over the ordered set of λ_is. One could think of an iterative procedure to maximize this function (see Exercise 2.16).

A situation where this type of data is encountered is in luminescence dating (see, e.g., Preusser et al., 2008). This is a technique that can be used to estimate the time since a sediment of soil was last exposed to sunlight. When a soil sample is exposed to light in a laboratory, the sample emits photons according to a process that can be modeled as Poisson process. For time intervals $(0, \delta], (\delta, 2\delta], \ldots, (T - \delta, T]$, independent Poisson counts are recorded for which the intensities can be assumed to be decreasing. The right panel of Figure 2.11 shows a scatter plot of measured counts, for a situation with $\delta = 0.04$ and $T = 40$ (both in seconds).

Example 2.6 (*Geometric extremum problem*) Let $X_1, X_2, \ldots, X_n$ be n independent random variables, geometrically distributed with parameter p_i such that the p_is are increasing. This means that for $1 \le i \le n$,

$$P(X_i = k) = p_i(1 - p_i)^{k-1} \text{ for } k = 1, 2, \ldots \text{ and } 0 \le p_1 \le p_2 \le \cdots \le p_n \le 1.$$

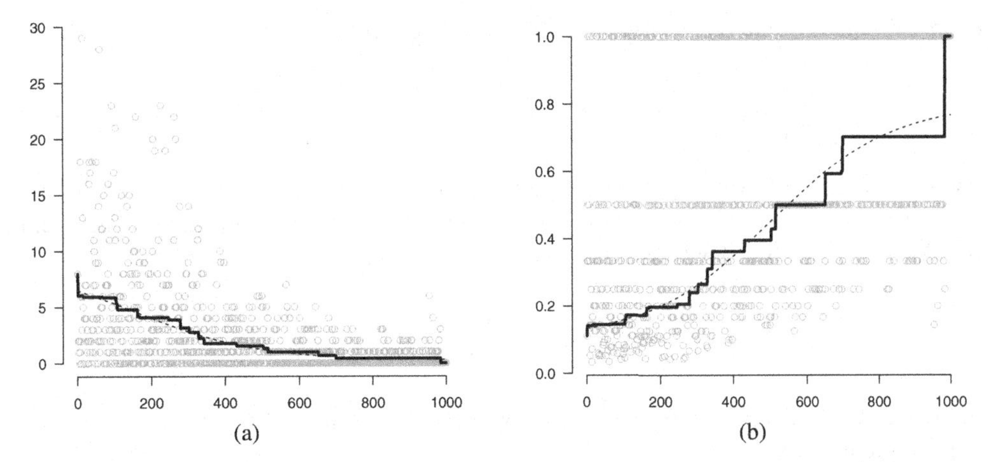

Figure 2.12 (a) A dataset of size $n = 1000$ from the Geometric extremum problem, with dashed true underlying values of $r_i = 1/p_i - 1$. The thick line indicates the maximum likelihood estimates of r_i as defined in Exercise 2.20. (b) The nonrestricted maximum likelihood estimates of p_i, $1/x_i$, as dots and the restricted maximum likelihood estimates of p_i as thick line.

Denote the observed data by $x_1, \ldots, x_n$. See Figure 2.12 for a scatter plot of a generated sample. The maximum likelihood estimator of parameter vector $(p_1, \ldots, p_n)$ can be obtained by maximizing the function

$$\ell(p_1, \ldots, p_n) = \sum_{i=1}^{n} ((x_i - 1)\log(1 - p_i) + \log p_i) \tag{2.22}$$

over the set of all n-tuples $0 \le p_1 \le p_2 \le \cdots \le p_n \le 1$.

Log likelihoods (2.21) and (2.22) do not immediately show that the maximizer can be constructed by computing the least convex minorant of a single cumulative sum diagram. To see that it is the case for these (and also other) examples, we identify a common structure in the optimization problems.

First, let Φ be a convex function on an interval $\mathbb{R}$, taking finite values on an interval I. Let ϕ be a function on I such that

$$\Phi(u) - \Phi(v) - \phi(v)(u - v) \ge 0$$

for all u and v in I. If Φ is differentiable at v, $\phi(v) = \Phi'(v)$; otherwise it can take any value between the left- and right derivative of Φ at v. Next, define the following function on $\mathbb{R}^2$

$$\Delta_\Phi(u, v) = \begin{cases} \Phi(u) - \Phi(v) - (u - v)\phi(v) & \text{if } u, v \in I \\ \infty & \text{if } u \notin I \text{ or } v \notin I. \end{cases} \tag{2.23}$$

Note that $\Delta_\Phi(u, v) \ge 0$ for all $u, v \in \mathbb{R}$ (see Exercise 2.17). Now we can prove the following theorem.

Theorem 2.1 *Consider the vector $y = (y_1, \ldots, y_n) \in \mathbb{R}^n$ and let $w = (w_1, \ldots, w_n) \in \mathbb{R}^n$ be a positive (weight) vector. Suppose that for all i, $y_i \in I$, the interval where Φ is finite. Let $r = (r_1, \ldots, r_n) \in \mathbb{R}^n$ be an isotonic vector, so with $r_1 \le \cdots \le r_n$, with $r_i \in I$ for all i.*

Denote by $\hat{r}$ the isotonic regression of y with weights given by w. Then

$$\sum_{i=1}^{n} \Delta_\Phi(y_i, r_i)w_i \geq \sum_{i=1}^{n} \Delta_\Phi(y_i, \hat{r}_i)w_i + \sum_{i=1}^{n} \Delta_\Phi(\hat{r}_i, r_i)w_i. \tag{2.24}$$

Before giving a proof of this theorem, note that it has some nice consequences. In view of Exercise 2.17, the second term at the right hand side of (2.24) is nonnegative. This yields the following inequality: for all isotonic vectors $r \in \mathbb{R}^n$ such that $r_i \in I$ for all i,

$$\sum_{i=1}^{n} \Delta_\Phi(y_i, \hat{r}_i)w_i \leq \sum_{i=1}^{n} \Delta_\Phi(y_i, r_i)w_i.$$

In other words: the isotonic regression $\hat{r}$ of y with weight function w does not only minimize the quadratic function

$$r \mapsto \frac{1}{2}\sum_{i=1}^{n} (r_i - y_i)^2 \, w_i$$

over all isotonic vectors $r \in \mathbb{R}^n$, but at the same time it minimizes

$$r \mapsto \sum_{i=1}^{n} \Delta_\Phi(y_i, r_i)w_i.$$

over the set of isotonic vectors r with $r_i \in I$ for all i.

A second, directly related and useful consequence is that $\hat{r}$ maximizes the function

$$r \mapsto \sum_{i=1}^{n} (\Phi(r_i) + (y_i - r_i)\phi(r_i)) \, w_i \tag{2.25}$$

over all isotonic $r \in \mathbb{R}^n$ with $r_i \in I$ for all i (see Exercise 2.18). Applying this result in the Poisson extremum problem in Example 2.5 with $\Phi(u) = u \log u$ on $I = (0, \infty)$ shows that the maximum likelihood estimator of $(\lambda_1, \ldots, \lambda_n)$ in the Poisson extremum problem is given by the isotonic regression of $(x_1, \ldots, x_n)$ with weights identically equal to one. (See Exercise 2.19). This observation was used to compute the estimates given in Figure 2.11. Similarly, Exercise 2.20 shows that maximizing (2.22) can be done by first computing the antitonic regression $\hat{r}$ of the vector $(x_1 - 1, \ldots, x_n - 1)$ with weights equal to one, and then setting $\hat{p}_i = 1/(\hat{r}_i + 1)$. The thick lines in Figure 2.11 indicate the ML estimators of the λ_is in the Poisson extremum problems. The left panel of Figure 2.12 shows the restricted (decreasing) ML estimates of the parameters $r_i = 1/p_i - 1$ in the geometric extremum problem. The right panel shows the corresponding ML estimates of the p_is as well as the ML estimates without using the monotonicity constraint as points.

Proof of Theorem 2.1 Subtracting the right-hand side of (2.24) from the left-hand side, we obtain

$$\sum_{i=1}^{n} (y_i - \hat{r}_i)(\phi(\hat{r}_i) - \phi(r_i))w_i.$$

Note that the vector $(\phi(r_1), \ldots, \phi(r_n))$ is isotonic because r is isotonic and ϕ is monotone. From Corollary 2.1 it then follows that

$$\sum_{i=1}^{n} (y_i - \hat{r}_i)\phi(r_i)w_i \le 0.$$

On the other hand, denote by $i_1 < i_2 < \cdots < i_{m-1}$ the indexes for which $\hat{r}_{i+1} > \hat{r}_i$ and take $i_0 = 0$ and $i_m = n$. Then

$$\sum_{i=1}^{n} (y_i - \hat{r}_i)\phi(\hat{r}_i)w_i = \sum_{j=1}^{m} \sum_{i=i_{j-1}+1}^{i_j} (y_i - \hat{r}_i)\phi(\hat{r}_i)w_i = \sum_{j=1}^{m} \phi(\hat{r}_{i_j}) \sum_{i=i_{j-1}+1}^{i_j} (y_i - \hat{r}_i)w_i.$$

The equality part of (2.3) shows that for $1 \le j \le m$,

$$\sum_{i=i_{j-1}+1}^{i_j} (y_i - \hat{r}_i)w_i = \sum_{i=1}^{i_j} (y_i - \hat{r}_i)w_i - \sum_{i=1}^{i_{j-1}+1} (y_i - \hat{r}_i)w_i = 0. \qquad \square$$

2.6 Estimating a Monotone Hazard Rate

In survival analysis and reliability theory, the hazard rate (also known as failure rate) is a natural function to model the distribution of data. It describes the probability of instantaneous failure at time x, given the subject has functioned until x. Let Z be a random variable with distribution function G_0 having density g_0 with respect to Lebesgue measure. Then at each point x where g_0 is continuous, the hazard rate h_0 of Z is defined by

$$h_0(x) = \lim_{\epsilon \downarrow 0} \epsilon^{-1} P(Z \in (x, x+\epsilon] | Z > x) = \lim_{\epsilon \downarrow 0} \frac{G_0(x+\epsilon) - G_0(x)}{\epsilon(1 - G_0(x))} = \frac{g_0(x)}{1 - G_0(x)}.$$

Note that this hazard rate indeed characterizes the distribution of Z, since

$$h_0(x) = -\frac{d}{dx} \log(1 - G_0(x)), \text{ implying } G_0(x) = 1 - \exp\left(-\int_0^x h_0(y)\,dy\right). \tag{2.26}$$

The exponential distributions are the only distributions with constant hazard rate, which is related to the memoryless property of this distribution. Other shapes of the hazard rate indicate whether the object suffers aging (increasing hazard rate) or is getting more reliable having survived longer (decreasing hazard rate).

Suppose we have a sample $Z_1, \ldots, Z_n$ from a distribution function G_0 on $[0, \infty)$, with density g_0 and hazard rate h_0. If one wants to estimate the hazard h_0 under the restriction that it is nondecreasing on the interval $[0, a]$, one of the simplest estimates is the least squares estimate $\hat{h}_n$, which minimizes the quadratic criterion

$$\tfrac{1}{2} \int_0^a h(x)^2\,dx - \int_{[0,a]} h(x)\,d\mathbb{H}_n(x), \tag{2.27}$$

under the restriction that h is monotone. Here $\mathbb{H}_n$ is the empirical cumulative hazard function

$$\mathbb{H}_n(x) = -\log\{1 - \mathbb{G}_n(x)\}, \ x < \max_i Z_i,$$

and $\mathbb{G}_n$ is the empirical distribution function of the sample $Z_1, \ldots, Z_n$. The rationale behind this criterion function is that $\mathbb{H}_n$ will be close to H_0 (defined as $\int_0^x h_0(y)\,dy$) asymptotically and $h \mapsto \frac{1}{2}\int_0^a h(x)^2\,dx - \int_0^a h(x)\,dH_0(x)$ is minimized by taking $h = h_0$ (which can be seen by completing the square).

Let $Z_{(1)}, \ldots, Z_{(n)}$ be the order statistics of the sample and let $Z_{(m)}$ be the largest order statistic $\le a$. To solve the minimization problem, we can first argue that, if $a < Z_{(n)}$, we can (Exercise 2.22) restrict the minimization problem to the problem of minimizing

$$\frac{1}{2}\sum_{i=1}^{m-1} h(Z_{(i)})^2 \left(Z_{(i+1)} - Z_{(i)}\right) + \frac{1}{2}h(Z_{(m)})^2 \left(a - Z_{(m)}\right) - \sum_{i=1}^{m} h(Z_{(i)})\Delta\mathbb{H}_{n,i}, \tag{2.28}$$

where $\Delta\mathbb{H}_{n,i} = \mathbb{H}_n(Z_{(i)}) - \mathbb{H}_n(Z_{(i)}-)$, and h is a nondecreasing and right-continuous function, constant on the intervals between successive jumps of the empirical distribution function. This is the isotonic regression problem of minimizing

$$\sum_{i=1}^{m}\left(h(Z_{(i)}) - \frac{\Delta\mathbb{H}_n(Z_{(i)})}{Z_{(i+1)} - Z_{(i)}}\right)^2 \left(Z_{(i+1)} - Z_{(i)}\right) + \left(h(Z_{(m)}) - \frac{\Delta\mathbb{H}_n(Z_{(m)})}{a - Z_{(m)}}\right)^2 \left(a - Z_{(m)}\right)$$

over nondecreasing functions h. So the solution at the points $Z_{(i)}$, $i = 1, \ldots, m$ is given by the left derivative of the cusum diagram, consisting of the points

$$\begin{aligned}
&(0,0), \left(Z_{(2)} - Z_{(1)}, \Delta\mathbb{H}_n(Z_{(1)})\right), \left(Z_{(3)} - Z_{(1)}, \Delta\mathbb{H}_n(Z_{(1)}) + \Delta\mathbb{H}_n(Z_{(2)})\right), \ldots, \left(a - Z_{(1)}, \mathbb{H}_n(a)\right)\\
&= (0,0), \left(Z_{(2)} - Z_{(1)}, \mathbb{H}_n(Z_{(1)})\right), \left(Z_{(3)} - Z_{(1)}, \mathbb{H}_n(Z_{(2)})\right), \ldots, \left(a - Z_{(1)}, \mathbb{H}_n(a)\right)
\end{aligned} \tag{2.29}$$

We can make the solution right continuous by letting it have a jump at the point immediately to the right of the location of a point of touch of the cusum diagram and its greatest convex minorant (see Exercise 2.22).

It is interesting that a maximum likelihood (ML) estimator of the increasing hazard rate can be defined and constructed. The ML estimator of g_0, assuming it has nondecreasing hazard rate on $[0, \infty)$, is formally defined as the maximizer of

$$g \mapsto \sum_{i=1}^{n} \log g(z_i)$$

over all density functions g such that its corresponding hazard rate h is nondecreasing. Here we write (as usual) lower case z for the observed data. In view of (2.26),

$$1 - G(z) = \exp\left\{-\int_0^z h(x)\,dx\right\} \text{ and } g(z) = h(z)\exp\left\{-\int_0^z h(x)\,dx\right\}$$

so that maximizing the log likelihood boils down to maximizing the function

$$\ell(h) = \sum_{i=1}^{n}\left(\log h(z_i) - \int_0^{z_i} h(x)\,dx\right) \tag{2.30}$$

over all nondecreasing hazard rates. Taking h such that its value tends to infinity in a left neighborhood of the largest observation, $z_{(n)}$, shows that this function is unbounded over the class of nondecreasing hazard rates on $[0, \infty)$ (see also Exercise 2.24). Hence, maximizing ℓ over the class of all nondecreasing hazard rates is not sensible. However, for fixed $c > 0$, it makes sense to maximize ℓ over all increasing hazard rates bounded above by c. From

Exercise 2.25 it follows that whenever such a maximizer exists, it can be taken constant between successive observed z_is and right continuous. Moreover, on the interval $[0, z_{(1)})$, it can be taken equal to zero. Now parameterize

$$h(z) = \begin{cases} 0 & \text{for } z \in [0, z_{(1)}) \\ h_i & \text{for } z \in [z_{(i)}, z_{(i+1)}),\ 1 \le i \le n-1 \\ c & \text{for } z \in [z_{(n)}, \infty) \end{cases}$$

and note that (2.30), neglecting the term $\log c$, can be rewritten as

$$\sum_{i=1}^{n-1} \left(\log h_i - (n-i)(z_{(i+1)} - z_{(i)})h_i\right)$$

$$= \sum_{i=1}^{n-1} \left(\frac{1}{(n-i)(z_{(i+1)} - z_{(i)})} \log h_i - h_i\right)(n-i)(z_{(i+1)} - z_{(i)}). \tag{2.31}$$

This function has the same structure as the log likelihood function in the Poisson extremum problem discussed in Example 2.5. Using Theorem 2.1, it follows that the maximizer $(\hat{h}_1, \ldots, \hat{h}_{n-1})$ of (2.31) is the isotonic regression of the vector y with weights w given by

$$y_i = \frac{1}{(n-i)(z_{(i+1)} - z_{(i)})} \text{ and } w_i = (n-i)(z_{(i+1)} - z_{(i)}) \quad \text{for } 1 \le i \le n-1.$$

This means that $\hat{h}_i$ can be computed as the left-hand slope of the greatest convex minorant of the diagram consisting of the points

$$P_0 = (0,0) \text{ and } P_j = \left(\sum_{k=1}^{j} (n-k)(z_{(k+1)} - z_{(k)}), j\right), \quad \text{for } j = 1, \ldots, n-1$$

evaluated at P_i. Figure 2.13 shows the maximum likelihood estimate $\hat{h}$ and the minimizer of

$$\tfrac{1}{2} \sum_{i=1}^{n-1} h(z_{(i)})^2 \left(z_{(i+1)} - z_{(i)}\right) - \sum_{i=1}^{n-1} h(z_{(i)}) \Delta \mathbb{H}_{n,i}, \tag{2.32}$$

where the criterion function is the left limit at $z_{(n)}$ of (2.28), as $a \uparrow z_{(n)}$, based on a sample of size $n = 50$ and $n = 5000$, respectively, from the Weibull density with hazard rate

$$h(z) = \frac{1}{9} z^2 1_{[0,\infty)}(z). \tag{2.33}$$

We can extend this to the situation where the data are right censored. In this case our observations are of the form

$$(T_1, \Delta_1), \ldots, (T_n, \Delta_n),$$

where $T_i = X_i \wedge C_i$, and Δ_i is an indicator which is equal to one if $T_i = X_i$ and zero if $T_i = C_i$; here X_i is the variable of interest and C_i is a censoring variable, which may censor the observation of X_i (if $C_i < X_i$). The T_i and X_i are assumed to be independent and drawn from distributions F and G, with densities f and g, respectively.

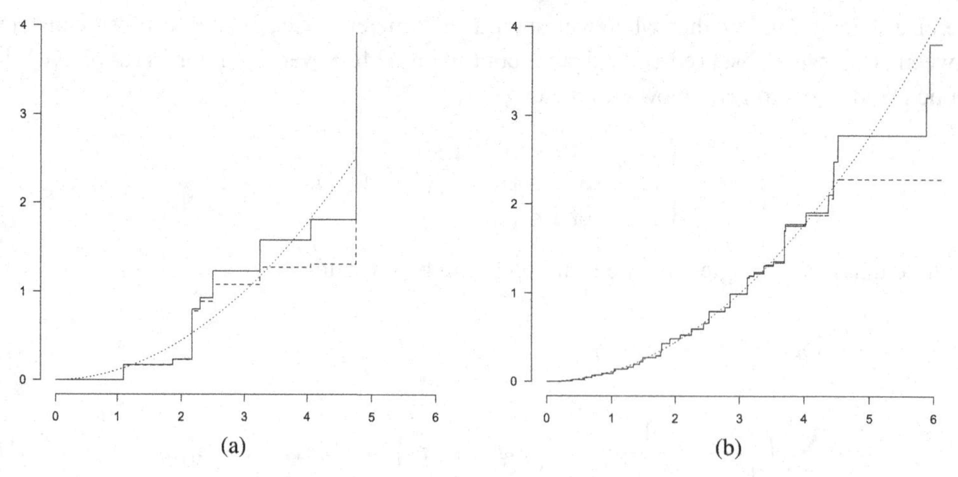

Figure 2.13 Two estimates of the hazard rate based on a sample of size $n = 50$ (a) and $n = 5000$ (b) from the Weibull density with hazard rate (2.33). The solid step function is the maximum likelihood estimate $\hat{h}$, the dashed step function is the least squares estimate $\tilde{h}$ minimizing (2.32). The dotted curve is the underlying hazard rate.

The likelihood can be written

$$\prod_{i=1}^{n} \{f(T_i)\{1 - G(T_i)\}\}^{\Delta_i} \{g(T_i)\{1 - F(T_i)\}\}^{1-\Delta_i}$$

$$= \prod_{i=1}^{n} f(T_i)^{\Delta_i}\{1 - F(T_i)\}^{1-\Delta_i} \prod_{i=1}^{n}\{1 - G(T_i)\}^{\Delta_i} g(T_i)^{1-\Delta_i}.$$

Let $\mathbb{P}_n$ be the empirical measure of the observations (T_i, Δ_i). The part of the log likelihood involving f and F, divided by n, can then be written in the form

$$\int (\delta \log f(t) + (1 - \delta)\log\{1 - F(t)\})\, d\mathbb{P}_n(t, \delta).$$

Rewriting this as a likelihood for the hazard $\lambda(x) = f(x)/\{1 - F(x)\}$ and cumulative hazard $\Lambda(x) = -\log\{1 - F(x)\}$, this turns into:

$$\int (\delta \log \lambda(t) - \Lambda(t))\, d\mathbb{P}_n(t, \delta).$$

We want to maximize this over hazard functions λ, which are nondecreasing. Hence the maximization problem has the same structure as the maximization problem for (2.30), with λ instead of h and also with extra indicators Δ_i, which were all equal to 1 in (2.30).

Let $T_{(1)} \le \cdots \le T_{(n)}$ be the observed order statistics and let $\Delta_{(1)}, \ldots, \Delta_{(n)}$ be the corresponding indicators. We can write

$$\int_{t<T_{(n)}} \Lambda(t)\, d\mathbb{P}_n(t, \delta) = \int_{t<T_{(n)}} \int_0^t \lambda(s)\, ds\, d\mathbb{P}_n(t, \delta) = \int_{t<T_{(n)}} \lambda(t) \int_{t\le s<T_{(n)}} d\mathbb{P}_n(s, \delta)\, dt.$$

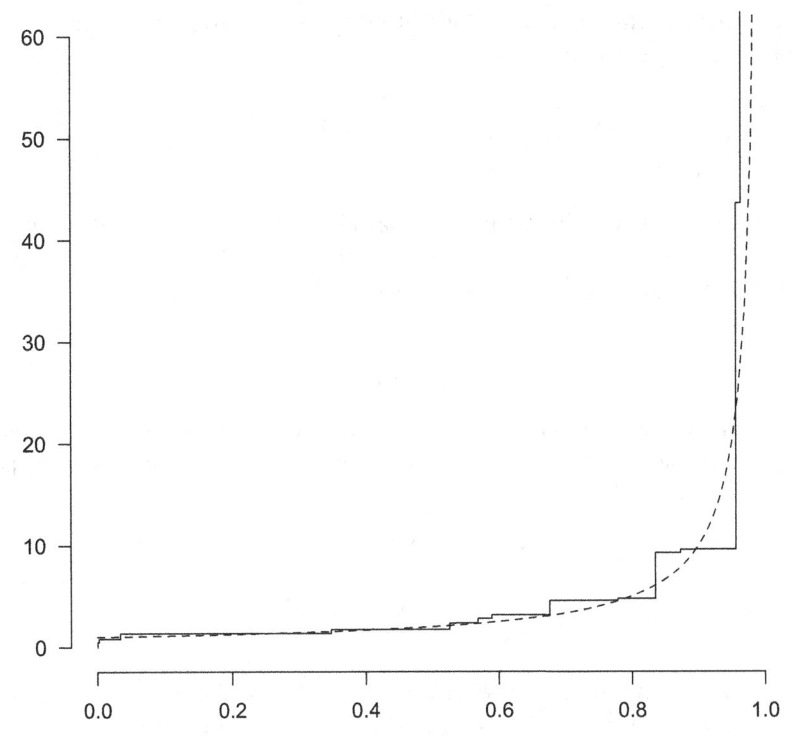

Figure 2.14 The MLE of the hazard (solid) and the real hazard (dashed) for the right-censoring model when both the distribution of the X_i and the distribution of the censoring variables C_i in uniform on the interval $[0, 1]$. The sample size is $n = 1000$.

Moreover, since λ, just as in the treatment of (2.30), can be assumed to be right continuous and piecewise constant on the intervals between the order statistics, we get:

$$\int_{t<T_{(n)}} \lambda(t) \int_{t\le s<T_{(n)}} d\mathbb{P}_n(s,\delta)\,dt$$
$$= \lambda(T_{(1)})\{T_{(2)} - T_{(1)})\} \int_{s\ge T_{(2)}} d\mathbb{P}_n(s,\delta) + \lambda(T_{(2)})\{T_{(3)} - T_{(2)}\} \int_{s\ge T_{(3)}} d\mathbb{P}_n(s,\delta) + \cdots$$

So the expression to maximize is:

$$\sum_{i=1}^{n-1} \left\{ \Delta_{(i)} \log \lambda(T_{(i)})) - n\lambda(T_{(i)})\{T_{(i+1)} - T_{(i)})\} \int_{s>T_{(i)}} d\mathbb{P}_n(s,\delta) \right\}. \tag{2.34}$$

This is an isotonic regression problem, with weights:

$$w_i = n\{T_{(i+1)} - T_{(i)})\} \int_{s\ge T_{(i+1)}} d\mathbb{P}_n(s,\delta),\ i = 1,\ldots,n-1,$$

and the solution is the left-continuous slope of the greatest convex minorant of the cusum diagram with points $P_0 = (0,0)$ and

$$P_i = \left(\sum_{j=1}^{i} w_j, \sum_{j=1}^{i} w_j g_j\right), \qquad g_i = \frac{\Delta_{(i)}}{w_i} \qquad i = 1, \dots, n-1. \tag{2.35}$$

Figure 2.14 shows the picture the MLE of the hazard, obtained in this way, for a sample of size $n = 1000$ in a situation where both the distribution of the X_i and the distribution of the censoring variables C_i are uniform on the interval $[0,1]$.

Exercises

2.1 Suppose the noise terms in (2.1) are independent and identically normally distributed with $\mu = 0$ and $\sigma^2 > 0$ not known. Show that the maximum likelihood estimate of r is then equal to the least squares estimate characterized in Lemma 2.1.

2.2 Denote the least squares objective function of Lemma 2.1 by

$$Q(r) = \frac{1}{2}\sum_{i=1}^{n}(r_i - y_i)^2 w_i.$$

Denote by $\mu^{(i)}$ the vector in $\mathbb{R}^n$ starting with i zeros followed by $n-i$ ones. Here $0 \le i \le n-1$. Write, for arbitrary $r \in C$, $r = \sum_{i=0}^{n-1} \alpha_i \mu^{(i)}$ with $\alpha_i \ge 0$ for all $i > 0$ and $\alpha_0 \in \mathbb{R}$. For $\hat{r} \in C$ satisfying (2.3), write $\hat{r} = \sum_{i=0}^{n-1} \hat{\alpha}_i \mu^{(i)}$ also with $\hat{\alpha}_i \ge 0$ for $i > 0$ and $\hat{\alpha}_0 \in \mathbb{R}$. Note that for $1 \le i \le n-1$, $\hat{\alpha}_i > 0$ if and only if $\hat{r}_{i+1} > \hat{r}_i$. Show that $\phi(\hat{r}) \le \phi(r)$. Hint: Define the convex function $\tau\colon [0,1] \to \mathbb{R}$ by $\tau(s) = Q(\hat{r} + s(r - \hat{r}))$ and use the inequality $Q(r) - Q(\hat{r}) = \tau(1) - \tau(0) \ge \tau'(0)$.

2.3 We can sample X_is from a distribution function F, with $F(0) = 1 - F(1) = 0$. Independently of that we can sample U_is from the uniform distribution on $(0,1)$. Define the random variables

$$Z_i = \begin{cases} X_i & \text{if } X_i > U_i \\ 0 & \text{if } X_i < U_i \end{cases}$$

and then only consider those Z_is that are nonzero. Derive the distribution of the resulting random variables.

2.4 Suppose $U \sim$Unif$(0,1)$ and independently Y has distribution function $\bar{F}$, where $\bar{F}$ is given by

$$\bar{F}(x) = \frac{1}{m_F}\int_0^x y\,dF(y), \qquad \text{where} \qquad 0 < m_F = \int y\,dF(y) < \infty.$$

a) Show that the random variable $Z = UY$ has survival function

$$P(Z > z) = \int_z^{\infty}\left(1 - \frac{z}{y}\right)\,d\bar{F}(y), \; z \ge 0.$$

b) Prove that Z has Lebesgue density g, given in (2.5).
c) Deduce from (b) that the survival function $x \mapsto 1 - F(x)$ is given by $g(x)/g(0)$ (see Keiding et al., 2012).

2.5 Observe that the minimizer of the function Q given by (2.10) with f_n given in (2.11) is constant between successive x_is. Show that it is the Grenander estimator that minimizes over all vectors g

satisfying $g_1 \geq g_2 \geq \cdots \geq g_n \geq 0$ such that $\sum_{i=1}^n g_i(x_i - x_{i-1}) = 1$ (implying that the solution is in fact a probability density function). Hint: note that

$$Q(g) = \sum_{i=1}^{n} \left(\frac{w(x_i)}{n(x_i - x_{i-1})} - g_i \right)^2 (x_i - x_{i-1})$$

and use Lemmas 2.1 and 2.2.

2.6 Verify that the minimizer of (2.10) over any function class also minimizes the function

$$g \mapsto \int g(x)^2\, dx - 2 \int g(x)\, dF_n(x),$$

where F_n denotes the distribution function with density f_n. This expression also motivates this criterion function in situations where the initial estimator F_n does not have a density with respect to Lebesgue measure. If F_n is taken to be $\mathbb{G}_n$, the empirical distribution function of the data, which estimator emerges?

2.7 (*Inconsistency of Grenander estimator at zero.*) Denote the order statistics of a sample of size n from a decreasing density g by $Z_{(1:n)} < Z_{(2:n)} < \cdots < Z_{(n:n)}$. Observe that by definition of the concave majorant of the empirical distribution function, $\hat{g}_n(0+) = \max_{1 \leq i \leq n} i/(nZ_{(i:n)}) \geq 1/(nZ_{(1:n)})$. Deduce from this that $\hat{g}_n(0+)$ is inconsistent for $g(0)$, e.g. by showing that

$$\liminf_{n \to \infty} P\left(\hat{g}_n(0+) \geq g(0)\right) \geq 1 - e^{-1}.$$

2.8 Let (b_n) be a vanishing sequence of positive numbers and consider the following estimator for $g(0)$ in the context of Exercise 2.7:

$$T_n = \frac{\mathbb{G}_n(b_n)}{b_n},$$

where $\mathbb{G}_n$ is the empirical distribution of $Z_1, \ldots, Z_n$. Assume $0 < g(0) < \infty$ and that g has a strictly negative (finite) right derivative at zero.

a) Show that for $n \to \infty$ the variance of T_n behaves like $g(0)/(nb_n)$.
b) Show that $E_g T_n - g(0)$ behaves like $b_n g'(0)/2$ for $n \to \infty$.
c) Combine the results in (a) and (b) to determine the asymptotic MSE optimal choice for b_n, i.e., determine c (depending on g) and $\alpha > 0$ so that $b_n = cn^{-\alpha}$ minimizes the asymptotic mean squared error.

2.9 (*Observation density in Current Status model.*) Let X be a random variable with distribution function F on $(0, \infty)$.

a) Let $t > 0$ be fixed and verify that for $\Delta = 1_{\{X \leq t\}}$ and $\delta \in \{0, 1\}$

$$P(\Delta = \delta) = \delta F(t) + (1-\delta)(1 - F(t)) \text{ and } \log P(\Delta = \delta) = \delta \log F(t) + (1-\delta)\log(1 - F(t)).$$

b) Let T be a random variable in $(0, \infty)$ with density g and independent of X. Show that

$$P(T \leq t, \Delta = \delta) = \delta \int_0^t F(s)g(s)\, ds + (1-\delta) \int_0^t (1 - F(s))g(s)\, ds, \quad \delta \in \{0, 1\}.$$

Conclude that (T, Δ) has density

$$h(t, \delta) = (\delta F(t) + (1-\delta)(1 - F(t)))\, g(t) \text{ w.r.t. } \lambda \times \mu \text{ on } (0, \infty) \times \{0, 1\}$$

where λ is Lebesgue measure on $(0, \infty)$ and μ counting measure on $\{0, 1\}$.

c) Verify log likelihood function (2.13).

2.10 In this book we adopt a nonparametric approach to estimating functions under shape constraints. In the spirit of Exercise 1.1, one can also estimate these functions parametrically. Consider the current status model of Section 2.3. Suppose the X_is are exponentially distributed with (unknown) parameter θ. Based on data as given in (2.12), find the (parametric) maximum likelihood estimator for F, i.e., the maximizer F_θ of (2.13) over the class of functions $\{F_\theta : \theta > 0\}$ with $F_\theta(x) = (1 - \exp(-x/\theta))1_{[0,\infty)}(x)$. If you don't get an explicit solution, can you show that there is a unique solution and can you describe a method to approximate it?

2.11 In the context of the proof of Lemma 2.3, show that for $i_{j-1} < m \le i_j$,

$$\sum_{i=m}^{i_j}(\delta_i - \hat{F}(t_i)) \le 0.$$

Hint: Interpret the sum of the δ_is as the increase of the cumulative sum diagram between location m and i_j and the sum of $\hat{F}(t_i)$s as increase of the convex minorant between location m and i_j in the diagram.

2.12 If all random variables are positive, (2.15) reduces to

$$g(z) = \int_0^z k(z-x)\,dF(x).$$

For a positive random variable, the Laplace transform of its distribution is well defined as Ee^{-tX}.

a) Show from the definition of Z that

$$Ee^{-tX} = \frac{Ee^{-tZ}}{Ee^{-tY}}.$$

b) Show, using the fact that the Laplace transform of a convolution is the product of the Laplace transform of the convolution factors, that

$$\int_0^\infty p(x)e^{-tx}\,dx = \frac{1}{t^2 Ee^{-tY}}.$$

if p satisfies (2.17).

c) Combine (a) and (b) to conclude that

$$\frac{1}{t^2}\cdot Ee^{-tX} = Ee^{-tZ}\cdot\int_0^\infty p(x)e^{-tx}\,dx$$

and obtain (2.18).

2.13 Show that the type 1 resolvent p associated with the uniform density $k = 1_{[0,1]}$ is given by $p(x) = 1 + \lfloor x \rfloor$ on $[0,\infty)$, by showing that it solves (2.17). Here $\lfloor x \rfloor$ denotes the integer part of x, i.e., the largest integer smaller than or equal to x.

2.14 Consider the setting of Section 2.4, with $X, Y \ge 0$. Show that if the function k is bounded, decreasing and smooth on $(0,\infty)$ in the sense that it can be represented as (2.16) with l a nonnegative function on $[0,\infty)$, the convolution density can be written as

$$g(z) = k(0)F(x) - k(0)\int_0^z F(z-u)l(u)\,du.$$

2.15 The estimator defined in Example 2.4 is sometimes called a projected inverse estimator, as an L_2 projection of a certain nonmonotone function on the class of monotone functions. Formulate, in the spirit of Exercise 2.5, the projection problem solved by the projected inverse estimator of Example 2.4, i.e., the function that is projected as well as the weights.

2.16 Suppose you would not recognize the problem of maximizing the log likelihood function (2.21) in the Poisson extremum problem as a generalized isotonic regression problem. One could then think of an iterative scheme to maximize ℓ given in (2.21).

a) Let $\mu_1 \le \mu_2 \le \cdots, \le \mu_n$ be fixed. Show that for each i,

$$x_i \log \lambda_i - \lambda_i = x_i \log \mu_i - \mu_i + x_i \log \left(1 + \frac{\lambda_i - \mu_i}{\mu_i}\right) - (\lambda_i - \mu_i). \tag{2.36}$$

b) Use the quadratic approximation $\log(1+u) \approx u - \frac{1}{2}u^2$ to get that the problem of maximizing the sum (over i) of terms (2.36) over the λ_is (keeping the μ_is fixed) can be approximated by minimizing the following quadratic function in the λ_is:

$$\phi_q(\lambda_1, \ldots, \lambda_n) = \sum_{i=1}^{n} \left[\lambda_i - \mu_i \left(2 - \frac{\mu_i}{x_i}\right)\right]^2 \frac{x_i}{\mu_i^2}.$$

Hint: terms that do not depend on λ_is can be discarded when it comes to minimization over the λ_is.

c) Construct an iterative algorithm based on the approximation based on (b). This algorithm is a special instance of the iterative convex minorant algorithm to be discussed in Section 7.3.

2.17 Show that the function Δ_Φ as defined in (2.23) is nonnegative on $\mathbb{R}^2$.

2.18 Show that the function (2.25) is maximized by the isotonic regression $\tilde{g}$ of g with weights w, using Theorem 2.1.

2.19 Let $\Phi\colon [0,\infty) \to \mathbb{R}$ be defined by $\Phi(u) = u \log u$.

a) Show that Φ is convex on $I = [0,\infty)$.

b) Show that choosing this function in Theorem 2.1 gives that the solution of maximization problem (2.21) is given by the isotonic regression of the vector $(x_1, \ldots, x_n)$ with weights $w_i \equiv 1$.

2.20 Consider the geometric extremum problem of Example 2.6. Define $r_i = 1/p_i - 1$ and $y_i = x_i - 1$.

a) Show that the log likelihood (2.22) can be rewritten in terms of the parameter vector r as follows:

$$\ell(r_1, \ldots, r_n) = \sum_{i=1}^{n} (y_i \log r_i - (y_i + 1)\log(r_i + 1)). \tag{2.37}$$

b) Consider the function $\Phi(u) = u \log u - (u+1)\log(u+1)$. Show that this function is convex on $[0,\infty)$ and that the resulting function (2.25) coincides with log likelihood (2.37).

c) Use Theorem 2.1 to show that the maximizer of (2.37) over all vectors $r = (r_1, \ldots, r_n)$ satisfying $r_1 \ge r_2 \ge \cdots \ge r_n$ is the antitonic regression of $(y_1, \ldots, y_n)$ with weights $w_i \equiv 1$.

d) Conclude that the ML estimate in the geometric extremum problem is given by $\hat{p}_i = 1/(\hat{r}_i + 1)$, where $\hat{r}$ is the antitonic regression described under (c).

2.21 The log likelihood in the current status model (2.13) is of the form

$$r \mapsto \sum_{i=1}^{n} (y_i \log r_i + (1 - y_i)\log(1 - r_i))$$

with $y_i \in [0, 1]$ for all i. This function is to be maximized over $(r_1, \dots, r_n)$ with $0 \le r_1 \le \cdots \le r_n \le 1$.

a) Argue that from a computational point of view, one may assume that $y_1 > 0$ and $y_n < 1$ in the sense that whenever the sequence of y_is starts with a number of zeros or ends with a number of ones, the corresponding optimal r_is can be determined independently of the values of the intermediate r_is.

b) Use Theorem 2.1 to give an alternative proof of Lemma 2.3. Hint: the function $\Phi\colon [0, 1] \to \mathbb{R}$ defined by $\Phi(u) = u \log u + (1-u)\log(1-u)$, defining it to be zero at 0 and 1, is convex on $[0, 1]$.

2.22 Show that the minimizer $\hat{h}_n$ of (2.27) is given at the points $Z_{(1)}, Z_{(2)}, \dots$ by the left-continuous slope of the greatest convex minorant of the cusum diagram given in (2.29). Show that we can extend this solution to a right-continuous, on the whole, interval, constant on the intervals $[Z_{(i)}, Z_{(i+1)})$, $1 \le i < m$, and $[Z_{(m)}, a]$.

2.23 Show that the estimator $\hat{h}_n$ of h in Exercise 2.22 is inconsistent at the boundary points of the interval of monotonicity, 0 and a.

2.24 Verify that log likelihood function (2.30) can be made arbitrarily high by choosing an appropriate nondecreasing hazard on $[0, \infty)$.

2.25 Show that a maximizer of ℓ defined in (2.30), if it exists, should be piecewise constant with jumps only possible at the observed data points Z_i and right continuous.

Bibliographic Remarks

Early references for what is now called isotonic regression are Ayer et al., 1955, and Van Eeden, 1956. In the textbooks Barlow et al., 1972, and Robertson et al., 1988, many results on this and related problems can be found. These books also contain many historical references on the subject of isotonic regression. The freezing data of Lake Monona were taken from http://www.aos.wisc.edu/~sco/lakes/Monona-ice.html.

The Grenander estimator for a decreasing density was introduced in Grenander, 1956. The model of Section 2.2 occurs in various applications, see, e.g., Watson, 1971, Patil and Rao, 1978, and Vardi, 1982. The current durations data were made available to us by Niels Keiding. The inconsistency of the Grenander estimator at zero (and a remedy) is discussed in Woodroofe and Sun, 1993. The current status model is closely related to monotone binary regression and monotone dose response models as studied in Ayer et al., 1955, and Van Eeden, 1956. This model as well as the deconvolution model was studied thoroughly in Groeneboom and Wellner, 1992. For other (mainly Fourier inversion based) nonparametric estimation methods in deconvolution problems, see Meister, 2009. Isotonic inverse estimators using the resolvent of the convolution kernel can be found in Van Es and van Zuijlen, 1996 (uniform noise), Jongbloed, 1998a (exponential noise), and Van Es et al., 1998 (more general noise density). Section 1.5 in Robertson et al., 1988, gives more details and examples on generalized isotonic regression problems, also in connection with general exponential families. The luminescence dating data used in Example 2.5 were kindly made available by the Netherlands Centre for Luminescence Dating. In Marshall and Proschan, 1965, the ML estimator for an increasing hazard rate was derived.

3

Asymptotic Theory for the Basic Monotone Problems

In Chapter 2 a number of estimation problems were introduced where monotonicity of functions can be taken into account in the estimation process. In this chapter asymptotic properties of monotone estimators will be derived. In Section 3.1 various methods for proving consistency of monotone estimators will be described. In Section 3.2, the pointwise limit behavior of the Grenander estimator will be derived heuristically. In particular, the typical rate of convergence of the estimator, $n^{-1/3}$, will emerge from heuristic calculations. In order to make the heuristics rigorous in concrete problems, properties of (the derivative of) convex minorants of functions are needed. Important properties, especially the switch relation, will be reviewed in Section 3.3. The empirical process theory needed to make the convergence of certain processes precise is introduced in Section 3.4. In Section 3.5, empirical process theory is applied to derive the asymptotic distribution of the isotonic inverse estimator in a deconvolution problem defined in Section 2.4. Using the switch relation of Section 3.3 and empirical process results from Section 3.4, the pointwise asymptotic distribution of the Grenander estimator is rigorously derived in Section 3.6. An alternative approach to settle the asymptotic distribution theory proceeds via the theory of martingales. Section 3.7 states some important results from that theory. Using these results, in Section 3.8 local asymptotic properties of the maximum likelihood estimator in the current status model are derived. Various limit distributions are encountered in this chapter, related to convex minorants of processes related to Brownian motion. One of these, the Chernoff distribution, is discussed in Section 3.9. In Section 3.10 results on the concave majorant of Brownian motion and Brownian bridge are stated and discussed.

3.1 Consistency

In this section, some general methods are discussed that can be used to prove consistency of nonparametric estimators in monotone estimation problems. The first is rather direct. It is based on the explicit construction of estimators as derivative of a convex minorant of a random set of points. This method will be illustrated using the Grenander estimator of a decreasing density. In models where an explicit construction of the estimator is not available, other methods are needed. The second method we describe is based on useful inequalities that can be inferred from the fact that a certain estimator minimizes a criterion function together with some empirical process theory. We will illustrate this method using a general class of deconvolution problems.

Recall that the maximum likelihood estimator of a decreasing density in the context of Section 2.2 is given by the left continuous derivative of the concave majorant of the

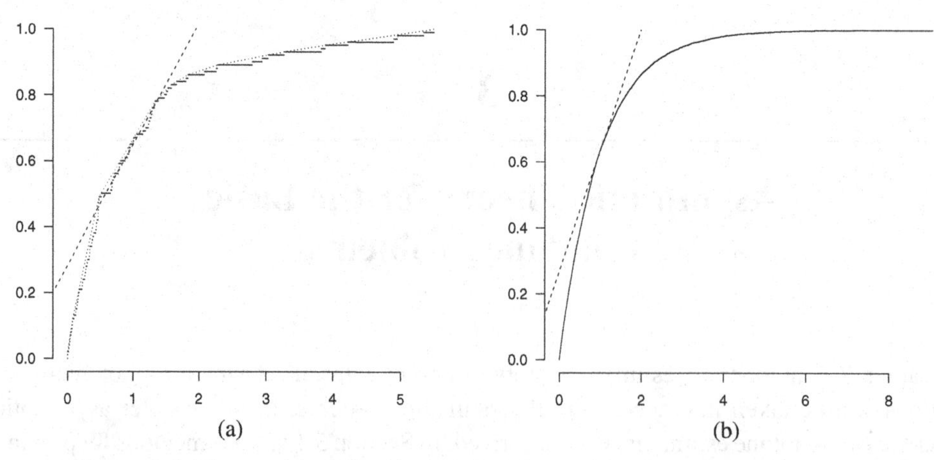

Figure 3.1 The empirical distribution function and its (least) concave majorant (dotted) based on a sample of size $n = 100$ (a) and $n = 10000$ (b) from the standard exponential distribution. The dashed lines are related to Figure 3.3.

empirical distribution function of the data. Denote by $X_1, \ldots, X_n$ the observed sample from the distribution with (decreasing) density g on $[0, \infty)$ and by $\mathbb{G}_n$ the empirical distribution of the sample. Then the classical Glivenko-Cantelli theorem states that $\|\mathbb{G}_n - G\|_\infty \to 0$ almost surely as $n \to \infty$. Here, G denotes the (concave) distribution function corresponding to g.

Now, for the dataset at hand, define $\epsilon = \|\mathbb{G}_n - G\|_\infty$. Then from Exercise 3.1 it follows that for all $x \geq 0$

$$-\epsilon \leq \mathbb{G}_n(x) - G(x) \leq \hat{G}_n(x) - G(x) \leq \epsilon,$$

implying Marshall's inequality:

$$\|\hat{G}_n - G\|_\infty \leq \|\mathbb{G}_n - G\|_\infty.$$

Combining this inequality with the Glivenko Cantelli theorem yields consistency for $\hat{G}_n$ as estimator for G:

$$\sup_{[0,\infty)} |\hat{G}_n(x) - G(x)| \to 0 \text{ a.s. as } n \to \infty. \tag{3.1}$$

See Figure 3.1 for an illustration of this result using samples of size $n = 100$ and $n = 10000$, respectively.

In order to strengthen this consistency of the distribution function $\hat{G}_n$ to that of its derivative, the following elementary lemma is quite useful.

Lemma 3.1 *Let the functions ϕ and ϕ_n, $n = 1, 2, \ldots$ be concave functions on an interval I such that $\sup_{x\in I} |\phi_n(x) - \phi(x)| \to 0$ as $n \to \infty$. Denote for x in the interior of I by $\phi^{(l)}(x)$ and $\phi^{(r)}(x)$ the left and right derivative of ϕ evaluated at x. Then*

$$\phi^{(l)}(x) \geq \limsup_{n\to\infty} \phi_n^{(l)}(x) \geq \liminf_{n\to\infty} \phi_n^{(r)}(x) \geq \phi^{(r)}(x).$$

In particular, if ϕ is differentiable at x, $\phi_n'(x) \to \phi'(x)$, where ϕ_n' is interpreted as either the left or the right derivative of ϕ_n at x.

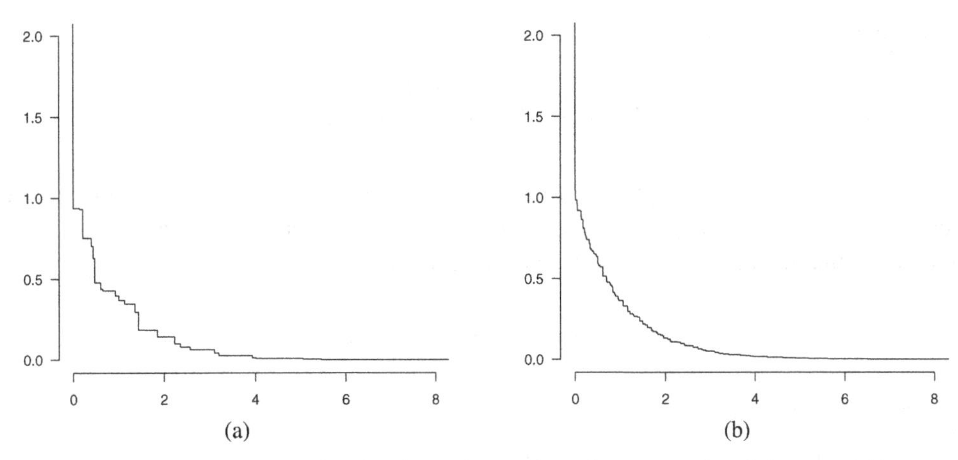

Figure 3.2 The Grenander density estimator based on a sample of size $n = 1000$ (a) and $n = 50000$ (b) from the standard exponential distribution.

Proof Fix $x \in I^o$, the interior of I, arbitrarily. Then for all $h > 0$ such that $[x - h, x + h] \subset I$,

$$\frac{\phi_n(x) - \phi_n(x - h)}{h} \geq \phi_n^l(x) \geq \phi_n^r(x) \geq \frac{\phi_n(x + h) - \phi_n(x)}{h}$$

by concavity of ϕ_n. Letting n to infinity, we see that for all these h

$$\frac{\phi(x) - \phi(x - h)}{h} \geq \liminf_{n \to \infty} \phi_n^{(l)}(x) \geq \limsup_{n \to \infty} \phi_n^{(r)}(x) \geq \frac{\phi(x + h) - \phi(x)}{h}.$$

The result follows upon letting $h \downarrow 0$. □

Corollary 3.1 *Let $\hat{g}_n$ be the Grenander estimator based on a sample of size n from the decreasing density g on $[0, \infty)$. Then for each $x > 0$, $\hat{g}_n(x) \to g(x)$ almost surely, as $n \to \infty$. Moreover, if g is continuous on $(0, \infty)$, then for each $c > 0$, $\sup_{x \geq c} |\hat{g}_n(x) - g(x)| \to 0$ almost surely for $n \to \infty$.*

Proof The pointwise result follows immediately from Lemma 3.1 and (3.1). The uniform consistency on intervals away from zero follows from the pointwise result and Exercise 3.2. □

Figure 3.2 illustrates consistency of the Grenander estimator for arguments bounded away from zero. It also illustrates the inconsistency problem identified in Exercise 2.7.

Another method to prove consistency for (nonparametric) M estimators in general is based on empirical process theory. In contrast to the method based on Marshall's inequality, which first yields a local result that can be used to derive more global results, this method starts with a global result that can later be used to obtain local results. We illustrate this method for ML estimators of a density over a specific class of densities.

Consider an i.i.d. sequence of random variables $Z_1, Z_2, \ldots$ having density g_0 belonging to a class of densities $\mathcal{G}$. The maximum likelihood estimator for g (based on the first n Z_is in the sequence) is the maximizer of

$$\ell(g) = \int \log g(z)\, d\mathbb{G}_n(z) = \frac{1}{n}\sum_{i=1}^{n} \log g(Z_i)$$

over all densities g that belong to $\mathcal{G}$. Without the need for an explicit representation of the solution of the optimization problem, a necessary condition for any maximizer $\hat{g}_n$ of the log likelihood is that its log likelihood exceeds that of the underlying density $g_0 \in \mathcal{G}$:

$$\ell(g_0) \le \ell(\hat{g}_n) \iff \int \log\left(\frac{\hat{g}_n(z)}{g_0(z)}\right) d\mathbb{G}_n(z) \ge 0.$$

This can be rewritten as

$$\int \log\left(\frac{\hat{g}_n(z)}{g_0(z)}\right) g_0(z)\, dz + \int \log\left(\frac{\hat{g}_n(z)}{g_0(z)}\right) d\left(\mathbb{G}_n - G_0\right)(z) \ge 0. \tag{3.2}$$

The first term in this expression can be bounded by using the inequality $\log x = 2\log\sqrt{x} \le 2(\sqrt{x}-1)$. Also using that $\int \hat{g}_n(z)\, dz = \int g_0(z)\, dz = 1$, this yields

$$\int \log\left(\frac{\hat{g}_n(z)}{g_0(z)}\right) g_0(z)\, dz \le 2\int \left(\sqrt{\hat{g}_n(z)g_0(z)} - g_0(z)\right) dz = -\int \left(\sqrt{\hat{g}_n(z)} - \sqrt{g_0(z)}\right)^2 dz. \tag{3.3}$$

The latter quantity is minus the squared L_2 distance between the square roots of both densities. Defining the squared Hellinger distance between $\hat{g}_n$ and g_0 by

$$h^2(\hat{g}_n, g_0) = \frac{1}{2}\left\|\sqrt{\hat{g}_n} - \sqrt{g_0}\right\|_{L_2}^2 = \frac{1}{2}\int \left(\sqrt{\hat{g}_n(z)} - \sqrt{g_0(z)}\right)^2 dz,$$

we therefore obtain from (3.2) and (3.3) that

$$0 \le h^2(\hat{g}_n, g_0) \le -\frac{1}{2}\int \log\left(\frac{\hat{g}_n(z)}{g_0(z)}\right) g_0(z)\, dz \le \frac{1}{2}\int \log\left(\frac{\hat{g}_n(z)}{g_0(z)}\right) d\left(\mathbb{G}_n - G_0\right)(z). \tag{3.4}$$

For a fixed density g rather than $\hat{g}_n$, the latter quantity will converge to zero by the law of large numbers if the expectation of $\log(g(Z)/g_0(Z))$ is finite. One of the central issues in empirical process theory is to study quantities such as

$$\sup_{\phi\in\Phi}\left|\int \phi(z)\, d\left(\mathbb{G}_n - G_0\right)(z)\right| \tag{3.5}$$

for sets of functions Φ. If for $\Phi = \{\phi\colon \phi(z) = \log(g(z)/g_0(z))\colon g \in \mathcal{G}\}$ this supremum would converge to zero, this would imply convergence to zero of the Hellinger distance between $\hat{g}_n$ and g_0 and thus to Hellinger consistency. Essentially, the class Φ should not be too large in some well defined sense (see also Section 3.4). An adaptation of this general method will be used in Section 4.6 for deconvolution problems. A simple example where this method can be applied is to the class of exponential densities:

$$\mathcal{G} = \{g_\theta : \theta \ge 1\} \text{ with } g_\theta(z) = \frac{1}{\theta}e^{-z/\theta}1_{[0,\infty)}(z).$$

Denoting the true parameter by θ_0, it follows that

$$\Phi = \{\phi_\theta : \theta \geq 1\} \text{ with } \phi_\theta(z) = \log\left(\frac{\theta_0}{\theta}\right) + z\left(\frac{1}{\theta} - \frac{1}{\theta_0}\right),$$

giving

$$\int \phi_\theta(z)\, d(\mathbb{G}_n - G_{\theta_0})(z) = \left(\frac{1}{\theta} - \frac{1}{\theta_0}\right) \int z\, d(\mathbb{G}_n - G_{\theta_0})(z).$$

Hence, for all $\theta \geq 1$,

$$\left| \int \phi_\theta(z)\, d\left(\mathbb{G}_n - G_0\right)(z) \right| \leq \frac{\theta_0 + 1}{\theta_0} \left| \int z\, d(\mathbb{G}_n - G_{\theta_0})(z) \right|.$$

Since the right hand side does not depend on θ, this bound is uniform in $\theta \geq 1$ and the strong law of large numbers determines that the right hand side tends to zero with probability one. Hence, the Hellinger distance of the ML estimator within the model of exponential densities (with expectation bounded away from zero) to the underlying exponential density converges to zero with probability one. Formulated differently, the ML estimator is strongly consistent in the Hellinger metric. Of course, there are more straightforward ways to prove this result, but this example illustrates the value of the general argument described without using empirical process theory to show convergence to zero of a random quantity of the type (3.5).

Yet another method to prove consistency of a maximum likelihood estimator over a convex class of densities is based on the following necessary condition for optimality of $\hat{g}_n$:

$$\lim_{\epsilon \downarrow 0} \epsilon^{-1} \left(\ell(\hat{g}_n + \epsilon(g - \hat{g}_n)) - \ell(\hat{g}_n)\right) \leq 0. \tag{3.6}$$

Here convexity of $\mathcal{G}$ and concavity of ℓ are used. Writing out the limit in (3.6) yields

$$0 \geq \lim_{\epsilon \downarrow 0} \frac{1}{n\epsilon} \sum_{i=1}^{n} \log\left(1 + \epsilon \frac{g(Z_i) - \hat{g}_n(Z_i)}{\hat{g}_n(Z_i)}\right) = \int \frac{g(z)}{\hat{g}_n(z)}\, d\mathbb{G}_n(z) - 1. \tag{3.7}$$

To infer consistency of the estimator from these inequalities, it makes sense to choose for g the underlying, true density g_0. It is a way of quantifying that the MLE has higher likelihood than g_0. In specific models inequality (3.6) can then be used to obtain a limiting inequality along subsequences. One can then, on a set of probability one, derive that any subsequence contains a further subsequence converging to a subdensity $\tilde{g}$, satisfying

$$1 \geq \int \frac{g_0(z)}{\tilde{g}(z)}\, dG_0(z) = \int \frac{g_0(z)^2}{\tilde{g}(z)}\, dz.$$

Since there can only be one such subdensity $\tilde{g}$ (namely the density g_0) satisfying this inequality, one can argue that with probability one, $\hat{g}_n$ should converge to g_0 pointwise. See Exercise 3.6.

3.2 Heuristic Asymptotics for the Grenander Estimator

Before starting a more rigorous asymptotic analysis of monotone estimators, we will give some heuristics in this section, leading to the typical $n^{-1/3}$-rate of convergence and the

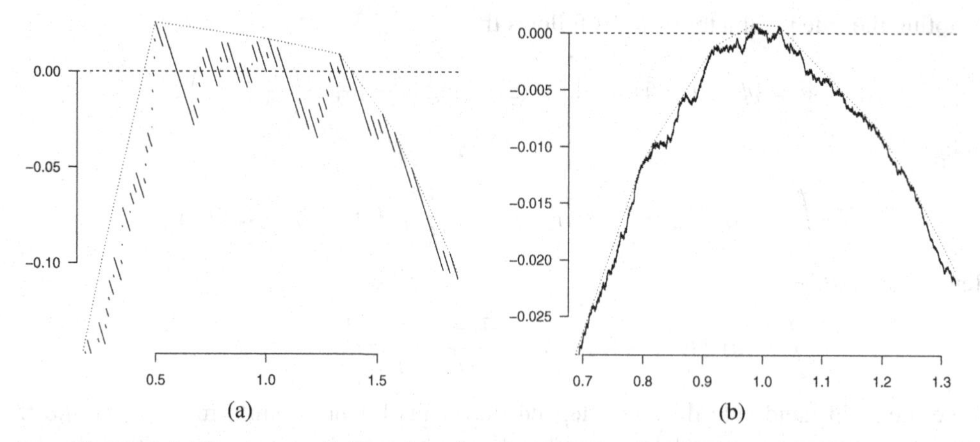

Figure 3.3 The process (3.9) and its (least) concave majorant (dotted) based on a sample of size $n = 100$ (a) and $n = 10000$ (b) from the standard exponential distribution. The samples coincide with those used in Figure 3.1.

asymptotic Chernoff distribution for the Grenander estimator at a fixed point x_0 whenever this underlying density has a strictly negative derivative at x_0. We will also consider the situation where the underlying density is constant in a neighborhood of x_0, leading to totally different asymptotic distribution. As seen in Corollary 3.1, the Grenander estimator is consistent in either situation.

For the first situation, let $x_0 > 0$ be a point where the underlying density g is continuously differentiable. Now consider the empirical distribution function and its concave majorant near this point x_0, and note that adding a linear drift in the picture shows that $x \mapsto \hat{G}_n(x) - \mathbb{G}_n(x_0) - g(x_0)(x - x_0)$ is the concave majorant of $x \mapsto \mathbb{G}_n(x) - \mathbb{G}_n(x_0) - g(x_0)(x - x_0)$ and the derivative of the concave majorant of the latter function at x_0 equals $\hat{g}_n(x_0) - g(x_0)$. We use the term "slocom" for the slope of the least concave majorant and write

$$\hat{g}_n(x_0) - g(x_0) = \text{slocom}\,[x \mapsto \mathbb{G}_n(x) - \mathbb{G}_n(x_0) - g(x_0)(x - x_0)]_{x=x_0}\,. \tag{3.8}$$

The process

$$x \mapsto \mathbb{G}_n(x) - \mathbb{G}_n(1) - g(1)(x - 1) \tag{3.9}$$

is shown in Figure 3.3 for the data sets of Figure 3.1. The dashed lines added to the empirical distribution functions in Figure 3.1 are $x \mapsto \mathbb{G}_n(1) + g(1)(x - 1)$.

Let us now localize and zoom in near x_0 on the process at the right hand side of (3.8). More specifically, let (δ_n) be a vanishing sequence of positive numbers and consider points $x_{n,t} = x_0 + \delta_n t$. Then, by the chain rule,

$$\delta_n\left(\hat{g}_n(x_0) - g(x_0)\right) = \text{slocom}\,[t \mapsto \mathbb{G}_n(x_0 + \delta_n t) - \mathbb{G}_n(x_0) - g(x_0)\delta_n t]_{t=0}\,.$$

Moreover, multiplying the function on the right by δ_n^{-2} results in the same multiplication for the derivative of the concave majorant, so

$$\delta_n^{-1}\left(\hat{g}_n(x_0) - g(x_0)\right) = \text{slocom}\left[t \mapsto \delta_n^{-2}\left(\mathbb{G}_n(x_0 + \delta_n t) - \mathbb{G}_n(x_0) - g(x_0)\delta_n t\right)\right]_{t=0}\,.$$

Observe that for $t \geq 0$ and $\xi_n \in [x_0, x_0 + \delta_n t]$

$$\delta_n^{-2} (\mathbb{G}_n(x_0 + \delta_n t) - \mathbb{G}_n(x_0) - g(x_0)\delta_n t) = \delta_n^{-2} \int 1_{(x_0, x_0+\delta_n t]} \, d(\mathbb{G}_n - G)(x) + \frac{1}{2} g'(\xi_n) t^2. \tag{3.10}$$

For $t \leq 0$ a similar expression holds.

By the central limit theorem, the first (random) term in (3.10) behaves (for t fixed) as $\sqrt{g(x_0)t/(n\delta_n^3)} N(0, 1)$. In fact, we will see that as a process,

$$\sqrt{\frac{n}{g(x_0)\delta_n}} \int 1_{(x_0, x_0+\delta_n t]} \, d(\mathbb{G}_n - G)(x) \to^D W(t),$$

where W is standard two-sided Brownian motion. Hence,

$$\delta_n^{-2} \int 1_{(x_0, x_0+\delta_n t]} \, d(\mathbb{G}_n - G)(x) \approx \frac{\sqrt{g(x_0)}}{\sqrt{n\delta_n^3}} W(t).$$

The second (deterministic) term in (3.10) behaves like $\frac{1}{2} g'(x_0) t^2$, so

$$\delta_n^{-1} (\hat{g}_n(x_0) - g(x_0)) \approx \text{slocom} \left[t \mapsto \frac{\sqrt{g(x_0)}}{\sqrt{n\delta_n^3}} W(t) + \frac{1}{2} g'(x_0) t^2 \right]_{t=0}.$$

Now suppose that we choose δ_n such that $\delta_n n^{1/3} \to \infty$. Then the random term in the function converges to zero. The slope of the concave majorant of the (concave) function $t \mapsto \frac{1}{2} g'(x_0) t^2$ at zero is zero, suggesting that

$$\delta_n^{-1} (\hat{g}_n(x_0) - g(x_0)) \to^P 0.$$

In other words: $\hat{g}_n(x_0)$ is rate-δ_n consistent at any rate converging to zero slower than $n^{-1/3}$.

Now suppose we take δ_n converging to zero at a faster rate than $n^{-1/3}$, i.e., with $\delta_n n^{1/3} \to 0$. Then the random part of (3.10) is dominant. The concave majorant of Brownian motion on $\mathbb{R}$ is not well defined, and will be infinite almost surely.

If we take $\delta_n = n^{-1/3}$, then both terms in (3.10) are of the same order and we see that

$$n^{1/3} (\hat{g}_n(x_0) - g(x_0)) \approx \text{slocom} \left[t \mapsto \sqrt{g(x_0)} W(t) + \frac{1}{2} g'(x_0) t^2 \right]_{t=0}.$$

This suggests that the concave majorant of these processes converge to the concave majorant of the limiting process. Finally, we reduce this asymptotic distribution to a canonical distribution that does not include constants depending on the density g. Using the scaling property of Brownian Motion (see Exercise 3.18), we obtain

$$\sqrt{g(x_0)} W(t) + \frac{1}{2} g'(x_0) t^2 \stackrel{\mathcal{D}}{=} \frac{g(x_0)^{2/3}}{(-\frac{1}{2} g'(x_0))^{1/3}} \left(W\left(\frac{(-\frac{1}{2} g'(x_0))^{2/3}}{g(x_0)^{1/3}} t \right) - \left(\frac{(-\frac{1}{2} g'(x_0))^{2/3}}{g(x_0)^{1/3}} t \right)^2 \right). \tag{3.11}$$

Therefore, again using the chain rule and the fact that multiplying a function by a constant will also multiply the slope of its concave majorant at zero by the same constant, we obtain

$$n^{1/3} (\hat{g}_n(x_0) - g(x_0)) \approx \frac{g(x_0)^{2/3}}{(-\frac{1}{2} g'(x_0))^{1/3}} \frac{(-\frac{1}{2} g'(x_0))^{2/3}}{g(x_0)^{1/3}} \text{slocom} \left[t \mapsto W(t) - t^2 \right]_{t=0}$$

implying

$$n^{1/3}\left|\frac{1}{2}g'(x_0)g(x_0)\right|^{-1/3}(\hat{g}_n(x_0)-g(x_0))\approx \text{slocom}\left[t\mapsto W(t)-t^2\right]_{t=0}, \tag{3.12}$$

an approximation that turns out to be right asymptotically. The distribution of the latter random variable is related to Chernoff's distribution to be studied in Section 3.9 (see also Exercise 3.12).

There are some important aspects we conveniently skipped in this derivation, aspects that need to be taken into account when deriving the asymptotics of the Grenander estimator rigorously. The first is the localization we use. In fact, the concave majorant of a process on $[0,\infty)$ near a point x_0 need not be determined locally near this point. Exercise 3.17 shows that there is something to be proved here in general. We also were a bit sloppy in applying continuous mapping ideas ("the concave majorants of the processes defined in (3.10) converge to the concave majorant of the limiting process"). In later sections these matters will be addressed more rigorously.

Now consider the situation where $x_0\in(a,b)$ such that $g(x)\equiv g(x_0)$ for $x\in[a,b]$. In words, x_0 belongs to a flat part of g. As an extreme case of this situation, assume the density g to be the uniform density on $[0,1]$ and let $x_0\in(0,1)$. Then we have

$$\begin{aligned}\sqrt{n}\,(\hat{g}_n(x_0)-g(x_0)) &= \text{slocom}\left[x\mapsto\sqrt{n}\,(\mathbb{G}_n(x)-g(x_0)x)\right]_{x=x_0}\\ &= \text{slocom}\left[x\mapsto\sqrt{n}\,(\mathbb{G}_n(x)-G(x))\right]_{x=x_0}\approx \text{slocom}\left[x\mapsto\mathbb{B}(x)\right]_{x=x_0}\end{aligned}$$

where $\mathbb{B}$ is the Brownian Bridge process on $[0,1]$. The limiting distribution will be further discussed in Section 3.10. As mentioned, taking g the uniform density on $[0,1]$ is a canonical choice when considering estimating a decreasing density at a point in an interval where the density is flat. Let us now look at the more general situation where g is constant on a nondegenerate interval $[a,b]$ and $x_0\in(a,b)$. Then,

$$\begin{aligned}\sqrt{n}\,(\hat{g}_n(x_0)-g(x_0)) &= \text{slocom}\left[x\mapsto\sqrt{n}\,(\mathbb{G}_n(x)-\mathbb{G}_n(a)-g(x_0)(x-a))\right]_{x=x_0}\\ &= \text{slocom}\left[x\mapsto\sqrt{n}\int_a^x d\,(\mathbb{G}_n-G)\,(x)\right]_{x=x_0},\end{aligned}$$

where the concave majorant is taken over $[0,\infty)$. Actually, it turns out (Carolan and Dykstra, 1999, Theorem 6.3) that the error made by taking the concave majorant over $[a,b]$ is of lower order, so that

$$\begin{aligned}\sqrt{n}\,(\hat{g}_n(x_0)-g(x_0)) &\approx \text{slocom}_{[a,b]}\left[x\mapsto\sqrt{n}\int_a^x d\,(\mathbb{G}_n-G)\,(x)\right]_{x=x_0}\\ &\approx \text{slocom}_{[a,b]}\left[x\mapsto\mathbb{B}(G(x))-\mathbb{B}(G(a)\right]_{x=x_0}\\ &= g(x_0)\text{slocom}_{[G(a),G(b)]}\left[t\mapsto\mathbb{B}(t)-\mathbb{B}(G(a))\right]_{t=G(x_0)}.\end{aligned}$$

Here we use the fact that the process $x\mapsto\sqrt{n}(\mathbb{G}_n-G)(x)$ on $[a,b]$ behaves like $x\mapsto\mathbb{B}(G(x))$ and, for $x\in[a,b]$, $G(x)=G(a)+g(x_0)(x-a)$. For the latter quantity (apart from the factor $g(x_0)$), note that

$$\text{slocom}_{[G(a),G(b)]}\left[t \mapsto \mathbb{B}(t) - \mathbb{B}(G(a))\right]_{t=G(x_0)} = \frac{\mathbb{B}(G(b)) - \mathbb{B}(G(a))}{(G(b) - G(a))}$$

$$+\,\text{slocom}_{[G(a),G(b)]}\left[t \mapsto \mathbb{B}(t) - \mathbb{B}(G(a)) - \frac{\mathbb{B}(G(b)) - \mathbb{B}(G(a))}{(G(b) - G(a))}(t - G(a))\right]_{t=G(x_0)}. \quad (3.13)$$

The two terms are independent random variables (see Exercise 3.7) and for the first term we have

$$\frac{\mathbb{B}(G(b)) - \mathbb{B}(G(a))}{G(b) - G(a)} \sim N\left(0, \frac{1 - g(x_0)(b-a)}{g(x_0)(b-a)}\right).$$

See Exercise 3.8.

For the second term in (3.13), the linearly compensated Brownian bridge on $[G(a), G(b)]$, it can be shown that this is itself a rescaled Brownian Bridge (see Exercise 3.10). Therefore, (apart from the factor $g(x_0)$) it has the same distribution as

$$\sqrt{G(b) - G(a)}\,\text{slocom}_{[G(a),G(b)]}\left[t \mapsto \mathbb{B}\left(\frac{t - G(a)}{G(b) - G(a)}\right)\right]_{t=G(x_0)},$$

which in turn has the same distribution as

$$\frac{1}{\sqrt{G(b) - G(a)}}\text{slocom}_{[0,1]}\left[u \mapsto \mathbb{B}(u)\right]_{u=\frac{G(x_0)-G(a)}{G(b)-G(a)}}.$$

This leads to the following result (where we include the factor $g(x_0)$ in the denominator at the left hand side):

$$\sqrt{n}\frac{\hat{g}_n(x_0) - g(x_0)}{g(x_0)} \approx^d \alpha Z + \beta Y \quad (3.14)$$

where Z is standard normally distributed,

$$Y = \text{slocom}_{[0,1]}\left[u \mapsto \mathbb{B}(u)\right]_{u=\frac{G(x_0)-G(a)}{G(b)-G(a)}}$$

independent of Z and

$$\alpha = \sqrt{\frac{1 - g(x_0)(b-a)}{g(x_0)(b-a)}} \text{ and } \beta = \frac{1}{\sqrt{g(x_0)(b-a)}}.$$

Although we will not return to more rigorous derivation for the flat density case, the slope process of the least concave majorant of Brownian bridge will be briefly considered in Section 3.10.

3.3 Convex Minorants: Basic Properties

As we saw, convex minorants and concave majorants play a special role in the setting of estimating a monotone function. Many estimators are characterized as derivative of the convex minorant of a certain random set of points in $\mathbb{R}^2$. In this section we derive some results that are particularly useful when the asymptotics of these estimators at a fixed point are to be assessed.

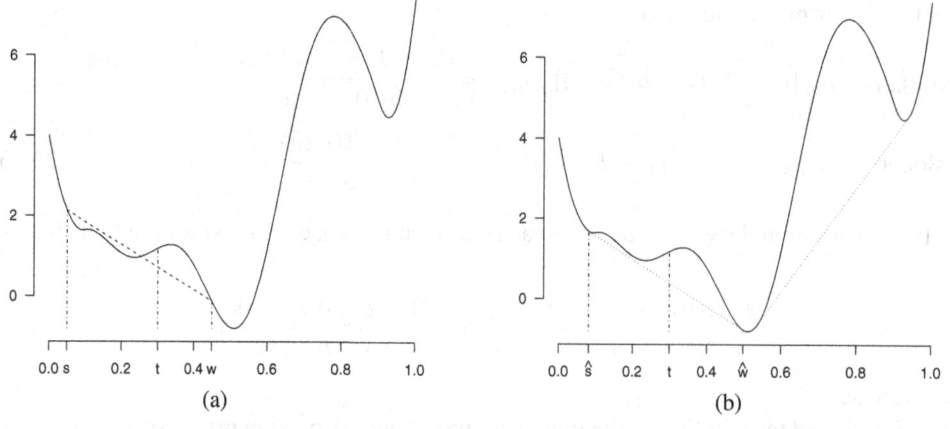

Figure 3.4 A (nonconvex) function Φ. (a) For fixed $t = 0.4$, the segment connecting $(s, \Phi(s))$ and $(w, \Phi(w))$ for chosen $s < t < w$. By first fixing s, searching for the $w \geq t$ that minimizes the slope of the segment, followed by varying $s < t$ to make the slope maximal, the left derivative of the convex minorant of Φ at t is found according to (3.15). (b) The solution and the $\hat{s} < t \leq \hat{w}$ solving the max-min problem for t.

Denote by Φ a lower semicontinuous function on an interval I and by Φ^* its convex minorant on I. Denote by $\Phi^{*,l}$ and $\Phi^{*,r}$ the left and right hand derivative of Φ^*. A first important observation is that for each $t \in I^o$,

$$\Phi^{*,l}(t) = \sup_{s<t} \inf_{w \geq t} \frac{\Phi(w) - \Phi(s)}{w - s} = \inf_{w \geq t} \sup_{s<t} \frac{\Phi(w) - \Phi(s)}{w - s} \tag{3.15}$$

and

$$\Phi^{*,r}(t) = \sup_{s \leq t} \inf_{w>t} \frac{\Phi(w) - \Phi(s)}{w - s} = \inf_{w>t} \sup_{s \leq t} \frac{\Phi(w) - \Phi(s)}{w - s} \tag{3.16}$$

See Figure 3.4 for an illustration of (3.15).

Of course, the suprema and infima are also restricted to I. Now define, for lower semicontinuous functions f,

$$\text{argmin}^+_{s \in I} f(s) = \sup\{s \in I: \ f(s) = \min_{u \in I} f(u)\}$$

and

$$\text{argmin}^-_{s \in I} f(s) = \inf\{s \in I: \ f(s) = \min_{u \in I} f(u)\}.$$

Lemma 3.2 relates the derivative of the convex minorant to an argmin functional applied to the function Φ with added drifts. The relations obtained are sometimes called the switch relations.

Lemma 3.2 *For all $t \in I^o$ and $v \in \mathbb{R}$,*

$$\Phi^{*,l}(t) > v \iff \text{argmin}^+_{s \in I} (\Phi(s) - vs) < t$$

and

$$\Phi^{*,r}(t) < v \iff \text{argmin}^-_{s \in I} (\Phi(s) - vs) > t.$$

Proof From (3.15) it follows that

$$\begin{aligned}\Phi^{*,l}(t) > v &\iff (\exists s < t)(\forall w \geq t)\colon \frac{\Phi(w) - \Phi(s)}{w - s} > v \\ &\iff (\exists s < t)(\forall w \geq t)\colon \Phi(w) - wv > \Phi(s) - sv \\ &\iff \operatorname{argmin}^{+}_{s \in I} (\Phi(s) - vs) < t\end{aligned}$$

For $\Phi^{*,r}$ the argument is similar. □

Let $\tilde{F}_n$ be defined as the right derivative of the convex minorant of the function U_n. Using the fact that the argmax function is invariant under multiplication by a positive number and addition of arbitrary constants, we see, using Lemma 3.2, that for each sequence δ_n of positive numbers,

$$\begin{aligned}\delta_n^{-1}\left(\tilde{F}_n(x) - F(x)\right) < v &\iff \operatorname{argmin}_s^{-}(U_n(s) - (F(x) + \delta_n v)s) > x \\ &\iff \operatorname{argmin}_s^{-}(U_n(x + \delta_n s) - U_n(x) - F(x)\delta_n s - \delta_n^2 vs) > 0 \\ &\iff \operatorname{argmin}_s^{-} Z_n(s) > 0\end{aligned}$$

where

$$\begin{aligned}Z_n(s) &= \delta_n^{-2}\left(U_n(x + \delta_n s) - U_n(x) - U(x + \delta_n s) + U(x)\right) \\ &\quad + \delta_n^{-2}\left(U(x + \delta_n s) - U(x) - F(x)\delta_n s\right) - vs.\end{aligned}$$

Therefore, the asymptotic behaviour of the rescaled difference between a convex minorant estimator and the estimate is intimitely related to the asymptotic behavior of the argmin of a sequence of stochastic processes Z_n. In particular, for each $v \in \mathbb{R}$,

$$P\left(\delta_n^{-1}\left(\tilde{F}_n(x) - F(x)\right) < v\right) = P\left(\operatorname{argmin}_s^{-} Z_n(s) > 0\right).$$

(Note that v is part of the definition of Z_n, just as x, but this is not stressed in the notation.) The process Z_n can be decomposed in an asymptotically deterministic part that usually can be handled using a Taylor approximation based on $\delta_n \downarrow 0$, and an intrinsically random part. To derive the asymptotic distribution of the estimator $\tilde{F}_n$, we will need some type of weak convergence of Z_n to an asymptotic stochastic process Z, and after that a suitable continuous mapping theorem for the argmin functional. Some background on these matters is provided in Section 3.4.

3.4 Some Empirical Process Theory

Empirical process theory has a rich history within the field of statistics. It is impossible to do justice to this branch of statistics in only one section of this book. We will state and explain only briefly some results that prove to be very useful when studying the asymptotic probabilistic behavior of isotonic inverse estimators at a fixed point. We borrow the theory from Van der Vaart and Wellner, 1996, and Van de Geer, 2000.

To state the results, we need to define some concepts and terminology. Let $\mathcal{F}$ be a class of real-valued functions on a subset I of the real line. Moreover, let $\|\cdot\|$ be a seminorm on $\mathcal{F}$. Given two functions $l \leq u$ on I, call the subset $[l, u] := \{f \in \mathcal{F} : l \leq f \leq u\}$ of functions a bracket in $\mathcal{F}$. A set of brackets covers $\mathcal{F}$ if for each $f \in \mathcal{F}$ there is a bracket $[l, u]$ such that $f \in [l, u]$. If, additionally, $\|u - l\| < \epsilon$, call $[l, u]$ an ϵ-bracket. The bracketing number $N_{[]}(\epsilon, \mathcal{F}, \|\cdot\|)$ is then defined as the minimum number of ϵ-brackets needed to cover the whole class $\mathcal{F}$. The ϵ-entropy with bracketing $H_{[]}(\epsilon, \mathcal{F}, \|\cdot\|)$ is defined as the logarithm of $N_{[]}(\epsilon, \mathcal{F}, \|\cdot\|)$. The bracketing number (and equivalently the entropy) of $\mathcal{F}$ as a function of ϵ measures the massiveness of the class.

A first basic result is a sufficient condition for a class of functions $\mathcal{F}: \mathbb{R} \to \mathbb{R}$ to be Glivenko Cantelli, i.e., that

$$\sup_{f \in \mathcal{F}} \left| \int f(z)\, d(\mathbb{G}_n - G_0)(z) \right| \to^{a.s.} 0 \tag{3.17}$$

where $\mathbb{G}_n$ denotes the empirical distribution function of an independent sample from the distribution with distribution function G_0. This result generalizes the classical Glivenko Cantelli theorem that $\sup_{z \in \mathbb{R}} |\mathbb{G}_n(z) - G_0(z)| \to 0$ a.s. and its proof is close to the proof of this classical result. Note that the classical result is indeed a special case of (3.17), taking

$$\mathcal{F} = \{f_z : z \in \mathbb{R}\}, \text{ where } f_z = 1_{(-\infty, z]}.$$

Lemma 3.3 *Suppose that $\mathcal{F} \subset L_1(G_0)$ and*

$$H_{[]}(\epsilon, \mathcal{F}, L_1(G_0)) < \infty \textit{ for all } \epsilon > 0.$$

Then $\mathcal{F}$ is Glivenko Cantelli, i.e., satisfies (3.17).

The proof of this result is outlined in Exercise 3.14. For our purposes, an important result from Birman and Solomjak, 1967, (see also Van de Geer, 2000, Lemma 3.8) is the following.

Lemma 3.4 *Let $\mathcal{F} = \{f : \mathbb{R} \to [0, 1],$ increasing$\}$. Then there exists a constant $A > 0$ such that*

$$H_{[]}(\epsilon, \mathcal{F}, L_1(G_0)) \le \frac{A}{\delta} \textit{ for all } \delta > 0.$$

Using the result of Lemma 3.4 as a building block, entropy bounds for many function classes can be constructed. For example, consider the class of functions on $\mathbb{R}$ of bounded variation, i.e., functions that can be expressed as the difference of two bounded monotone functions on $\mathbb{R}$. Indeed, for the class

$$\mathcal{H} = \{h = f - g : f, g \in \mathcal{F}\}$$

it follows that

$$H_{[]}(\epsilon, \mathcal{F}, L_1(G_0)) \le \frac{4A}{\delta}$$

for all $\delta > 0$, with the same A as in Lemma 3.4 (see Exercise 3.15). Combined with Lemma 3.3, this leads to a preservation theorem for the Glivenko Cantelli property under the transformation taking the class $\mathcal{F}$ to the class $\mathcal{H}$ of all pairwise differences in $\mathcal{F}$.

Now we state some results on the asymptotic behavior of the argmin functional applied to a sequence of stochastic processes. The main theorem that we will use is the following. It is a type of continuous mapping theorem for the argmin functional.

Theorem 3.1 (Theorem 3.2.2 in Van der Vaart and Wellner, 1996) *Let Z_n and Z be stochastic processes indexed by $\mathbb{R}$ such that $Z_n \rightsquigarrow Z$ in $\ell^\infty([-K, K])$ for each $K > 0$. Moreover,* *suppose that almost all sample paths of Z are continuous and have a unique minimizer* $\hat{h}$*, assumed to be bounded in probability. If the sequence* $\hat{h}_n$ *is uniformly tight, and satisfies* $Z_n(\hat{h}_n) \le \inf_h Z_n(h) + o_P(1)$*, then* $\hat{h}_n \to^{\mathcal{D}} \hat{h}$ *in* $\mathbb{R}$.

Some clarification is needed. The space $\ell^\infty([-K, K])$ consists of all functions $f: [-K, K] \to \mathbb{R}$ such that $\|f\|_\infty = \sup_{t\in[-K,K]} |f(t)| < \infty$. Weak convergence in this space, denoted by $\rightsquigarrow$, to a Borel measurable Z, entails

$$E^* f(Z_n) \to Ef(Z)$$

for all bounded continuous functions $f: \ell^\infty([-K, K]) \to \mathbb{R}$. The E^* stands for outer expectation. This is needed since we do not require Z_n to be Borel measurable as mapping from a probability space into ℓ^∞. To verify the conditions in this theorem, and the nature of, e.g., the limit process Z, we will need some more concepts and results.

An important concept is that of an envelope F of the class $\mathcal{F}$ of functions defined on the set I. That is any function F on I such that for each $z \in I$ $|f(z)| \le F(z)$ for all $f \in \mathcal{F}$. A possible candidate as envelope is the minimal (or natural) envelope $F(z) = \sup_{f\in\mathcal{F}} |f(z)|$ for $z \in I$.

To establish convergence of Z_n to Z in our applications, we can use Theorem 3.2, giving sufficient conditions for this convergence.

Theorem 3.2 (Theorem 2.11.23 in Van der Vaart and Wellner, 1996) *For each n, let* $\mathcal{F}_n = \{f_{n,t}: t \in [-K, K]\}$ *be a class of measurable functions on* $\mathbb{R}$ *with measurable natural envelope* $F_n(z) = \sup_{|t|\le K} |f_{n,t}(z)|$ *satisfying the following two conditions:*

$$\|F_n\|_{2,g}^2 = \int F_n(z)^2 g(z)\, dz = O(1) \quad as\ n \to \infty \tag{3.18}$$

and

$$\int_{\{z:\ F_n(z) > \eta\sqrt{n}\}} F_n(z)^2 g(z)\, dz \to 0 \quad as\ n \to \infty. \tag{3.19}$$

Moreover, let the classes $\mathcal{F}_n$ *be such that for all* $\delta_n \downarrow 0$

$$\sup_{\{s,t\in[-K,K]:\ |s-t|<\delta_n\}} \int (f_{n,s}(z) - f_{n,s}(z))^2 g(z)\, dz \to 0 \quad as\ n \to \infty \tag{3.20}$$

and

$$\int_0^{\delta_n} \sqrt{\log N_{[\,]}(\epsilon\|F_n\|_{2,g}, \mathcal{F}_n, \|\cdot\|_{2,g})}\, d\epsilon \to 0.$$

Then the sequence of stochastic processes

$$t \to \sqrt{n} \int f_{n,t}(z)\, d(\mathbb{G}_n - G)(z)$$

converges in distribution to a Gaussian process, provided the sequence of covariance functions converges pointwise on $[-K, K]^2$

As part of the conditions of Theorem 3.1, we need the sequence of argmins $\hat{h}_n$ to be uniformly tight. Theorem 3.3 can be used to establish this tightness.

Theorem 3.3 (Theorem 3.2.5 in Van der Vaart and Wellner, 1996) *Let $\mathbb{M}_n$ be stochastic processes on an interval $I \subset \mathbb{R}$ and $\mathbb{M}$ a deterministic function on I such that for all θ in a neighborhood of θ_0*

$$\mathbb{M}(\theta) - \mathbb{M}(\theta_0) \gtrsim |\theta - \theta_0|^2 . \tag{3.21}$$

Moreover, suppose that

$$E^* \sup_{|\theta-\theta_0|<\delta} |(\mathbb{M}_n - \mathbb{M})(\theta) - (\mathbb{M}_n - \mathbb{M})(\theta_0)| \le \frac{\phi_n(\delta)}{\sqrt{n}} \tag{3.22}$$

for functions such that $\delta \mapsto \phi_n(\delta)/\delta^\alpha$ is decreasing for some $\alpha < 2$ (not depending on n). Let r_n be a sequence of positive numbers such that

$$r_n^2 \phi_n\left(\frac{1}{r_n}\right) \lesssim \sqrt{n} \text{ for all } n.$$

If the sequence $\hat{\theta}_n$ satisfies $\mathbb{M}_n(\hat{\theta}_n) \le \mathbb{M}_n(\theta_0) + O_P(r_n^{-2})$ and converges to θ_0 in probability, then $\hat{\theta}_n = \theta_0 + O_P(r_n^{-1})$.

To verify condition (3.22) in practical situations, the following theorem is useful. It uses the concept bracketing integral of the class $\mathcal{F}$ with $\|\cdot\|$, which is defined by

$$J_{[]}(\delta, \mathcal{F}, \|\cdot\|) = \int_0^\delta \sqrt{1 + \log N_{[]}(\epsilon \|F\|, \mathcal{F}, \|\cdot\|)}\, d\epsilon.$$

Theorem 3.4 (Theorem 2.14.2 in Van der Vaart and Wellner, 1996)

$$E_g^* \sup_{f \in \mathcal{F}} |\sqrt{n} \int f(z)\, d(\mathbb{G}_n - G)(z)| \le c J_{[]}(1, \mathcal{F}, \|\cdot\|_{2,g}) \|F\|_{2,g}.$$

3.5 Asymptotic Distribution in Exponential Deconvolution Model

In this section we will derive the asymptotic distribution of the nonparametric estimator of F defined in Section 2.4 in the exponential deconvolution problem. Remember that in the exponential deconvolution problem we have

$$U(x) = \int_0^x F(y)\, dy = G(x) + \int_0^x G(y)\, dy \text{ and } U_n(x) = \mathbb{G}_n(x^-) + \int_0^x \mathbb{G}_n(y)\, dy,$$

where $\mathbb{G}_n$ is the empirical distribution function based on the first n random variables of an i.i.d. sequence generated by the density

$$g(z) = \int_0^z e^{x-z}\, dF(x)$$

with corresponding distribution function G. (The choice for $\mathbb{G}_n(x^-) = \lim_{y \uparrow x} \mathbb{G}_n(y)$ instead of $\mathbb{G}_n(x)$ does not influence the convex minorant and its derivative. It is only chosen like this to make U_n lower semicontinuous, making the argmin functionals well defined.) The estimator $\tilde{F}_n(x)$ is defined as the right derivative of the convex minorant of U_n evaluated at the point x. Fix a point $x > 0$ such that $F'(y) = f(y)$ for y in a neighborhood of x for some continuous strictly positive function f.

In view of Section 3.3, we know that for each $v \in \mathbb{R}$,

$$\delta_n^{-1}\left(\tilde{F}_n(x) - F(x)\right) < v \iff \text{argmin}_s^- Z_n(s) > 0,$$

where

$$\begin{aligned} Z_n(s) &= \delta_n^{-2}\left(U_n(x+\delta_n s) - U_n(x) - U(x+\delta_n s) + U(x)\right) \\ &\quad + \delta_n^{-2}\left(U(x+\delta_n s) - U(x) - F(x)\delta_n s\right) - vs \\ &= W_n(s) + D_n(s) - vs. \end{aligned}$$

Here, for $s > 0$,

$$\begin{aligned} W_n(s) &= \delta_n^{-2}\left(U_n(x+\delta_n s) - U_n(x) - U(x+\delta_n s) + U(x)\right) \\ &= \delta_n^{-2}\int_{(x,x+\delta_n s]} d(\mathbb{G}_n - G)(z) + \delta_n^{-2}\int_{(x,x+\delta_n s]} (\mathbb{G}_n(z) - G(z))\,dz \end{aligned}$$

and (we assumed F to have a continuous strictly positive derivative f in a neighborhood of x),

$$\begin{aligned} D_n(s) &= \delta_n^{-2}\left(U(x+\delta_n s) - U(x) - F(x)\delta_n s\right) \\ &= \delta_n^{-2}\int_x^{x+\delta_n s} (F(y) - F(x))\,dy = \frac{1}{2}f(x)s^2 + R_n(s), \end{aligned}$$

where (the nonrandom) R_n tends to zero uniformly on compacta, whenever $\delta_n \downarrow 0$. The exact rescaling rate δ_n is not relevant for this. The case $s < 0$ proceeds completely analogously. One way of dealing with the two cases at the same time is to take the integral of the difference between two indicator functions, but for notational convenience we will not pursue this here.

What can be said about the rate of δ_n needed to get nondegenerate stochastic behavior for the process W_n? For a fixed finite value of s, the second term of W_n is of order $\delta_n^{-1}n^{-1/2}$, which follows from the fact that $\sup_{x \le z \le x+\delta_n s} |\mathbb{G}_n(z) - G(z)| = O_p(n^{-1/2})$. The first term of W_n is an increment of the empirical process over an interval of length $\delta_n s$, multiplied by $n^{-1/2}\delta_n^{-2}$. This will therefore be $O_p(\delta_n^{-3/2}n^{-1/2})$. In particular, this term dominates the second term, and to make it asymptotically nondegenerate, we need $\delta_n \sim n^{-1/3}$. This will be the choice for this section.

We now proceed studying the behavior of

$$\begin{aligned} W_n(s) &= n^{2/3}\int_{(x,x+n^{-1/3}s]} d(\mathbb{G}_n - G)(z) + n^{2/3}\int_{(x,x+n^{-1/3}s]} (\mathbb{G}_n(z) - G(z))\,dz \\ &= n^{2/3}\int_{(x,x+n^{-1/3}s]} d(\mathbb{G}_n - G)(z) + R_n^{(2)}(s), \end{aligned}$$

where for each $K < \infty$

$$\sup_{0 \le s \le K} R_n^{(2)}(s) \le n^{2/3}n^{-1/3}K\|\mathbb{G}_n - G\|_\infty = O_p(n^{-1/6})$$

as $n \to \infty$. Using Theorem 3.2 (see Exercise 3.21), we see that for each $K < \infty$

$$W_n \rightsquigarrow \sqrt{g(x)}W \text{ in } \ell^\infty([-K,K]), \tag{3.23}$$

where W is standard two-sided Brownian motion on $\mathbb{R}$. Therefore,

$$Z_n \rightsquigarrow Z \text{ in } \ell^\infty([-K, K]), \text{ where } Z(s) = \sqrt{g(x)}W(s) + \frac{1}{2}f(x)s^2.$$

Since $\mathrm{Var}(Z(s) - Z(t)) \neq 0$ for all $s \neq t$, the process Z has almost surely a unique minimizer. (See Kim and Pollard, 1990, Lemma 2.6.)

The next step in establishing the asymptotic distribution of $\tilde{F}(x)$, is to show that the sequence of argmins is tight. To this end we observe that we can apply Theorem 3.3. Here

$$\begin{aligned} I\!\!M_n(\theta) &= U_n(x+\theta) - U_n(x) - \theta F(x) - v\theta n^{-1/3} \\ &= \int_{[0,x+\theta]} (1 + x - \theta - z)\, d\mathbb{G}_n(z) - \int_{[0,x]} (1 + x - z)\, d\mathbb{G}_n(z) - \theta F(x) - v\theta n^{-1/3} \end{aligned}$$

and

$$I\!\!M(\theta) = U(x+\theta) - U(x) - \theta F(x).$$

Note that in the notation of Theorem 3.3, $\theta_0 = 0$, and also that $I\!\!M_n(0) = I\!\!M(0) = 0$. Moreover,

$$I\!\!M(\theta) \sim \frac{1}{2} f(x)\theta^2 \text{ as } \theta \to 0$$

taking care of condition (3.21).

From Exercise 3.25, it follows that $\hat{\theta}_n = \mathrm{argmin}_\theta^- I\!\!M_n(\theta) \to^P 0$ as $n \to \infty$. Now note that

$$(I\!\!M_n - I\!\!M)(\theta) = \int (1 + x + \theta - z)1_{[0,x+\theta]}(z) - (1 + x - z)1_{[0,x]}(z)\, d(\mathbb{G}_n - G)(z) - \theta v n^{-1/3}$$

so that

$$\sup_{|\theta|\le\delta} |(I\!\!M_n - I\!\!M)(\theta)| \le n^{-1/2}\left(\sup_{|\theta|\le\delta} \sqrt{n} \left| \int m_\theta(z)\, d(\mathbb{G}_n - G)(z) \right| + \delta v n^{1/6} \right) \tag{3.24}$$

where we write (see also Figure 3.5)

$$m_\theta(z) = (1 + x + \theta - z)1_{[0,x+\theta]}(z) - (1 + x - z)1_{[0,x]}(z). \tag{3.25}$$

To determine a function ϕ_n such that (3.22) holds, we will apply Theorem 3.4 to the first term in (3.24) applied to the function class

$$\mathcal{M}_\delta = \{m_\theta \colon |\theta| \le \delta\} \tag{3.26}$$

To this end, we need a bound on the bracketing number of the class as well as a bound on the second moment (under g) of the envelope of this class. By Exercise 3.23, we get that for each $N > 1$ the set of brackets

$$\{[m_{i\delta/N}, m_{(i+1)\delta/N}] \colon i = -N, -N+1, \ldots, N-1\}$$

covers the class $\mathcal{M}_\delta$. To determine the bracketing number as a function of ϵ, we have to determine how big N must be in order for the brackets to become ϵ-brackets. Exercise 3.24 gives that

$$N_{[\,]}(\epsilon, \mathcal{M}_\delta, \|\cdot\|_{2,g}) \sim \lceil c\delta\epsilon^{-2} \rceil.$$

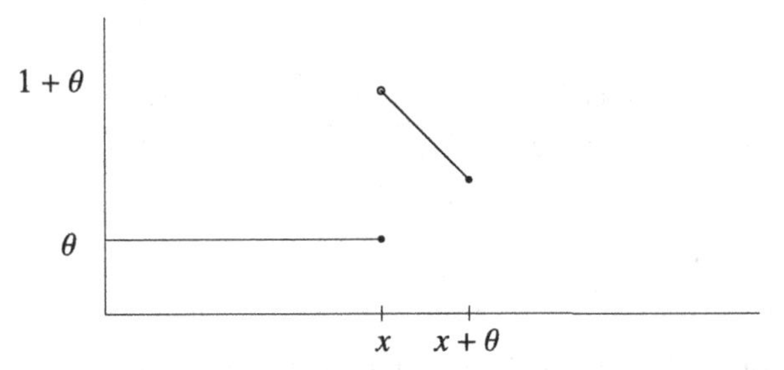

Figure 3.5 The function m_θ for x fixed as defined in (3.25).

Using the envelope of $\mathcal{M}_\delta$ and its norm provided in Exercise 3.26, we get for the bracketing integral at 1

$$J_{[\,]}(1, \mathcal{M}_\delta, \|\cdot\|_{2,g}) = \int_0^1 \sqrt{1 + \log N_{[\,]}(\epsilon\|M_\delta\|, \mathcal{M}_\delta, \|\cdot\|_{2,g})}\, d\epsilon$$
$$\leq \int_0^1 \sqrt{1 + \log c/\epsilon^2}\, d\epsilon < \infty.$$

By Theorem 3.4, and using (3.24), we get that we may take

$$\phi_n(\delta) = c\sqrt{\delta} + \delta v n^{1/6}$$

in (3.22). Hence, since $\phi_n(\delta)/\delta$ is certainly decreasing and $\hat\theta_n \to^P 0$, we obtain the rate $r_n = n^{1/3}$ and $\hat\theta_n = \operatorname{argmin}_\theta \mathbb{M}_n(\theta) = O_p(n^{-1/3})$. Since

$$Z_n(t) = n^{2/3}\mathbb{M}_n(n^{-1/3}t),$$

this means that

$$\hat h_n = \operatorname{argmin}_h Z_n(h) = O_p(1),$$

establishing the uniform tightness of $\hat h_n$. Together with the weak convergence of Z_n to Z, Theorem 3.1 can be used to conclude

$$\hat h_n \to^{\mathcal{D}} \hat h = \operatorname{argmin}_s \sqrt{g(x)}W(s) + \frac{1}{2}f(x)s^2$$

Using the scaling property of the Wiener process, we see that for each $a > 0$ and $b \in \mathbb{R}$

$$\operatorname{argmin}_{t\in\mathbb{R}}\left(aW(t) + (t-b)^2\right) =^{\mathcal{D}} a^{2/3}\operatorname{argmin}_{t\in\mathbb{R}}\left(W(t) + t^2\right) + b$$

(see also Exercise 3.18). This implies that

$$\operatorname{argmin}_{t\in\mathbb{R}}\left(\sqrt{g(x)}W(t) + \frac{1}{2}f(x)t^2 - vt\right) =^{\mathcal{D}}$$
$$\frac{g(x)^{1/3}2^{2/3}}{f(x)^{2/3}}\operatorname{argmin}_{t\in\mathbb{R}}\left(W(t) + t^2\right) + \frac{v}{f(x)}.$$

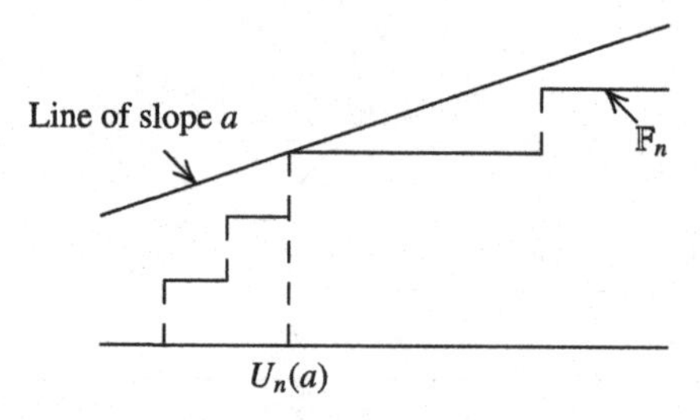

Figure 3.6 The switch relation for the Grenander estimator.

Hence,

$$\lim_{n\to\infty} P\left(n^{1/3}\left(\hat{F}_n(x) - F(x)\right) < v\right)$$
$$= P\left(\text{argmin}_{t\in\mathbb{R}}\left(W(t) + t^2\right) > -vf(x)^{-1/3}g(x)^{-1/3}2^{-2/3}\right)$$

showing that

$$n^{1/3}f(x)^{-1/3}g(x)^{-1/3}2^{-2/3}\left(\hat{F}_n(x) - F(x)\right) \to^{\mathcal{D}} \text{argmin}_{t\in\mathbb{R}}\left(W(t) + t^2\right). \tag{3.27}$$

This limiting random variable has the Chernoff distribution that will be discussed further in Section 3.9.

3.6 Limit Distribution of the Grenander Estimator

In Section 3.2, a heuristic derivation of the pointwise asymptotic distribution of the Grenander estimator was given, leading to (3.12). In this section we rigorously prove the result, originally proved by Prakasa Rao. The setting is the same as in Section 3.2, so $\hat{g}_n$ is the Grenander estimator (MLE) of the decreasing density $g\colon [0,\infty) \to [0,\infty)$ based on a sample $X_1, \ldots, X_n$ from this density.

Theorem 3.5 (Prakasa Rao, 1969) *Suppose that g has a strictly negative derivative g' at the interior point x_0 and $g(x_0) > 0$. Then*

$$n^{1/3}\left\{\hat{g}_n(x_0) - g(x_0)\right\} / \left|4g(x_0)g'(x_0)\right|^{1/3} \xrightarrow{\mathcal{D}} Z,\ n \to \infty,$$

where $\xrightarrow{\mathcal{D}}$ denotes convergence in distribution, and $Z = \text{argmax}_t\{W(t) - t^2\}$, that is: Z is the (almost surely unique) location of the maximum of two-sided Brownian motion minus the parabola $y(t) = t^2$.

Define, for $a > 0$, $U_n(a)$,

$$U_n(a) = \sup\left\{t \geq 0 : \mathbb{G}_n(t) - at \text{ is maximal}\right\}.$$

Then the switch relation encountered in Section 3.3 (here for the least concave majorant rather than the greatest convex minorant) gives that with probability one

$$\hat{g}_n(x_0) \leq a \iff U_n(a) \leq x_0.$$

See Figure 3.6 for an illustration of this relation.

Let δ_n be defined by

$$\delta_n = n^{-1/3}\left|4g'(x_0)g(x_0)\right|^{1/3}.$$

Then we get:

$$\mathbb{P}\{\hat{g}_n(x_0) - g(x_0) \le x\delta_n\} = \mathbb{P}\left\{U_n\left(g(x_0) + x\delta_n\right) \le x_0\right\}.$$

Let $a = g(x_0)$. Then we have to study the limit behavior of

$$\mathbb{P}\left\{U_n\left(a + x\delta_n\right) - x_0 \le 0\right\},$$

where

$$U_n(a + x\delta_n) = \sup\left\{t \ge 0 : \mathbb{G}_n(t) - \left(a + x\delta_n\right)t \text{ is maximal}\right\}.$$

From this point, we can take several different approaches, for example the Hungarian approximation, as in Groeneboom, 1985; the martingale approach, as will be used in Section 3.8 for the current status model; or the empirical process approach, as in Section 3.4. We take the last approach.

First note that

$$\begin{aligned}\delta_n^{-1}\{U_n(a + x\delta_n) - x_0\} &= \delta_n^{-1}\{\sup\{t : \mathbb{G}_n(t) - (a + x\delta_n)t \text{ is maximal}\} - x_0\}\\ &= \delta_n^{-1}\sup\{t : \mathbb{G}_n(x_0 + t) - (a + x\delta_n)(x_0 + t) \text{ is maximal}\}\\ &= \sup\{t : \mathbb{G}_n(x_0 + \delta_n t) - (a + x\delta_n)(x_0 + \delta_n t) \text{ is maximal}\}.\end{aligned}$$

Since the location of the maximum of a process does not change if the process is multiplied by a positive constant or a constant is added, the location of the maximum of the process

$$t \mapsto \mathbb{G}_n(x_0 + t\delta_n) - (a + x\delta_n)(x_0 + t\delta_n)$$

is the same as the location of the maximum of the process

$$\begin{aligned}t \mapsto\ &\delta_n^{-2}\{\mathbb{G}_n(x_0 + \delta_n t) - \mathbb{G}_n(x_0) - G_0(x_0 + \delta_n t) + G_0(x_0)\}\\ &+ \delta_n^{-2}\{G_0(x_0 + \delta_n t) + G_0(x_0) - at\delta_n\} - xt, \qquad t \ge -x_0/\delta_n. \end{aligned} \tag{3.28}$$

The process (3.28) converges, for every $K > 0$, in the space $D[-K, K]$ with the Skorohod topology, to the process

$$t \mapsto \left|4g(x_0)g'(x_0)\right|^{-2/3} W\left(4^{1/3}\left|g'(x_0)\right|^{1/3} g(x_0)^{4/3} t\right) + \tfrac{1}{2}g'(x_0)t^2 - xt, \qquad t \in [-K, K],$$

where W is two-sided Brownian motion, originating from zero. After the rescaling $u = t|g'(x_0)|$, the process changes (in distribution) to

$$u \mapsto \frac{1}{2|g'(x_0)|}\left\{W(u) - u^2 - 2xu\right\},\ u \in \left[-K|g'(x_0)|, K|g'(x_0)|\right],$$

where we use Brownian scaling to simplify the Brownian motion part. The location of the maximum of this process has the same distribution as the location of the maximum of the process

$$u \mapsto W(u) - (u + x)^2,\ u \in \left[-K|g'(x_0)|, K|g'(x_0)|\right],$$

So we obtain:

$$\mathbb{P}\left\{U_n(a+x\delta_n)-x_0\le 0\right\}=\mathbb{P}\left\{\delta_n^{-1}|g'(x_0)|\left\{U_n(a+x\delta_n)-x_0\right\}\le 0\right\},$$

where

$$\delta_n^{-1}|g'(x_0)|\left\{U_n(a+x\delta_n)-x_0\right\}\xrightarrow{\mathcal{D}}\operatorname{argmax}\left\{W(u)-(u+x)^2:u\in\mathbb{R}\right\},$$

provided we can prove tightness of the process on the left-hand side. Note that, since we only have to consider the probability

$$\mathbb{P}\left\{U_n(a+x\delta_n)-x_0\le 0\right\},$$

we can multiply $U_n(a+x\delta_n)-x_0$ by an arbitrary scaling factor to obtain the desired convergence.

Let $V(a)$ be the location of the maximum of the process $W(u)-(u-a)^2$, $u\in\mathbb{R}$, where W is two-sided Brownian motion on $\mathbb{R}$. Then $a\mapsto V(a)-a$ is a stationary process, and hence:

$$\mathbb{P}\{V(a)\le y\}=\mathbb{P}\{V(0)\le y-a\}.$$

Hence:

$$\mathbb{P}\left\{\operatorname{argmax}\left\{W(u)-(u+x)^2\right\}\le 0\right\}=\mathbb{P}\{V(-x)\le 0\}=\mathbb{P}\{V(0)\le x\}.$$

The tightness can be derived from Theorem 3.3, where we take $\theta_0=0$, $\theta=t$ and

$$\mathbb{M}(t)=-\left\{G(x_0+t)-G(x_0)-at\right\},\ \ \mathbb{M}_n(t)=-\left\{\mathbb{G}_n(x_0+t)-\mathbb{G}_n(x_0)-(a+\delta_nx)t\right\}.$$

Suppose $\hat{t}_n$ minimizes $t\mapsto\mathbb{M}_n(t)$, then we want $\delta_n^{-1}\hat{t}_n$ to be bounded in probability. First of all, we have by the assumptions of Theorem 3.5:

$$\mathbb{M}(t)=-\tfrac{1}{2}g'(x_0)t^2+o(t^2),\ t\to 0.$$

We now take $\gamma_n(\delta)=\sqrt{\delta}+\delta n^{1/6}$ and $r_n=n^{1/3}$. Then:

$$E\sup_{|t-x_0|<\delta}\left|(\mathbb{M}_n-\mathbb{M})(t)-(\mathbb{M}_n-\mathbb{M})(x_0)\right|\lesssim\frac{\gamma_n(\delta)}{\sqrt{n}}\quad\text{and}\quad n^{2/3}\gamma_n\left(n^{-1/3}\right)\lesssim\sqrt{n}.\tag{3.29}$$

Furthermore, $\mathbb{M}_n(\hat{t}_n)\le 0=\mathbb{M}_n(0)$, and we therefore only still have to check the property $\hat{t}_n\xrightarrow{p}0$.

Since for all fixed $x\in\mathbb{R}$, $a+\delta_nx>0$ for $a>0$ and sufficiently large n, we get for all large n:

$$\mathbb{M}_n(t)\to\infty,\ t\to\infty,$$

so $\hat{t}_n\in[-x_0,B]$ for sufficiently large $B=B(x)$ and all large n. For $t\ge\epsilon>0$, we get

$$\begin{aligned}&-\left\{\mathbb{G}_n(x_0+t)-\mathbb{G}_n(x_0)-(a+\delta_nx)t\right\}\ge-\tfrac{1}{2}g'(x_0)\{1+o(1)\}\epsilon^2+\delta_nxt+O_p\left(n^{-1/2}\right)\\&\ge-\tfrac{1}{4}g'(x_0)\epsilon^2+\delta_nxt+O_p\left(n^{-1/2}\right),\end{aligned}$$

if $\epsilon > 0$ is sufficiently small, using $\|\mathbb{G}_n - G\|_\infty = O_p(n^{-1/2})$ and the convexity and nonnegativity of the function

$$t \mapsto -\{G(x_0+t) - G(x_0) - at\}.$$

We similarly get, for $t \le -\epsilon$ and small $\epsilon > 0$,

$$-\{\mathbb{G}_n(x_0+t) - \mathbb{G}_n(x_0) - (a+\delta_n x)t\} \ge -\tfrac{1}{4}g'(x_0)\epsilon^2 + \delta_n xt + O_p\big(n^{-1/2}\big).$$

Since $\delta_n \asymp n^{-1/3}$ and $f_0'(x_0) < 0$, this means that $\mathbb{M}_n(t)$ is, with probability tending to one, strictly positive outside the neighborhood $(-\epsilon, \epsilon)$, if $t \in [-x_0, B]$. Since $\mathbb{M}_n(0) = 0$, this means that $\hat{t}_n \in (-\epsilon, \epsilon)$ with probability tending to one and hence $\hat{t}_n \xrightarrow{p} 0$. The result now follows from Theorem 3.3.

3.7 Some Martingale Theory

In Section 3.8 the asymptotic distribution of the MLE in the current status model will be derived using techniques from martingale theory. In this section we briefly introduce the key ideas and techniques related to this elegant area of probability theory for reference.

Definition 3.1 For a fixed $a \ge 0$, a stochastic process $\{Z(t) : t \ge a\}$ in $D[0,\infty)$ is a martingale with respect to an increasing family of σ-algebras $\{\mathcal{F}_t : t \ge a\}$, if it is adapted to this family, i.e., $Z(t)$ is $\mathcal{F}_t$-measurable for each $t \ge a$, and if

$$E|Z(t)| < \infty,\ t \ge a,\ \text{ and } E\left\{Z(t) \,\middle|\, \mathcal{F}_s\right\} = Z(s),\ t > s \ge a,\ a.s. \tag{3.30}$$

The increasing family of σ-algebras $\{\mathcal{F}_t : t \ge a\}$ is called a filtration of σ-algebras. Z is called an L_2-martingale, if $EZ_t^2 < \infty$, for all $t \ge a$.

A related concept is that of a submartingale. In fact, it has all properties of a martingale except that the equality in (3.30) is replaced by an inequality:

$$E\left\{Z(t) \,\middle|\, \mathcal{F}_s\right\} \ge Z(s),\ t > s \ge a,\ a.s.$$

For nonnegative submartingales there is a very useful inequality, Doob's (maximal) inequality, that bounds the expectation of the supremum of the process over an interval by the expectation of the nonnegative submartingale taken at the right end point of the interval. A version of this inequality reads:

$$E \sup_{s \in [a,t]} Z_n(s) \le EZ_n(t). \tag{3.31}$$

Another result to be used, is a martingale central limit theorem, given in Rebolledo, 1980. To state the theorem, we need the quadratic variation process. For an L_2-martingale Z, the quadratic variation or square brackets process $t \mapsto [Z](t)$ is defined as the limit in probability, over partitions $0 = t_0 < t_1 < \cdots < t_n = t$ of

$$\sum_{i=1}^n \{Z(t_i) - Z(t_{i-1})\}^2.$$

Theorem 3.6 (Rebolledo's theorem) *Let* (Z_n) *be a sequence of* L_2 *martingales in* $D(0,\infty)$ *with corresponding quadratic variation processes* $[Z_n]$. *Let* H *be a continuous increasing function on* $[0,\infty)$, *such that* $H(0)=0$. *Furthermore, suppose*

(i) $Z_n(0)\to 0$ *in probability,*
(ii) $[Z_n](t)\to H(t)$ *in probability, for each fixed* $t>0$,
(iii) $E\max_{u\in[0,t]}\{Z_n(u)-Z_n(u-)\}^2\to 0$, *for each fixed* $t>0$.

Then the sequence (Z_n) *converges in* $D(0,\infty)$ *to the process*

$$t\mapsto W(H(t)),\ t\ge 0,$$

where $t\mapsto W(t)$ *is standard Brownian motion on* $[0,\infty)$,

This follows from Theorem 2, p. 273, Rebolledo, 1980.

3.8 Asymptotic Distribution of the MLE in the Current Status Model

We now turn to the local limit distribution of the MLE for current status data, using the same set up as in Section 2.3. That is: $X_1, X_2,\dots,$ are independent nonnegative random variables with distribution function F_0 and $T_1, T_2,\dots$ are independent random variables with distribution function G. Moreover, the T_is and X_js are also independent. For a sample size n, the data available to us are given by

$$(T_1,\Delta_1),\dots,(T_n,\Delta_n),$$

where $\Delta_i=1_{\{X_i\le T_i\}}$. The following result will be proved.

Theorem 3.7 *Let* t_0 *be such that* $0<F_0(t_0), G(t_0)<1$, *and let* F_0 *and* G *be differentiable at* t_0, *with strictly positive derivatives* $f_0(t_0)$ *and* $g(t_0)$, *respectively. Furthermore, let* $\hat F_n$ *be the MLE of* F_0. *Then we have, as* $n\to\infty$,

$$n^{1/3}\left\{\hat F_n(t_0)-F_0(t_0)\right\}/\left\{4F_0(t_0)(1-F_0(t_0))f_0(t_0)/g(t_0)\right\}^{1/3}\to^{\mathcal D} Z, \tag{3.32}$$

where $\to^{\mathcal D}$ *denotes convergence in distribution, and where* Z *is the last time where standard two-sided Brownian motion plus the parabola* $y(t)=t^2$ *reaches its minimum.*

In order to prove this result, we first give a road map. Let $\mathbb G_n$ be the empirical distribution function of $T_1,\dots,T_n$, and define V_n by

$$V_n(t)=n^{-1}\sum_{i=1}^n\Delta_i 1_{\{T_i\le t\}}=\int\delta 1_{(-\infty,t]}(t')\,d\mathbb P_n(t',\delta),$$

where $\mathbb P_n$ is the empirical distribution function of $(T_1,\Delta_1),\dots,(T_n,\Delta_n)$. Then we know from Lemma 2.3 that the MLE of F_0 at $T_{(i)}$ can be obtained by taking the left derivative of the greatest convex minorant of the diagram consisting of the points $(0,0)$ and $\bigl(i/n, V_n(T_{(i)})\bigr)$, $i=1,\dots,n$. Note that this cumulative sum diagram can also be defined by

$$(\mathbb G_n(t), V_n(t)),\ t\in\mathbb R, \tag{3.33}$$

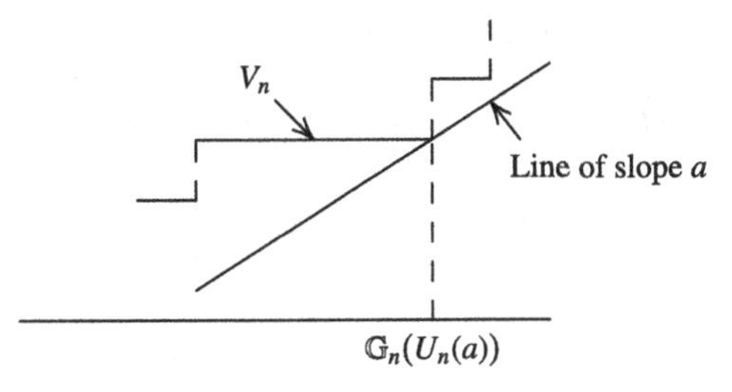

Figure 3.7 The switch relation.

since (3.33) runs through the values (0, 0) and $\big(i/n,\, V_n(T_{(i)})\big), i = 1, \ldots, n$. This observation will also be of use in Section 8.5, defining the maximum smoothed likelihood estimator in the current status model.

Furthermore, let

$$U_n(a) = \operatorname{argmin}\{t \in \mathbb{R} : V_n(t) - a\mathbb{G}_n(t)\}. \tag{3.34}$$

Then, in view of Section 3.3, we have the following switch relation:

$$\hat{F}_n(t) \geq a \iff \mathbb{G}_n(t) \geq \mathbb{G}_n(U_n(a)) \iff t \geq U_n(a). \tag{3.35}$$

If $a = F_0(t)$, this implies

$$\mathbb{P}\left\{n^{1/3}\{\hat{F}_n(t) - F_0(t)\} \geq x\right\} = \mathbb{P}\left\{\hat{F}_n(t) \geq a + n^{-1/3}x\right\} = \mathbb{P}\left\{U_n(a + n^{-1/3}x) \leq t\right\}. \tag{3.36}$$

Figure 3.7 shows how the idea applies to the cumulative sum diagram (3.33).

In view of (3.36), the inverse process $a \mapsto U_n(a)$ can be considered instead of the original process $t \mapsto \hat{F}_n(t)$. We have the following tightness condition for the process U_n. It is similar to Lemma 5.3 on p. 93 in Groeneboom and Wellner, 1992, and indeed proved in the same way.

Lemma 3.5 *Assume that the conditions of Theorem 3.7 are satisfied and let* $a_0 = F_0(t_0)$. *Then for each* $\epsilon > 0$ *and* $M_1 > 0$ *an* $M_2 > 0$ *can be found such that*

$$\mathbb{P}\left\{\sup_{a\in[-M_1,M_1]} n^{1/3}\{U_n(a_0 + n^{-1/3}a) - t_0\} > M_2\right\} < \epsilon,$$

and

$$\mathbb{P}\left\{\inf_{a\in[-M_1,M_1]} n^{1/3}\{U_n(a_0 + n^{-1/3}a) - t_0\} < -M_2\right\} < \epsilon,$$

for all large n.

Proof We only prove the first inequality, since the second inequality can be proved analogously. First note that

$$\mathbb{P}\left\{\sup_{a\in[-M_1,M_1]} n^{1/3}\{U_n(a_0+n^{-1/3}a) - t_0\} > M_2\right\} = \mathbb{P}\left\{n^{1/3}\{U_n(a_0+n^{-1/3}M_1) - t_0\} > M_2\right\},$$

since the process U_n is nondecreasing. Furthermore,

$$\mathbb{P}\left\{n^{1/3}\{U_n(a_0+n^{-1/3}M_1)-t_0\}>M_2\right\}$$
$$\leq \mathbb{P}\left\{\exists t>M_2 : W_n(t)-n^{2/3}\{a_0+n^{-1/3}M_1\}\{\mathbb{G}_n(t_0+n^{-1/3}t)-\mathbb{G}_n(t_0)\}\leq 0\right\},$$

where

$$W_n(t)=n^{2/3}\left\{V_n(t_0+n^{-1/3}t)-V_n(t_0)\right\},\ t\in\mathbb{R}.$$

For $t>0$, the process W_n can be written as

$$W_n(t)=n^{-1/3}\sum_{i:T_i\in[t_0,t_0+n^{-1/3}t]}\Delta_i,$$

and this process can be turned into a martingale $\widetilde{W}_n$, by defining

$$\widetilde{W}_n(t)=n^{-1/3}\sum_{i:T_i\in[t_0,t_0+n^{-1/3}t]}\{\Delta_i-F_0(T_i)\}=n^{2/3}\int_{u\in[t_0,t_0+n^{-1/3}t]}\{\delta-F_0(u)\}\,d\mathbb{P}_n(u,\delta). \tag{3.37}$$

The process $\widetilde{W}_n$ is a martingale with respect to the family of σ-algebras $\{\mathcal{F}_{n,t}:t\geq 0\}$ defined by

$$\mathcal{F}_{n,t}=\sigma\left\{(T_i,\Delta_i):T_i\in[t_0,t_0+n^{-1/3}t]\right\},t\geq 0. \tag{3.38}$$

We also condition on $\mathbb{G}_n(t_0)$. We now get:

$$\mathbb{P}\left\{\exists t>M_2 : W_n(t)-n^{1/3}M_1\{\mathbb{G}_n(t_0+n^{-1/3}t)-\mathbb{G}_n(t_0)\}\leq 0\right\}$$
$$=\mathbb{P}\left\{\exists t>M_2 : \widetilde{W}_n(t)+n^{2/3}\int_{u\in[t_0,t_0+n^{-1/3}t]}\{F_0(u)-F_0(t_0)\}\,d\mathbb{G}_n(u)\right.$$
$$\left.-n^{1/3}M_1\{\mathbb{G}_n(t_0+n^{-1/3}t)-\mathbb{G}_n(t_0)\}\leq 0\right\}$$
$$=\mathbb{P}\left\{\exists t>M_2 : \widetilde{W}_n(t)+n^{2/3}\int_{u\in[t_0,t_0+n^{-1/3}t]}\{F_0(u)-F_0(t_0)-n^{-1/3}M_1\}\,d\mathbb{G}_n(u)\leq 0\right\}.$$

By the fact that $\widetilde{W}_n$ is a martingale and Doob's submartingale inequality we now get for $\epsilon>0$ and $A>0$:

$$\mathbb{P}\left\{\exists u\in\left[(j-1)n^{-1/3},jn^{-1/3}\right):\left|\widetilde{W}_n(u)\right|>\epsilon(j-1)^2+A\right\}$$
$$\leq n^{4/3}\mathbb{E}\widetilde{W}_n\big(jn^{-1/3}\big)^2/\big(\epsilon(j-1)^2+A\big)^2\leq cj/\big(\epsilon(j-1)^2+A\big)^2,$$

for $j\geq 1$, where c does not depend on $j\in\mathbb{N}$.

The arguments of Lemma 4.1 in Kim and Pollard, 1990, imply that there exists a tight sequence of random variables (A_n) such that

$$n^{-2/3}\left|\widetilde{W}_n(u)\right|\leq\epsilon u^2+n^{-2/3}A_n,\ u\geq 0$$

for all $\epsilon > 0$. Moreover, by the conditions of Theorem 3.7 we have for each $M > 0$ and $\eta > 0$ an $M_2 > 0$ and $\epsilon > 0$ such that

$$\forall t \ge M_2 n^{-1/3} : \int_{u \in [t_0, t_0 + n^{-1/3} t]} \{F_0(u) - F_0(t_0)\}\, d\mathbb{G}_n(u) \ge M n^{-2/3} \vee \epsilon t^2,$$

with probability larger than $1 - \eta$. The statement now follows. □

We also need the following result.

Lemma 3.6 *Let W be two-sided Brownian motion on $\mathbb{R}$, originating from zero. Let $D(\mathbb{R})$ be the space of right continuous functions, with left limits (cadlag functions) on $\mathbb{R}$, equipped with the metric of uniform convergence on compact sets, and let $t_0 > 0$. Let $\widetilde{W}_n$ be defined by (3.37), that is:*

$$\widetilde{W}_n(t) = \begin{cases} n^{-1/3} \sum_{i: T_i \in [t_0, t_0 + n^{-1/3} t]} \{\Delta_i - F_0(T_i)\}, & t \ge 0, \\ n^{-1/3} \sum_{i: T_i \in [t_0 + n^{-1/3} t, t_0]} \{\Delta_i - F_0(T_i)\}, & t < 0,\ t_0 + n^{-1/3} t > 0, \\ 0, & \text{otherwise.} \end{cases}$$

Then $\widetilde{W}_n$ converges in $D(\mathbb{R})$ in distribution to the process V, defined by

$$V(t) = \sqrt{g(t_0) F_0(t_0)(1 - F_0(t_0))}\, W(t),\ t \in \mathbb{R}. \tag{3.39}$$

Proof It was noted in the proof of Lemma 3.5 that $t \mapsto \widetilde{W}_n(t)$, $t \ge 0$, is a martingale. The quadratic variation process is given by

$$\left[\widetilde{W}_n\right](t) = n^{-2/3} \sum_{T_i \in [t_0, t_0 + n^{-1/3} t]} \{\Delta_i - F_0(T_i)\}^2,\ t \ge 0. \tag{3.40}$$

We check the conditions of Theorem 3.6 for the sequence $(\widetilde{W}_n)$, restricting these processes to the positive halfline. Condition (i) is clearly satisfied. To check (ii), we note that, by (3.40), $\left[\widetilde{W}_n\right](t)$ can be written

$$\left[\widetilde{W}_n\right](t) = \sum_{i=1}^n Z_{n,i},$$

where

$$Z_{n,i} = n^{-2/3} \{\Delta_i - F_0(T_i)\}^2 1_{[t_0, t_0 + n^{-1/3} t]}(T_i),\ i = 1, \dots, n.$$

The $Z_{n,i}$, $i = 1, \dots, n$ satisfy

$$\begin{aligned} \sum_{i=1}^n E Z_{n,i} &= n^{1/3} \int_{t_0}^{t_0 + n^{-1/3} t} F_0(u)\{1 - F_0(u)\}\, dG(u) \\ &= F_0(t_0)\{1 - F_0(t_0)\} g(t_0) t + o(1),\ n \to \infty, \end{aligned}$$

and

$$\operatorname{Var}\left(\sum_{i=1}^{n} Z_{n,i}\right) \le nEZ_{n,1}^2 = n^{-1/3}\int_{t_0}^{t_0+n^{-1/3}t} \{\{1-F_0(u)\}^4 F_0(u)+F_0(u)^4\{1-F_0(u)\}\}dG(u)$$

$$\le n^{-1/3}\int_{t_0}^{t_0+n^{-1/3}t} F_0(u)\{1-F_0(u)\}\,dG(u) \to 0,\ n\to\infty.$$

Hence, by Chebyshev's inequality,

$$\left[\widetilde{W}_n\right](t) \to F_0(t_0)\{1-F_0(t_0)\}\,g(t_0)t, \text{ in probability, as } n\to\infty.$$

Since the jumps of V_n are bounded above by $n^{-1/3}$, condition (iii) is clearly satisfied.

Thus the conditions of Theorem 3.6 are satisfied with

$$H(t) = F_0(t_0)\{1-F_0(t_0)\}\,g(t_0)t,\ t\ge 0.$$

The scaling property of Brownian motion gives that

$$t \mapsto \sigma^{-1/2}W(\sigma t) \stackrel{\mathcal{D}}{=} t\mapsto W(t), \text{ for all } \sigma>0. \tag{3.41}$$

This yields that

$$t\mapsto W(H(t)),\ t\ge 0,$$

has the same distribution as

$$t\mapsto \sqrt{F_0(t_0)\{1-F_0(t_0)\}\,g(t_0)}\,W(t),\ t\ge 0.$$

The statement for the process on the negative halfline follows in a similar way. □

We can now easily prove Theorem 3.7. By the switch relation (3.36), we have:

$$\mathbb{P}\left\{n^{1/3}\{\hat{F}_n(t_0)-F_0(t_0)\}\ge x\right\} = \mathbb{P}\left\{U_n(a_0+n^{-1/3}x)\le t_0\right\}.$$

Furthermore,

$$\mathbb{P}\left\{U_n(a_0+n^{-1/3}x)\le t_0\right\} = \mathbb{P}\left\{U_n(a_0+n^{-1/3}x)-t_0\le 0\right\}$$

$$= \mathbb{P}\left\{n^{1/3}\{U_n(a_0+n^{-1/3}x)-t_0\}\le 0\right\}.$$

By Lemma 3.5, the sequence $n^{1/3}\{U_n(a_0+n^{-1/3}x)-t_0\}$ is tight, and by Theorem 3.1, applied to the process

$$t\mapsto \tilde{W}_n(t)+n^{2/3}\int_{t_0}^{t_0+n^{-1/3}t)}\{F_0(u)-F_0(t_0)\}\,dG(u),$$

it converges in distribution in the Skorohod topology to the process

$$t\mapsto V(t)+\tfrac{1}{2}f_0(t_0)g(t_0)t^2,\ t\in\mathbb{R},$$

where V is defined as in Lemma 3.6. The result can now be deduced from Brownian scaling property (3.41). See Exercise 3.27, using the fact that the time scale of the cusum diagram converges locally to $t\mapsto g(t_0)t$.

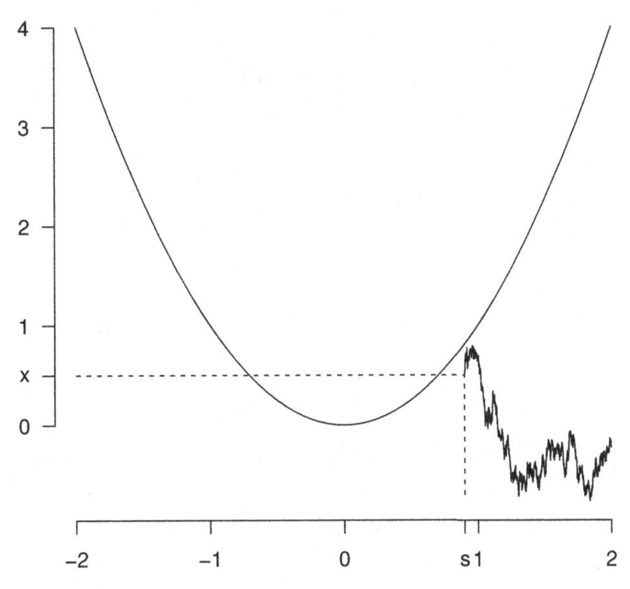

Figure 3.8 The process W conditioned to have value x at $t = s$ on $[s, 2]$ and the function $t \mapsto t^2$ on $[-2, 2]$.

3.9 Chernoff's Distribution

Let $\{W(t) : t \in \mathbb{R}\}$ be standard two-sided Brownian motion, originating from zero. The determination of the distribution of the (almost surely unique) location of the maximum of $\{W(t) - t^2 : t \in \mathbb{R}\}$, which occurs as limit distribution in Theorem 3.5, Theorem 3.7, and other places in this book, has a long history, which probably started with the paper by Chernoff, 1964, in a study of the limit distribution of an estimator of the mode of a distribution. In that paper, the density of the location of the maximum of $\{W(t) - t^2 : t \in \mathbb{R}\}$, which we will denote by

$$Z = \operatorname{argmax}_t\{W(t) - t^2,\ t \in \mathbb{R}\}, \tag{3.42}$$

is characterized in a way we will now describe. Define for $s, x \in \mathbb{R}$ the quantity $u(s, x)$ by

$$u(s, x) = \mathbb{P}\left\{W(t) > t^2 \text{ for some } t > s \,|\, W(s) = x\right\}.$$

See Figure 3.8. Then we have the convolution equation:

$$u(s, x) = \mathbb{E}\left\{u\left(s + \epsilon,\, W(s + \epsilon)\right)\right\} = u(s, x) + \frac{\partial}{\partial s}u(s, x)\epsilon + \frac{1}{2}\frac{\partial^2}{\partial y^2}u(s, x)\epsilon + o(\epsilon),$$

if $x < s^2$. Hence, the function u satisfies the following partial differential equation on the region $\{(s, x) \in \mathbb{R}^2 \colon x < s^2\}$

$$\frac{\partial}{\partial s}u(s, x) = -\frac{1}{2}\frac{\partial^2}{\partial x^2}u(s, x)$$

and

$$u(s, x) = 1,\ x \ge s^2, \qquad u(s, x) \to 0,\ x \to -\infty.$$

Define

$$M_h = \max_{t\in[s-h,s]} W(t)$$

and

$$u_2(t) = \lim_{x\uparrow t^2} \frac{\partial}{\partial x} u(t,x). \tag{3.43}$$

Then:

$$\mathbb{P}\left\{\max_{t\ge s}\left\{W(t)-t^2\right\} < M_h - s^2 \mid W(s), M_h\right\} = -\left\{W(s)-M_h\right\}\partial_2 u(s,s^2) + O_p(h).$$

Similarly:

$$\mathbb{P}\left\{\max_{t\le s-h}\left\{W(t)-t^2\right\} < M_h - (s-h)^2 \mid W(s-h), M_h\right\}$$
$$= -\left\{W(s-h)-M_h\right\}\partial_2 u(-s,s^2) + O_p(h).$$

The conclusion is that:

$$\mathbb{P}\{Z\in[s-h,s]\} \sim \mathbb{P}\left\{\max_{x\notin[s-h,s]}\left\{W(t)-t^2\right\} < M_h - s^2\right\}$$
$$\sim \mathbb{E}\left\{M_h - W(s)\right\}\left\{M_h - W(s-h)\right\}\partial_2 u(s,s^2)\partial_2 u(-s,s^2)$$
$$= h\mathbb{E}\left(\max_{x\in[0,1]}\mathbb{B}(x)\right)^2 \partial_2 u(s,s^2)\partial_2 u(-s,s^2) = \tfrac{1}{2}h\,\partial_2 u(s,s^2)\partial_2 u(-s,s^2),\ h\downarrow 0.$$

where $\mathbb{B}$ is standard Brownian bridge as described in Exercise 3.9. This gives the following lemma.

Lemma 3.7 *Let f_Z be the density of the (almost surely unique) location of the maximum for standard Brownian motion minus $y(t)=t^2$. Let $u(t,x)$ be the solution of the heat equation*

$$\frac{\partial}{\partial t}u(t,x) = -\frac{1}{2}\frac{\partial^2}{\partial x^2}u(t,x),$$

for $x\le t^2$, under the boundary conditions

$$u(t,x)\ge 0,\quad u(t,t^2) := \lim_{x\uparrow t^2} u(t,x) = 1,\quad (t,x)\in\mathbb{R}^2,\quad \lim_{x\downarrow -\infty} u(t,x) = 0,\quad t\in\mathbb{R}.$$

Then

$$f_Z(t) = \tfrac{1}{2}u_2(-t)u_2(t),\ t\in\mathbb{R}. \tag{3.44}$$

where the function u_2 is defined by (3.43).

The original attempts to compute the density f_Z were based on numerically solving the heat equation, but it soon became clear that this method did not produce a very accurate solution, mainly because of the rather awkward boundary conditions. However, around 1984 the connection with Airy functions was discovered and this connection was exploited to give analytic solutions in the papers Daniels and Skyrme, 1985, Temme, 1985, and Groeneboom, 1989, which were all written in 1984, although the last paper appeared much later.

There seems to be a recent revival of interest in this area of research; see, e.g., Janson et al., 2010, Groeneboom, 2010, Groeneboom, 2011, Groeneboom and Temme, 2011, Pimentel, 2014, and Janson, 2013. These recent papers (except Pimentel, 2014) rely heavily on the results in Daniels and Skyrme, 1985, and Groeneboom, 1989, but the derivation of the results in these papers is not a simple matter. The most natural approach still seems to use the Cameron-Martin formula for making the transition from Brownian motion with drift to Brownian motion without drift, and next to use the Feynman-Kac formula for determining the distribution of the Radon-Nikodym derivative of the Brownian motion with parabolic drift with respect to the Brownian motion without drift from the corresponding second order differential equation. This is the approach followed in Groeneboom, 1989, and taken up again in Groeneboom et al., 2013. We now give a description of the latter approach.

This method starts with the following theorem.

Theorem 3.8 (Theorem 2.1 in Groeneboom, 1989) *Let, for $s \in \mathbb{R}$ and $x < 0$, $Q^{(s,x)}$ be the probability measure on the Borel σ-field of $C([s,\infty):\mathbb{R})$, corresponding to the process $\{X(t): t \ge s\}$, where $X(t) = W(t) - t^2$, starting at position x at time s, and where $\{W(t): t \ge s\}$ is Brownian motion, starting at $x + s^2$ at time s. Let the first passage time τ_0 of zero of the process X be defined by*

$$\tau_0 = \inf\{t \ge s : X(s) = 0\},$$

where, as usual, we define $\tau_0 = \infty$, if $\{t \ge s : X(t) = 0\} = \emptyset$. Then

(i)

$$Q^{(s,x)}\{\tau_0 \in dt\} = e^{-\frac{2}{3}\left(t^3 - s^3\right) + 2sx}\psi_x(t-s)E^0\left\{\exp\left\{-2\int_0^{t-s} B(u)\,du\right\}\Big|\,B(t-s) = -x\right\}dt,$$

where B is a Bessel(3) *process, starting at zero at time* 0*, with corresponding expectation E^0, and where $\psi_z(u) = \left(2\pi u^3\right)^{-1/2} z\exp\left(-z^2/(2u)\right)$, $u, z > 0$, is the value at u of the density of the first passage time through zero of Brownian motion, starting at z at time* 0*.*

(ii)

$$Q^{(s,x)}\{\tau_0 \in dt\} = e^{-\frac{2}{3}\left(t^3 - s^3\right) + 2sx}h_x(t-s)\,dt, \tag{3.45}$$

where the function $h_x : \mathbb{R}_+ \to \mathbb{R}_+$ has Laplace transform

$$\hat{h}_x(\lambda) = \int_0^\infty e^{-\lambda u}h_x(u)\,du = \mathrm{Ai}(\xi - 4^{1/3}x)/\mathrm{Ai}(\xi), \quad \xi = 2^{-1/3}\lambda > 0, \tag{3.46}$$

and Ai *denotes the Airy function* Ai.

Remark Note that the function h_x in the definition of the density of the stopping time τ_0 has by part (ii) of Theorem 3.8 the representation

$$h_x(t) = \frac{1}{2\pi}\int_{v=-\infty}^{\infty} e^{itv}\frac{\mathrm{Ai}(i2^{-1/3}v - 4^{1/3}x)}{\mathrm{Ai}(i2^{-1/3}v)}\,dv, \; t > 0. \tag{3.47}$$

This representation is obtained by inverting the Laplace transform.

The big jump forward with respect to Lemma 3.7 is that we have the distribution of the hitting time τ_0 of Brownian motion minus a parabolic drift, expressed by (3.45), where the

Table 3.1 *Most Frequently Used Quantiles of the Chernoff Distribution*

Probability p	0.90	0.95	0.975	0.99	0.995	0.999
Quantile $F^{-1}(p)$	0.6642	0.8451	0.9982	1.1716	1.2867	1.5167

only part that is giving real trouble is expressed by (3.46) or (3.47). Theorem 3.8 should in principle be sufficient to derive the density f_Z of (3.44), since, defining

$$q(s) = \lim_{x\uparrow 0} \frac{\partial}{\partial x} Q^{(s,x)}\left\{X_t < 0,\ \forall t \ge s\right\} = \lim_{x\uparrow 0} \frac{\partial}{\partial x} Q^{(s,x)}\left\{\tau_0 = \infty\right\},$$

we find:

$$f_Z(s) = \tfrac{1}{2} q(s) q(-s),$$

following a line of reasoning similar to the derivation of (3.44) in Chernoff, 1964.

Theorem 3.8 is proved in Groeneboom et al., 2013, by the Feynman-Kac formula with a stopping time, which, in turn, is essentially Ito's formula plus a martingale argument. We have:

$$\lim_{x\uparrow 0} \frac{\partial}{\partial x} Q^{(s,x)}\left\{\tau_0 < \infty\right\} = \lim_{x\uparrow 0} \frac{\partial}{\partial x} e^{2sx+\frac{2}{3}s^3} \frac{1}{2\pi} \int_{t=0}^{\infty} e^{ivt-\frac{2}{3}(s+t)^3} \int_{v=-\infty}^{\infty} \frac{\mathrm{Ai}(i2^{-1/3}v - 4^{1/3}x)}{\mathrm{Ai}(i2^{-1/3}v)}\, dv\, dt.$$

It follows from Theorem 3.1 in Groeneboom et al., 2013, that

$$\begin{aligned} &e^{2sx+\frac{2}{3}s^3} \frac{1}{2\pi} \int_{t=0}^{\infty} e^{ivt-\frac{2}{3}(s+t)^3} \int_{v=-\infty}^{\infty} \frac{\mathrm{Ai}(i2^{-1/3}v - 4^{1/3}x)}{\mathrm{Ai}(i2^{-1/3}v)}\, dv\, dt \\ &= 1 - e^{2sx+\frac{2}{3}s^3} g(s,x) = 1 - \frac{e^{\frac{2}{3}s^3}}{2\pi} \int_{u=-\infty}^{\infty} \frac{\int_{y=0}^{-4^{1/3}x} e^{-2^{1/3}s(iu+y)} \mathrm{Ai}(iu+y)\, dy}{\mathrm{Ai}(iu)^2}\, du, \end{aligned}$$

and hence:

$$q(s) = \lim_{x\uparrow 0} \frac{\partial}{\partial x} Q^{(s,x)}\left\{\tau_0 < \infty\right\} = \frac{e^{\frac{2}{3}s^3}}{2^{1/3}\pi} \int_{u=-\infty}^{\infty} \frac{e^{-i2^{1/3}su}}{\mathrm{Ai}(iu)}\, du = \frac{e^{\frac{2}{3}s^3}}{2^{2/3}\pi} \int_{u=-\infty}^{\infty} \frac{e^{-isu}}{\mathrm{Ai}(i2^{-1/3}u)}\, du.$$

This leads to the following recapitulation of Lemma 3.7.

Theorem 3.9 *Let f_Z be the density of the (almost surely unique) location of the maximum for standard Brownian motion minus $y(t) = t^2$. Then:*

$$f_Z(s) = \tfrac{1}{2} q(s) q(-s),\ t \in \mathbb{R}, \tag{3.48}$$

where

$$q(s) = \frac{1}{4^{2/3}\pi} \int_{u=-\infty}^{\infty} \frac{e^{-isu}}{\mathrm{Ai}(i2^{-1/3}u)}\, du. \tag{3.49}$$

Figure 3.9 shows pictures of the density function and distribution function associated with the Chernoff distribution. Also the 90% quantile (0.664) is indicated in the pictures. The most frequently used quantiles of the Chernoff distribution are given in Table 3.1.

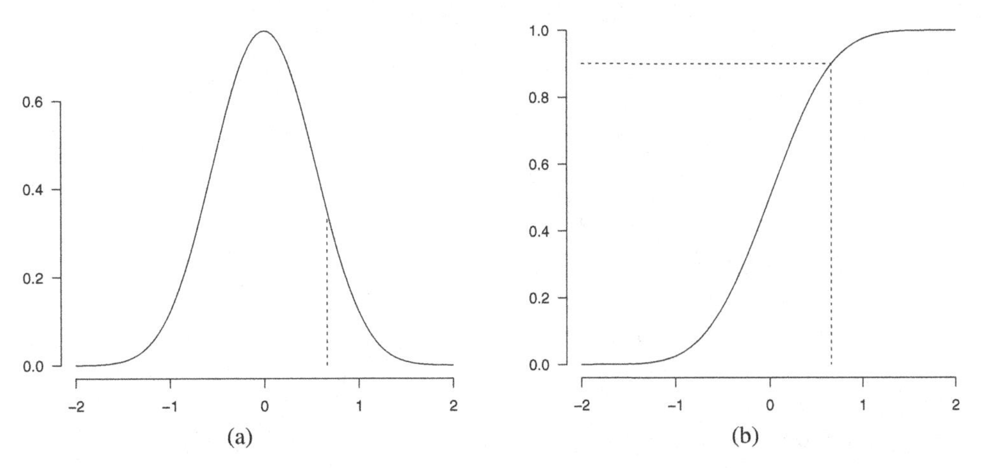

Figure 3.9 The density (a) and distribution (b) function of the Chernoff distribution. In both pictures the 90% quantile is indicated.

3.10 The Concave Majorant of Brownian Motion and Brownian Bridge

In this section we consider two processes. The first is the concave majorant of Brownian motion on $[0, \infty)$ and the second is the concave majorant of Brownian bridge on $[0, 1]$. See Figure 3.10.

First consider the concave majorant of one-sided Brownian motion without drift, starting at zero, and define

$$\tau(1/a) = \sigma(a) = \sup\{t \geq 0 : W(t) - at \text{ is maximal}\}.$$

In order to let time increase if we go to the right, we make the shift $\tau(1/a) = \sigma(a)$. Let S_t be the left continuous slope of the least concave majorant of Brownian motion at t. We have

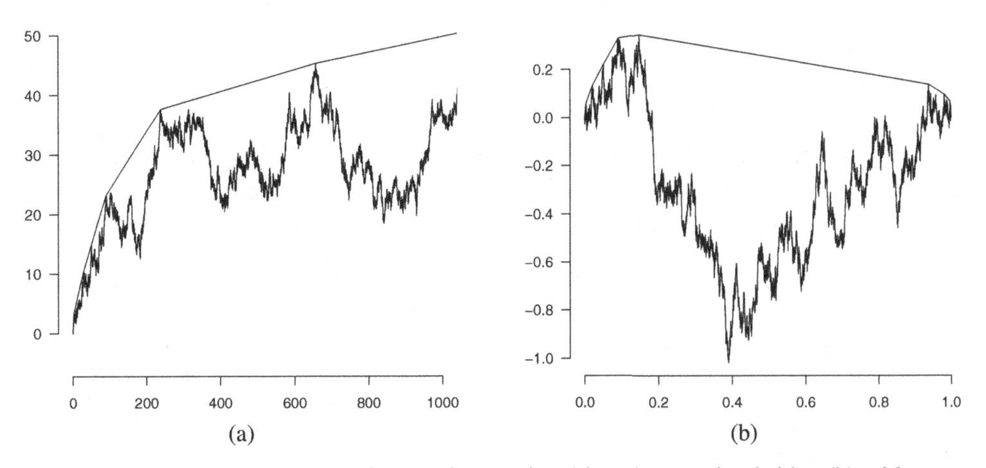

Figure 3.10 Realizations of Brownian motion (a) and Brownian bridge (b) with their (least) concave majorants. For the Brownian motion, the concave majorant is computed on [0, 10.000] and plotted on [0, 1000].

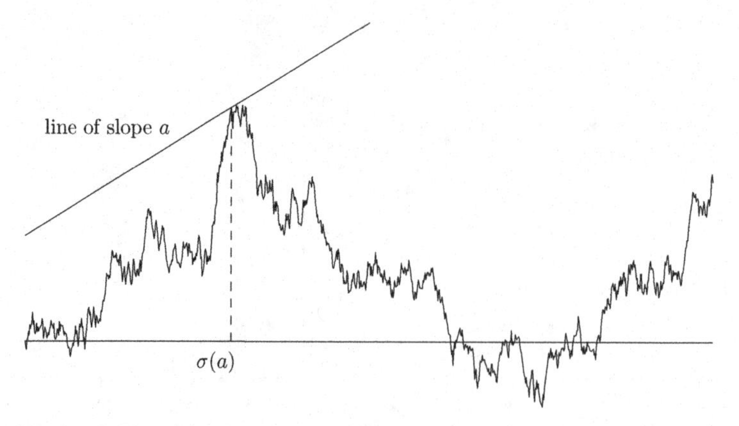

Figure 3.11 $\sigma(a) = \text{argmax}_x\{W(x) - ax : x \geq 0\}$, $\tau(a) = \sigma(1/a)$.

again the switch relation:

$$S_t \geq a \iff t \leq \sigma(a) = \tau(1/a).$$

See also Figure 3.11.

The argmax process $a \mapsto \tau(a)$ for Brownian motion is a jump process with independent increments. It can be considered to be the inverse of the process of slopes of the least concave majorant of Brownian motion, as the first item of the next theorem shows.

Theorem 3.10 (Groeneboom, 1983) *(i) The argmax process $a \mapsto \tau(a)$ is a time inhomogeneous process with independent increments, and, for $u > 0$,*

$$\frac{\mathbb{P}\left\{\tau(a+h) - \tau(a) \in du \mid \tau(a) = t\right\}}{h} \sim \frac{e^{-u/(2a^2)}}{a^2\sqrt{2\pi u}}\, du, \; h \downarrow 0.$$

(ii) Let $N(a, b)$ be the number of jumps of τ in $[a, b]$. Then

$$N(a, b) \stackrel{\mathcal{D}}{=} \text{Poisson}\,(\log(b/a))$$

and hence:

$$\{N(a, b) - \log(b/a)\} / \sqrt{\log(b/a)} \xrightarrow{\mathcal{D}} N(0, 1).$$

As a consequence of this we can generate Brownian motion by first generating the greatest concave majorant, and next generating Brownian excursions between successive points of change of slope of the concave majorant.

Corollary 3.2 (Groeneboom, 1983) *(i) Brownian motion on $[0, \infty)$ can be decomposed into the argmax process τ and Brownian excursions.*

(ii) If S_n is the slope of the concave majorant of the uniform empirical process

$$U_n(t) = \sqrt{n}\{\mathbb{F}_n(t) - F(t)\}, \; t \in [0, 1],$$

then

$$\left\{\int_0^1 S_n(t)^2\, dt - \log n\right\} / \sqrt{3 \log n} \xrightarrow{\mathcal{D}} N(0, 1).$$

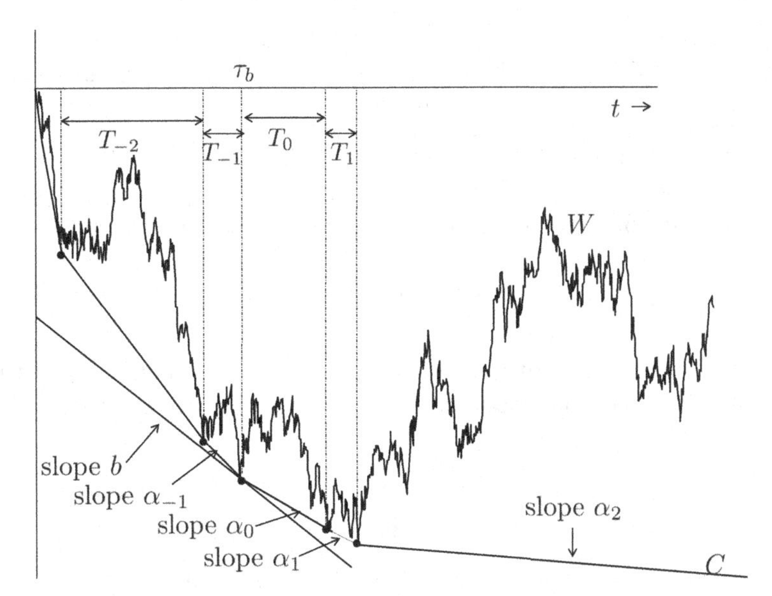

Figure 3.12 A realization of one-sided Brownian motion W with its convex minorant C on $[0, \infty)$ as considered in Theorem 3.12.

Part 2 uses Doob's transformation (to go from Brownian motion to Brownian bridge) and Hungarian embedding.

Similar methods yield for the number of jumps N_n of the concave majorant of the uniform empirical process $U_n = \sqrt{n}\{\mathbb{F}_n - F\}$:

Theorem 3.11 (Sparre Andersen, 1954)

$$\{N_n - \log n\} / \sqrt{\log n} \xrightarrow{\mathcal{D}} N(0, 1).$$

A very short proof of this result, using a conditional (Poisson) representation of the concave majorant of a uniform distribution function, given in Groeneboom and Pyke, 1983, can be found in Groeneboom and Lopuhaä, 1993. In Pitman, 1983, an interpretation of the results of Groeneboom, 1983, in terms of Bessel processes is given. It describes the embedded Markov chain of the greatest convex minorant of the Brownian motion and therefore the slopes start at $-\infty$ instead of 0. This is another method for letting the time variable increase as we go to the right. Another way to do this was used earlier for the least concave majorant, where we made the change of variables $\sigma(a) \rightarrow \tau(1/a)$ to make this happen.

Theorem 3.12 (Pitman, 1983) *Fix $b < 0$ and let $\tau_b = \text{argmin}\{W(x) - bx\}$; see Figure 3.12. Then:*

(i) The next slope α_0 of the convex minorant is uniformly distributed on $(b, 0)$, and conditionally on $\alpha_0, \ldots, \alpha_n$, the next slope α_{n+1} is uniform on $(\alpha_n, 0)$.
(ii) The preceding slope α_{-1} has density $|b|x^{-2}$ on the interval $(-\infty, b)$, and, conditional on $\alpha_{-n}, \ldots, \alpha_{-1}$, α_{-n-1} has density $|\alpha_{-n}|x^{-2}$ on $(-\infty, \alpha_{-n})$.
(iii) The sequences $\{\alpha_i,\ i < 0\}$ and $\{\alpha_i,\ i \geq 0\}$ are independent.

(iv) Conditional on all the slopes α_i, the lengths of the segments T_i between successive touches of W and C are independent, and T_i has a Gamma$\left(\frac{1}{2}, \frac{1}{2}\alpha_i^2\right)$ distribution:

$$\mathbb{P}\left\{T_i \in dt \,\middle|\, \alpha_i = a,\, \alpha_j, j \neq i\right| = \frac{|a|}{\sqrt{2\pi t}} e^{-\frac{1}{2}a^2 t} dt.$$

In Exercise 3.36 to 3.38 it is deduced from an alternative form of Theorem 3.12 that

$$\tilde{f}_n(0+) \xrightarrow{\mathcal{D}} \sup_{t>0} \frac{N_t}{t} \stackrel{\mathcal{D}}{=} 1/U,$$

where $\{N_t : t \geq 0\}$ is the standard Poisson process on $\mathbb{R}^+$, U is uniformly distributed on $[0, 1]$, and $\tilde{f}_n(0+)$ is the Grenander estimator at zero for a sample of n uniformly distributed random variables on $[0, 1]$.

Exercises

3.1 Let (F_n) be a sequence of functions on an interval I and $\hat{F}_n$ the least concave majorant of F_n taken over I. Moreover, let F be a concave function on I and define $\epsilon = \sup_{x\in I} |F_n(x) - F(x)|$.
   a) Argue that for all $x \in I$, $F_n(x) \leq \hat{F}_n(x) \leq F(x) + \epsilon$. Hint: use that $x \mapsto F(x) + \epsilon$ is concave on I.
   b) Infer that $-\epsilon \leq F_n(x) - F(x) \leq \hat{F}_n(x) - F(x) \leq \epsilon$ for all $x \in I$.
   c) Conclude that $\sup_{x\in I} |\hat{F}_n(x) - F(x)| \leq \sup_{x\in I} |F_n(x) - F(x)|$. This inequality is known as Marshall's inequality.

3.2 Let (f_n) be a sequence of bounded, decreasing functions on an interval $I \subset \mathbb{R}$ and f be bounded, decreasing and continuous on I. Suppose that for all $x \in I$, $f_n(x) \to f(x)$ a.s. as $n \to \infty$. Moreover, if I is unbounded, suppose that $f(\pm\infty) \to f(\pm\infty)$ a.s. Show that then $\sup_{x\in I} |f_n(x) - f(x)| \to 0$ a.s.

3.3 Consider an exponential class of densities on $\mathbb{R}$:

$$g_\theta(z) = \exp\left(a(\theta)t(z) + s(z) + b(\theta)\right) 1_A(z).$$

Follow the argument used to prove consistency of the ML estimator in the class of exponential densities (Section 3.1) to derive sufficient conditions for the ML estimator to be consistent in this model. The essence is that within this more general class of densities the behavior of (3.5) can also be studied easily because the integral reduces to a deterministic factor involving the unknown parameter and a random factor that does not depend on this parameter.

3.4 Let f and g be two probability densities on $\mathbb{R}$. Use that $u - v = (\sqrt{u} - \sqrt{v})(\sqrt{u} + \sqrt{v})$ for non-negative u and v to show the following inequality for the (squared) Hellinger distance between f and g:

$$\frac{1}{2}\int \frac{(f(z) - g(z))^2}{f(z) + g(z)}\, dz \geq \frac{1}{2}\int \left(\sqrt{f(z)} - \sqrt{g(z)}\right)^2 dz = h^2(f, g).$$

3.5 The second method of proof discussed in Section 3.1 can be applied to rather general classes of densities. If the class of densities has the property of being convex (the class of exponential densities is not convex), the method can be adapted in a useful way; useful in the sense that it leads to a more tractable function class Φ in view of studying (3.5).

a) Use the fact that $\ell((\hat{g}+g_0)/2) \le \ell(\hat{g})$ in case $\mathcal{G}$ is convex and that $\log u \le u-1$ to show that

$$0 \le \int \frac{2\hat{g}_n(z)}{\hat{g}_n(z)+g_0(z)}\, d\left(\mathbb{G}_n - G_0\right)(z) - \int \frac{g_0(z)-\hat{g}_n(z)}{\hat{g}_n(z)+g_0(z)}\, dG_0(z).$$

b) For the last term at the right hand side of (a), show that

$$\int \frac{g_0(z)-\hat{g}_n(z)}{\hat{g}_n(z)+g_0(z)}\, dG_0(z) = \int \frac{(\hat{g}_n(z)-g_0(z))^2}{\hat{g}_n(z)+g_0(z)}\, dz.$$

Conclude with Exercise 3.4 that

$$h^2(\hat{g}_n, g_0) \le \int \frac{2\hat{g}_n(z)}{\hat{g}_n(z)+g_0(z)} d\left(\mathbb{G}_n - G_0\right)(z).$$

3.6 Consider a function $\tilde{g}$ on $\mathbb{R}$ such that $\tilde{g} \ge 0$ and $\int \tilde{g}(x)\,dz \le 1$ ($\tilde{g}$ is a subdensity). Let furthermore g_0 be a continuous probability density on $\mathbb{R}$ and suppose that

$$\int \frac{g_0(z)^2}{\tilde{g}(z)} \le 1.$$

Show that then, necessarily, $\tilde{g} = g_0$ almost everywhere with respect to Lebesgue measure.

3.7 Show that the two random variables at the right hand side of (3.13) are independent.

3.8 Show that the first term at the right hand side of (3.13) is normally distributed with expectation zero and variance

$$\frac{1-(G(b)+G(a)}{G(b)-G(a)} = \frac{1-g(x_0)(b-a)}{g(x_0)(b-a)}.$$

3.9 The standard Wiener process (or Brownian motion) on $[0,\infty)$ is a Gaussian process $(W_t)_{t=0}^{\infty}$ with

$$EW_t = 0 \text{ and } EW_sW_t = s \wedge t$$

for all $0 \le s, t < \infty$. Brownian bridge (also called tied-down Brownian motion) can be defined in terms of the process W by

$$\mathbb{B}(t) = \mathbb{B}_t = W_t - tW_1, \text{ for } t \in [0,1].$$

Show that $(\mathbb{B})_{t=0}^{1}$ is a Gaussian process with

$$E\mathbb{B}_t = 0 \text{ and } E\mathbb{B}_s\mathbb{B}_t = s \wedge t - st.$$

for $0 \le s, t \le 1$.

3.10 Let W and $\mathbb{B}$ be Brownian bridge as defined in Exercise 3.9. Let $0 < u < v < 1$ be fixed and define the process (S_t) on $[u, v]$ by

$$S_t = \mathbb{B}_t - \mathbb{B}_u - \frac{\mathbb{B}_v - \mathbb{B}_u}{v-u}(t-u).$$

a) Show that

$$S_t = W_t - W_u - \frac{W_v - W_u}{v-u}(t-u).$$

b) Show that, for $s, t \in (u, v)$,

$$ES_t = 0 \text{ and } ES_sS_t = s \wedge t - u - \frac{(s-u)(t-u)}{v-u}$$
$$= (v-u)\left(\frac{s-u}{v-u} \wedge \frac{t-u}{v-u} - \frac{s-u}{v-u}\frac{t-u}{v-u}\right).$$

c) Define the process (U_t) on $[u, v]$ by

$$U_t = \sqrt{v-u}\mathbb{B}\left(\frac{t-u}{v-u}\right).$$

Show that the processes S and U have the same distribution by showing they are both Gaussian with the same expectation and covariance function.

3.11 In (3.14) the asymptotic distribution of the Grenander estimator at x_0 is given, where $x_0 \in (a, b)$ and g is constant on $[a, b]$ (and not on an interval strictly containing $[a, b]$).

a) Express the asymptotic bias and variance of $\sqrt{n}(\hat{g}_n(x_0) - g(x_0))$ in terms of a, b, g and a universal constant.

Suppose it is known that a and b were known. A natural estimator for $g(x_0)$ would be

$$T_n = \frac{\mathbb{G}_n(b) - \mathbb{G}_n(a)}{b-a}.$$

b) Compute the asymptotic bias and variance of $\sqrt{n}(T_n - g(x_0))$, depending on a, b and g.

3.12 Let W be two-sided standard Brownian Motion. Consider the random variable Y, defined as the derivative of the convex minorant of the process $t \mapsto W(t) + t^2$, evaluated at zero (see (3.12)). Define Z as in Section 3.9, so $Z = \text{argmax}_t\{W(t) - t^2\}$ (see (3.42)). In this exercise we derive that $Y =^d 2Z$.

a) Show that Z has the same distribution as $Z' = \text{argmin}_t\{W(t) + t^2\}$ and that its distribution is symmetric (i.e., $P(Z < -z) = P(Z > z)$ for all z).

b) Use Lemma 3.2 to show that $Y < y$ if and only if $\text{argmin}_t\{W(t) + t^2 - yt\} > 0$.

c) Show that $\text{argmin}_t\{W(t) + t^2 - yt\} = \text{argmin}_t\{W(t) + (t - y/2)^2\} = y/2 + Z'$.

d) Combine (a), (b) and (c) to show that for each $y \in \mathbb{R}$, $P(Y < y) = P(Z' > -y/2) = P(Z < y/2)$.

3.13 Let $\hat{F}_n$ be the nonparametric estimator of the distribution function in the exponential deconvolution setting, introduced in Section 2.4. Prove that for each $x > 0$, $\hat{F}_n(x) \to^P F(x)$ almost surely. Moreover, assuming that $F'(y) = f(y)$ is strictly positive in a neighborhood of x, show that for some $\nu > 0$, $\sup_{\{y:\, |y-x|<\nu\}} |\tilde{F}_n(y) - F(y)| \to 0$ almost surely.

3.14 This exercise outlines the proof of Lemma 3.3.

a) Choose $\epsilon > 0$ and a minimal set of ϵ-brackets $\{[l_j, u_j]: 1 \le j \le N\}$. Then show that for any $f \in \mathcal{F}$, there exists a $j \in \{1, 2, \ldots, N\}$ such that

$$\int f(z)\, d(\mathbb{G}_n - G_0)(z) \le \int u_j(z)\, d(\mathbb{G}_n - G_0)(z) + \epsilon$$

and

$$\int f(z)\, d(\mathbb{G}_n - G_0)(z) \ge \int l_j(z)\, d(\mathbb{G}_n - G_0)(z) - \epsilon.$$

b) Use the fact that $\mathcal{F} \subset L_1(G_0)$ and that $[l_j, u_j]$ are brackets in the seminorm $L_1(G_0)$ to show that for each j

$$\int l_j(z)\, d(\mathbb{G}_n - G_0)(z) \to^{a.s.} 0 \text{ and } \int u_j(z)\, d(\mathbb{G}_n - G_0)(z) \to^{a.s.} 0$$

for $n \to \infty$ and conclude that therefore almost surely

$$\max_{1 \le j \le N} \left| \int l_j(z)\, d(\mathbb{G}_n - G_0)(z) \right| < \epsilon \text{ and } \max_{1 \le j \le N} \left| \int u_j(z)\, d(\mathbb{G}_n - G_0)(z) \right| < \epsilon$$

for n sufficiently large.

c) Conclude (3.17).

3.15 Consider two function classes $\mathcal{F}_1$ and $\mathcal{F}_2$ containing (measurable) real valued functions on $\mathbb{R}$. Suppose that for some $A_1, A_2 > 0$

$$H_{[]}(\epsilon, \mathcal{F}_1, L_1(G_0)) \le \frac{A_1}{\delta} \text{ and } H_{[]}(\epsilon, \mathcal{F}_2, L_1(G_0)) \le \frac{A_2}{\delta} \text{ for all } \delta > 0.$$

Define $\mathcal{H} = \{f_1 - f_2 \colon f_1 \in \mathcal{F}_1,\ f_2 \in \mathcal{F}_2\}$. Show that

$$H_{[]}(\epsilon, \mathcal{H}, L_1(G_0)) \le H_{[]}(\epsilon/2, \mathcal{F}_1, L_1(G_0)) + H_{[]}(\epsilon/2, \mathcal{F}_2, L_1(G_0)) \le \frac{2(A_1 + A_2)}{\delta}.$$

Hint: Choose a minimal set of $\epsilon/2$-brackets of $\mathcal{F}_1$ and $\mathcal{F}_2$. Denote these by $[l_i^1, u_i^1]$ $(1 \le i \le N_{[]}(\epsilon/2, \mathcal{F}_1, L_1(G_0)))$ and $[l_j^2, u_j^2]$ $(1 \le j \le N_{[]}(\epsilon/2, \mathcal{F}_2, L_1(G_0)))$ respectively. Then define the brackets $u_{ij} = u_i^1 - l_j^2$ and $l_{ij} = l_i^1 - u_j^2$ for the appropriate values of i and j. Show that these functions are ϵ-brackets for $\mathcal{H}$ and that for all $h \in \mathcal{H}$ there is one of these brackets with $l_{ij} \le h \le u_{ij}$.

3.16 In Exercise 2.14 an expression was derived for the convolution densities for deconvolution problems with bounded smooth decreasing kernels on $(0, \infty)$. Show that the class of densities $\mathcal{G}$ that can allow this expression for a distribution function F on $[0, \infty)$ is a Glivenko Cantelli class. Hint: note that the expression in Exercise 2.14 is the difference of two monotone bounded functions and use Exercise 3.15.

3.17 Give an example of a sequence of functions (f_n) on $\mathbb{R}$ such that for all $x \in \mathbb{R}$, $f_n(x) \to f(x)$ for some convex function f on $\mathbb{R}$ and such that the sequence of greatest convex minorants of f_n converges to a function different from f.

3.18 Use (3.41) to show that for any $\alpha, \beta > 0$,

$$t \mapsto \alpha\left(W(\beta t) - (\beta t)^2\right) \stackrel{\mathcal{D}}{=} t \mapsto \alpha\sqrt{\beta}\, W(t) - \alpha\beta^2 t^2.$$

Choose α and β appropriately to obtain (3.11).

3.19 Show that $\hat{g}_n(0+)/g(0) \to^d \sup_{t>0} t^{-1} N(t)$, where N is a standard Poisson process on $[0, \infty)$. See also Exercise 2.7 and Figure 3.2. Hint: observe that

$$\hat{g}_n(0+) = \max_{1 \le i \le n} \frac{i}{n Z_{(i)}} = \left(\min_{1 \le i \le n} n Z_{(i)}/i \right)^{-1}.$$

3.20 Show that for a standard Poisson process on $[0, \infty)$, $\sup_{t>0} t^{-1} N(t) =^d 1/U$ where U is uniformly distributed on $(0, 1)$. Conclude from this, and Exercise 3.19, that

$$\lim_{n \to \infty} P\left(g_n(0+) \ge g(0)\right) = 1$$

and relate this to Exercise 2.7.

3.21 Verify the conditions of Theorem 3.2 needed to conclude (3.23), noting that

$$f_{n,t}(z) = n^{1/6}\left(1_{[0,x+n^{-1/3}t]}(z) - 1_{[0,x]}(z)\right),$$

and that $\mathrm{Cov}(W(s), W(t)) = s \wedge t$.

3.22 Show that $\hat{\theta}_n = \mathrm{argmin}_\theta^- \mathbb{M}_n(\theta) \to^P 0$ as $n \to \infty$. Use the consistency result of Exercise 3.13 and note that in view of Lemma 3.2

$$\hat{\theta}_n > \epsilon \iff \tilde{F}_n(x+\epsilon) < F(x) + vn^{-1/3} \text{ and } \hat{\theta}_n \le -\epsilon \iff \tilde{F}_n(x-\epsilon) \ge F(x) + vn^{-1/3}.$$

3.23 Draw a picture of a few of some of the functions in $\mathcal{M}_\delta$ as defined in (3.26) (see also Figure 3.5) to verify that for $\theta_1 \le \theta_2$ the following holds:

$$m_{\theta_1}(z) \le m_{\theta_2}(z) \text{ for all } z.$$

3.24 Consider the functions defined in (3.25). Show that

$$\|m_{\theta_1} - m_{\theta_2}\|_{2,g}^2 = \int (m_{\theta_1}(z) - m_{\theta_2}(z))^2 g(z)\,dz \sim c|\theta_1 - \theta_2|$$

as $|\theta_1 - \theta_2| \to 0$ and $|\theta_1|, |\theta_2| < \delta$.

3.25 Show in the context of Section 3.5 that $\hat{\theta}_n = \mathrm{argmin}_\theta^- \mathbb{M}_n(\theta) \to^P 0$ as $n \to \infty$. Use the consistency result that for some $v > 0$, $\sup_{\{y:\,|y-x|<v\}} |\tilde{F}_n(y) - F(y)| \to 0$ a.s. as derived in Section 3.1, and note that in view of Lemma 3.2

$$\hat{\theta}_n > \epsilon \iff \tilde{F}_n(x+\epsilon) < F(x) + vn^{-1/3} \text{ and } \hat{\theta}_n \le -\epsilon \iff \tilde{F}_n(x-\epsilon) \ge F(x) + vn^{-1/3}.$$

3.26 Consider the functions defined in (3.25). Assume that F' is continuous in a neighborhood of x. Show that for each $|\theta| \le \delta$,

$$|m_\theta(z)| \le M_\delta(z) := \delta 1_{[0,x+\delta]}(z) + 1_{[x-\delta,x+\delta]}(z)$$

so that $\|M_\delta\|_{2,g} \sim g(x)\sqrt{\delta}$.

3.27 Let W be (standard) two-sided Brownian motion. For any positive constants c_1 and c_2, the location of the minimum of the process

$$t \mapsto c_1 W(t) + c_2 t^2,\ t \in \mathbb{R}$$

is almost surely unique. Denote this location of the minimum by

$$\mathrm{argmin}_{t\in\mathbb{R}} \left\{c_1 W(t) + c_2 t^2\right\}.$$

Use (3.41) to show that $\mathrm{argmin}_{t\in\mathbb{R}} \left\{c_1 W(t) + c_2 t^2\right\}$ has the same distribution as

$$(c_1/c_2)^{2/3} \mathrm{argmin}_{t\in\mathbb{R}} \left\{W(t) + t^2\right\}.$$

Use this property to deduce the statement of Theorem 3.7.

3.28 Show that if the process $\{Z(t)\colon t \ge 0\}$ in $D[0,\infty)$ is an L_2 martingale, the process $\{Z(t)^2\colon t \ge 0\}$ in $D[0,\infty)$ is a submartingale.

3.29 Suppose that F_0 is strictly increasing in the point x and that $F_0(x) = a \in (0,1)$. Show, using the consistency result of the preceding section, that

$$T_n(a) \xrightarrow{\text{a.s.}} F_0^{-1}(a). \tag{3.50}$$

3.30 Show that the process Z_n as defined in (3.37) is a martingale with respect to the filtration $\{\mathcal{F}_t\colon t \ge t_0\}$, where $\mathcal{F}_t = \sigma\{(T_i, \Delta_i)\colon t_0 \le T_i \le t\}$.

3.31 Show that the quadratic variation process of the process $\widetilde{W}_n$ in the proof of Lemma 3.6 is given by

$$\left[\widetilde{W}_n\right](t) = n^{-2/3} \sum_{T_i \in [t_0, t_0 + n^{-1/3} t]} \{\Delta_i - F_0(T_i)\}^2 \, , \; t \geq 0.$$

3.32 Let $\mathbb{F}_n$ be the empirical distribution function of sample $X_1, \ldots, X_n$ from a distribution on $[0, \infty)$ with continuous distribution function F, and let $a = \sup\{t \geq 0 : F(t) < 1\}$. Show that

$$\{\mathbb{F}_n(t) - F(t)\} / \{1 - F(t)\}, \, t \in [0, a)$$

is a martingale with respect to the filtration $\{\sigma\{\mathbb{F}_n(s)\} : s \in [0, t]\}, \; t \in [0, a)\}$.

Exercises 3.33 to 3.35 give an alternative route to Lemma 3.6. We assume that the conditions of Theorem 3.7 are satisfied.

3.33 Let, for t_0 as in Theorem 3.7, the function $\langle \widetilde{W}_n \rangle$ be defined by

$$\langle \widetilde{W}_n \rangle(t) = n^{1/3} \int_{t_0}^{t_0 + n^{-1/3} t} F_0(u) \{1 - F_0(u)\} \frac{1 - \mathbb{G}_n(u)}{1 - G(u)} \, dG(u), \; t \geq 0. \tag{3.51}$$

Show that the process

$$t \mapsto \widetilde{W}_n(t)^2 - \langle \widetilde{W}_n \rangle(t), \; t \geq 0$$

is a martingale with respect to the filtration $\{\mathcal{F}_{n,t} : t \geq 0\}$, defined by (3.38), also conditioning on $\mathbb{G}_n(t_0)$.

3.34 The process $\langle \widetilde{W}_n \rangle$ of Exercise 3.33 is called the *conditional variance* process or *predictable variation* process. Show that, in probability, for each (fixed) $t > 0$,

$$\langle \widetilde{W}_n \rangle(t) \longrightarrow F_0(t_0) \{1 - F_0(t_0)\} \, g(t_0) t, \; n \to \infty.$$

3.35 Use theorem 13, in chapter 8, p. 179 of Pollard, 1984, or the $\langle\rangle$ instead of the [] condition of theorem 2, p. 273, Rebolledo, 1980, to derive Lemma 3.6, using the result of Exercise 3.34 instead of part (ii) of Theorem 3.6.

3.36 Prove the following variant of the theorem of Pitman, 1983, for the concave majorant of Brownian motion. Fix $b \in (0, \infty)$ and let $T_0 = \tau_b = \text{argmax}\{W(x) - bx\}$. Then:

a) The next slope α_0 of the convex minorant is uniformly distributed on $(0, b)$, and conditionally on $\alpha_0, \ldots, \alpha_n$, the next slope α_{n+1} is uniform on $(0, \alpha_n)$.
b) The preceding slope α_{-1} has density bx^{-2} on the interval (b, ∞), and, conditional on $\alpha_{-n}, \ldots, \alpha_{-1}, \alpha_{-n-1}$ has density $\alpha_{-n} x^{-2}$ on $(-\infty, \alpha_{-n})$.
c) The sequences $\{\alpha_i, \, i < 0\}$ and $\{\alpha_i, \, i \geq 0\}$ are independent.
d) Conditional on all the slopes α_i, the lengths of the segments T_i are independent, and T_i has a gamma$\left(\frac{1}{2}, \frac{1}{2}\alpha_i^2\right)$ distribution:

$$\mathbb{P}\left\{T_i \in dt \;\middle|\; \alpha_i = a, \, \alpha_j, \, j \neq i\right\} = \frac{a}{\sqrt{2\pi t}} e^{-\frac{1}{2}a^2 t} dt.$$

3.37 Deduce from part (b) of Exercise 3.36 that the slope of the concave majorant $\tilde{f}_n(0+)$ at zero of the empirical distribution function for the uniform empirical process satisfies

$$\tilde{f}_n(0+) \xrightarrow{\mathcal{D}} 1/U,$$

where U is uniform on $[0, 1]$.

3.38 Deduce from Exercises 3.36 and 3.37 that

$$1/U \stackrel{\mathcal{D}}{=} \sup_{t>0} \frac{N_t}{t},$$

where $\{N_t : t \geq 0\}$ is the standard Poisson process on $\mathbb{R}_+$.

Bibliographic Remarks

More on the basic inequalities used to derive consistency can be found in Van de Geer, 2000. The approach to prove consistency by using the fact that the derivative of the criterion function in the direction of the underlying distribution is nonnegative was used in Jewell, 1982. Marshall's inequality was derived in Marshall, 1969. Thorough introductions to empirical process theory are Pollard, 1984, Van der Vaart and Wellner, 1996, and Kosorok, 2008b. The asymptotic distribution for the Grenander estimator was first established in Prakasa Rao, 1969, whereas for the isotonic estimator in the exponential deconvolution model it was derived in Jongbloed, 1998a, and (for more general convolution kernels) in Van Es et al., 1998, using the approach of Kim and Pollard, 1990. The asymptotic theory of the Grenander estimator at flat regions can be found in Carolan and Dykstra, 1999. For the maximum likelihood estimator in the current status model, the asymptotic distribution was derived in Groeneboom and Wellner, 1992. The approach to proving Theorem 3.5 follows the proof in Groeneboom, 1985. The Chernoff distribution was introduced in Chernoff, 1964. Approximate quantiles of the distribution can be found in Groeneboom and Wellner, 2001. Recent papers on this distribution are Janson et al., 2010, Groeneboom, 2010, Groeneboom, 2011, Groeneboom and Temme, 2011, Pimentel, 2014, and Janson, 2013.

4

Other Univariate Problems Involving Monotonicity Constraints

In Chapter 2, various models were introduced where monotonicity constraints are clearly involved. In this chapter, more problems will be described where monotonicity constraints play an important role. These constraints can be related to the interpretation of these problems as inverse problems, as also seen in Section 1.5. Some distribution function F (by definition monotone) in the background is to be estimated based on data from an induced distribution function G. Monotonicity of F induces more or less explicit shape constraints on the sampling distribution function G.

First a classical problem from stereology is considered: Wicksell's problem. This is concerned with estimating the distribution of radii of spheres randomly scattered in an opaque medium based on radii of circular profiles obtained by intersecting the medium with a plane. The second problem is that of estimating a concave regression function based on noisy data. Related to this, a simple model from ornithology is introduced. It concerns estimating the distribution of sojourn times of birds at an oasis based on observed times when specific birds were caught at the oasis. As in Wicksell's problem, imposing certain assumptions, the sampling distribution can be expressed in terms of the underlying distribution of interest. Also the estimation of log concave densities, star shaped distribution functions and distribution functions in deconvolution models more general than that of Section 2.4 and in the interval censoring case 2 will be considered.

For the problems discussed in this chapter, estimation procedures will be introduced, characterizations of these estimators will be given and some estimators will also be studied asymptotically. Estimation paradigms as plug-in inverse estimators, least squares estimators and maximum likelihood estimators will be illustrated and studied.

4.1 Wicksell's Corpuscle Problem

In the early 1920s, the Swedish mathematician Sven D. Wicksell at the University of Lund was confronted with an interesting problem from the medical sciences. Anatomist T. Helman tried to get hold of the distribution of the size of so-called follicles in human spleens. Postmortem examinations were executed, during which spleens were sliced at several places. Profiles of follicles present on the slices were observed.

Of course, estimating the size of these follicles based on this type of data is an ill-specified problem. A model is needed to describe the situation. It was Wicksell who built a model and already proposed a first nonparametric estimator for the distribution function.

Consider the following model. In $\mathbb{R}^3$, points are distributed according to a homogeneous Poisson process. Centered at those points, there are spheres of varying sizes. All squared radii

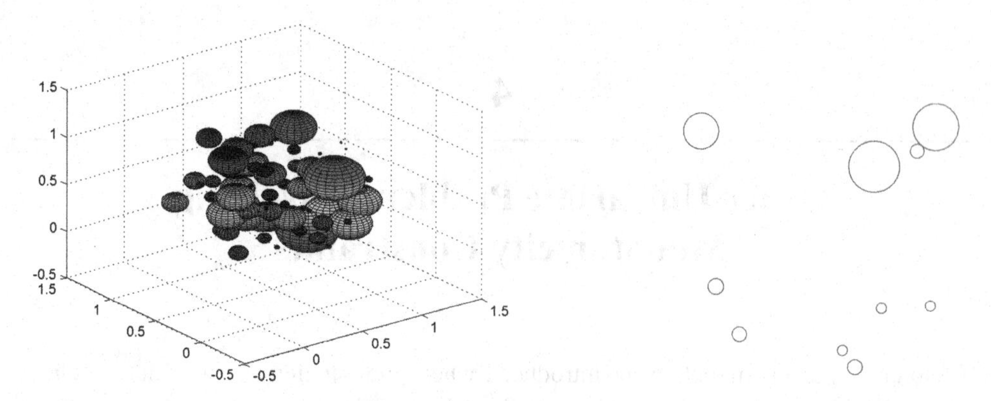

Figure 4.1 The three-dimensional medium with spheres and an illustration of the visible circular profiles in a cross-section.

are assumed to be independent and distributed according to a distribution function F. This is the object we want to estimate. However, the sphere radii cannot be observed. What can be observed are radii of circular profiles on a cross-section of the medium, say the xy-plane in $\mathbb{R}^3$; see Figure 4.1 for an illustration of this model.

It is not immediately obvious whether this model describes the practical situation well. However, when the follicles are approximately spherical, the distance between the places where the spleen is cut is big compared with the radii of the large follicles and the number of overlapping spheres is negligible (Poisson process has low intensity). The model seems to fit the real situation fairly well.

In the model, two phenomena are working in opposite directions. First, the radius of an observable circle clearly cannot exceed the radius of its associated sphere, and second, spheres that have a large radius are more likely to be cut than the small ones. This second point can be formalized by saying that the spheres are sampled proportionally to their radii. This means that the distribution of the spheres actually cut is not F, but a size biased version of F. In this case, the biased distribution is given by

$$F^b(x) = \frac{1}{m_F} \int_0^x \sqrt{y}\, dF(y),$$

where $m_F = \int_0^\infty \sqrt{y}\, dF(y)$, the expected sphere radius under F. Note the resemblance with (2.4). Now, given the fact that a sphere with squared radius X is cut by the plane, what can be said about the distribution of the observable squared circle radius? This behavior can be deduced from the fact that the distance of the cutting plane to the center of the sphere is uniformly distributed on $[0, \sqrt{X}]$. Pythagoras then shows (see Figure 4.2) that the observable squared circle radius can be seen as a random fraction of X:

$$Z = (1 - U^2)X = YX,$$

where U is uniformly distributed on $(0, 1)$ and, consequently, Y has Beta$(1, 1/2)$ density

$$k_Y(y) = \frac{1}{2}(1 - y)^{-1/2} 1_{(0,1)}(y).$$

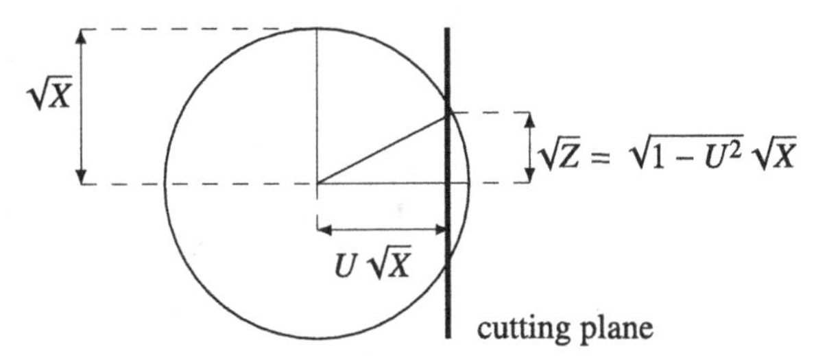

Figure 4.2 View on the sphere in the direction of the vertical cutting plane. The sphere radius is $\sqrt{X}$, the position of the cut is uniformly distributed on the right half of the sphere.

Therefore, using Exercise 4.1, the density g of Z can, for $z \geq 0$, be written as

$$g(z) = \int_{(0,\infty)} x^{-1} k_Y(z/x)\, dF^b(x) = \frac{1}{m_F} \int_{(z,\infty)} x^{-1/2} k_Y(z/x)\, dF(x) = \frac{1}{2m_F} \int_{(z,\infty)} \frac{dF(x)}{\sqrt{x-z}}.$$

Having derived the density of the observables in terms of the distribution function of interest, the question of whether we can also explicitly express F in terms of g emerges. In the current situation, there is an explicit inverse relation, expressing F in terms of g. Indeed (see Exercise 4.2),

$$F(x) = 1 - \frac{\int_x^\infty (z-x)^{-1/2} g(z)\, dz}{\int_0^\infty z^{-1/2} g(z)\, dz}. \tag{4.1}$$

Let us pause and take a look at what we have done. We used the mechanics behind the process of collecting the data in our model to derive the distribution of the observables in terms of the distribution of interest. This process of determining the sampling distribution in terms of the object F is what we called the direct problem. After that we used mathematical manipulations to get (or check) some inverse relation to express the quantity of interest in terms of the sampling distribution. This is often called the inverse problem. The next step, to estimate F based on data generated by g, could be called the statistical inverse problem (see also Section 1.5).

As can be seen from Exercise 4.3, not all densities g will give a distribution function when plugged into (4.1). This shows that the transformation taking F to g is injective, but not surjective with respect to the class of densities on $(0, \infty)$.

A natural estimator for F is obtained using (4.1). Having a sample $Z_1, \ldots, Z_n$ from density g, one can use the empirical distribution function based on this, $\mathbb{G}_n$, and substitute $d\mathbb{G}_n(z)$ for $g(z)\, dz$ in (4.1). Formally,

$$\hat{F}_n(x) = 1 - \frac{\int_{(x,\infty)} (z-x)^{-1/2}\, d\mathbb{G}_n(z)}{\int_0^\infty z^{-1/2}\, d\mathbb{G}_n(z)}.$$

For fixed values of x this so-called plug-in estimator is consistent and asymptotically normally distributed (see Exercise 4.5). Despite the acceptable pointwise behavior of this estimator, as a function of x it really exhibits some undesirable features. See Figure 4.3 for $\hat{F}_n$ based

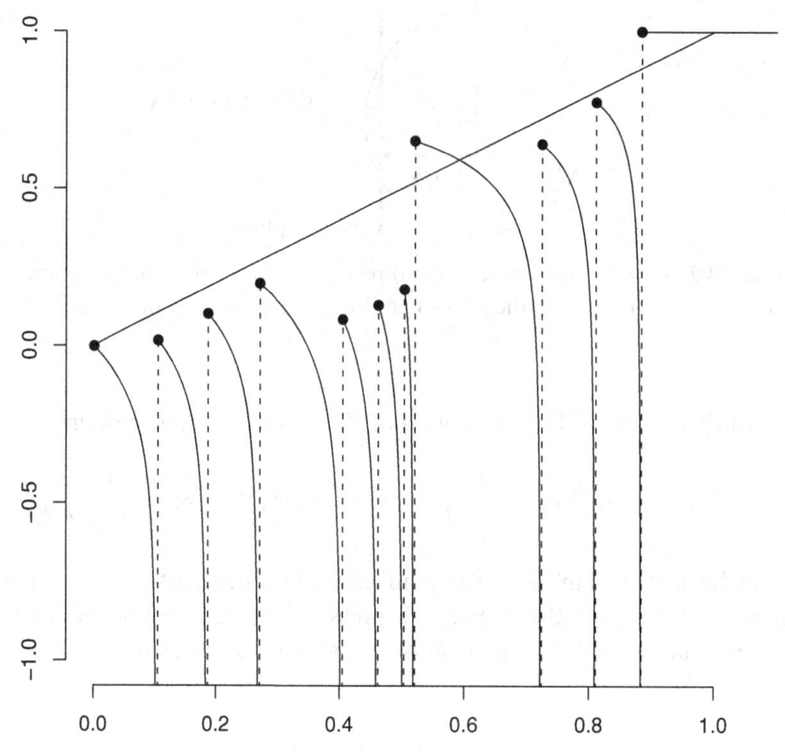

Figure 4.3 Plug-in estimator of F based on a sample of size 10. The underlying uniform distribution function is the solid straight line.

on a sample of size 10 from the density

$$g(z) = \frac{3}{2}\sqrt{1-z}\,1_{[0,1]}(z)$$

corresponding to the standard uniform distribution for X. It is clear that the estimator is piecewise decreasing and has infinite discontinuities at the observed data points.

There are various possibilities to construct a monotone estimator in this setting. One option is to define a likelihood of the observations, and maximize this over densities g corresponding proper distribution functions F. See Exercise 4.8 and 4.9. Another option is to turn the estimate of Figure 4.3 into a monotone function using the techniques of Chapter 2. In view of (4.1), this can be done by writing

$$1 - F(x) = \frac{T(x)}{T(0)} \quad \text{with} \quad T(x) = \int_x^\infty \frac{g(z)}{\sqrt{z-x}}\,dz,$$

where T is decreasing and defining its (nonmonotone) estimator

$$T_n(x) = \int_{(x,\infty)} \frac{d\mathbb{G}_n(z)}{\sqrt{z-x}} = \frac{1}{n}\sum_{i=1}^n (Z_i - x)^{-1/2}\,1_{(x,\infty)}(Z_i).$$

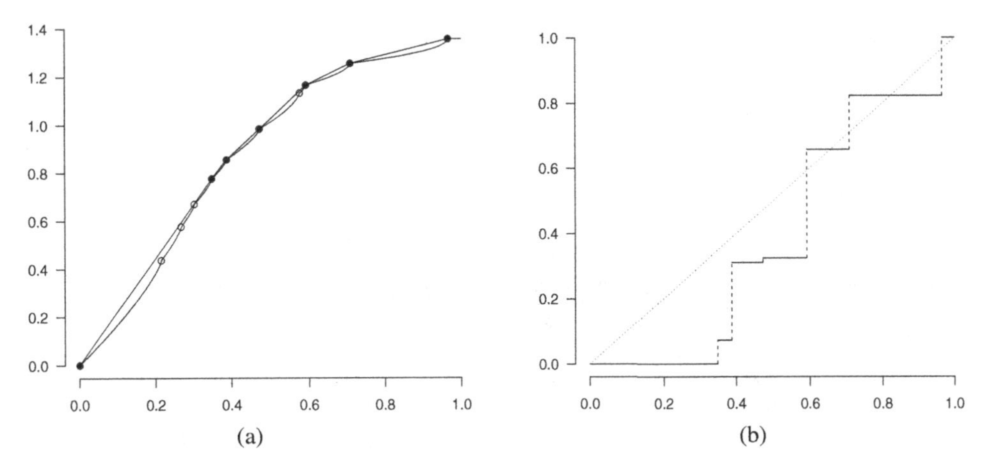

Figure 4.4 (a) The function U_n defined in (4.2) with its least concave majorant based on the same data set of size $n = 10$ as Figure 4.3. The bend points of the concave majorant are the solid points. (b) The estimate $\tilde{F}_n$ based on the right derivative of this concave majorant via (4.3) with the dotted underlying (uniform) distribution.

Then T_n can be isotonized by taking the (right) derivative of the least concave majorant of its primitive

$$U_n(x) = \int_0^x T_n(y)\,dy = \frac{2}{n}\sum_{i=1}^{n}\sqrt{Z_i} - \frac{2}{n}\sum_{i=1}^{n}\sqrt{Z_i - x}\,1_{(x,\infty)}(Z_i). \tag{4.2}$$

(See Exercise 4.6.) Denoting this derivative by $\tilde{T}_n$, the following estimator for F is obtained

$$\tilde{F}_n(x) = 1 - \frac{\tilde{T}_n(x)}{\tilde{T}_n(0)}. \tag{4.3}$$

For the same data set as used in Figure 4.3, this function U_n and its concave majorant are given in the Figure 4.4a. The resulting estimate for F is shown in Figure 4.4b. Using the approach described in Section 3.1, $\tilde{T}_n$ can be shown to be consistent (see Exercise 4.7).

What did Wicksell do in his original paper? He constructed a system of linear equations matching observed cell frequencies with their expectations and solved this system. In Section 8.2 another estimator based on Wicksell's original (binned) data will be computed.

4.2 Convex Regression

The Canadian income data have been analyzed by several authors, see, e.g., Ruppert et al., 2003, and Meyer, 2008. They represent a sample of 205 Canadian workers. The relationship between log(income) and age can be expected to be concave, and we can estimate this relationship nonparametrically, only using the concavity restriction.

A picture of the data, together with the nonparametric regression estimate, under the concavity restriction, is shown in Figure 4.5, and compared there with a concave estimate, as discussed in Meyer, 2008, using cubic splines. For computing this estimate we used Meyer's `R` function `cspl`, using 5 knots at equal quantile distances. The nonparametric regression

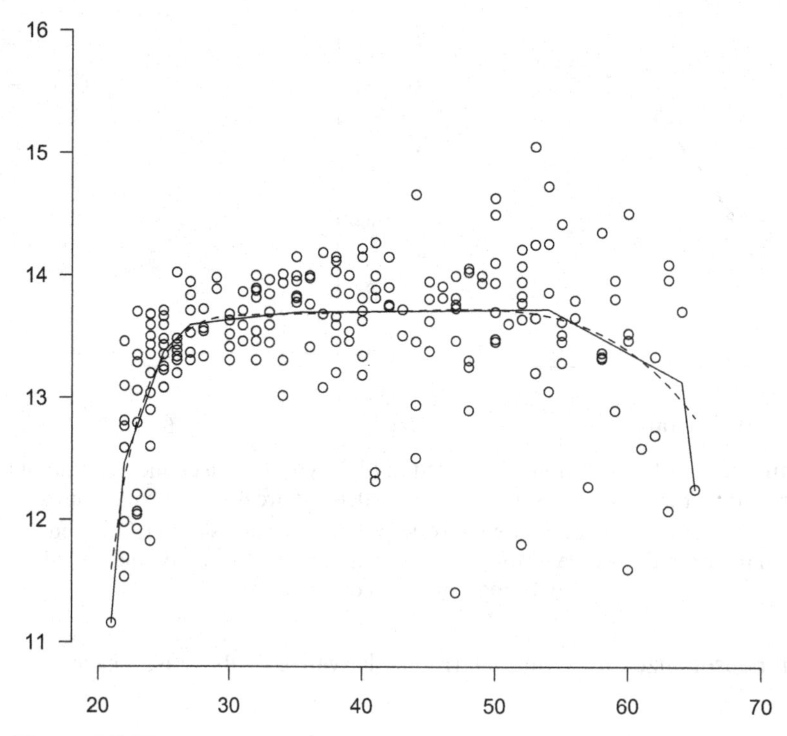

Figure 4.5 The nonparametric concave regression function for the Canadian income data (solid curve), together with a concave cubic spline estimate with 5 knots, as defined in Meyer, 2008 (dashed curve). The data represent log(income) (y-coordinates) plotted against age (x-coordinates), for 205 persons.

estimate, under the concavity restriction, was computed using the support reduction algorithm to be discussed in Section 7.4. This algorithm is very fast. It was used on the present data by fitting the minus log(income) data to a convex curve.

It is seen that the two estimates are very close, except at the very right end. One of the differences between the estimates is that for the estimate of Meyer, 2008, the number of knots of the cubic spline has to be specified, whereas no such choice is needed for the nonparametric concave regression estimate; the latter estimate chooses the locations of the bend points of the piecewise linear convex estimate automatically. Both estimates will probably have a local $n^{-2/5}$ rate of convergence at interior points. This is in fact proved for the nonparametric concave regression estimate in Groeneboom et al., 2001a, and a characterization of the limit distribution is given in Groeneboom et al., 2001b. The latter characterization is in terms of the second derivative of the invelope of integrated Brownian motion $+\,t^4$, which is a cubic spline touching the integrated Brownian motion $+\,t^4$ and lying above this process between points of touch, whereas its second derivative is convex. For further details, see Section 11.1.

The formal setting for this problem runs as follows. Consider given the following data set of size n: $\{(x_i, y_i): i = 1, \dots, n\}$ where $x_1 < x_2 < \cdots < x_n$ and where y_i is a realization of the random variable

$$Y_i = r_0(x_i) + \epsilon_i \tag{4.4}$$

for a convex function r_0 on $\mathbb{R}$. Here $\{\epsilon_i : i = 1, \ldots, n\}$ are i.i.d. random variables. Writing $\mathcal{K}$ for the set of all convex functions on $\mathbb{R}$, the first suggestion for a least squares estimate of r_0 is

$$\text{argmin}_{r \in \mathcal{K}} \phi(r), \quad \text{where} \quad \phi(r) = \frac{1}{2} \sum_{i=1}^{n} (y_i - r(x_i))^2 . \tag{4.5}$$

It is clear that this definition needs more specification. For instance, any solution to the minimization problem can be extended quite arbitrarily (although convex) outside the range of the x_is. Also, between the x_is there is some arbitrariness in the way a function can be chosen. We therefore confine ourselves to minimizing ϕ over the subclass $\mathcal{K}_n$ of $\mathcal{K}$ consisting of the functions that are linear between successive x_is, as well as to the left and the right of the range of the x_is. Hence, we define

$$\hat{r}_n = \text{argmin}_{r \in \mathcal{K}_n} \phi(r)$$

where ϕ is given in (4.5). Note that $r \in \mathcal{K}_n$ can be parameterized naturally by $(r_1, \ldots, r_n) = (r(x_1), \ldots, r(x_n)) \in \tilde{\mathcal{K}}_n \subset \mathbb{R}^n$ where

$$\tilde{\mathcal{K}}_n = \left\{ r \in \mathbb{R}^n : \frac{r_i - r_{i-1}}{x_i - x_{i-1}} \leq \frac{r_{i+1} - r_i}{x_{i+1} - x_i} \text{ for all } i = 2 \ldots, n-1 \right\} .$$

Using the identification $\mathcal{K}_n = \tilde{\mathcal{K}}_n$, we have the following lemma, ensuring uniqueness of the estimator.

Lemma 4.1 *There is a unique function $\hat{r}_n \in \mathcal{K}_n$ that minimizes ϕ over $\mathcal{K}_n$.*

Proof Follows immediately from the strict convexity of $\phi : \mathcal{K}_n \to \mathbb{R}$ and the fact that $\phi(r) \to \infty$ as $\|r\|_2 \to \infty$. □

Next step is to characterize the least squares estimator.

Lemma 4.2 *Define $\hat{R}_k = \sum_{i=1}^{k} \hat{r}_i$ and $S_k = \sum_{i=1}^{k} Y_i$. Then $\hat{r}_n = \text{argmin}_{r \in \mathcal{K}_n} \phi(r)$ if and only if $\hat{R}_n = S_n$ and*

$$\sum_{k=1}^{j-1} \hat{R}_k (x_{k+1} - x_k) \begin{cases} \geq \sum_{k=1}^{j-1} S_k (x_{k+1} - x_k) & j = 2, 3, \ldots, n \\ = \sum_{k=1}^{j-1} S_k (x_{k+1} - x_k) & \text{if } \hat{r}_n \text{ has a kink at } x_j \text{ or } j = n. \end{cases} \tag{4.6}$$

The lemma is proved in Section 2.3 of Groeneboom et al., 2001a. This characterization was used in the routine used in computing the convex regression function for the Canadian income data. It is also the basis of the R package `conreg`, written by Martin Mächler. See Figure 4.6 for a visualization of the necessary and sufficient optimality conditions for the Canadian income data.

4.3 Convex Density Estimation

An example of a nonparametric density estimator under a convexity constraint is given in Hampel's bird migration problem. Consider a population of birds of a certain type that cross the desert individually and stop over at an oasis. Ornithologists are interested in the time

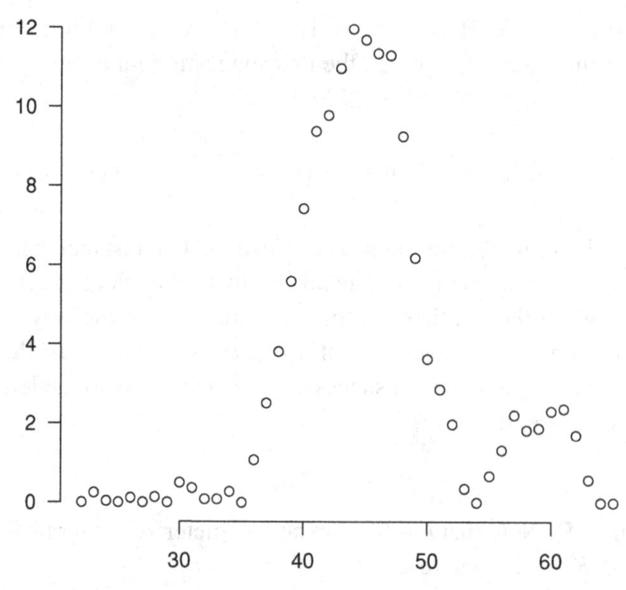

Figure 4.6 Visualization of the conditions given in (4.6) for the Canadian income data. The differences between the left and right hand side of (4.6) are plotted against the x_is.

spent by a generic bird at this oasis. The following model can be used to investigate these sojourn times.

To each bird, a positive random variable X with distribution function F is attached, denoting the time spent at the oasis. This quantity cannot be observed. Also, independent of X, the bird has a homogeneous Poisson process $\{N(t)\colon t \geq 0\}$ attached to it, with intensity λ, assumed to be small. The times the bird is caught at the oasis are then those jump points of the Poisson process that occur before X. Note that $N(X)$ is the number of times the bird is caught while staying over at the oasis (see also Figure 4.7).

The data for one bird consist of those catching times that occurred before X. Of course, it is conditional on the fact that $N(X) \geq 1$. In Hampel's model only those observations are used that correspond to birds that have been caught exactly twice. We will derive the distribution of the difference in time between the two catches in terms of the unknown distribution function X of the sojourn time.

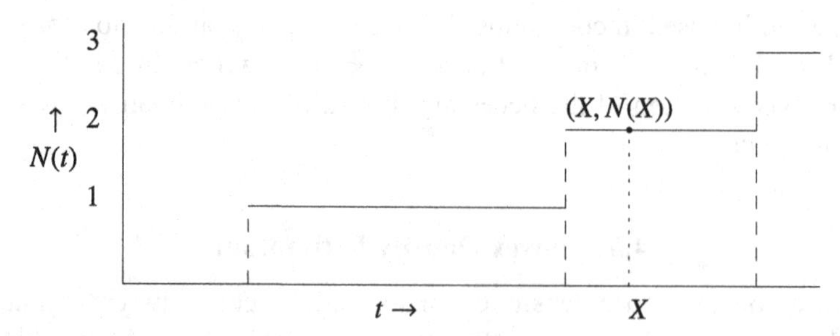

Figure 4.7 Actual sojourn time of the bird X and the (Poisson) process of catches. In this example, the bird is caught twice during its stay at the oasis.

The first question is: what is the distribution of the sojourn time of a bird given it is caught exactly twice?

$$\tilde{F}(x) = P(X \le x | N(X) = 2) = \frac{P(X \le x \wedge N(X) = 2)}{P(N(X) = 2)} =$$

$$= \frac{\int_{y\in[0,x]} P(N(X) = 2 | X = y) dF(y)}{\int_{y\in[0,\infty)} P(N(X) = 2 | X = y) dF(y)} = \frac{\int_{y\in[0,x]} \frac{1}{2}(\lambda y)^2 e^{-\lambda y} dF(y)}{\int_{y\in[0,\infty)} \frac{1}{2}(\lambda y)^2 e^{-\lambda y} dF(y)} \approx \frac{\int_{y\in[0,x]} y^2 dF(y)}{\int_{y\in[0,\infty)} y^2 dF(y)}$$

for small λ (and, for example, if F has bounded support). In other words, the conditional distribution is (for small λ) a size biased (in this case squared length biased) distribution associated with F.

The second question is: what is the distribution of Z (time between the two catches) given that $X = x$ and there are exactly two catches? Given X and $N(X) = 2$, the jumps of N in $[0, x]$ are uniformly distributed on this interval. Therefore, writing $U_{(1)}$ and $U_{(2)}$ for the order statistics of two uniformly distributed random variables in $(0, 1)$,

$$P(Z > z | X = x, N(X) = 2) = P(U_{(2)} - U_{(1)} > z/x) = (1 - z/x)^2 \quad \text{for } 0 < z < x.$$

Consequently,

$$P(Z > z | N(X) = 2) = \int_0^\infty P(Z > z | X = x, N(X) = 2)\, d\tilde{F}(x) = \frac{\int_z^\infty (x - z)^2\, dF(x)}{\int_0^\infty x^2\, dF(x)}. \tag{4.7}$$

We see that we explicitly have to require the second moment of F to exist. Of course, if we take F to have bounded support, this assumption certainly holds.

Differentiating (4.7) with respect to z, we get for the density g of Z that

$$g(z) = \frac{2 \int_z^\infty (x - z)\, dF(x)}{\int_0^\infty x^2\, dF(x)}$$

and differentiating once again gives

$$g'(z) = -\frac{2(1 - F(z))}{\int_0^\infty x^2\, dF(x)}.$$

Monotonicity and positivity of $z \mapsto 1 - F(z)$ force the density g to be convex and decreasing on $(0, \infty)$. Moreover, the inverse relation expressing F in terms of g is given by

$$F(x) = 1 - \frac{g'(x)}{g'(0)}.$$

We now turn to the problem of estimating a convex decreasing density on $[0, \infty)$ based on an i.i.d. sample of size n. One way of doing this is maximum likelihood (see Exercise 4.12). Here we consider a least squares approach. Given a dataset $z_1, \ldots, z_n$ from the density g, with empirical distribution function $\mathbb{G}_n$, define the quadratic objective function Q by

$$Q(g) = \frac{1}{2} \int g(z)^2\, dz - \int g(z)\, d\mathbb{G}_n(z).$$

The intuition behind this function is that the true underlying g_0 minimizes the function

$$g \mapsto \frac{1}{2} \int (g(z) - g_0(z))^2\, dz = \frac{1}{2} \int g(z)^2\, dz - \int g(z)\, dG_0(z) + \frac{1}{2} \int g_0(z)^2\, dz$$

where the last term does not depend on g and G_0 represents the distribution function corresponding to g_0. Replacing the unknown G_0 in the latter expression by the estimate $\mathbb{G}_n$ yields Q. This reasoning also explains that the resulting estimator that minimizes Q over all convex decreasing estimators is named the least squares estimator; see Exercise 1.1 for an estimator of this type in a parametric setting and Section 2.6 for a similar reasoning behind a least squares estimator of a monotone hazard. Note that we do not restrict the minimization to the class of convex decreasing densities, but instead take the larger class consisting of positive multiples of these densities,

$$\mathcal{K} = \{g\colon [0,\infty) \to [0,\infty)\colon g \text{ convex, decreasing, integrable}\}.$$

This estimator will be shown to be piecewise linear with at most n changes of slope occurring between observation points. Moreover, the minimizer of Q over $\mathcal{K}$ will be shown to be a density automatically.

Lemma 4.3 *There exists a unique $\hat{g}_n \in \mathcal{K}$ that minimizes Q over $\mathcal{K}$. This solution is piecewise linear, and has at most one change of slope between two successive ordered observations $z_{(i)}$ and $z_{(i+1)}$ and no changes of slope at observation points. The first change of slope is to the right of the first order statistic and the last change of slope is to the right of the largest order statistic.*

Proof For existence, we first show that there is a bounded convex decreasing function $\bar{g}$ with bounded support such that the minimization can be restricted to the subset

$$\{g \in \mathcal{K}\colon g \le \bar{g}\} \tag{4.8}$$

of $\mathcal{K}$.

First note that there is a $c_1 > 0$ such that any candidate to be the minimizer of Q should have a left derivative at $z_{(1)}$ bounded above in absolute value by c_1. Indeed, if g is a function in $\mathcal{K}$, then

$$g(z) \ge g(z_{(1)}) + g'(z_{(1)}-)(z - z_{(1)}) \quad \text{for } x \in [0, z_{(1)}],$$

and

$$\begin{aligned}
Q(g) &\ge \frac{1}{2}\int_0^{z_{(1)}} g(z)^2\,dz - g(z_{(1)})\\
&\ge \frac{1}{2}\int_0^{z_{(1)}} \left(g(z_{(1)}) + g'(z_{(1)}-)(z - z_{(1)})\right)^2\,dz - g(z_{(1)})\\
&\ge \frac{1}{2} z_{(1)} g(z_{(1)})^2 + \frac{1}{6} z_{(1)}^3 g'(z_{(1)}-)^2 - g(z_{(1)})\\
&\ge -(2z_{(1)})^{-1} + \frac{1}{6} z_{(1)}^3 g'(z_{(1)}-)^2,
\end{aligned}$$

showing that $Q(g)$ tends to infinity as the left derivative of g at $z_{(1)}$ tends to minus infinity. In the last inequality we use that $u \mapsto \frac{1}{2}z_{(1)}u^2 - u$ attains its minimum at $u = 1/z_{(1)}$. This same argument can be used to show that the right derivative at $z_{(n)}$ of any solution candidate g is bounded below in absolute value by some $c_2 > 0$, whenever $g(z_{(n)}) > 0$.

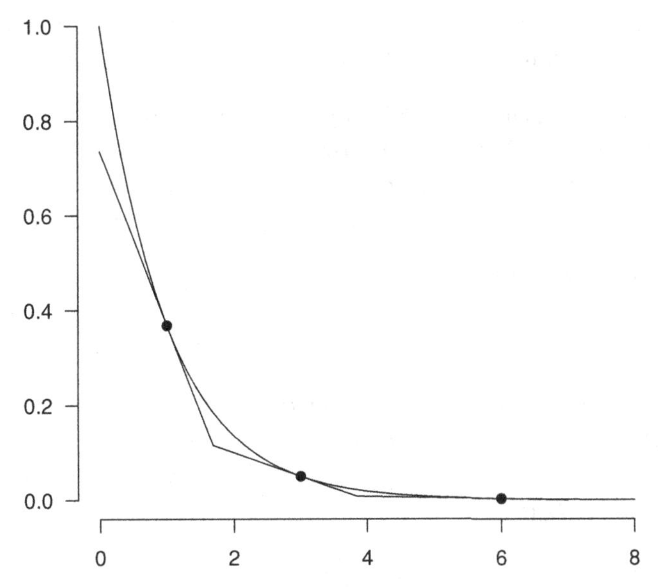

Figure 4.8 A piecewise linear convex function coinciding with g at the "observations" 1, 3 and 6 and lying below g throughout.

Additionally, it is clear that $g(z_{(1)})$ is bounded by some constant c_3. This follows from the fact that

$$Q(g) \geq \frac{1}{2} g(z_{(1)})^2 z_{(1)} - g(z_{(1)}),$$

which tends to infinity as $g(z_{(1)})$ tends to infinity.

To conclude the existence argument, observe that we may restrict attention to functions in $\mathcal{K}$ that are linear on the interval $[0, z_{(1)}]$. Indeed, any element g of $\mathcal{K}$ can be modified to a $\hat{g} \in \mathcal{K}$, which is linear on $[0, z_{(1)}]$ as follows:

$$\hat{g}(z) = \begin{cases} g(z_{(1)}) + g'(z_{(1)}+)(z - z_{(1)}) & \text{for } z \in [0, z_{(1)}] \\ g(z) & \text{for } z > z_{(1)}, \end{cases}$$

and if $g \neq \hat{g}$, $Q(g) > Q(\hat{g})$ (only first term is influenced by going from g to $\hat{g}$). For the same reason, attention can be restricted to functions that behave linearly between the point $z_{(n)}$ and the point where it hits zero, by extending a function using its left derivative at the point $z_{(n)}$. In fact, this argument can be adapted to show that a solution of the minimization problem has at most one change of slope between successive observations. Indeed, let g be a given convex decreasing function, and fix its values at the observation points. Then one can construct a piecewise linear function that lies entirely below g, and has the same values at the observation points. This shows that Q is decreased when going from g to this piecewise linear version, since the first term of Q decreases and the second term stays the same. See Figure 4.8.

Hence, defining the function

$$\bar{g}(z) = \begin{cases} c_3 + c_1(z_{(1)} - z) & \text{for } z \in [0, z_{(1)}] \\ (c_3 - c_2(z - z_{(1)})) \vee 0 & \text{for } z > z_{(1)}, \end{cases}$$

we see that the minimization of Q over $\mathcal{K}$ may be restricted to the compact set (4.8). Uniqueness of the solution follows from the strict convexity of Q on $\mathcal{K}$. □

The lemma below gives a characterization of $\hat{g}_n$. In contrast to the situation of estimating a decreasing density (Section 2.2), this characterization cannot be used immediately to compute the estimator. For that it is too implicit. However, it can be used to check whether a candidate function minimizes Q over $\mathcal{K}$ and to derive further qualitative properties of $\hat{g}_n$.

Lemma 4.4 *Let Y_n be defined by*

$$Y_n(x) = \int_0^x \mathbb{G}_n(z)\,dz, \qquad x \ge 0.$$

Then the piecewise linear function $\hat{g}_n \in \mathcal{K}$ minimizes Q over $\mathcal{K}$ if and only if the following conditions are satisfied for $\hat{g}_n$ and its second integral $H_n(x) = \int_{0<t<u<x} \hat{g}_n(t)\,dt\,du$:

$$H_n(x) \begin{cases} \ge Y_n(x), & \text{if } x \ge 0, \\ = Y_n(x) & \text{if } \hat{g}_n'(x+) > \hat{g}_n'(x-). \end{cases} \tag{4.9}$$

Proof Let $\hat{g}_n \in \mathcal{K}$ satisfy (4.9), and note that this implies

$$\int_{(0,\infty)} \{H_n(x) - Y_n(x)\}\, d\hat{g}_n'(x) = 0. \tag{4.10}$$

Choose $g \in \mathcal{K}$ arbitrarily. Then we get, using integration by parts,

$$Q_n(g) - Q_n(\hat{g}_n) \ge \int_{(0,\infty)} \left\{\hat{H}_n(x) - Y_n(x)\right\} d(g' - \hat{g}_n')(x).$$

But using (4.10) and (4.9), we get

$$\int_{(0,\infty)} \{H_n(x) - Y_n(x)\}\, d(g' - \hat{g}_n')(x) = \int_{(0,\infty)} \{H_n(x) - Y_n(x)\}\, dg'(x) \ge 0.$$

Hence $\hat{g}_n$ minimizes Q over $\mathcal{K}$.

Conversely, suppose that $\hat{g}_n$ minimizes Q over $\mathcal{K}$. Consider, for $x > 0$, the function $g_x \in \mathcal{K}$, defined by

$$g_x(t) = (x - t)_+, \qquad t \ge 0. \tag{4.11}$$

Then necessarily

$$\lim_{\epsilon \downarrow 0} \frac{Q(\hat{g}_n + \epsilon g_x) - Q_n(\hat{g}_n)}{\epsilon} = H_n(x) - Y_n(x) \ge 0.$$

This yields the inequality part of (4.9). Furthermore,

$$\lim_{\epsilon \to 0} \frac{Q((1+\epsilon)\hat{g}_n) - Q_n(\hat{g}_n)}{\epsilon} = \int_{(0,\infty)} \{H_n(x) - Y_n(x)\}\, d\hat{g}_n'(x) = 0,$$

which is (4.10). This can, however, only hold if the equality part of (4.9) also holds. □

See Figure 4.9 for a picture of the LS estimator based on the data set

$$\{0.0357, 0.0434, 0.0906, 0.0989, 0.2049, 0.4583, 0.6035, 1.2783, 2.0637, 2.3275\}$$

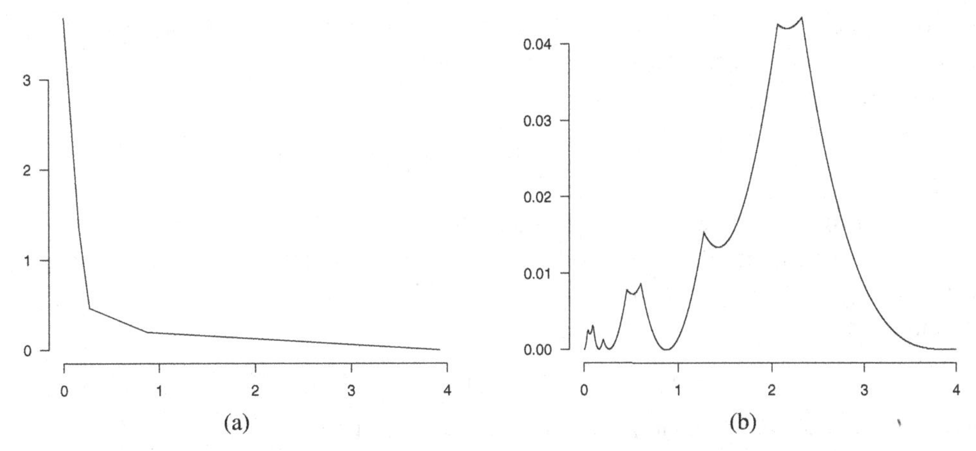

Figure 4.9 (a) The LS estimator of a convex decreasing density g based on a sample of size $n = 10$. (b) The process $x \mapsto H_n(x) - Y_n(x)$ of Lemma 4.4 based on this data set.

and in the right panel the corresponding function $Y_n - H_n$ of Lemma 4.4. The solution is obtained using the support reduction algorithm to be discussed in Section 7.4.

Recall that in the definition of $\hat{g}_n$, the constraint of having integral equal to one is not included. Using the characterization of $\hat{g}_n$, it can be deduced that $\hat{g}_n$ actually is a probability density. This is one of the statements in the next corollary.

Corollary 4.1 *Let H_n satisfy condition (4.9) of Lemma 4.4 and let $\hat{g}_n = H_n''$. Then:*

(i) $\hat{G}_n(x) = \mathbb{G}_n(x)$ *for each x such that $\hat{g}_n'(x-) < \hat{g}_n'(x+)$, where $\hat{G}_n(x) = \int_0^x \hat{g}_n(t)\,dt$.*
(ii) $\hat{g}_n(z_{(n)}) > 0$*, where $z_{(n)}$ is the largest order statistic of the sample.*
(iii) The least squares estimator is a density, so $\int \hat{g}_n(x)\,dx = 1$.
(iv) Let $0 < t_1 < \cdots < t_m$ be the points of change of slope of H_n'' and let $t_0 = 0$. Then $\hat{g}_n$ and H_n have the following "midpoint properties":

$$\hat{g}_n(\bar{t}_k) = \frac{1}{2}\{\hat{g}_n(t_{k-1}) + \hat{g}_n(t_k)\} = \frac{\mathbb{G}_n(t_k) - \mathbb{G}_n(t_{k-1})}{t_k - t_{k-1}}, \tag{4.12}$$

and

$$H_n(\bar{t}_k) = \frac{1}{2}\{Y_n(t_{k-1}) + Y_n(t_k)\} - \frac{1}{8}\{\mathbb{G}_n(t_k) - \mathbb{G}_n(t_{k-1})\}(t_k - t_{k-1}), \tag{4.13}$$

for $k = 1, \ldots, m$, where $\bar{t}_k = (t_{k-1} + t_k)/2$.

Proof For proving (i), note that at each point z such that $\hat{g}_n'(z-) < \hat{g}_n'(z+)$ (note that such a point cannot be an observation point by Lemma 4.3) we have by (4.9) that $Y_n(z) = H_n(z)$. Since $H_n \geq Y_n$ throughout and both Y_n and H_n are differentiable at z, we have that $\hat{G}_n(x) = \mathbb{G}_n(x)$.

For (ii), we will prove that the upper support point of the piecewise linear density $\hat{g}_n$, $z(\hat{g}_n)$, satisfies $z(\hat{g}_n) > z_{(n)}$. From Lemma 4.3 we already know that $z(\hat{g}_n) \neq z_{(n)}$. Now suppose that $z(\hat{g}_n) < z_{(n)}$. Then for all $x > z_{(n)}$

$$H_n'(x) = \hat{G}_n(x) = \hat{G}_n(z(\hat{g}_n)) \stackrel{(i)}{=} \mathbb{G}_n(z(\hat{g}_n)) < 1\,.$$

However, since $Y_n'(x) = \mathbb{G}_n(x) = 1$ for all $x > z_{(n)}$, inevitably the inequality part of (4.9) would be violated eventually. Hence $z(\hat{g}_n) > z_{(n)}$ and (ii) follows.
For (iii), combine (i) and (ii) to get

$$\int \hat{g}_n(x)\,dx = \hat{G}_n(z(\hat{g}_n)) = \mathbb{G}_n(z(\hat{g}_n)) = 1.$$

The first part of (iv) is an easy consequence of the fact that $\hat{G}_n(t_i) = \mathbb{G}_n(t_i)$, $i = 0, \ldots, m$ (part (i)), combined with the property that $\hat{g}_n$ is linear on the intervals $[t_{i-1}, t_i]$. Again by the fact that $\hat{g}_n$ is linear on $[t_{k-1}, t_k]$, we get that $\tilde{H}_n$ is a cubic polynomial on $[t_{k-1}, t_k]$, determined by

$$H_n(t_{k-1}) = Y_n(t_{k-1}),\ H_n(t_k) = Y_n(t_k),\ H_n'(t_{k-1}) = \mathbb{G}_n(t_{k-1}),\ H_n'(t_k) = \mathbb{G}_n(t_k),$$

using that H_n is tangent to Y_n at t_{k-1} and t_k. This implies (4.13). □

Having a well defined estimator for the convex decreasing density and knowing some of its properties, we will now consider consistency of $\hat{g}_n$.

Theorem 4.1 *(Consistency of LS density estimator) Suppose that $Z_1, Z_2, \ldots$ are i.i.d. random variables with bounded convex decreasing density g_0. Then the least squares estimator $\hat{g}_n$ based on $Z_1, \ldots, Z_n$ is uniformly consistent on closed intervals bounded away from 0: i.e., for each $c > 0$, we have, with probability one,*

$$\sup_{c \le x < \infty} |\hat{g}_n(x) - g_0(x)| \to 0. \tag{4.14}$$

Proof The proof is based on the characterization of the estimator given in Lemma 4.4. We let $\mathcal{T}_n$ denote the set of locations of change of slope of H_n'', where H_n is defined as in Lemma 4.4.

Fix $\delta > 0$, such that $[0, \delta]$ is contained in the interior of the support of g_0, and let $\tau_{n,1} \in \mathcal{T}_n$ be the last point of change of slope in $(0, \delta]$, or zero if there is no such point. Since, with probability one,

$$\liminf_{n\to\infty} Z_{(n)} > \delta$$

and, by Lemma 4.3, the last point of change of slope is to the right of $Z_{(n)}$, we may assume that there exists a point of change of slope $\tau_{n,2}$ strictly to the right of δ. Let $\tau_{n,2}$ be the first point of change of slope that is strictly to the right of δ. Then the sequence $(\hat{g}_n(\tau_{n,1}))$ is uniformly bounded. This is seen in the following way. Let $\tau_n = \{\tau_{n,1} + \tau_{n,2}\}/2$. Then $\tau_n \ge \delta/2$ and hence, by Exercise 4.10,

$$\hat{g}_n(\tau_n) \le \hat{g}_n(\delta/2) \le 1/\delta.$$

This implies that we have an upper bound for $\hat{g}_n(\tau_{n,1})$ that only depends on δ. Indeed, if $\tau_{n,1} > \delta/2$, $\hat{g}_n(\tau_{n,1}) \le \hat{g}_n(\delta/2) \le 1/\delta$ by (4.10). If $\tau_{n,1} \le \delta/2$, we can use linearity of $\hat{g}_n$ on $[\tau_{n,1}, \delta]$ to get

$$1 \ge \int_{\tau_{n,1}}^{\delta} \hat{g}_n(x)\,dx = \frac{1}{2}(\delta - \tau_{n,1})(\hat{g}_n(\delta) + \hat{g}_n(\tau_{n,1})) \ge \frac{1}{4}\delta\hat{g}_n(\tau_{n,1}),$$

giving $\hat{g}_n(\tau_{n,1}) \le 4/\delta$. Moreover, the right derivative of $\hat{g}_n$ has a uniform absolute upper bound at $\tau_{n,1}$, also only depending on δ. This can be verified analogously.

On the interval $[\tau_{n,1}, \infty)$, we have:

$$\frac{1}{2}\int_{[\tau_{n,1},\infty)} \hat{g}_n(x)^2\,dx - \int_{[\tau_{n,1},\infty)} \hat{g}_n(x)\,d\mathbb{G}_n(x) \le \frac{1}{2}\int_{[\tau_{n,1},\infty)} g_0(x)^2\,dx - \int_{[\tau_{n,1},\infty)} g_0(x)\,d\mathbb{G}_n(x).$$

This follows from writing $g_0^2 - \hat{g}_n^2 = (g_0 - \hat{g}_n)^2 + 2\hat{g}_n(g_0 - \hat{g}_n)$, implying, using integration by parts,

$$\frac{1}{2}\int_{[\tau_{n,1},\infty)} g_0(x)^2\,dx - \int_{[\tau_{n,1},\infty)} g_0(x)\,d\mathbb{G}_n(x) - \frac{1}{2}\int_{[\tau_{n,1},\infty)} \hat{g}_n(x)^2\,dx + \int_{[\tau_{n,1},\infty)} \hat{g}_n(x)\,d\mathbb{G}_n(x)$$

$$\ge \int_{[\tau_{n,1},\infty)} \hat{g}_n(x)\{g_0(x) - \hat{g}_n(x)\}\,dx - \int_{[\tau_{n,1},\infty)} \{g_0(x) - \hat{g}_n(x)\}\,d\mathbb{G}_n(x)$$

$$= \int_{[\tau_{n,1},\infty)} \{H_n(x) - Y_n(x)\}\,d(g_0' - \hat{g}_n')(x) = \int_{[\tau_{n,1},\infty)} \{H_n(x) - Y_n(x)\}\,dg_0'(x) \ge 0.$$

This argument was used in the proof of Lemma 4.4 on the interval $(0, \infty)$.

Since $\tau_{n,1} \in [0, \delta]$, for each subsequence of (ℓ) there must be a further subsequence converging to a point $\tau_1 \in [0, \delta]$. Using a Helly argument, there will be a further subsequence (n_k) so that, for each $x \in (\tau_1, \infty)$, $\hat{g}_{n_k}(x) \to \hat{g}(x) = \hat{g}(x, \omega)$, where $\hat{g}$ is a convex function on $[\tau_1, \infty)$, satisfying $\hat{g}(\tau_1) < \infty$. The function $\hat{g}$ satisfies:

$$\frac{1}{2}\int_{[\tau_1,\infty)} \hat{g}(x)^2\,dx - \int_{[\tau_1,\infty)} \hat{g}(x)\,dG_0(x) \le \frac{1}{2}\int_{[\tau_1,\infty)} g_0(x)^2\,dx - \int_{[\tau_1,\infty)} g_0(x)\,dG_0(x), \tag{4.15}$$

where the integrals on the right side are finite, also if $\tau_1 = 0$, since $g_0(0) < \infty$. But this implies

$$\int_{[\tau_1,\infty)} \{\hat{g}(x) - g_0(x)\}^2\,dx \le 0, \tag{4.16}$$

and hence $\hat{g}(x) = g_0(x)$, for $x \ge \tau_1$. Since $\delta > 0$ can be chosen arbitrarily small, we get that for any $c > 0$, each subsequence $\hat{g}_\ell$ has a subsequence that converges to g_0 at each point $x \ge c$. By the monotonicity of g_0, the convergence has to be uniform. □

The uniform consistency cannot be extended to the interval $[0, \infty)$. Exercise 4.11 shows that $\hat{g}_n$ is inconsistent at zero. The next challenge is to derive the asymptotic (pointwise) distribution of $\hat{g}_n$. This problem will be addressed in Section 11.1.

4.4 Log Concave Densities

A density g on $\mathbb{R}$ is said to be log concave if its logarithm $f = \log g$ is concave, i.e., for all $x, y \in \mathbb{R}$ and $\lambda \in [0, 1]$

$$f(\lambda x + (1-\lambda)y) \ge \lambda f(x) + (1-\lambda)f(y).$$

The best known log concave densities on $\mathbb{R}$ are the normal densities. However, many more densities are log concave, including the Laplace densities the, Gumbel density and uniform densities on an interval. In the last case, $f = \log g$ is constant on the interval where g is positive and $-\infty$ outside this interval. Log normal densities are automatically unimodal, in the

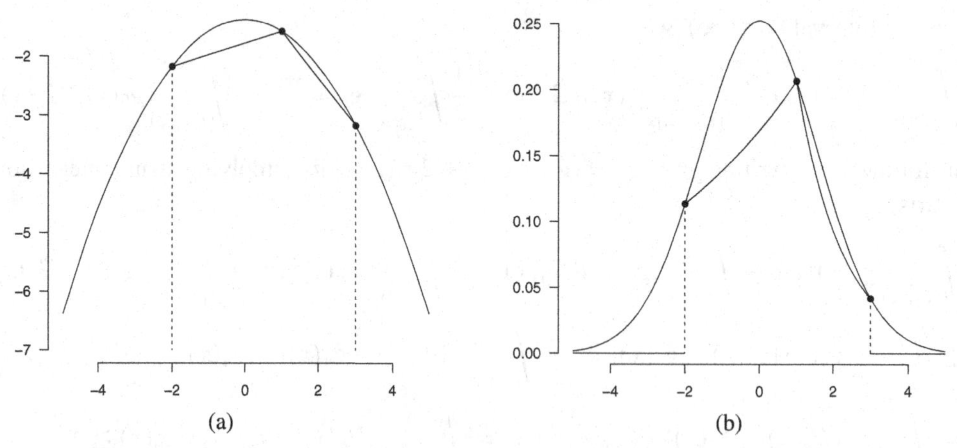

Figure 4.10 (a) A concave log-density f and a piecewise linear function that coincides with f at the three observation points. (b) The corresponding log concave density and piecewise log linear subdensity. Inflating the latter to obtain a true probability density will lead to a log concave density with higher log likelihood than $\exp(f)$.

sense that there exists an $x \in \mathbb{R}$ such that g is nondecreasing on $(-\infty, x]$ and nonincreasing on $[x, \infty)$ (see Exercise 4.14).

Consider the problem of estimating a log concave density g based on an i.i.d. sample of size n from g: $Z_1, Z_2, \ldots, Z_n$. Denoting a typical dataset of this type by $z_1, \ldots, z_n$, a natural approach would be maximum likelihood, since the log likelihood function is particularly simple. Indeed, writing $g = \exp(f)$ with f concave, we have

$$\ell(g) = \sum_{i=1}^{n} \log g(z_i) = \sum_{i=1}^{n} f(z_i).$$

This function is to be maximized over all concave functions f such that $\int \exp(f(x))\,dx = 1$.

Also in this problem, the maximization over the function class of all concave functions can be reduced to a maximization problem over finitely many (in fact n) parameters. Indeed, the solution to the optimization problem (if it exists) can be seen to be a log concave density g for which $f = \log g$ is piecewise linear on $[z_{(1)}, z_{(n)}]$ with changes of slope at the observed data points z_i and $f(z) = -\infty$ for $z < z_{(1)}$ and $z > z_{(n)}$. Here $z_{(1)} < \cdots < z_{(n)}$ denote the ordered observations.

To see this, first note that there exist log concave densities g such that $\ell(g) > -\infty$ (take an arbitrary normal density). Then consider an arbitrary concave function f with $\ell(\exp(f)) > -\infty$ that satisfies $\int \exp(f(x))\,dx = 1$. Now define the function $\tilde{f}$ by $\tilde{f}(z_i) = f(z_i)$ for all i, by linear interpolation between successive points z_i and $f(z) = -\infty$ for $z < z_{(1)}$ and $z > z_{(n)}$. Then concavity of f implies that $\tilde{f}$ is concave and $\tilde{f} \le f$. Hence, $\tilde{g} = \exp \tilde{f} \le \exp f = g$, implying that $\int \tilde{g}(x)\,dx \le 1$. Since $g(x_i) = \tilde{g}(x_i)$ for all i, $\ell(g) = \ell(\tilde{g})$. If $\int \tilde{g}(x)\,dx < 1$, so $\tilde{f}$ is really different from f, one can define the log concave density $\bar{g} = \tilde{g} / \int \tilde{g}(x)\,dx$ with $\bar{g}(z_i) > \tilde{g}(z_i) = g(z_i)$ for all i. This implies $\ell(\bar{g}) > \ell(g)$. In other words: for each log concave density g that is not of the specified type, one can construct a log concave density $\bar{g}$ of the specified type such that $\ell(\bar{g}) > \ell(g)$. See also Figure 4.10.

Let us now consider the ML estimator for the log concave density g. For (fixed) data set $z_{(1)} < \cdots < z_{(n)}$, write $f_j = f(z_{(j)})$ for $1 \le j \le n$ and define the log likelihood function with Lagrange relaxation term (see Exercise 4.16) by

$$\ell(f) = \frac{1}{n}\sum_{i=1}^{n} f_i - \int_{z_{(1)}}^{z_{(n)}} \exp\{f(z)\}\,dz = \frac{1}{n}\sum_{i=1}^{n} f_i - \sum_{i=2}^{n} \frac{e^{f_i} - e^{f_{i-1}}}{f_i - f_{i-1}}(z_{(i)} - z_{(i-1)}). \quad (4.17)$$

For the last equality, see Exercise 4.15. The ML estimator for g is $\exp(\hat{f})$ where $\hat{f}$ maximizes (4.17) over all concave, piecewise linear functions f on $[z_{(1)}, z_{(n)}]$ with changes of slope restricted to the points $z_{(i)}$, $1 \le i \le n$. Existence and uniqueness of this estimator is established in Exercise 4.17.

A characterization and interesting properties of the ML estimator $\hat{g}$ and its logarithm $\hat{f}$ can be obtained by a by now familiar variational argument. Let Δ be a function on $\mathbb{R}$ such that for some $\lambda > 0$ the function $\hat{f} + \lambda\Delta$ is concave on $\mathbb{R}$. Then the following inequality holds:

$$0 \ge \lim_{\epsilon \downarrow 0} \frac{\ell(\hat{f} + \epsilon\Delta) - \ell(\hat{f})}{\epsilon} = \frac{1}{n}\sum_{i=1}^{n} \Delta(z_i) - \int e^{\hat{f}(z)}\Delta(z)\,dz = \frac{1}{n}\sum_{i=1}^{n} \Delta(z_i) - \int \Delta(z)\hat{g}(z)\,dz. \quad (4.18)$$

This necessary condition for optimality has some immediate consequences. Taking $\Delta(z) = \pm 1$, it yields $\int \hat{g}(z)\,dz = 1$. Taking $\Delta(z) = \pm z$ leads to $\int z\,\hat{g}(z)\,dz = \bar{z}_n$, so that the expected value with respect to the MLE equals the mean of the sample. Finally, taking $\Delta(z) = -z^2$ we obtain $\int z^2 \hat{g}(z)\,dz \le \frac{1}{n}\sum_{i=1}^{n} z_i^2$. Hence, the variance corresponding to the MLE is smaller than or equal to the sample variance of the z_is, so the MLE is more concentrated than the empirical distribution function of the data. Considerations in this line also lead to a characterization of the MLE.

Lemma 4.5 *Let $\tilde{f}$ be a concave function on $\mathbb{R}$ with $\tilde{f}(z) = -\infty$ for $z \notin [z_{(1)}, z_{(n)}]$ and linear on intervals of the type $[z_{(i-1)}, z_{(i)}]$ for $2 \le i \le n$. Moreover, assume that $\int \exp(\tilde{f}(z))\,dz = 1$. Then $\tilde{f}$ maximizes ℓ over all concave functions of the same type if and only if*

$$\int_{z_{(1)}}^{z} \tilde{G}(x)\,dx \begin{cases} \le \int_{z_{(1)}}^{z} \mathbb{G}_n(x)\,dx & \text{for all } z \in [z_{(1)}, z_{(n)}] \\ = \int_{z_{(1)}}^{z} \mathbb{G}_n(x)\,dx & \text{for all } z \in [z_{(1)}, z_{(n)}] \text{ with } \tilde{f}'(z-) > \tilde{f}'(z+) \end{cases} \quad (4.19)$$

where $\mathbb{G}_n$ is the empirical distribution function of the z_i's and $\tilde{G}$ the distribution function with density $\exp(\tilde{f})$.

Proof Necessity follows from (4.18). Indeed, suppose $\tilde{f}$ maximizes ℓ over all concave functions as stated and define for $z \in [z_{(1)}, z_{(n)}]$ the function

$$\Delta^{(z)}(x) = -(z - x)_+ = -\int_{-\infty}^{z} 1_{[x,\infty)}(y)\,dy.$$

Note that any positive multiple of this function can be added to $\tilde{f}$ without destroying its concavity and that

$$\frac{1}{n}\sum_{i=1}^{n} \Delta^{(z)}(z_i) = -\frac{1}{n}\sum_{i=1}^{n} \int_{-\infty}^{z} 1_{[z_{(i)},\infty)}(y)\,dy = -\int_{-\infty}^{z} \mathbb{G}_n(y)\,dy \quad (4.20)$$

and

$$-\int e^{\tilde{f}(x)}\Delta^{(z)}(x)\,dx = \int_{-\infty}^{z} \tilde{G}(x)\,dx. \tag{4.21}$$

Using (4.18) this leads to

$$\int_{-\infty}^{z} \mathbb{G}_n(y)\,dy \geq \int_{-\infty}^{z} \tilde{G}(x)\,dx.$$

Because for z such that $\tilde{f}'(z-) > \tilde{f}'(z+)$ inequality (4.18) can also be applied to the function $-\Delta^{(z)}$, the equality part of (4.19) follows.

For sufficiency, use the following representations of $\tilde{f}$ and an arbitrary concave f, piecewise linear between successive z_is:

$$\tilde{f}(z) = \tilde{\alpha}_0 + \sum_{i=1}^{n-1} \tilde{\alpha}_i \Delta^{(z_{(i)})}(z) \text{ and } f(z) = \alpha_0 + \sum_{i=1}^{n-1} \alpha_i \Delta^{(z_{(i)})}(z)$$

where $\alpha_0, \tilde{\alpha}_0 \in \mathbb{R}$ and for $1 \leq i \leq n-1$, $\alpha_i, \tilde{\alpha}_i \geq 0$. Using that $e^{\tilde{f}} - e^{f} \leq (\tilde{f} - f)e^{\tilde{f}}$, it follows that

$$\begin{aligned}
\ell(\tilde{f}) - \ell(f) &= \int (\tilde{f}(z) - f(z))\,d\mathbb{G}_n(z) - \int \left(e^{\tilde{f}(z)} - e^{f(z)}\right) dz \\
&\geq \int (\tilde{f}(z) - f(z))\,d\mathbb{G}_n(z) - \int (\tilde{f}(z) - f(z))e^{\tilde{f}(z)}\,dz \\
&= (\tilde{\alpha}_0 - \alpha_0)\left(1 - \int e^{\tilde{f}(z)}\,dz\right) \\
&\quad + \sum_{i=1}^{n-1} (\tilde{\alpha}_i - \alpha_i)\left(\int \Delta^{(z_{(i)})}(z)\,d\mathbb{G}_n(z) - \int \Delta^{(z_{(i)})}(z)e^{\tilde{f}(z)}\,dz\right) \\
&= \sum_{i=1}^{n-1} (\tilde{\alpha}_i - \alpha_i)\left(-\int_{-\infty}^{z_{(i)}} \mathbb{G}_n(z)\,dz + \int_{-\infty}^{z_{(i)}} \tilde{G}(z)\,dz\right)
\end{aligned} \tag{4.22}$$

where we use the integral assumption on $\exp \tilde{f}$ in the last step, as well as (4.20) and (4.21). Furthermore, defining $\tilde{I} = \{0 \leq i \leq n-1 \colon \tilde{\alpha}_i > 0\}$ as the set of indices i where $\tilde{f}'(z_{(i)}-) > \tilde{f}'(z_{(i)}+)$, it follows that the summation in (4.22) can be restricted to the terms $i \notin \tilde{I}$ in view of the equalities in (4.19) for $i \in \tilde{I}$. For $i \notin \tilde{I}$, $\tilde{\alpha}_i = 0$ so that the lower bound reduces to

$$\sum_{i \notin \tilde{I}} \alpha_i \left(\int_{-\infty}^{z_{(i)}} \mathbb{G}_n(z)\,dz - \int_{-\infty}^{z_{(i)}} \tilde{G}(z)\,dz\right) \geq 0$$

by the inequalities in (4.19) and the fact that $\alpha_i \geq 0$ for $i \geq 1$. □

Figure 4.11 shows the functions $\hat{f}$ and $\hat{g}$ respectively based on a sample of size $n = 250$ from the standard normal density. Figure 4.12a shows the estimator $\hat{G}$ based on this data set along with the empirical distribution function $\mathbb{G}_n$. Figure 4.12b visualizes the optimality conditions given in (4.19) by showing $z \mapsto \int_{-\infty}^{z}(\mathbb{G}_n(x) - \hat{G}_n(x))\,dx$. The pictures

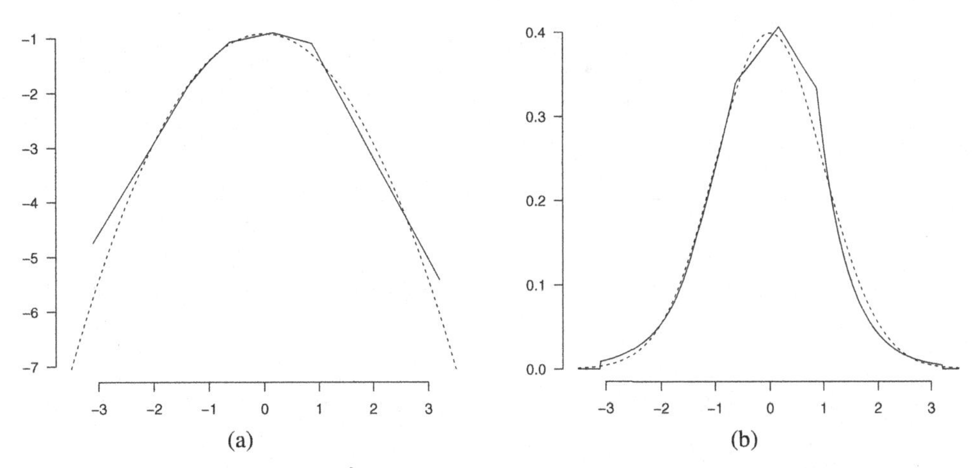

Figure 4.11 (a) The MLE $\hat{f}$ of the log density f based on a sample of size $n = 250$ from the standard normal density. (b) The associated MLE $\hat{g}$ of g. In both pictures the dashed curve denotes the underlying function to be estimated.

were produced using the `logConDens` R-package developed by Lutz Dümbgen and Kaspar Rufibach.

4.5 Star Shaped Distributions on [0, 1]

Let X be a random variable taking values in the interval $[0, 1]$ and let U be a random variable independent of X and uniformly distributed on $[0, 1]$. Define $Z = \max\{X, U\}$ and consider the problem of estimating the distribution function F of X based on n independent copies of the random variable Z. It is clear that the distribution function of Z is given by

$$G(z) = \mathbb{P}(X \leq z \text{ and } U \leq z) = zF(z), \quad z \in [0, 1].$$

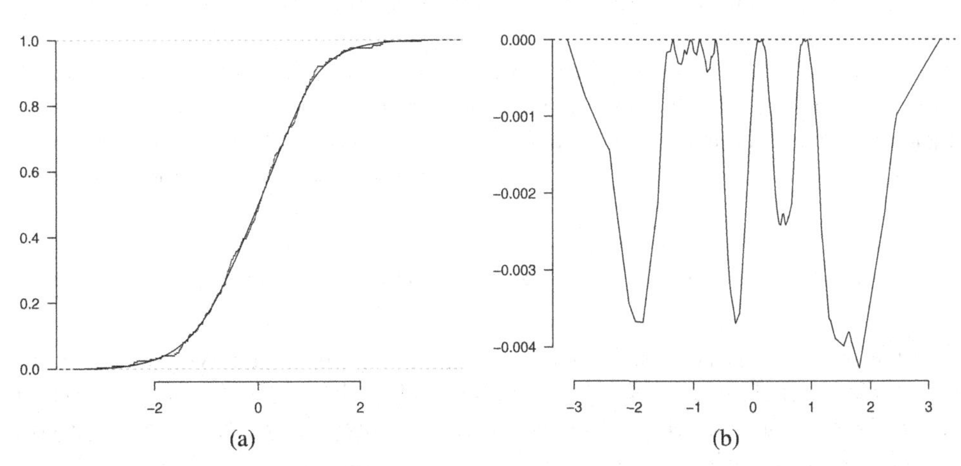

Figure 4.12 (a) The MLE $\hat{G}$ of the distribution function G based on the sample of Figure 4.11. (b) The optimality conditions of (4.19).

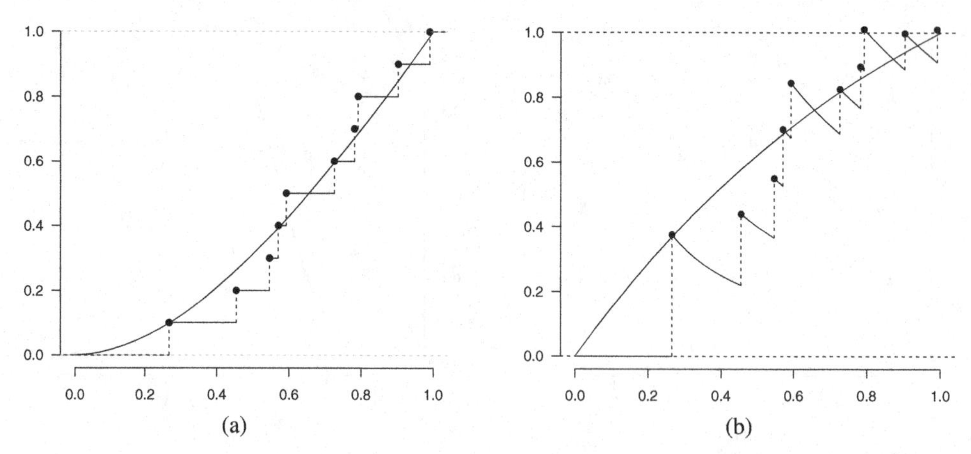

Figure 4.13 The empirical distribution function $\mathbb{G}_n$ (a) and resulting estimator $\tilde{F}_n$ (b) based on a sample of size $n = 10$ from the (star shaped) distribution function $G(z) = z(1 - e^{-z})/(1 - e^{-1})$ on $[0, 1]$.

This expression imposes a specific shape constraint on G: a star shaped constraint. This terminology comes from the fact that from every point in the set $\{(x, y) \in [0, 1]^2 : y \geq G(x)\}$ (the epigraph of G on $[0, 1]$) the segment connecting $(0, 0)$ to this point is also contained in this set. In terms of monotonicity constraints it entails that the function $z \mapsto G(z)/z$ is nondecreasing on $[0, 1]$.

Note that in view of the relation between F and G, a natural estimator for F based on the ordered set of observed data $z_{(1)} < \cdots < z_{(n)}$ is given by

$$\tilde{F}_n(z) = \frac{1}{nz} \sum_{i=1}^{n} 1_{[z_{(i)},1]}(z) = \frac{\mathbb{G}_n(z)}{z}, \quad z \in (0, 1]$$

where $\mathbb{G}_n$ denotes the empirical distribution function of the z_is. It is clear from Figure 4.13 that this estimator does not satisfy the natural monotonicity constraint of distribution functions. On the other hand, the law of large numbers does show that for each $z \in (0, 1]$, $\tilde{F}_n(z) \to F(z)$ as $n \to \infty$ with probability one.

One way to obtain a star shaped estimator of G is to first construct an estimator for the primitive of F, and regularizing this estimator by taking its convex minorant. This convex estimator can be differentiated to obtain a monotone estimator for F, leading to a star shaped estimator for G via multiplication by z. Using $\tilde{F}_n$ we obtain an estimator for the primitive of F:

$$U_n(z) = \int_0^z \tilde{F}_n(y)\, dy = \frac{1}{n} \sum_{i=1}^{n} 1_{[z_i,1]}(z) \log\left(\frac{z}{z_i}\right), \quad z \in (0, 1].$$

Clearly, this is an unbiased estimator for $U(z) = \int_0^z F(y)\, dy$ and by the law of large numbers it is pointwise strongly consistent. Moreover, since U_n and U are all nondecreasing, zero at zero and one at one and continuous, the pointwise consistency can be strengthened to uniform consistency. Taking as estimator for F

$$\hat{F}_n(z) = \hat{U}_n^{(r)}(z) \wedge 1,$$

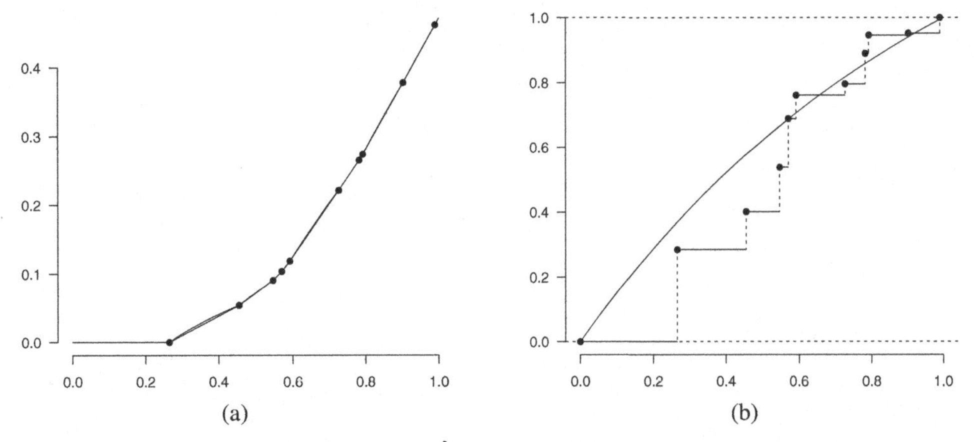

Figure 4.14 (a) The functions U_n, $\hat{U}_n$ and underlying (true) U based on the same data as used in Figure 4.13. (b) The resulting $\hat{F}_n$.

where $\hat{U}_n$ is the greatest convex minorant of U_n on [0, 1] and the superscript (r) denotes the right derivative, consistency can be derived using Lemma 3.1 (see Exercise 4.20). In Figure 4.14 U_n, $\hat{U}_n$ and the resulting estimator $\hat{F}_n$ are shown based on the same data as used for Figure 4.13.

Yet another interesting estimator in this model is the ML estimator. A first problem to be solved is to define the likelihood properly. As seen in Exercise 4.18, unlike concave distribution functions, star shaped distribution functions are not necessarily absolutely continuous with respect to Lebesgue measure. A density of G with respect to Lebesgue measure can therefore not be used in the definition of the likelihood. Analogous to how this problem is dealt with when the ML estimator of a nonrestricted distribution function is defined (see Exercise 4.19), the following definition of the log likelihood function seems natural

$$\ell(G) = \sum_{i=1}^{n} \log(G(z_{(i)}) - G(z_{(i)}-) = \sum_{i=1}^{n} \log(F(z_{(i)}) - F(z_{(i)}-) + \sum_{i=1}^{n} \log z_{(i)}. \qquad (4.23)$$

Since this last sum does not involve the G (or equivalently F), maximizing this function over all star shaped distribution functions G boils down to maximizing the first sum over all distribution functions F on [0, 1], without further restrictions. The solution to this problem is the discrete distribution function with masses $1/n$ on each observed value $z_{(i)}$. See Exercise 4.19. For the ML estimator of G, this means that it is defined by

$$\hat{G}_n(z) = z\mathbb{G}_n(z), \;\; z \in (0, 1].$$

Note that this estimator does not consistently estimate G: it converges almost surely uniformly to $z \mapsto zG(z)$ on [0, 1]. For the dataset of size 10 used in this section as well as for another dataset of size $n = 1000$, Figure 4.15 shows the ML estimates for G.

One could object against the definition of the log likelihood in the sense that a star shaped distribution function on [0, 1] can never be purely discrete (except for the degenerate distribution on $\{1\}$), something that is not taken into account. One alternative definition

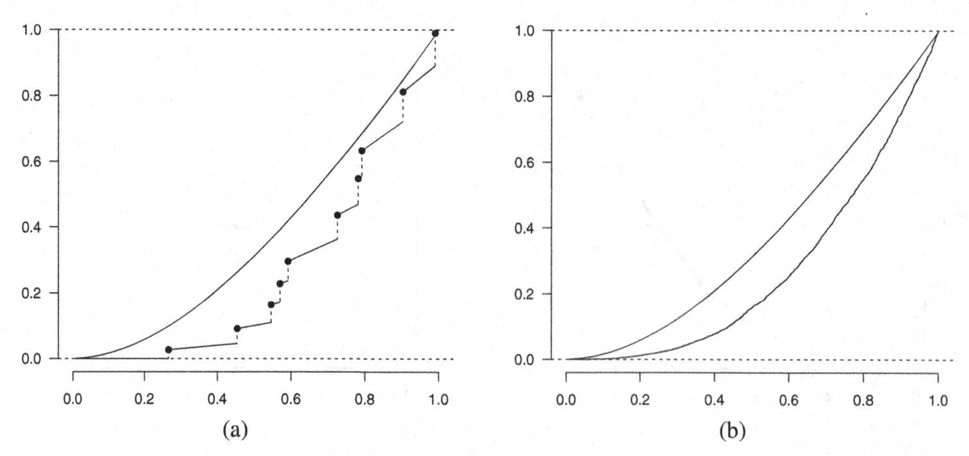

Figure 4.15 (a) The mle $\hat{G}_n$ of G and the underlying (true) G based on the same data as used in Figure 4.13. (b) The estimator based on a sample of size $n = 1000$ from the same distribution.

would be

$$\ell(G) = \sum_{i=1}^{n} \log \left(G(z_{(i)}) - G(z_{(i-1)}) \right),$$

which for star shaped distribution functions is really different from (4.23). It can be shown, however, that the maximizer of this function will not result in a consistent estimator for G. In particular, for G the uniform distribution function on [0, 1], the estimator will converge almost surely to $z \mapsto z^{1/e+1} = z^{1/e}G(z)$ uniformly on [0, 1].

Consistent likelihood-based nonparametric estimators for G can be obtained by using a smoothing idea in the spirit of what will be seen in Chapter 8. Given a partition of [0, 1] in $k_n \ll n$ intervals of length $1/k_n$, say $(y_{i-1}, y_i]$ for $1 \le i \le k_n$, the following smoothed log likelihood can be defined:

$$G \mapsto \sum_{i=1}^{k_n} n_i \log (G(y_i) - G(y_{i-1})) \quad \text{with} \quad n_i = \{j : z_j \in (y_{i-1}, y_i]\}.$$

If k_n tends to infinity sufficiently slowly, the maximizer of this function can be shown to be consistent for G.

4.6 Deconvolution Problems

As in Section 2.4, consider two independent samples $X_1, \ldots, X_n$ and $Y_1, \ldots, Y_n$ generated by an unknown distribution function F and a known probability density k. Instead of observing these samples directly, a sample $Z_1, \ldots, Z_n$ is observed where $Z_i = X_i + Y_i$ for each i. The Z_i then constitute a sample from the convolution density given by

$$g(z) = g_F(z) = [k * dF](z) = \int_{-\infty}^{\infty} k(z - x)\, dF(x). \tag{4.24}$$

Using the observable Z_is, the density g can be estimated directly, but needs to be deconvolved with k to obtain an estimator for F.

If the characteristic function of Y has no zeroes, one can determine the characteristic function of X by dividing the characteristic function of Z by that of Y, so that F can be found by Fourier inversion.

Example 4.1 *(Gaussian deconvolution)* The density of Y_i is given by

$$k(y) = \frac{1}{\sqrt{2\pi}} e^{-\frac{1}{2}y^2}$$

so that Z_i has the density

$$g(z) = \int_{-\infty}^{\infty} \frac{1}{\sqrt{2\pi}} e^{-\frac{1}{2}(z-x)^2}\, dF(x).$$

The standard normal density has characteristic function

$$Ee^{itY} = \int_{-\infty}^{\infty} k(y)e^{ity}\, dy = \exp\left(-\tfrac{1}{2}t^2\right).$$

Hence, the characteristic function of X is given by

$$Ee^{itX} = e^{t^2/2} Ee^{itZ}.$$

A natural approach to estimating F is to take the empirical characteristic function based on the observable Z_is, divide it by the (known, computable) characteristic function of Y and apply the inverse Fourier transform to this function to obtain an estimator for F. Quite some work has been done in this direction. A general problem faced when this approach is used is the poor quality of the estimator of the characteristic function of Z for large values of the argument. Usually a truncation is used in the integral to compute the inverse Fourier transform. As seen in Section 2.4, there are situations where the inverse relation can be formulated more explicitly, without using characteristic functions, if the type one resolvent of k exists. This is in particular the case when $X, Y \geq 0$ and k is smooth on $(0, \infty)$ with a finite discontinuity at zero. We have seen plug-in type estimators for this situation. These estimators can be constructed explicitly. For ML estimators this is, in general, not the case.

In this section we consider a smooth decreasing density k on $(0, \infty)$ (although the results hold also for more general noise densities) and introduce the nonparametric maximum likelihood estimator in that case. Also, for this method, inverse Fourier transforms are not required. The class of densities is parameterized by the class $\mathcal{F}$ of all distribution functions via (4.24).

Based on a realization $z_1, \ldots, z_n$ of a sample $Z_1, \ldots, Z_n$ from a density g of type (4.24), the log likelihood of a particular distribution function F is given by

$$\ell(F) = \sum_{i=1}^{n} \log g_F(z_i) = \sum_{i=1}^{n} \log\left(\int k(z_i - x)\, dF(x)\right). \tag{4.25}$$

Using the arguments seen before (e.g., in Lemma 4.4), the following characterization can be given for the ML estimator.

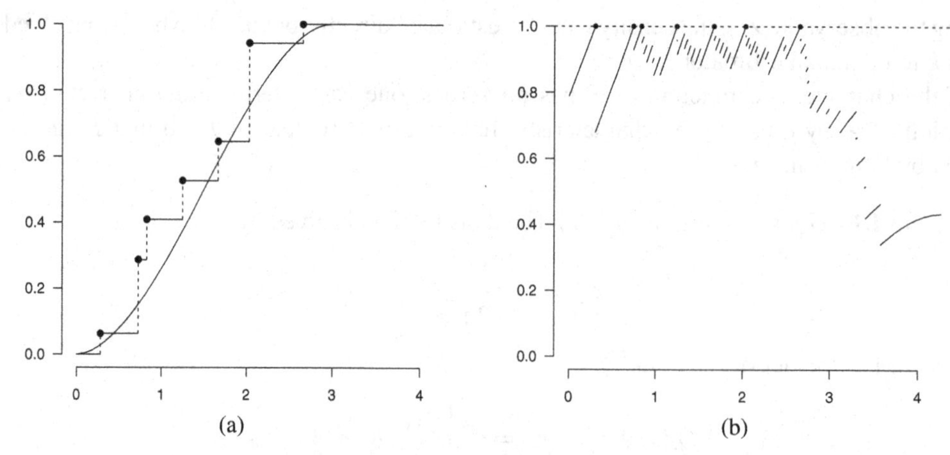

Figure 4.16 (a) The ML estimate of F based on a sample of size $n = 50$ from the density $g = f * k$ with f and k defined in (4.26). (b) The characterizing (in-)equalities of Lemma 4.6 are visualized.

Lemma 4.6 *Let $F \in \mathcal{F}$ and g_F be as in (4.24). Then F is the ML estimator (maximizing (4.25) over $\mathcal{F}$) if and only if*

$$C_F(x) := \int \frac{k(z-x)}{g_F(z)}\, d\mathbb{G}_n(z) \begin{cases} \leq 1 & \text{for all } x \in \mathbb{R} \\ = 1 & \text{for all } x \in \mathbb{R} \text{ with } F(x) - F(x-) > 0 \end{cases}$$

This characterizing lemma can be used to check whether a given F maximizes (4.25), but it does not give a constructive way to obtain this maximizer. In Chapter 7 computational issues for this type of implicitly characterized estimators are addressed. We generated a dataset of size $n = 50$ from the convolution of the following densities

$$f(x) = \frac{2}{9}x(3-x)1_{[0,3]}(x) \text{ and } k(x) = \sqrt{\frac{2}{\pi}}e^{-x^2/2}1_{[0,\infty)}(x) \tag{4.26}$$

(rescaled Beta(2, 2) density and that of the absolute value of a standard normal random variable) and computed the ML estimator using the EM algorithm to be described in Section 7.2; see also Exercise 7.19. In Figure 4.16 the ML estimate $\hat{F}$ is given with the underlying distribution function. Moreover, for this estimate, the function

$$x \mapsto C_{\hat{F}}(x) = \int \frac{k(z-x)}{g_F(z)}\, d\mathbb{G}_n(z)$$

is shown together with the horizontal line at level one. In view of Lemma 4.6 is clear that $\hat{F}$ is indeed the ML estimator.

To prove Hellinger consistency of the ML estimator of observation density g, we use Exercise 3.5. It suffices to show that the class of functions

$$\Phi = \left\{\phi_F : \phi_F(z) = \frac{g_F(z)}{g_F(z) + g_{F_0}(z)} : F \in \mathcal{F}\right\}$$

is Glivenko Cantelli, i.e., that for $n \to \infty$

$$\sup_{\phi\in\Phi}\left|\int \phi(z)\, d(\mathbb{G}_n - G_0)(z)\right| \to^{a.s.} 0$$

As seen in Section 3.4, a sufficient condition for this is that for each $\delta > 0$,

$$H_{[]}(\delta, \Phi, L_1(G_0)) = \log N_{[]}(\delta, \Phi, L_1(G_0)) < \infty.$$

To see that this entropy with bracketing is finite, fix $\delta > 0$. For fixed $\epsilon > 0$, define the set

$$I_\epsilon = \{z \geq 0\colon g_{F_0}(z) \leq \epsilon\} \text{ such that } I_\epsilon \downarrow \{z \geq 0\colon g_{F_0}(z) = 0\} \text{ as } \epsilon \downarrow 0$$

giving (by dominated convergence) that $\int_{I_\epsilon} g_{F_0}(z)\, dz \downarrow 0$ as $\epsilon \downarrow 0$. Choose $\epsilon > 0$ sufficiently small such that

$$\int_{I_\epsilon} g_{F_0}(z)\, dz \leq \frac{1}{2}\delta.$$

Moreover, given this δ and corresponding ϵ, construct the (in view of Exercise 3.16 finitely many) $\epsilon\delta^2/2$-brackets $[g_l, g_u]$ of

$$\mathcal{G} = \{g_F\colon F \text{ a distribution function on } (0, \infty)\}.$$

Then observe (Exercise 4.22) that these brackets $[g_l, g_u]$ give rise to brackets $[h_l, h_u]$ in Φ where

$$h_l(z) = \frac{g_l(z)}{g_l(z) + g_{F_0}(z)} \quad \text{and} \quad h_u(z) = \frac{g_u(z)}{g_u(z) + g_{F_0}(z)}$$

and, using Exercise 4.21, that

$$\begin{aligned}
\int (h_u(z) - h_l(z)) g_{F_0}(z)\, dz &= \left(\int_{I_\epsilon} + \int_{I_\epsilon^c}\right)(h_u(z) - h_l(z)) g_{F_0}(z)\, dz \\
&\leq \int_{I_\epsilon} g_{F_0}(z)\, dz + \frac{1}{\epsilon}\int_{I_\epsilon^c}(g_u(z) - g_l(z)) g_{F_0}(z)\, dz \\
&\leq \frac{1}{2}\delta + \frac{1}{\epsilon}\int (g_u(z) - g_l(z)) g_{F_0}(z)\, dz \leq \frac{1}{2}\delta + \frac{1}{\epsilon}\frac{1}{2}\epsilon\delta^2 = \delta.
\end{aligned}$$

Hence, the constructed brackets $[h_l \cdot h_u]$ are actually δ-brackets. This shows that $N_{[]}(\delta, \Phi, L_1(G_0)) \leq N_{[]}(\epsilon\delta^2/2, \mathcal{G}, L_1(G_0)) < \infty$.

Having Hellinger consistency of the ML estimator of the observation density, i.e.

$$\int \left(\sqrt{g_{\hat{F}_n}(z)} - \sqrt{g_{F_0}(z)}\right)^2 dz \to 0 \text{ a.s.}, \tag{4.27}$$

the next question of interest is whether the ML estimator of the underlying distribution function $\hat{F}_n$ is also consistent for F_0.

To prove this, note that in the current situation we have the following relation expressing the distribution function F in terms of its convolution density g:

$$F(x) = \frac{d}{dx}\int_0^x p(x - z) g(z)\, dz \text{ so also } \int_0^x F(z)\, dz = \int_0^x p(x - z) g(z)\, dz$$

where p is the type one resolvent of k, solving the integral equation $\int_0^x p(z-x)k(x)\,dx = x$ on $[0,\infty)$. This function is nonnegative and nondecreasing on $[0,\infty)$ (see Section 2.4). Using these properties of p, the Cauchy Schwarz inequality, the inequality

$$g_F(z) = \int_0^z k(z-x)\,dF(x) \le k(0)F(z) \le k(0)$$

and the equality $u - v = (\sqrt{u}-\sqrt{v})(\sqrt{u}+\sqrt{v})$, this relation implies that for each $x > 0$,

$$\begin{aligned}\int_0^x \left|\hat{F}_n(z) - F_0(z)\right| dz &= \int_0^x p(x-z)|g_{\hat{F}_n}(z) - g_{F_0}(z)|\,dz \\ &\le \sqrt{\int_0^x p(x-z)^2\,dz}\sqrt{\int_0^x (g_{\hat{F}_n}(z) - g_{F_0}(z))^2\,dz} \\ &\le \sqrt{x}p(x)\sqrt{\int_0^x \left(\sqrt{g_{\hat{F}_n}(z)} + \sqrt{g_{F_0}(z)}\right)^2 \left(\sqrt{g_{\hat{F}_n}(z)} - \sqrt{g_{F_0}(z)}\right)^2 dz} \\ &\le 2\sqrt{k(0)}\sqrt{x}p(x)\sqrt{\int_0^\infty \left(\sqrt{g_{\hat{F}_n}(z)} - \sqrt{g_{F_0}(z)}\right)^2 dz}.\end{aligned}$$

The right hand side tends to zero almost surely because of (4.27). This means that for all $x > 0$,

$$\int_0^x \left|\hat{F}_n(z) - F_0(z)\right| dz \to 0 \text{ a.s.}$$

Using Exercise 4.23, this implies that for each $x > 0$,

$$F_0(x-) \le \liminf_{n\to\infty} \hat{F}_n(x) \le \limsup_{n\to\infty} \hat{F}_n(x) \le F_0(x) \text{ a.s.}$$

Assuming continuity of F_0, pointwise consistency of $\hat{F}_n$ follows. This can be strengthened to uniform consistency along the lines of Exercise 3.2.

4.7 Interval Censoring Case 2

We now discuss an extension of the current status model introduced in Section 2.3, where one considers more observation times per unobservable "hidden" variable X. The simplest model of this type is called interval censoring case 2. Let X be an unobservable random variable from an unknown distribution function F_0 on $[0,\infty)$. Moreover, let (T, U) be a random vector with density h, independent of X, where $U > T \ge 0$ with probability one. Instead of observing X directly, the pair (T, U) together with the indicators

$$\Delta_1 = 1_{\{X\le T\}},\ \Delta_2 = 1_{\{T<X\le U\}} \text{ and } \Delta_3 = 1 - \Delta_1 - \Delta_1$$

are observed. These observations provide information on the position of the random variables X with respect to the corresponding observation times T and U. This setting is known as interval censoring case 2, where the 2 refers to the number of inspection times for each X.

From Exercise 4.24 we get that the density (with respect to an appropriate dominating measure) of the observable vector $(T, U, \Delta_1, \Delta_2)$ is given by

$$g_F(t, u, \delta_1, \delta_2) = F(t)^{\delta_1}(F(u) - F(t))^{\delta_2}(1 - F(u))^{1-\delta_1-\delta_2}g(t, u).$$

Now consider the situation that data $(t_i, u_i, \delta_{1,i}, \delta_{2,i})$ $(1 \le i \le n)$ are observed as independent realizations of this random vector. The log likelihood for F (apart from a term not depending on F) is given by

$$\ell(F) = \sum_{i=1}^{n} \delta_{1,i} \log F(t_i) + \delta_{2,i} \log(F(u_i) - F(t_i)) + (1 - \delta_{1,i} - \delta_{2,i}) \log(1 - F(u_i)). \tag{4.28}$$

A maximum likelihood (ML) estimator for F is a maximizer of this log likelihood. First note that the log likelihood only depends on the function F through its values at the points $\{t_i, u_i : 1 \le i \le n\}$. As long as monotonicity of F is preserved, changing F at intermediate values of the argument does not change the value of ℓ. Therefore, in finding a maximizer of ℓ over all distribution functions, attention can be restricted to those that are piecewise constant with all jumps concentrated on the points $\{t_i, u_i : 1 \le i \le n\}$ and possibly one point to the right of the maximal u_i. Even a further reduction of possible jump points can be made (see Exercise 4.25). There are also other estimators one can consider in this model. Examples of smooth estimators in this situation are the SMLE (smoothed maximum likelihood estimator) and MSLE (maximum smoothed likelihood estimator). Smooth estimators in this model will be discussed in Section 8.6.

Figure 4.17 visualizes a generated dataset in the spirit of Figure 2.7. The data were generated using sample size $n = 50$, $F(x) = \sqrt{x}$ on $(0, 1)$ and (T, U) jointly distributed as the minimum and maximum of two independent standard uniformly distributed random variables.

For the ML estimator, a characterizing lemma can be proved using the variational argument used before. This leads to the following lemma (see also Exercise 4.26).

Lemma 4.7 *The (sub-) distribution function $\hat{F}$ maximizes ℓ defined by (4.28) over all piecewise constant (sub-) distribution functions F with jumps concentrated on the set $\mathcal{T} = \{t_i, u_i : 1 \le i \le n\}$ if and only if*

$$\frac{1}{n}\sum_{i=1}^{n}\left(\frac{\delta_{1,i}1_{[0,t_i]}(x)}{\hat{F}(t_i)} + \frac{\delta_{2,i}1_{(t_i,u_i]}(x)}{\hat{F}(u_i) - \hat{F}(t_i)} + \frac{(1-\delta_{1,i}-\delta_{2,i})1_{(u_i,\infty)}(x)}{1-\hat{F}(u_i)}\right)\begin{cases} \le 1 & \text{for all } x \in \mathcal{T} \\ = 1 & \text{if } \hat{F}(x) > \hat{F}(x-) \end{cases} \tag{4.29}$$

The ML estimator will be considered from a computational point of view in Section 7.2. Using the iterative convex minorant algorithm described there, the estimator given Figure 4.18a is found as maximizer of (4.28) over the class of piecewise constant distribution functions described earlier. Figure 4.18b shows the picture corresponding to the characterizing (in-)equalities of (4.29) showing that indeed the distribution function found maximizes the log likelihood.

Interestingly, Lemma 4.7 also characterizes the the MLE for the distribution function of the hidden variable in the case that one has more observation times $T_i, U_i, V_i, \ldots$ "per hidden variable" X_i and not necessarily the same number of observation times for all i. The log

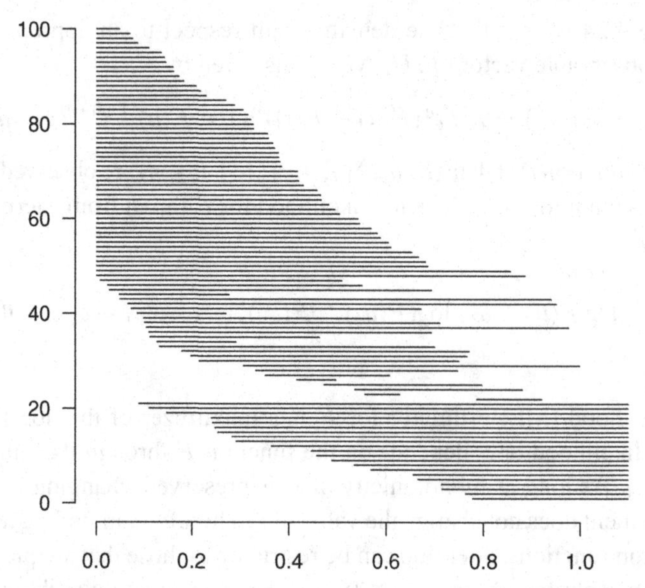

Figure 4.17 Visualization of a data set of size $n = 100$. The horizontal line at level i indicates the possible values of X_i given the observed inspection times and indicators for the ith subject. If $\delta_{1,i} = 1$, the line corresponds to $[0, t_i]$, if $\delta_{2,i} = 1$ to $(t_i, u_i]$ and if $\delta_{1,i} + \delta_{2,i} = 0$ it represents (u_i, ∞).

likelihood in this so-called mixed case interval censoring model has the same structure as in case of two inspection times per subject because at most two of the observation times of the set $\{T_i, U_i, V_i, \dots\}$ are relevant for the location of the hidden variable. If we know that the hidden variable is located between two observation times, while the other observation times for this hidden variable are either more to the right or more to the left, then these other observation times do not give extra information and do not play a role in characterizing the MLE. Likewise, if we know that the hidden variable lies to the right of all these observation

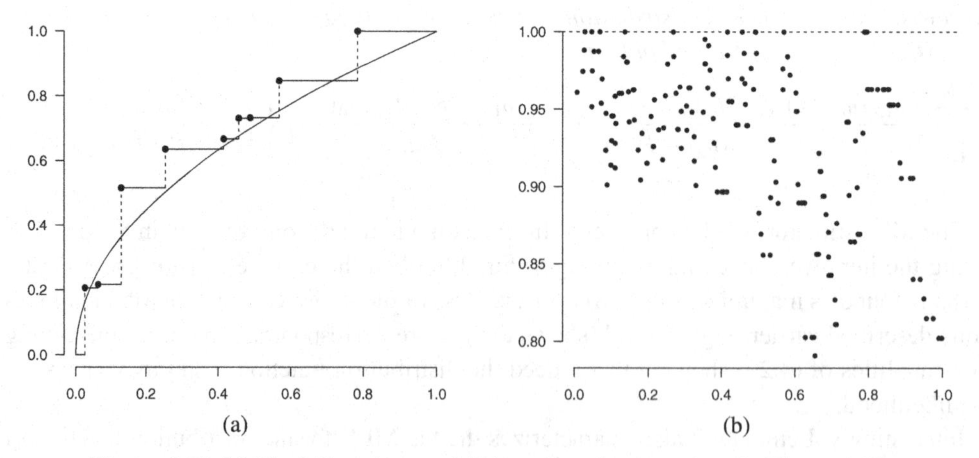

Figure 4.18 (a) The ML estimate of F based on the data given in Figure 4.17. (b) The characterizing (in-)equalities of Lemma 4.7.

times, all observation times smaller than the largest one do not give extra information. A similar situation occurs if the hidden event time lies to the left of the smallest observation time for this variable. In the last two cases only one observation time gives relevant information and the other ones can be discarded. But of course the distribution theory will be different if we have more observation times per unobservable, and higher accuracy of the MLE of F_0 can be expected than in the situation that we just have two observation times.

From an asymptotic point of view, there is an interesting distinction between models, depending on whether inspection times can be arbitrarily close to each other or not. For practical purposes, the separated case (where there is some strictly positive minimal distance ν between the two inspection times) is of more interest. In that case, the asymptotic behavior of the ML estimator is comparable to the behavior in the current status (interval censoring case 1) model (see Section 2.3). In case 2 one does get a bit more information for each subject, but there are no subjects for which the variable of interest X_i is more precisely known than by the separation threshold $\nu > 0$. The result that follows (showing the $n^{-1/3}$ rate as seen in Theorem 3.7 for the current status model) is proved in Groeneboom, 1996: under appropriate conditions,

$$n^{1/3}\{2\xi(t_0)/f_0(t_0)\}^{1/3}\{\hat{F}_n(t_0) - F_0(t_0)\} \xrightarrow{\mathcal{D}} 2Z, \tag{4.30}$$

where Z is the last time where standard two-sided Brownian motion minus the parabola $y(t) = t^2$ reaches its maximum (so it has the Chernoff distribution described in Section 3.9). Here

$$\xi(t_0) = \frac{g_1(t_0)}{F_0(t_0)} + k_1(t_0) + k_2(t_0) + \frac{g_2(t_0)}{1 - F_0(t_0)}$$

with

$$k_1(u) = \int_u^\infty \frac{g(u, v)}{F_0(v) - F_0(u)}\, dv, \quad \text{and} \quad k_2(v) = \int_0^v \frac{g(u, v)}{F_0(v) - F_0(u)}\, du.$$

See Exercise 4.27 to see that in the separated case, the ML estimator is more accurate than if data were obtained within the current status model (so with only one observation time per subject).

Intuitively, it seems reasonable to expect that in the nonseparated situation better estimation is possible than in the separated case, since with positive probability there are observation intervals (T_i, U_i) of arbitrarily short length. The presence of an X_i in a small observation interval $(T_i, U_i]$ gives more precise information on its location than in a larger observation interval. On the other hand, one cannot expect too much gain from this, since the probability of getting a small interval and an observation X_i contained in it is itself small. This better performance could materialize in a faster rate of convergence or merely in a lower asymptotic variance. In Birgé, 1999, an alternative estimator to the ML estimator is proposed, attaining rate of convergence $(n \log n)^{-1/3}$ in the nonseparated case. In Groeneboom and Ketelaars, 2011, this estimator is shown to be asymptotically normally distributed and a simulation study comparing this estimator to the ML estimator indicates that the ML estimator outperforms this estimator. The transition from the rate $n^{-1/3}$ to $(n \log n)^{-1/3}$ quantifies the (small) gain in precision, in going from the separated to the nonseparated case. In Groeneboom and Wellner,

1992, it is conjectured that under certain conditions on the densities f_0 and h (including that $0 < h(t_0, t_0) < \infty$),

$$(n \log n)^{1/3} \left\{ \hat{F}_n(t_0) - F_0(t_0) \right\} / \left\{ \tfrac{3}{4} f_0(t_0)^2 / g(t_0, t_0) \right\}^{1/3} \xrightarrow{\mathcal{D}} 2Z, \qquad (4.31)$$

where Z has the Chernoff distribution described in Section 3.9. In Groeneboom and Ketelaars, 2011, a simulation study can be found showing that the fit between the variance of the MLE and the value predicted by the asymptotic theory on the basis of (4.31) is rather good for the case in which F_0 is uniform. In the simulation study the uniform density on the upper triangle of the unit square for the pair (T_i, U_i) was chosen as an example of the observation density h in the nonseparated case. Other simulation results that show less convincing support of the conjecture can also be found in Groeneboom and Ketelaars, 2011, but that is conceivably due to slow convergence to the limit as similar behavior is found for Birgé's estimator in this case.

Exercises

4.1 Consider two independent, positive random variables X and Y, where X has distribution function F and Y density k on $(0, \infty)$. Then the random variable Z has probability density g on $(0, \infty)$ given by

$$g(z) = \int_{(0,\infty)} x^{-1} k(z/x)\, dF(x).$$

4.2 Verify inverse formula (4.1) in Wicksell's problem.

4.3 Argue that the sampling density g in Wicksell's problem cannot be standard uniform.

4.4 Show that if the squared sphere radii in Wicksell's problem are exponentially distributed, the same holds for the squared circle radii. Determine the relation between the parameters of the two distributions.

4.5 Substituting the empirical measure $d\mathbb{G}_n(z)$ based on an i.i.d. sample $Z_1, Z_2, \ldots, Z_n$ for $g(z)\,dz$ in (4.1), the following estimator for $F(x)$ is obtained:

$$\hat{F}_n(x) = 1 - \frac{T_n(x)}{T_n(0)} \quad \text{with} \quad T_n(x) = \frac{1}{n} \sum_{i=1}^{n} (Z_i - x)^{-1/2} 1_{(x,\infty)}(Z_i).$$

a) Show that for each $x \geq 0$, $\hat{F}_n(x) \to^P F(x)$.
b) Observe that $\hat{F}_n$ is not a monotone function (e.g., by first looking at the estimator for a sample of size $n = 1$).
c) Consider a point $x > 0$ where g is continuous and such that $0 < F(x) < 1$. Argue that $g(x) > 0$ and conclude that $\mathrm{Var}(T_n(x)) = \infty$.
d) Apply the central limit theorem for i.i.d. random variables with infinite variance to derive the asymptotic (normal) distribution of the estimator $\hat{F}_n(x)$.

4.6 Verify expression (4.2) for the primitive U_n of the function T_n in Wicksell's problem.

4.7 Consider the isotonized estimator $\tilde{T}_n$ defined as right derivative of the least concave majorant of the process U_n defined in (4.2). Use Lemma 3.1 to show that for $x > 0$, $\tilde{T}_n(x) \to T(x)$ almost surely as $n \to \infty$.

4.8 A natural method to apply in Wicksell's problem is maximum likelihood. Given a realization $z_1, \ldots, z_n$ of the sample $Z_1, Z_2, \ldots, Z_n$ from g, define the log likelihood function by

$$\ell(F) = \sum_{i=1}^{n} \log g_F(z_i) = \sum_{i=1}^{n} \log \left(\int_{(x,\infty)} (z_i - x)^{-1/2} \, dF(x) \right) - n \log(2m_F).$$

Show that for certain choices of F, this function can be made infinite. Hint: think of F with a peaked density at an observation point.

4.9 As seen in Exercise 4.8, taken over the class of all distribution functions F on $[0, \infty)$, ℓ is unbounded (it even takes infinite values). One way out is to restrict the class of distribution functions to be considered.

a) A natural class of distribution functions seems to be the one consisting of piecewise constant distribution functions, only having jumps at the observed data $z_1, \ldots, z_n$. Show that for this type of distribution function, $g_F(z_{(n)}) = 0$, implying that $\ell(F) = -\infty$ for all distribution functions of this type.

b) Show that allowing for mass to the right of $z_{(n)}$ (e.g., at $z_{(n)} + 1$) or excluding the term $\log g_F(z_{(n)})$ from the log likelihood, one obtains a log likelihood such that $\ell(F) > -\infty$ for some F and $\ell(F) < \infty$ for all F.

4.10 Show that all convex decreasing densities g on $[0, \infty)$ satisfy the inequality $g(z) \leq 1/(2z)$ for all $z \geq 0$. Hint: consider triangular densities $\{g_\theta : \theta > 0\}$ where

$$g_\theta(z) = \frac{2}{\theta^2}(\theta - z) 1_{[0,\theta]}(z).$$

4.11 Show that the least squares estimator $\tilde{g}_n$ of a convex decreasing density is inconsistent at zero.

4.12 Show that if the maximum likelihood estimator for a convex decreasing density based on a sample of size n exists, it will be a piecewise linear density with at most one change of slope between two successive observations, one change of slope to the right of the largest order statistic $z_{(n)}$ and no changes of slope at observed data points. Hint: starting from an arbitrary convex decreasing density g, construct a convex decreasing subdensity $\tilde{g}$ of the required type that has the same log likelihood; then define $\bar{g} = \tilde{g} / \int \tilde{g}$.

4.13 Consider estimating a convex decreasing density based on a sample of size $n = 1$. Write z_1 for the observed value and show that the least squares estimator considered in Section 4.3 and the ML estimator of Exercise 4.12 are given by, respectively,

$$\tilde{g}(z) = \frac{2(3z_1 - z)_+}{9z_1^2} \quad \text{and} \quad \tilde{g}(z) = \frac{2(2z_1 - z)_+}{4z_1^2}.$$

4.14 a) Show that all log concave densities are unimodal.

b) Show that a log concave density g has subexponential tails, in the sense that for some $c > 0$,

$$\int_{\mathbb{R}} g(x) e^{c|x|} \, dx < \infty.$$

c) Show that the Laplace, logistic and Beta(α, β) densities with $\alpha, \beta \geq 1$ are log concave, i.e.,

$$f(x) = \frac{1}{2} e^{-|x|}, \quad f(x) = \frac{x^{\alpha-1}(1-x)^{\beta-1}}{B(\alpha, \beta)} 1_{(0,1)}(x) \quad \text{and} \quad f(x) = \frac{e^x}{(1+e^x)^2}.$$

4.15 Let f be the function defined on $[z_{(i-1)}, z_{(i)}]$ as linear interpolation of $(z_{(i-1)}, f_{i-1})$ and $(z_{(i)}, f_i)$. Show that

$$\int_{z_{(i-1)}}^{z_{(i)}} \exp(f(z))\,dz = (z_{(i)} - z_{(i-1)})\frac{\exp(f_i) - \exp(f_{i-1})}{f_i - f_{i-1}}$$

provided $f_{i-1} \neq f_i$. In the case $f_{i-1} = f_i$ it is clear that the integral is given by $\exp(f_i)(z_{(i)} - z_{(i-1)})$.

4.16 Show that if $\hat{f}$ maximizes the relaxed log likelihood given in (4.17) over the class of log concave functions on $\mathbb{R}$, $\exp \hat{f}$ is a probability density in the sense that $\int \exp(f(z))\,dz = 1$. Hint: argue that

$$\lim_{\epsilon \to 0} \epsilon^{-1}\left(\ell(\hat{f} + \epsilon) - \ell(\hat{f})\right)$$

equals zero and and evaluate the limit.

4.17 Show that the function ℓ defined by the first equality in (4.17) is strictly concave on the class of concave functions. Conclude existence and uniqueness of the ML estimator $\hat{f}$ maximizing ℓ.

4.18 Construct a star shaped distribution function on [0, 1] that is not continuous.

4.19 Consider an i.i.d. sample $X_1, \ldots, X_n$ from a distribution with unknown distribution function F. Denote the class of all distribution functions on $\mathbb{R}$ by $\mathcal{F}$ and its subclass of distribution functions with continuous density functions by $\mathcal{F}_s$. On the subclass $\mathcal{F}_s$, the log likelihood function can be defined by

$$\ell(f) = \sum_{i=1}^{n} \log f(X_i).$$

a) Show that ℓ is unbounded on $\mathcal{F}_s$, so that the ML estimator of f over $\mathcal{F}_s$ is not well defined. Denote the subclass of all discrete distribution functions on R by $\mathcal{F}_d$. Define, for $F \in \mathcal{F}_s$, the log likelihood function based on $X_1, \ldots, X_n$ by

$$\ell(F) = \sum_{i=1}^{n} \log\left(F(X_i) - F(X_i-)\right), \tag{4.32}$$

where $F(X_i-) = \lim_{x \uparrow X_i} F(x)$.

b) Show that the maximizer of ℓ defined in (4.32) over all discrete distribution functions is given by the empirical distribution function of $X_1, \ldots, X_n$. Hint: first argue that if a maximizer exists, it should assign all its mass to the observed X_is. Then, denote the number of distinct values among the X_is by m and parameterize the optimization problem in terms of probability vectors $p = (p_1, \ldots, p_m)$ satisfying $p_i \geq 0$ for all i and $\sum_{i=1}^{m} p_i = 1$.

4.20 Fill in the details of the consistency result for the estimator $\hat{F}_n$ defined in Section 4.5.

4.21 Define, for $u > 0$, the function $\tau_u(v) = v/(v + u)$ on $[0, \infty)$. Show that τ_u is nondecreasing and that its derivative satisfies $\tau_u'(v) \leq 1/u$.

4.22 Show in the context of Section 4.6 that if $g_l \leq g \leq g_u$ for functions $g_u, g, g_u \in \mathcal{G}$, then also

$$\frac{g_l}{g_l + g_{F_0}} \leq \frac{g}{g + g_{F_0}} \leq \frac{g_u}{g_u + g_{F_0}}.$$

See also Exercise 4.21. Conclude that a bracket $[g_l, g_u]$ in $\mathcal{G}$ gives rise to a bracket $[g_l/(g_l + g_{F_0}), g_u/(g_u + g_{F_0})]$ in Φ.

4.23 Consider a sequence of (possibly sub-) distribution functions (F_n) and a distribution function F such that for all $x > 0$

$$\int_0^x |F_n(z) - F(z)|\, dz \to 0.$$

a) Show that the sequence (F_n) is asymptotically tight, in the sense that for each $\epsilon > 0$ there exists an $M > 0$ such that for all n sufficiently large $F_n(M) \geq 1 - \epsilon$.
b) Show that for all $x > 0$

$$F(x-) \leq \liminf_{n\to\infty} F_n(x) \leq \limsup_{n\to\infty} F_n(x) \leq F(x).$$

4.24 Show that, given $t < u$ in the interval censoring case 2 model, the distribution of the triple $(\Delta_1, \Delta_2, \Delta_3)$ is given by

$$P_F\left((\Delta_1, \Delta_2, \Delta_3) = (\delta_1, \delta_2, \delta_3) | (T, U) = (t, u)\right) = \begin{cases} F(t) & \text{if } (\delta_1, \delta_2, \delta_3) = (1, 0, 0) \\ F(u) - F(t) & \text{if } (\delta_1, \delta_2, \delta_3) = (0, 1, 0) \\ 1 - F(u) & \text{if } (\delta_1, \delta_2, \delta_3) = (0, 0, 1). \end{cases}$$

Note that this distribution can be written compactly as follows

$$P_F\left((\Delta_1, \Delta_2, \Delta_3) = (\delta_1, \delta_2, \delta_3) | (T, U) = (t, u)\right) = F(t)^{\delta_1}(F(u) - F(t))^{\delta_2}(1 - F(u))^{\delta_3}$$

for $(\delta_1, \delta_2, \delta_3) \in \{0, 1\}^3$ with $\delta_1 + \delta_2 + \delta_3 = 1$.

4.25 Argue that in maximizing the log likelihood for the interval censoring case 2 model, attention can be restricted to piecewise constant distribution functions having jumps restricted to the set

$$\{t_i : \delta_{1,i} + \delta_{2,i} = 1\} \cup \{u_i : \delta_{1,i} = 0\} \cup \{s\}$$

where $s > \max_i u_i$.

4.26 Prove Lemma 4.7 by taking directional derivatives of the log likelihood (4.28) in appropriate directions.

4.27 The asymptotic distribution of the ML estimator for $F(t_0)$ based on current status data is given in Theorem 3.7. Based on two inspection times that are strictly separated, (4.30) gives the asymptotic distribution of the ML estimator. Both asymptotic distributions are the same up to a scale change. Argue intuitively that the ML estimator in the current status setting should not have smaller asymptotic variance than the ML estimator in the interval censoring case 2 setting. Also compare the asymptotic variances.

Bibliographic Remarks

Wicksell's problem was originally posed in Wicksell, 1925. Smooth estimators were proposed, among others, in Hall and Smith, 1988, and Van Es and Hoogendoorn, 1990. Extensions of the model were also introduced and studied, e.g., in Wicksell, 1926, for ellipsoidal particles, Feuerverger and Hall, 2000, for thick slices and McGarrity et al., 2014, for a model involving circular cylinders. In Ohser and Mücklich, 2000, extensions to various nonspherical particles are considered. Convex regression is a widely applied model. Consistency of the least squares estimator is shown in Hanson and Pledger, 1976. Asymptotic distribution theory can be found in Groeneboom et al., 2001a. The bird-catching model was introduced in Hampel, 1987. Nonparametric estimators for this model were introduced and studied in Anevski, 2003, and Groeneboom et al., 2001a. The model itself was further developed in Anevski, 2007. Extensions of the model to k-monotone densities are studied in Balabdaoui and Wellner, 2007. In Walther, 2001, the problem of maximum likelihood estimation of a log concave density is studied. Basic properties of the estimators are derived in Dümbgen and Rufibach, 2009, and limit theory is established in Balabdaoui et al., 2009. The MLE of a log concave density was computed using the `logConDens` R-package; see also Dümbgen and Rufibach, 2011. Estimators (including the inconsistent ML estimator) for star shaped distributions were introduced in Barlow et al., 1972. Consistency of the maximum smoothed likelihood estimator (a type of estimator that will be studied in Chapter 8) of a star shaped distribution can be found in Jongbloed, 2009. Various approaches to estimation in the deconvolution problem can be found in Meister, 2009. For a restricted class of convolution kernels, isotonic estimators were introduced and studied in Van Es et al., 1998. For the same convolution kernels, concave estimators for the distribution function were defined and studied in Jongbloed and van der Meulen, 2009. Consistency of the ML estimator under mild conditions on the convolution density is established in Groeneboom, Jongbloed and Michael, 2012. The interval censoring case 2 problem is thoroughly studied in Groeneboom and Wellner, 1992. Under an until now unproven working hypothesis, the asymptotic distribution of the ML estimator with convergence rate $(n \log n)^{-1/3}$ in the unseparated case is established. Under the same working hypothesis, Wellner, 1995, proves (4.30) in the separated case. The working hypothesis is that the MLE has an asymptotic distribution which is the same as that of a one step toy estimator, obtained by doing one step of the iterative convex minorant algorithm (to be discussed in Section 7.3) starting from the underlying distribution function F_0. See p. 89 of Groeneboom and Wellner, 1992.

5

Higher Dimensional Problems

There are many situations in which one is interested in a bivariate or higher dimensional distribution function and direct observations from that distribution function are not available. Just like Chapter 4, this chapter is concerned with estimating such distribution functions based on observations that only carry partial information on the underlying random quantity of interest. The competing risk model with current status observations is introduced and studied in Section 5.1. This model is related to the current status model of Section 2.3, but apart from information on whether or not a subject already died at time of inspection, information on the death cause (belonging to a finite set of causes) is also available in case the subject has already died. Another model of interest concerns the natural two-dimensional generalization of the current status model. The unobservable random quantity of interest is a bivariate event time and, based on two-dimensional inspection times, a quadrant in $\mathbb{R}^2$ can be identified where the event took place. This problem and the more general bivariate case 2 interval censoring model are considered in Section 5.2. In the current status with continuous marks model, the quantity of interest is again a random vector (X, Y) in $\mathbb{R}^2$. Based on an inspection time T for X, its current status is observed, so whether or not $X \leq T$. In case $X \leq T$, also the corresponding value Y is observed. Otherwise Y is not observed. This model, to be considered in Section 5.3, is clearly related to the competing risks model, where the discrete variable indicating the risk factor is replaced by a possibly continuous random variable. Finally, the problem of estimating a multivariate log concave density is considered in Section 5.4. Unlike monotonicity, the concept of convexity is well established in higher dimensions and log concave densities on $\mathbb{R}^k$ can be estimated nonparametrically.

5.1 Competing Risks with Current Status Observations

Consider a situation where for a certain object there are several (say $K \in \mathbb{N}$) possible causes of failure and that at a random point T in time the object is inspected. It is observed whether or not this object broke down before time T or not (its current status). In case the object has broken down, it is also observed which of the K possible causes (competing risks) lead to the breakdown. Write X for the time of breakdown and $Y \in \{1, 2, \ldots, K\}$ for the corresponding cause. Together with inspection time T, the indicator vector

$$\Delta = (\Delta_1, \ldots, \Delta_K) \quad \text{with} \quad \Delta_k = 1_{[X \leq T, Y=k]}$$

is observed. Note that if all indicators are zero, this means that $X > T$. If $X \leq T$ also the breakdown cause is observed, so then exactly one of the K indicators equals 1. Assuming (X, Y) to have joint distribution function F and T to be independent of (X, Y), this model is

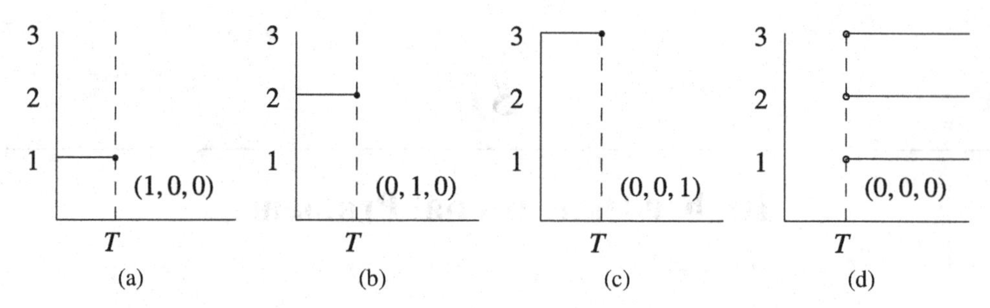

Figure 5.1 Graphical representation of the observed data (T, Δ) in an example with $K = 3$ competing risks. The horizontal line segments indicate the values of (X, Y) that are consistent with (T, Δ), for each of the four possible values of Δ also indicated in the pictures.

known as the competing risk model with current status observations. See Figure 5.1 for the various possibilities for (X, Y) given an observation (T, Δ) with $K = 3$. Data of this type arise naturally in cross-sectional studies with several failure causes. Moreover, similar data arise in HIV vaccine trials; see Hudgens et al., 2001. Clearly, the model is a generalization of the current status model introduced in Section 2.3 since that model corresponds to the situation $K = 1$.

Note that, given T, the vector

$$(\Delta_1, \Delta_2, \ldots, \Delta_K, \Delta_{K+1}) \text{ with } \Delta_{K+1} = 1 - \sum_{k=1}^{K} \Delta_k \tag{5.1}$$

has a multinomial distribution with parameters 1 and $(F_1(T), \ldots, F_K(T), 1 - F_X(T))$ where

$$F_k(t) = P(X \le t, Y = k), \quad t \ge 0, k = 1, 2, \ldots, K,$$

is the subdistribution function of X for risk level k and $F_X = \sum_k F_k$ is the marginal distribution function of X (see Exercise 5.1). Denoting by e_k the kth unit vector in $\mathbb{R}^K$, by # counting measure on $D = \{e_k : k = 1, \ldots, K + 1\}$ and G the distribution of T, we can define the measure $\mu = G \times \#$ on $\mathbb{R} \times D$. With respect to this (dominating) measure, the density of a single observation (T, Δ) is given by

$$p_F(t, \delta) = \prod_{k=1}^{K} F_k(t)^{\delta_k} (1 - F_X(t))^{1-\delta_+}, \tag{5.2}$$

where $\delta_+ = \sum_{k=1}^{K} \delta_k$.

Now consider an independent sample of size n, distributed as $(T, \Delta_1, \ldots, \Delta_K)$,

$$\left(T_i, \Delta^i\right) = \left(T_i, \Delta_1^i, \ldots, \Delta_K^i\right), \; i = 1, \ldots, n,$$

where, for $1 \le i \le n$,

$$\Delta^i = \left(\Delta_1^i, \ldots, \Delta_K^i\right) \quad \text{with} \quad \Delta_k^i = 1_{\{X_i \le T_i, \, Y = k\}}, \; k = 1, \ldots, K.$$

Also define

$$\Delta^i_{K+1} = 1 - \sum_{k=1}^{K} \Delta^i_k = 1_{\{X_i > T_i\}}.$$

Using (5.2) and independence of the observations, the log likelihood (divided by n) is given by

$$\begin{aligned} \ell(F) &= \int \log p_F(t,\delta)\, d\mathbb{P}_n(t,\delta) \\ &= \int \left\{ \sum_{k=1}^{K} \delta_k \log F_k(t) + (1-\delta_+)\log(1-F_+(t)) \right\} d\mathbb{P}_n(t,\delta), \end{aligned} \tag{5.3}$$

where $\mathbb{P}_n$ is the empirical distribution of (T_i, Δ^i), $i = 1, \dots, n$. An MLE $\hat{F}_n = (\hat{F}_{n1}, \dots, \hat{F}_{nK})$ can then be defined by the property

$$\ell(\hat{F}_n) = \max_{F \in \mathcal{F}_K} \ell(F) \tag{5.4}$$

where

$$\begin{aligned} \mathcal{F}_K = \{F = (F_1, \dots, F_K)\colon\ & F_1, \dots, F_K \text{ are subdistribution functions,} \\ & \text{such that for all } x \geq 0\colon \sum_{k=1}^{K} F_k(x) \leq 1\}. \end{aligned} \tag{5.5}$$

Before considering this MLE, we first follow an approach taken in Jewell et al., 2003. There, part of the information in the data is discarded, reducing the problem of estimating the subdistribution functions F_k to the known current status problem. For $1 \leq k \leq K$ the kth element of the vector $\tilde{F} = (\tilde{F}_1, \dots, \tilde{F}_K)$ is defined as maximizer of the kth marginal log likelihood for the reduced current status data (T_i, Δ^i_k), $i = 1, \dots, n$:

$$\ell_k(F) = \int \left\{ \delta_k \log F_k(t) + (1-\delta_k)\log(1-F_k(t)) \right\} d\mathbb{P}_n(t,\delta), \tag{5.6}$$

over the class $\mathcal{F}$ of all subdistribution functions on $\mathbb{R}$. Thus, $\tilde{F}_k$ uses only the kth entry of the Δ-vector. We see that the maximum marginal likelihood estimator splits the estimation problem into K well known univariate current status problems. Therefore, its computation and asymptotic theory follow straightforwardly from known results on current status data. But this simplification comes at a cost. For example, the constraint $\sum_{k=1}^{K} F_k(t) \leq 1$ may be violated (see Exercise 5.5).

To illustrate the estimator $\tilde{F}$ (and the ML estimator $\hat{F}$), we use the two data sets represented in Figure 5.2. These data sets are generated as follows. The number of risks is $K = 2$ and T is independent of (X, Y). The distribution function of T is $G(t) = P(T \leq t) = 1 - \exp(-t)$ on $[0, \infty)$ and Y and $X|Y$ are distributed as follows ($t \geq 0$): $P(Y = k) = k/3$ and $P(X \leq t | Y = k) = 1 - \exp(-kt)$ for $k = 1, 2$. This yields

$$F_k(t) = (k/3)\{1 - \exp(-kt)\} \text{ for } k = 1, 2. \tag{5.7}$$

Using the reduced datasets, we obtain the estimators $\tilde{F}_1$ for F_1 and $\tilde{F}_2$ for F_2 using the characterization of Lemma 2.3. Figure 5.3 shows the estimates based on the datasets of size

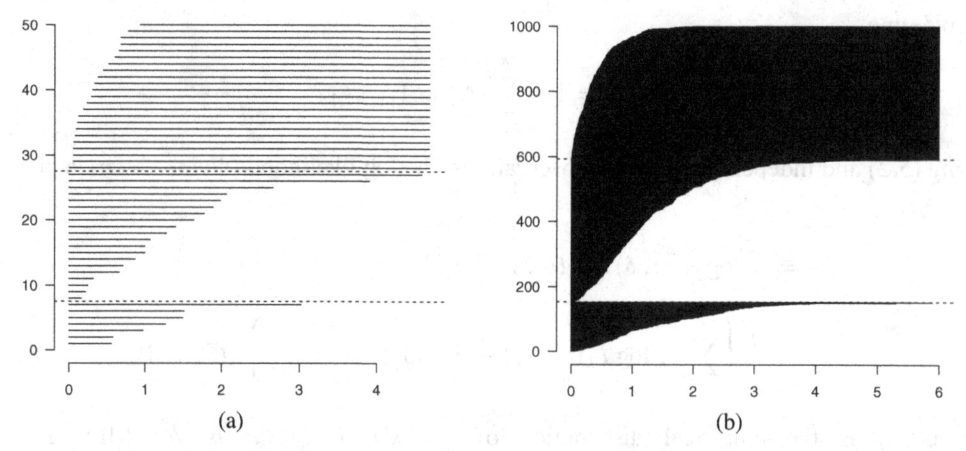

Figure 5.2 Graphical representation of a data set of size $n = 50$ and $n = 1000$ respectively, with $K = 2$ competing risks. The lines below the lowest dashed line indicate the intervals containing the X_is corresponding to observed T_is with $\Delta_1^i = 1$ (so with risk variable $Y = 1$). The lines, then, for $Y = 2$ ($\Delta_2^i = 1$), are given between the two dashed lines. The lines above the highest dashed line indicate the intervals containing those X_is with $\Delta_1^i = \Delta_2^i = 0$.

50 and 1,000. Note that indeed for both data sets one of the subdistribution functions is estimated by a genuine distribution function (implying in this case that the range constraint on F_X is violated by $\tilde{F}_X$).

We now return to the MLE. This estimator was first defined in Hudgens et al., 2001. To understand its form, let $F = (F_1, \ldots, F_K) \in \mathcal{F}_K$, where $\mathcal{F}_K$ is as defined in (5.5). Since only values of the subdistribution functions F_k at the observation times appear in the log likelihood ℓ, it makes sense to only estimate the subdistribution functions at these values. In

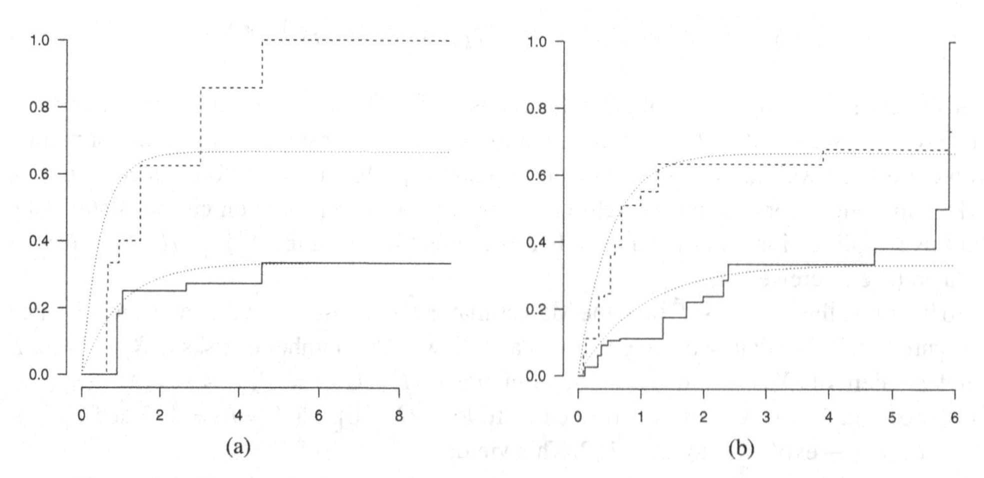

Figure 5.3 The marginal ML estimators based on two marginal data sets obtained from Figure 5.2. The solid line is $\tilde{F}_1$ and the dashed function the estimate $\tilde{F}_2$ as maximizer of (5.6) over all subdistribution functions. The dotted functions are the underlying subdistribution functions F_1 and F_2 given in (5.7).

doing so, optimization problem (5.4) reduces to a finite dimensional optimization problem. Its solution exists by corollary 38.10 in Zeidler, 1985.

Defining the sets $\mathcal{T}_k$ by

$$\mathcal{T}_k = \{T_i, i = 1, \ldots, n : \Delta_k^i + \Delta_{K+1}^i > 0\} \cup \{T_{(n)}\} \tag{5.8}$$

for $1 \le k \le K + 1$, the following lemma is proved in Groeneboom et al., 2008a.

Lemma 5.1 *For each $k = 1, \ldots, K + 1$, $\hat{F}_k(t)$ is unique at $t \in \mathcal{T}_k$. Moreover, $\hat{F}_k(\infty)$ is unique if and only if $\Delta_{K+1}^i = 0$ for all observations with $T_i = T_{(n)}$.*

See also Exercise 5.6. Comparing the optimization problem for the MLE with that of the marginal estimators, we note the following differences:

(a) The objective function ℓ for the MLE contains a term $1 - F_+$, involving the sum of the subdistribution functions, while the objective function $\tilde{\ell}(F) = \sum_k \ell_k$ for the marginal estimator $\tilde{F}$ only contains the individual components.
(b) The set $\mathcal{F}_K$ for the MLE contains the constraint $F_+ \le 1$, while the set $\mathcal{F}$ for the components of $\tilde{F}$ only involves range constraints on the individual components.

It turns out that the MLE also has an interpretation via the derivatives of greatest convex minorants. Since in practice there often will be ties, we denote the log likelihood by

$$\ell(F) = \sum_{i=1}^{p} \left\{ \sum_{k=1}^{K} N_{ik} \log F_k(T_{(i)}) + N_{i,K+1} \log\left(1 - F_+(T_{(i)})\right) \right\}, \tag{5.9}$$

where we assume that there are p strictly different order statistics $T_{(i)}$ in the sample that has a total of n observations, and where we assume that there are N_{ik} observations with $\Delta_k = 1$, $k = 1, \ldots, K + 1$, at the ith order statistic.

Then the MLE $\hat{F}_{nk}$, $k = 1, \ldots, K$, is the greatest convex minorant of the (self-induced) cusum diagram with points $P_0 = (0, 0)$ and

$$P_i = \left(n^{-1} \sum_{j=1}^{i} \frac{N_{j,K+1}}{1 - \hat{F}_{n,+}(T_{(j)})} + \lambda_n 1_{\{i=p\}}, n^{-1} \sum_{j=1}^{i} N_{jk} \right), \qquad i = 1, \ldots, p,$$

where

$$\lambda_n = 1 - n^{-1} \sum_{j=1}^{p} \frac{N_{K+1,i}}{1 - \hat{F}_{n,+}(T_{(j)})}.$$

This means that the (self-induced) cusum diagrams are given by

$$\left(\int_{u \le t} \frac{\delta_{K+1}}{1 - \hat{F}_{n+}(u)}\, d\mathbb{P}_n(u, \delta) + \lambda_n 1_{\{u = T_{(p)}\}}, \quad \int_{t_k \le t} \delta_k \, d\mathbb{P}_n(u, \delta) \right), \quad t \ge 0,\ k = 1, \ldots, K, \tag{5.10}$$

where λ_n is given by

$$\lambda_n = 1 - \int \frac{\delta_{K+1}}{1 - \hat{F}_{n+}(u)}\, d\mathbb{P}_n(u, \delta) \ge 0, \tag{5.11}$$

and $T_{(p)}$ is the largest of the strictly different order statistics (see earlier).

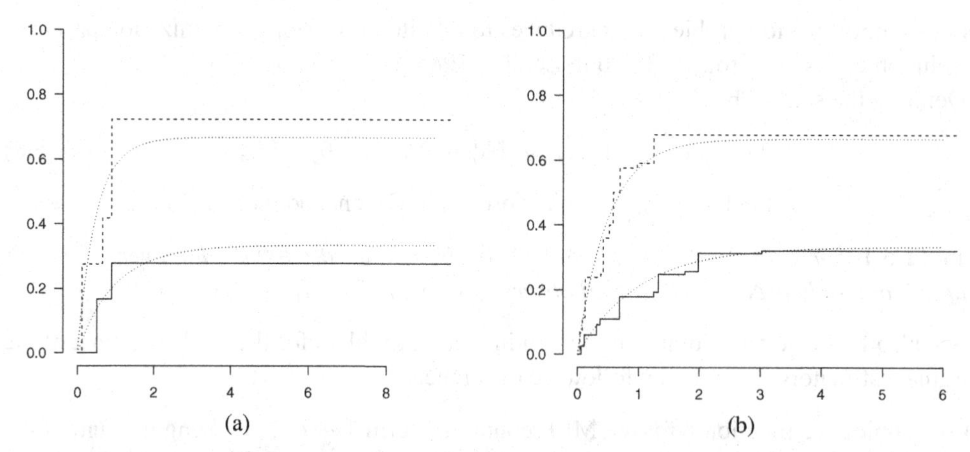

Figure 5.4 The ML estimators based on the data represented in Figure 5.1. The solid function is $\hat{F}_1$, the dashed function $\hat{F}_2$. The dotted functions are the underlying distribution functions F_1 and F_2 given in (5.7).

This convex minorant characterization is derived from corollary 2.10 in Groeneboom et al., 2008a (Exercise 5.4), which is given as follows.

Lemma 5.2 (Corollary 2.10, Groeneboom et al., 2008a) *Let λ_n be given by (5.11), then $\hat{F}_n = (\hat{F}_{n1}, \ldots, \hat{F}_{nK})$ is an MLE if for all $k = 1, \ldots, K$ and each point T_i such that $\Delta_{ik} = 1$*

$$\int_{u\in[T_i,s)} \left\{ \frac{\delta_k}{\hat{F}_{nk}} - \frac{\delta_{K+1}}{1-\hat{F}_{n+}(u)} \right\} d\mathbb{P}_n(u,\delta) \geq \lambda_n 1_{[\tau_{nk},s)}(T_{(p)}),\ s \in \mathbb{R},$$

where equality holds if T_i is a point of increase of $\hat{F}_{nk}$ and where $T_{(p)}$ is the largest of the strictly ordered order statistics.

Algorithms for computing the MLE are discussed in Section 7.5. The more complicated objective function for the MLE poses new challenges in the derivation of the local rate of convergence of the MLE. Moreover, it gives rise to a new limiting process for the local limiting distribution of the MLE, see Groeneboom et al., 2008b. The constraint $F_+ \leq 1$ on the space over which we maximize is important for smaller sample sizes, but its effect vanishes asymptotically. Figure 5.4 shows the ML estimators based on the data of Figure 5.1. The estimators are computed using the R-package `MLEcens`.

In a simulation study in Groeneboom et al., 2008a, properties of the MLE and the marginal estimator are compared empirically. The MLE and the marginal estimator seem to behave in a similar way for small values of t, but tend to diverge for larger values of t. Furthermore, the marginal estimator often violates the constraint for the marginal distribution of X, that $\sum_{k=1}^{K} \tilde{F}_k(x) \leq 1$.

We note that for the data sets of Figure 5.1 both $\hat{F}_1 + \hat{F}_2$ and $\tilde{F}_1 + \tilde{F}_2$ provide estimators for the overall (marginal) failure time distribution F_X of X. A third estimator for this distribution is given by the MLE for the reduced current status data $(T, \Delta_1 + \Delta_2)$, ignoring information on the failure causes. These three estimators are typically not the same; see Figure 5.5.

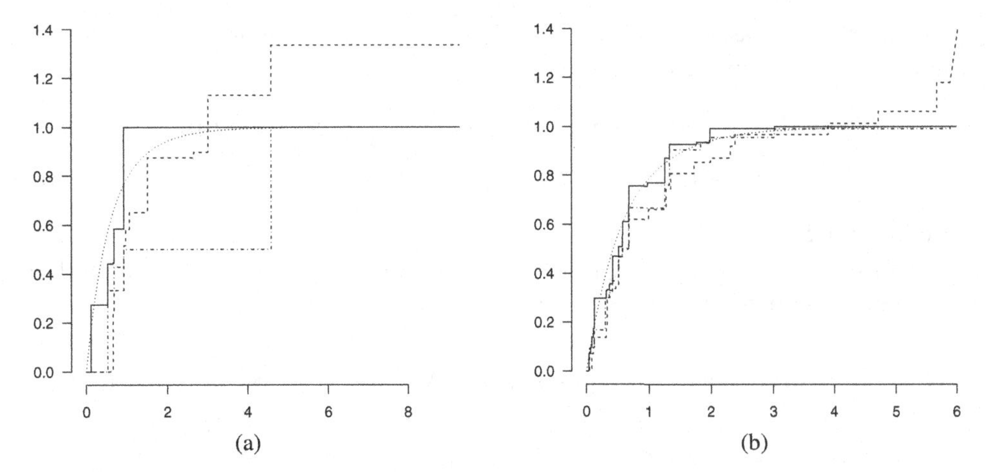

Figure 5.5 Various estimators of the marginal distribution function F_X based on the data of Figure 5.1: the ML estimate $\hat{F}_1 + \hat{F}_2$ (solid), the Jewel estimate $\tilde{F}_1 + \tilde{F}_2$ (dashed) and the ML estimate based on current status data $(T, \Delta_1 + \Delta_2)$ without taking into account the information on the risk causing the various failures (dash-dotted). The dotted function is the underlying distribution function $F_1 + F_2$, where F_1 and F_2 are given in (5.7).

5.2 Bivariate Interval Censoring

In Section 2.3 the current status model is introduced and in Section 4.7 the more general case 2 interval censoring model is introduced. Bivariate extensions of these models are considered in this section. We start with the bivariate current status model. The random vector of interest is (X, Y), assumed to have an unknown (joint) distribution function F. This random vector cannot be observed. Instead, a bivariate inspection time (T_1, T_2), independent of (X, Y) and having joint distribution function G, is observed, together with the information in which of the quadrants (relative to (T_1, T_2)) (X, Y) is located. More precisely, the observation is the quadruple $(T_1, T_2, \Delta_1, \Delta_2)$, where

$$\Delta_1 = 1_{\{X \le T_1\}}, \quad \Delta_2 = 1_{\{Y \le T_2\}}. \tag{5.12}$$

See also Figure 5.6, indicating for given values of (T_1, T_2) the possible regions for (X, Y) given the information in (Δ_1, Δ_2). Defining the measure $\mu_G \times \#$ on $(0, \infty)^2 \times \{0, 1\}^2$, where

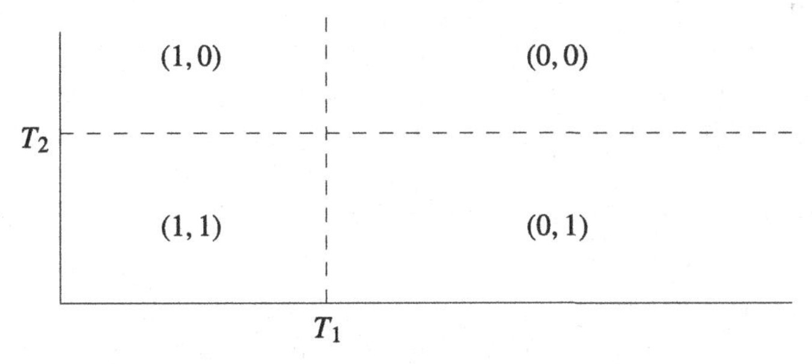

Figure 5.6 Given (T_1, T_2), the four quadrants that can possibly contain (X, Y) are indicated by the possible observable indicator vectors (Δ_1, Δ_2).

μ_G is the probability measure with distribution function G and # counting measure on $\{0,1\}^2$, the density of this random vector is given by

$$p_F(t_1,t_2,\delta_1,\delta_2) = F(t_1,t_2)^{\delta_1\delta_2}(F(t_1,\infty)-F(t_1,t_2))^{\delta_1(1-\delta_2)}(F(\infty,t_2)-F(t_1,t_2))^{(1-\delta_1)\delta_2}$$
$$\times(1-F(\infty,t_2)-F(t_1,\infty)+F(t_1,t_2))^{(1-\delta_1)(1-\delta_2)}. \tag{5.13}$$

See also Exercise 5.7.

Now consider having an i.i.d. sample $(T_{i1},T_{i2},\Delta_{i1},\Delta_{i2})$, $1\le i\le n$ of observations from the bivariate current status model. The log likelihood of F is then given by

$$\ell(F)=\sum_{i=1}^n\{\Delta_{i1}\Delta_{i2}\log F(T_{i1},T_{i2})+\Delta_{i1}(1-\Delta_{i2})\log\{F(T_{i1},\infty)-F(T_{i1},T_{i2})\}$$
$$+(1-\Delta_{i1})\Delta_{i2}\log\{F(\infty,T_{i2})-F(T_{i1},T_{i2})\}$$
$$+(1-\Delta_{i1})(1-\Delta_{i2})\log\{1-F(\infty,T_{i2})-F(T_{i1},\infty)+F(T_{i1},T_{i12})\}\}.$$

A maximum likelihood estimator $\hat F_n$ maximizes ℓ over all bivariate distribution functions F. Another formulation is that $\hat F_n$ has to maximize

$$\frac{1}{n}\ell(F)=\int\delta_1\delta_2\log F(u,v)\,d\,\mathbb{P}_n+\int\delta_1(1-\delta_2)\log\{F_1(u)-F(u,v)\}\,d\,\mathbb{P}_n$$
$$+\int(1-\delta_1)\delta_2\log\{F_2(v)-F(u,v)\}\,d\,\mathbb{P}_n$$
$$+\int(1-\delta_1)(1-\delta_2)\log\{1-F_1(u)-F_2(v)+F(u,v)\}\,d\mathbb{P}_n \tag{5.14}$$

over F, where F_1 and F_2 are the first and second marginal distribution functions of F, respectively, and $\mathbb{P}_n$ is the empirical distribution function of the observations $(\delta_{i1},\delta_{i2},u_i,v_i)$. Unlike the univariate current status model, a simple construction of the MLE via a greatest convex minorant is not known. What can be derived is a characterization of the MLE via (in-) equalities. Uniform consistency can be proved for this estimator, if the underlying distribution has compact support, and the distribution of the hidden variable is absolutely continuous with respect to the observation distribution. The reason that the MLE is consistent for the bivariate current status model, and inconsistent for the bivariate right censoring model (where one has, in fact, more information), is that in the latter case the MLE only uses the information on lines, if the observation is uncensored in one coordinate, and does not use the surrounding information for the uncensored coordinate. One would need information on the conditional distribution on these lines to distribute mass in such a way that a consistent estimate would result, but this conditional distribution is not available, since it is part of the estimation problem.

In this section, we look for a ML estimator over the class of bivariate distribution functions that can be represented as

$$F=\sum_{j=1}^m\alpha_j 1_{[\tau_j,\infty)},\quad \sum_{j=1}^m\alpha_j\le1,\ \alpha_j\ge0,\quad 1\le j\le m=m_n, \tag{5.15}$$

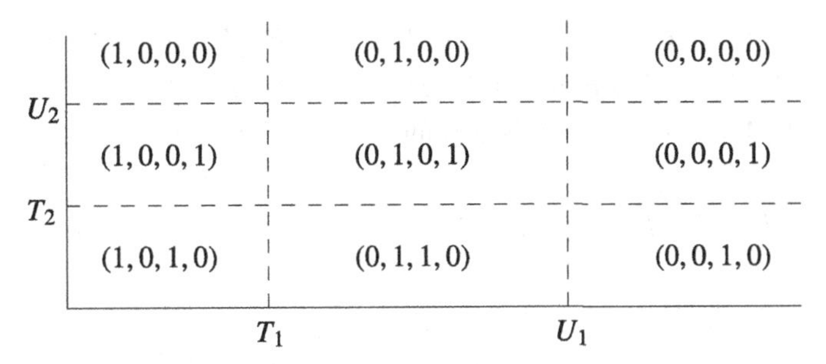

Figure 5.7 The nine regions that possibly contain (X, Y) indexed by the observable indicator vector $(\Delta_1^{(1)}, \Delta_2^{(1)}, \Delta_1^{(2)}, \Delta_2^{(2)})$.

where we denote by $[\tau_j, \infty)$ the quadrant $[t_{j1}, \infty) \times [t_{j2}, \infty)$, where $\tau_j = (t_{j1}, t_{j2})$. The optmality conditions are:

$$\int_{[\mathbf{t},\infty)} \frac{\delta_1 \delta_2}{F(u, v)}\, d\,\mathbb{P}_n + \int_{[t_1,\infty)\times[0,t_2)} \frac{\delta_1(1-\delta_2)}{F_1(u) - F(u, v)}\, d\,\mathbb{P}_n$$
$$+ \int_{[0,t_1)\times[t_2,\infty)} \frac{(1-\delta_1)\delta_2}{F_2(v) - F(u, v)}\, d\,\mathbb{P}_n$$
$$+ \int_{[0,t_1)\times[0,t_2)} \frac{(1-\delta_1)(1-\delta_2)}{1 - F_1(u) - F_2(v) + F(u, v)}\, d\,\mathbb{P}_n \le 1 \qquad (5.16)$$

for all $\mathbf{t} = (t_1, t_2) \in \mathbb{R}^2$. We must have equality in (5.16), if $\mathbf{t} = \tau_j$, $j = 1, \ldots, m_n$, where the rectangles $[\tau_j, \infty)$ are the generators of the solution, i.e., if the corresponding $\alpha_j > 0$ (see Exercise 5.8).

For the case 2 type bivariate interval censoring model, still the random vector of interest is given by $(X, Y) \sim F$, but now a typical observation is given by the following vector:

$$\left(T_1, U_1, T_2, U_2, \Delta_1^{(1)}, \Delta_2^{(1)}, \Delta_1^{(2)}, \Delta_2^{(2)}\right).$$

Here, independent of (X, Y), $(T_1, U_1, T_2, U_2) \sim G$ where G is a distribution function on $(0, \infty)^4$ such that under G, $T_1 < U_1$ and $T_2 < U_2$ with probability one. Moreover,

$$\Delta_1^{(1)} = 1_{\{X_1 \le T_1\}}, \quad \Delta_1^{(2)} = 1_{\{T_1 < X_1 \le U_1\}}, \quad \Delta_2^{(1)} = 1_{\{X_2 \le T_2\}}, \quad \Delta_2^{(2)} = 1_{\{T_2 < X_2 \le U_2\}}.$$

Given (T_1, U_1, T_2, U_2), Figure 5.7 shows the various possibilities for $(\Delta_1^{(1)}, \Delta_2^{(1)}, \Delta_1^{(2)}, \Delta_2^{(2)})$ indicating the region containing (X, Y). With respect to the measure $\mu_G \times \#$ on $\mathbb{R}^4 \times \{0, 1\}^4$, where μ_G is the distribution on $(0, \infty)^4$ with distribution function G and # counting measure on $\{0, 1\}^4$, this observation has density

$$p_F(t_1, u_1, t_2, u_2, \delta_1^{(1)}, \delta_2^{(1)}, \delta_1^{(2)}, \delta_2^{(2)}) = P_F\left((X, Y) \in A_{(t_1, u_1, t_2, u_2, \delta_1^{(1)}, \delta_2^{(1)}, \delta_1^{(2)}, \delta_2^{(2)})}\right) \qquad (5.17)$$

where $A_{(t_1,u_1,t_2,u_2,\delta_1^{(1)},\delta_2^{(1)},\delta_1^{(2)},\delta_2^{(2)})}$ denotes the area defined by (t_1, u_1, t_2, u_2) and the indicator vector $(\delta_1^{(1)}, \delta_2^{(1)}, \delta_1^{(2)}, \delta_2^{(2)})$ using Figure 5.7 (see also Exercise 5.9).

Now consider an independent sample obtained in this way,

$$\left(T_{i1}, U_{i1}, T_{i2}, U_{i2}, \Delta_{i1}^{(1)}, \Delta_{i2}^{(1)}, \Delta_{i1}^{(2)}, \Delta_{i2}^{(2)}\right), \; i = 1, \ldots, n, \tag{5.18}$$

where

$$\Delta_{i1}^{(1)} = 1_{\{X_{i1} \le T_{i1}\}}, \quad \Delta_{i1}^{(2)} = 1_{\{T_{i1} < X_{i1} \le U_{i1}\}}, \quad \Delta_{i2}^{(1)} = 1_{\{X_{i2} \le T_{i2}\}}, \quad \Delta_{i2}^{(2)} = 1_{\{T_{i2} < X_{i2} \le U_{i2}\}}.$$

The log likelihood of F is then given by

$$\ell(F) = \int \log p_F(t_1, u_1, t_2, u_2, \delta_1^{(1)}, \delta_2^{(1)}, \delta_1^{(2)}, \delta_2^{(2)})\, d\mathbb{P}_n, \tag{5.19}$$

where p_F is given by (5.17) and $\mathbb{P}_n$ denotes the empirical distribution of the observations given in (5.18). Defining

$$\Delta_{i1}^{(3)} = 1 - \Delta_{i1}^{(1)} - \Delta_{i1}^{(2)}, \qquad \Delta_{i2}^{(3)} = 1 - \Delta_{i2}^{(1)} - \Delta_{i2}^{(2)},$$

and the corresponding generic values $\delta_i^{(j)}$ similarly, the measure $dV_n^{(ij)}$ can be defined by

$$dV_n^{(ij)} = \delta_1^{(i)}\delta_1^{(j)}\delta_2^{(i)}\delta_2^{(j)}\, d\mathbb{P}_n\left(t, u, v, w, \delta_1^{(i)}, \delta_2^{(i)}, \delta_1^{(j)}, \delta_2^{(j)}\right), \; 1 \le i, j \le 3,$$

and an MLE of F is then obtained by maximizing

$$\begin{aligned}
\ell(F) = &\int \log F(t, v)\, dV_n^{(11)} + \int_{t \le u,\, v \le w} \log\{F(u, w) - F(t, w) - F(u, v) + F(t, v)\}\, dV_n^{(22)} \\
&+ \int_{u \ge t} \log\{F(u, v) - F(t, v)\}\, dV_n^{(21)} + \int_{w \ge v} \log\{F(t, w) - F(t, v)\}\, dV_n^{(12)} \\
&+ \int_{u \ge t} \log\{F_1(u) - F_1(t) - F(u, w) + F(t, w)\}\, dV_n^{(23)} \\
&+ \int_{w \ge v} \log\{F_2(w) - F_2(v) - F(u, w) + F(u, v)\}\, dV_n^{(32)} \\
&+ \int_{w \ge v} \log\{F_1(t) - F(t, w)\}\, dV_n^{(13)} + \int_{u \ge t} \log\{F_2(v) - F(u, v)\}\, dV_n^{(31)} \\
&+ \int \log\{1 - F_1(u) - F_2(w) + F(u, w)\}\, dV_n^{(33)}
\end{aligned} \tag{5.20}$$

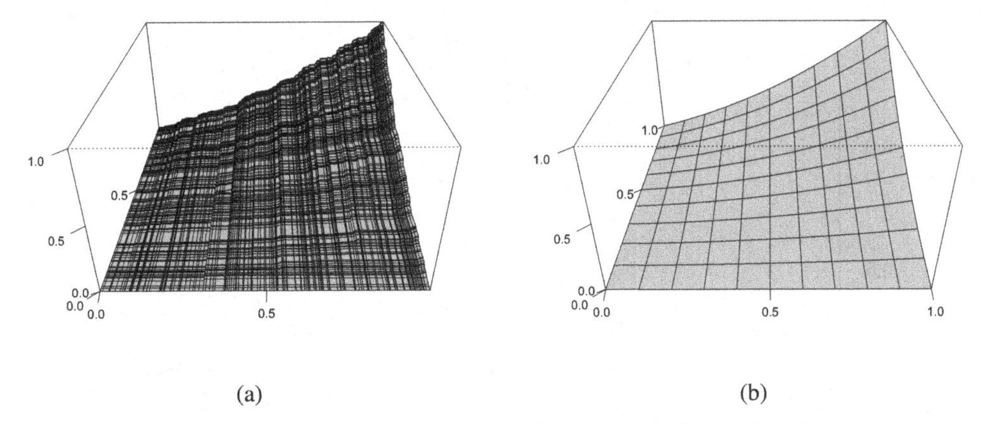

(a) (b)

Figure 5.8 Sieved MLE (a) and underlying distribution function $F_0(x, y) = \frac{1}{2}xy(x + y)$ (b) for a sample of size $n = 5000$ with a uniform observation distribution on $[0, 1]^2$.

over bivariate distribution functions F. The Fenchel optimality conditions become

$$
\begin{aligned}
&\int_{t \ge x,\, v \ge y} \frac{dV_n^{(11)}}{F(t, v)} + \int_{t < x \le u,\, v < y \le w} \frac{dV_n^{(22)}}{F(u, w) - F(t, w) - F(u, v) + F(t, v)} \\
&\quad + \int_{t < x \le u,\, v \ge y} \frac{dV_n^{(21)}}{F(u, v) - F(t, v)} + \int_{t \ge x,\, v < y \le w} \frac{dV_n^{(12)}}{F(t, w) - F(t, v)} \\
&\quad + \int_{t < x \le u,\, w \ge y} \frac{dV_n^{(23)}}{F_1(u) - F_1(t) - F(u, w) + F(t, w)} \\
&\quad + \int_{u \ge x,\, v < y \le w} \frac{dV_n^{(32)}}{F_2(w) - F_2(v) - F(u, w) + F(u, v)} \\
&\quad + \int_{t < x \le u,\, v \ge y} \frac{dV_n^{(13)}}{F_1(t) - F(t, w)} + \int_{t \ge x,\, v < y \le w} \frac{dV_n^{(31)}}{F_2(v) - F(u, v)} \\
&\quad + \int_{u < x,\, w < y} \frac{dV_n^{(33)}}{1 - F_1(u) - F_2(w) + F(u, w)} \le 1, \qquad (5.21)
\end{aligned}
$$

with equality if (x, y) is a point of mass of dF. The MLE is consistent; see, e.g., Song, 2001. In Section 8.7 a maximum smoothed likelihood estimator (MSLE) will be considered, obtained by maximizing the smoothed log likelihood, which arises if we smooth the measures $dV_n^{(ij)}$, $1 \le i, j \le 3$.

In Section 7.4, algorithmic issues for this model are discussed, in particular also for a sieved version of the ML estimator, where inequalities (5.21) are only attained at a fine grid of (x, y)-values. Figure 5.8 shows this approximate ML estimator together with a picture of the real underlying distribution function for a sample of size $n = 5000$ and underlying distribution with density $f(x, y) = x + y$ on the unit square.

A picture of the levels of the MLE and the underlying distribution function is shown in Figure 5.9. The estimate of the MLE seems pretty good. There is a conjecture that it converges at rate $n^{-1/3}$, just like the univariate MLE (in the separated case), but this conjecture has not

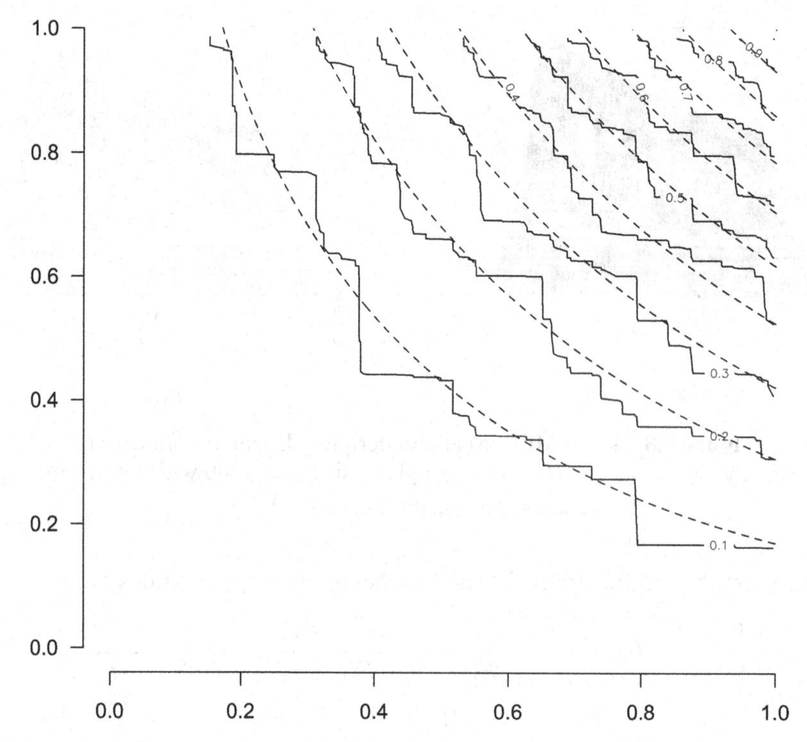

Figure 5.9 The level plots of the MLE and the underlying distribution function (dashed), for the same sample as in Figure 5.8.

been proved. We will say more about this issue in the chapter on asymptotics. A picture of the marginal distribution functions of the MLE are shown in Figure 5.10. In Section 7.4 also a real bivariate interval censored data set from Betensky and Finkelstein, 1999, is analyzed, where X denotes the time the opportunistic cytomegalovirus (CMV) appears in the blood of a patient and Y the time it appears in the urine.

5.3 Current Status with Continuous Marks

Consider a random vector (X, Y) in $[0, \infty)^2$ with distribution function F. Instead of observing this vector directly, we observe an inspection time T, independent of (X, Y), together with the indicator $\Delta = 1_{\{X \le T\}}$. Moreover, in case $\Delta = 1$, also the associated Y is observed. If $\Delta = 0$, Y is not observed. One can think of X as survival time and Y as associated continuous mark that is only observed if the patient has already died at time T (possibly some quantity that has been measured postmortem). An application where observations can be modeled by this model is the HIV vaccine trial studied by Hudgens et al., 2007. In these HIV vaccine trials, participants were injected with a vaccine and tested for infection with HIV during several follow ups. Efficacy of the vaccine might depend on the genetic sequence of the exposing virus, and the so-called viral distance Y between the DNA of the infecting virus and the virus in the vaccine could be considered as a continuous mark variable. In general, the time X to HIV infection is subject to interval censoring case k, with current status censoring (or interval censoring case 1) as a special instance. The current status continuous marks model

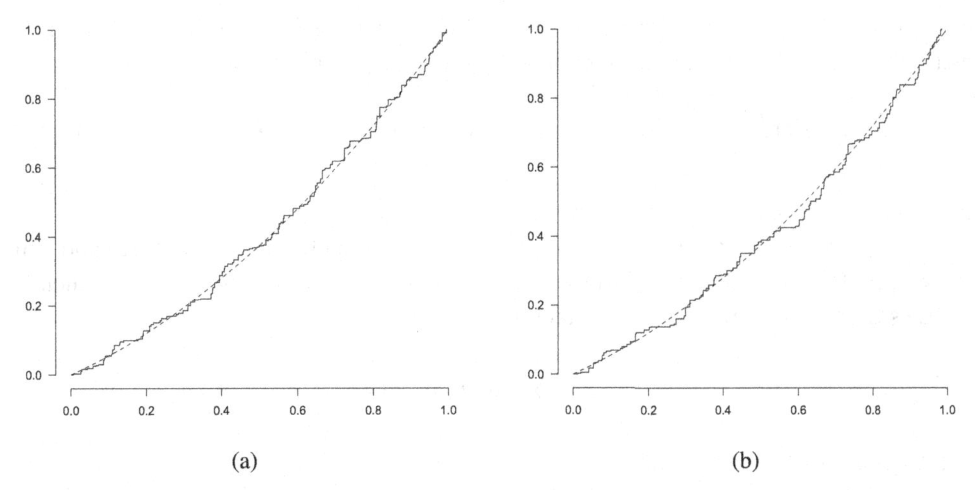

Figure 5.10 First (a) and second (b) marginals of the MLE (solid curves) for the same sample as in Figure 5.8. The real underlying marginal distribution functions are given by the dashed curves.

is clearly related to the competing risk with current status observations in the sense that then Y has a distribution concentrated on a finite number of labels (death causes).

Figure 5.11 shows the possible observation sets, either $(0, T] \times \{Y\}$ or $(T, \infty) \times [0, \infty)$. Now suppose (X, Y) has (joint) density f on $(0, \infty)^2$. Then $P_f(Y = 0) = 0$ and the observable data can be summarized as $(T, \Delta \cdot Y)$. Indeed, if the second component is zero, this means $\Delta = 0$ almost surely and if it is nonzero, it equals Y almost surely. The random vector $(T, \Delta \cdot Y)$ takes values in $(0, \infty) \times [0, \infty)$, where the second component has a point mass at zero. To derive the density of $(T, \Delta \cdot Y)$, consider the following dominating measure on $(0, \infty) \times [0, \infty)$: for B a Borel set in $(0, \infty) \times [0, \infty)$,

$$\mu(B) = \lambda_2(B) + \lambda_1(B_0) \text{ where } B_0 = \{x \in (0, \infty) : (x, 0) \in B\}$$

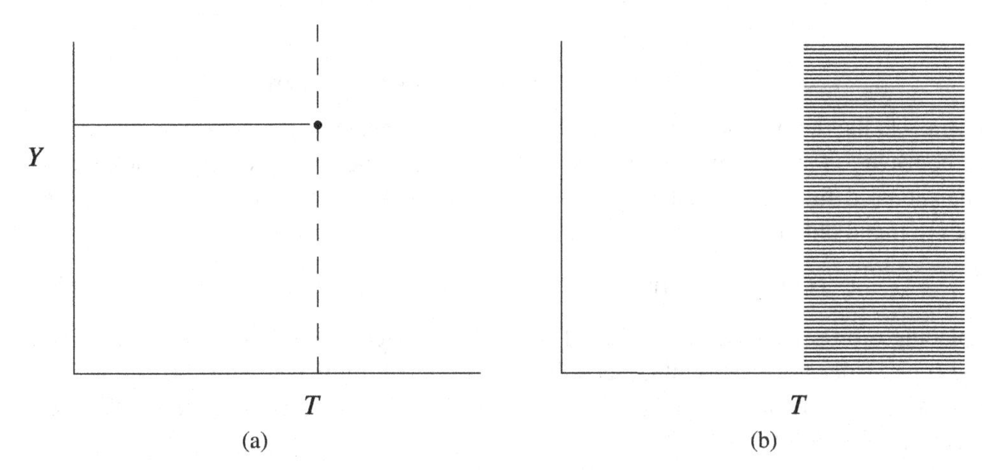

Figure 5.11 The two possible sets containing (X, Y) depending on whether $\Delta = 1$ (a) or $\Delta = 0$ (b).

and λ_i denotes Lebesgue measure on $[0, \infty)^i$. Now, denoting by g the density of T, the density of $(T, \Delta \cdot Y)$ with respect to dominating measure μ is given by

$$p_f(t, z) = 1_{\{0\}}(z)g(t)\int_{x=t}^{\infty}\int_{y=0}^{\infty} f(x, y)\,dy\,dx + 1_{(0,\infty)}(z)g(t)\int_0^t f(x, z)\,dx \text{ on } (0, \infty)\times[0,\infty). \tag{5.22}$$

See Exercise 5.10 for an alternative expression.

Now suppose we have a sample $(t_1, z_1), (t_2, z_2), \ldots, (t_n, z_n)$ from density (5.22) and consider the problem of estimating f from these. A natural estimator to define would then be the ML estimator, maximizing the log likelihood function

$$\ell(f) = \sum_{i=1}^{n} \log p_f(t_i, z_i)$$

over a large class of joint densities f.

The MLE in the current status continuous marks model is inconsistent and Maathuis and Wellner, 2008, obtain a consistent estimator by discretizing the mark variable to K levels. The resulting observations can then be viewed as observations from the current status K-competing risk model. Apart from consistency, global and local asymptotic distribution properties for the MLE in the latter model were proved in Groeneboom et al., 2008a, and Groeneboom et al., 2008b. Asymptotic results for $K \to \infty$ as $n \to \infty$ are not yet known.

In Groeneboom, Jongbloed and Witte, 2012, an alternative method is considered, the approach via maximum smoothed likelihood. This is a natural approach, as in other models where MLEs are inconsistent (Section 4.5), MSLEs (maximum smoothed likelihood estimators) also provide consistent estimators. The basic idea is to replace the empirical distribution function in the log-likelihood by a smooth estimator. It is proved in Groeneboom, Jongbloed and Witte, 2012, that, for a histogram-type smoothing of the observation distribution, the resulting MSLE is consistent under certain regularity conditions, so avoids the inconsistency of the raw MLE. A rather fast converging algorithm is provided for computing the MSLEs in the appendix of the cited paper. More on maximum smoothed likelihood estimation can be found in Chapter 8.

5.4 Multivariate Log Concave Densities

Contrary to the definition of monotonicity in higher dimensions, the definition of convexity and concavity is straightforward and unambiguous in higher dimensions. In fact, the one-dimensional definition immediately generalizes to saying that a function $f : \mathbb{R}^d \to \mathbb{R}$ is concave if and only if, for all $x, y \in \mathbb{R}^d$,

$$f(\lambda x + (1-\lambda)y) \geq \lambda f(x) + (1-\lambda)f(y) \text{ for all } \lambda \in [0, 1]. \tag{5.23}$$

Replacing $\geq$ by $\leq$ yields the definition of convexity. In Section 4.4 the problem of estimating a log concave density on $\mathbb{R}$ is introduced. In this section the multivariate problem is discussed.

Consider a probability density g on $\mathbb{R}^d$ such that $f = \log g$ is concave on $\mathbb{R}^d$, i.e., it satisfies (5.23) for all $x, y \in \mathbb{R}^d$ and

$$\int_{\mathbb{R}^d} g(x)\,dx = \int_{\mathbb{R}^d} \exp f(x)\,dx = 1.$$

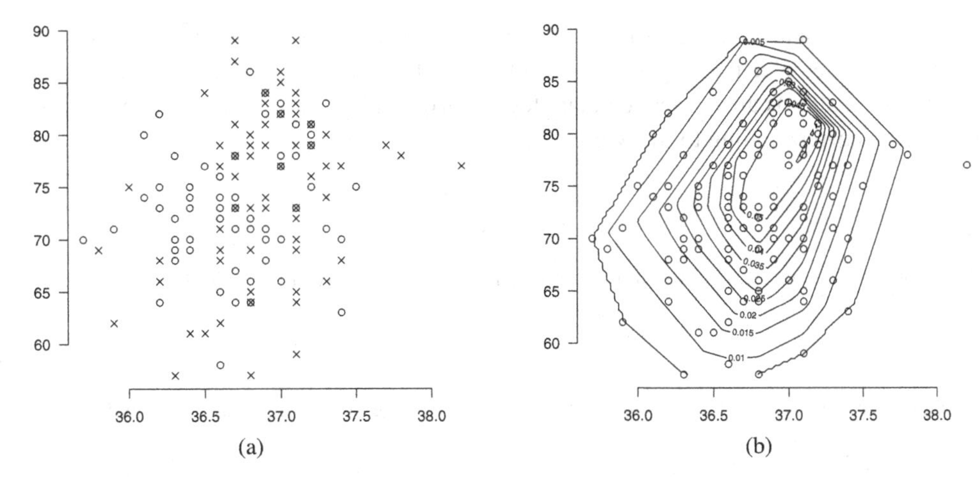

Figure 5.12 (a) Scatter plot of the body temperature (°C) and heart rate (beats per minute) data. The circles indicate measurements for men, the crosses measurements for women. (b) Level sets for the ML estimator of the log concave density based on all data.

This density g is then called log concave. There are many examples of log concave densities on $\mathbb{R}^d$, e.g., the class of Gaussian densities (see Exercise 5.11).

Now let $z_1, \ldots, z_n$ be the realization of a sample of size n from a log concave probability density on $\mathbb{R}^d$. Define the log likelihood by

$$\ell(g) = \frac{1}{n} \sum_{i=1}^{n} \log g(z_i) - \int_{\mathbb{R}^d} g(z)\, dz. \tag{5.24}$$

The ML estimator can be defined as maximizer of this function over all log concave densities on $\mathbb{R}^d$. From Exercise 5.14 it follows that whenever this ML estimator is well defined, it can be defined as maximizer of ℓ over all log concave functions (rather than only density functions) on $\mathbb{R}^d$. Another property of the ML estimator, if it exists, is that its support equals the convex hull C of the observed z_is in $\mathbb{R}^d$ (see Exercise 5.16). Moreover, it can be seen that with probability one the set C has positive Lebesgue measure in $\mathbb{R}^d$ as soon as $n > d$. Finally, another structural property of an ML estimator (if it exists) can be derived: log piecewise linearity. This follows from Exercise 5.17. Having noted all this, existence of the ML estimator can be established. Indeed, a candidate log concave function g is determined by its values taken at the observed data points $z_1, \ldots, z_n$, say $g_1, \ldots, g_n$. Moreover, log likelihood function (5.24) depends on these values continuously (see Exercise 5.18). In view of Exercise 5.19, maximization of ℓ may be restricted to a bounded subset of $\mathbb{R}^n$. Since a continuous function on a bounded subset of $\mathbb{R}^n$ attains its maximal value, existence of the ML estimator is guaranteed.

We illustrate the the ML estimator of a log concave function on $\mathbb{R}^2$ using a dataset studied in Schoemaker, 1996, derived from data reported in Mackowiak et al., 1992. For 65 men and 65 women the body temperature and heart rate are given. Figure 5.12 shows a scatter plot of the measurements.

Modeling the data to be a sample from a log concave density on $\mathbb{R}^2$, the same for both male and female, the R-package LogConcDEAD by Cule, Gramacy and Samworth (see Cule et al., 2009) can be used to obtain level sets of the ML estimator. Figure 5.12 shows the level sets corresponding to the ML estimator.

Exercises

5.1 Show that the conditional distribution, given T, of the vector given in (5.1) is

$$\text{Mult}(1, F_1(T), \cdots, F_K(T), 1 - F_X(T))).$$

5.2 Consider the following small dataset in the context of the competing risk model with current status observations:

$$\{(4, 0, 1), (3, 1, 0), (2, 0, 0), (5, 0, 1)\}.$$

Clearly, $n = 4$ and $K = 2$ in this example. Compute the estimators introduced in Section 5.3 based on these data.

5.3 In marginal log likelihood (5.6), $F(t)$ occurs for all observed values of t, for all k. This is somewhat peculiar, since for level $Y = 1$, the $F(T)$s corresponding to $\Delta_2 = 1$ (indicating that failure occurred before T, due to cause 2) enter the log likelihood *as if* it were known that the corresponding failure due to cause 1 happened after T, whereas it is only known that failure due to cause 1 happened (hypothetically because failure had already happened due to cause 2) after $X \leq T$. Construct a log likelihood that incorporates this information.

5.4 Derive, in the competing risk model, the convex minorant characterization of the MLE via a set of convex minorants of the cusum diagrams in (5.10) from Lemma 5.2.

5.5 Suppose in the competing risk current status model that the highest T observed corresponds to a value $\Delta_k = 1$ for some $k \in \{1, \ldots, K\}$. Argue that the naive estimate $\tilde{F}_k$ is then a genuine distribution function, with $\lim_{x\to\infty} \tilde{F}_k(x) = 1$. If there is also a T_i with corresponding $\Delta^i_j = 1$ for some $j \neq k$, argue that $\tilde{F}_j(t) > 0$ for $t \geq T_i$ and conclude that $\tilde{F}_X = \sum_\ell \tilde{F}_\ell$ will in general not satisfy the range constraint $\lim_{x\to\infty} \tilde{F}_X(x) = 1$.

5.6 Verify Lemma 5.1 for the case $K = 1$, the univariate current status model, e.g., using the characterization of the ML estimator in Lemma 2.3.

5.7 Show that the density of the observable vector in the bivariate current status model is given by (5.13).

5.8 Let $\hat{F}$ be a bivariate distribution function that can be represented as (5.15) and maximizes log likelihood (5.14) in the bivariate current status model over all bivariate distribution functions of that type. Show that $\hat{F}$ then necessarily satisfies inequalities (5.16).

5.9 For $t_1 < u_1$ and $t_2 < u_2$ given, show that

$$P_F\left((X, Y) \in A_{(t_1,u_1,t_2,u_2,0,1,0,0)}\right) = F(u_1, \infty) - F(u_1, u_2) - F(t_1, \infty) + F(t_1, u_2)$$

in (5.17).

5.10 Derive density (5.22) for the observable vector in the current status with continuous mark model and observe that this density can also be written as

$$p_f(t,z) = \begin{cases} g(t)(1 - F_X(t)) & \text{if } z = 0 \\ g(t)\frac{\partial}{\partial z}F(t,z) & \text{if } z > 0, \end{cases}$$

where F is the distribution function of (X, Y) corresponding with f and F_X the marginal distribution of X under f.

5.11 Let g be the density of a Gaussian random variable X in $\mathbb{R}^d$, with mean $\mu \in \mathbb{R}^d$ and nonsingular covariance matrix S. Show that g is log concave.

5.12 Let $\phi : \mathbb{R}^+ \to \mathbb{R}$ be concave and decreasing on $\mathbb{R}^+$. Define, for $x \in \mathbb{R}^d$, $f(x) = \phi(\|x\|)$. Show that f is concave on $\mathbb{R}^d$.

5.13 Let $g_1, \ldots, g_d$ be univariate log concave densities. The product density on $\mathbb{R}^d$ is given by

$$g(z_1, \ldots, z_d) = g_1(z_1) \cdots g_d(z_d).$$

Show that g is a log concave density on $\mathbb{R}^d$.

5.14 Suppose that g maximizes (5.24) over all log concave functions on $\mathbb{R}^d$. Show that then, necessarily, $\int_{\mathbb{R}^d} g(z)\,dz = 1$.

5.15 Let g be a log concave density on $\mathbb{R}^2$ such that its support has positive Lebesgue measure. Show that the convex hull C of a sample of size 3 from g has positive Lebesgue measure with probability one.

5.16 Let g be a log concave density on $\mathbb{R}^d$ and $z_1, \ldots, z_n \in \mathbb{R}^d$ be given. Denote by C the convex hull of $z_1, \ldots, z_n$ and define

$$\tilde{g}(z) = \frac{g(z)1_C(z)}{\int_C g(y)\,dy}.$$

For ℓ as defined in (5.24), argue that $\ell(\tilde{g}) \geq \ell(g)$. Moreover, argue that the inequality is strict in case $\int_C g(y)\,dy < 1$.

5.17 Let g be a log concave density on $\mathbb{R}^d$. Define $y_i = \log g(z_i)$ for $1 \leq i \leq n$ and

$$h(z) = \inf\left\{k(z)\colon k\colon \mathbb{R}^d \to \mathbb{R},\ \text{concave},\ k(z_i) \geq y_i,\ 1 \leq i \leq n\right\}.$$

Show that $\exp(h)$ is log concave and $\ell(\exp(h)) \geq \ell(g)$ where ℓ is as defined in (5.24).

5.18 Show that log likelihood function (5.24) depends continuously on the values $g(z_1), \ldots, g(z_n)$ of the log concave density g evaluated at the data.

5.19 Assume that the convex hull C of the points $z_1, \ldots, z_n$ has strictly positive volume (something that happens with probability one if the underlying density has support on $\mathbb{R}^d$ with positive Lebesgue measure and $n > d$).

a) Take $g(z) = 1_C(z)$ and observe that $\ell(g) > -\infty$.

b) Let g be a log concave function on C such that $\log g$ is piecewise linear in the spirit of Exercise 5.17. Argue that there exists a $\delta > 0$ such that $g(z_i) > c > 0$ implies that $g(z_i) > c/2$ on a subset of C with Lebesgue measure greater than or equal to δ.

c) Combine (a) and (b) to show that in maximizing (5.24) over all log concave functions, one can determine a value M such that maximization may be restricted to those densities g for which $g(z_i) \leq M$ for all i.

Bibliographic Remarks

The problem of nonparametric estimation in the competing risk problem with current status observations has been studied by Hudgens et al., 2001, Jewell et al., 2003, Jewell and Kalbfleisch, 2004, and Groeneboom et al., 2008a,b. These papers introduce various nonparametric estimators, provide algorithms to compute them and show simulation studies that compare them. Local asymptotic theory for the nonparametric maximum likelihood estimator (MLE) is developed in Groeneboom et al., 2008a,b, and compared there with the naive estimator of Jewell et al., 2003. Song, 2001, discusses minimax lower bounds for the estimation rate of the MLE and consistency of the MLE in the bivariate interval censoring model.

The (consistent) MLE for bivariate current status data is discussed in Section 5.2. In a clearly related model, where bivariate right-censored data are available, the MLE is not consistent, see Tsai et al., 1986. Modifications of the MLE to ensure consistency have been discussed by several authors, see, e.g., Dabrowska, 1988 and Van der Laan, 1996.

In the current status model or the interval censoring model with more observation times for the hidden variable, one only has information on the interval to which the hidden variable belongs, and the MLE therefore automatically uses the surrounding information. For this reason a reduction of the bivariate right-censoring model to the interval censoring model has been proposed to obtain consistent estimators of the bivariate distribution function. In this way the information of a whole set of lines is combined.

In Section 5.3 the continuous marks problem is considered with interval censored event time X. If instead X is subject to right-censoring and Y is only observed if X is observed, Huang and Louis, 1998, study a nonparametric estimator of the bivariate distribution function F_0. This estimator is uniformly strongly consistent and asymptotically normally distributed. Hudgens et al., 2007, study several estimators for the joint distribution of a survival time and a continuous mark variable, when the survival time is interval censored and the mark variable is possibly missing. In this paper a computational algorithm for the MLE is proposed, but since the MLE is inconsistent (Maathuis and Wellner, 2008), one would be inclined to recommend not to use this estimator.

The MLE for a multidimensional log-concave density is studied in Cule et al., 2010, Cule and Samworth, 2010, and Schuhmacher et al., 2011.

6

Lower Bounds on Estimation Rates

Consider the problem of estimating the distribution function F in the exponential deconvolution model described in Example 2.4. In that case, the sampling density is known to belong to the class

$$\mathcal{G} = \left\{ g : g(z) = e^{-z} \int_0^z e^x \, dF(x),\ z \geq 0, \text{ where } F \text{ is a distribution function on } [0, \infty) \right\}.$$

If one wants to estimate the distribution function F at a fixed point x, a natural functional in terms of F, one is interested in the following (rather unnatural) functional of the sampling density g:

$$Tg = T(g) = g(x) + \int_0^x g(y) \, dy.$$

See (2.20). The problem of estimating Tg based on a sample from the density g that is known to belong to a class $\mathcal{G}$ of densities possesses an intrinsic difficulty. In this chapter we consider the problem of quantifying this complexity in some detail.

Let $\mathcal{G}$ be a class of densities on a measure space $(\mathcal{X}, \mathcal{B})$ carrying a σ-finite measure λ and T a functional defined on $\mathcal{G}$. We are interested in the intrinsic difficulty of the problem of estimating Tg based on a sample from a density g, which is known to be contained in $\mathcal{G}$. In the next section we introduce the concept minimax risk as a way to measure how hard this estimation problem is. In Section 6.2 a theorem is proved stating that the minimax risk can asymptotically be bounded from below by a quantity involving a local modulus of continuity of T over $\mathcal{G}$. The proof of this theorem is quite elementary, having the triangle inequality at its heart. In Section 6.3 another frequently used tool for deriving lower bounds for the minimax risk is introduced: the Van Trees inequality. We will see that the modulus of continuity encountered in Section 6.2 also appears quite naturally when applying the Van Trees inequality. Finally, in Section 6.4, we apply the theory to some of the problems introduced in Chapters 2 and 4.

6.1 Global and Local Minimax Risk

In this section we derive a quantity depending on the class of densities $\mathcal{G}$, the functional T and the sample size n, which can be used to measure the difficulty of the problem of estimating Tg based on a sample of size n from a density g known to be contained in $\mathcal{G}$. The asymptotic behavior of this quantity for $n \to \infty$ depends on $\mathcal{G}$ and T only. That behavior gives information on the complexity of the estimation problem.

Let t_n $(n \geq 1)$ be an estimation procedure, i.e., a sequence of measurable functions (t_n) where $t_n\colon \mathcal{X}^n \to \mathbb{R}$. For each $n \geq 1$ the random variable $T_n = t_n(X_1, X_2, \ldots, X_n)$, where $X_1, X_2, \ldots$ is a sequence of independent identically distributed random variables with density $g \in \mathcal{G}$, is an estimator for Tg. For well behaved estimators we expect the random variable $T_n - Tg$ to be concentrated around zero in some sense. One way to formalize this is to define the risk of the estimation procedure t_n, evaluated at $g \in \mathcal{G}$, as follows:

$$R_l(n, t_n, g; T) = E_{g^{\otimes n}} l(|t_n(\mathcal{X}) - Tg|), \tag{6.1}$$

where l is an increasing loss function on $[0, \infty)$ with $l(t) = 0$ if and only if $t = 0$ and $g^{\otimes n}$ is the n-fold product density associated with g. To say that "at a point $g \in \mathcal{G}$" the estimator $T_n^{(1)}$ is better than the estimator $T_n^{(2)}$ means that $R_l(n, t_n^{(1)}, g) < R_l(n, t_n^{(2)}, g)$ for the corresponding estimation procedures. To be able to say that the estimator $T_n^{(1)}$ is better than the estimator $T_n^{(2)}$ "over $\mathcal{G}$," the behavior of the estimators at a fixed g is not enough: the awkward estimator $T_n \equiv Tg_0$ behaves very well at g_0, but for all $g \in \mathcal{G}$ with $Tg \neq Tg_0$ its risk cannot tend to zero as $n \to \infty$. Therefore, we define the max risk of t_n over $\mathcal{G}$ as follows:

$$\mathrm{MR}_l(n, t_n; T, \mathcal{G}) = \sup_{g \in \mathcal{G}} R_l(n, t_n, g; T), \tag{6.2}$$

meaning that the quality of t_n is measured by its worst case performance over $\mathcal{G}$. Taking the infimum in (6.2) over all possible procedures t_n, we get a quantity called the (global) minimax risk for estimating Tg based on a sample of size n known to be generated by a density belonging to $\mathcal{G}$. Explicitly:

$$\mathrm{MMR}_l(n; T, \mathcal{G}) = \inf_{t_n} \mathrm{MR}_l(n, t_n; T, \mathcal{G}) = \inf_{t_n} \sup_{g \in \mathcal{G}} E_{g^{\otimes n}} l(|t_n(\mathcal{X}) - Tg|). \tag{6.3}$$

For fixed n this quantity indicates how hard the estimation problem is.

It is clear that, if $\mathrm{MMR}_l(n_0; T, \mathcal{G})$ is finite for some $n_0 \geq 1$, $(\mathrm{MMR}_l(n; T, \mathcal{G}))_{n=n_0}^{\infty}$ is a decreasing sequence of positive numbers. In fact, usually $\mathrm{MMR}_l(n; T, \mathcal{G}) \downarrow 0$ for $n \to \infty$. The rate δ_n of this convergence is a measure for the ill-posedness of the estimation problem. If, for $n \to \infty$,

$$\mathrm{MMR}_l(n; T, \mathcal{G}) \asymp \delta_n,$$

we say that Tg is δ_n−estimable with loss function l.

It is trivially true that for each sequence $\mathcal{G}_n$ of subsets of $\mathcal{G}$

$$\mathrm{MMR}_l(n; T, \mathcal{G}) \geq \mathrm{MMR}_l(n; T, \mathcal{G}_n) = \inf_{t_n} \sup_{g \in \mathcal{G}_n} E_{g^{\otimes n}} l(|t_n(\mathcal{X}) - Tg|). \tag{6.4}$$

The quantity at the right hand side of (6.4) is called a local minimax risk. Apart from the fact that the global minimax risk can be bounded from below by any local minimax risk, the local minimax risk is an interesting quantity in its own right. If $\mathcal{G}_n$ is a shrinking neighborhood of one fixed $g \in \mathcal{G}$ (in some sense to be made explicit in Section 6.2), the dependence on g of $\mathrm{MMR}_l(n; T, \mathcal{G}_n)$ for $n \to \infty$ indicates the role played by the true underlying density g in the difficulty of the estimation problem. For instance, if it can be shown in a specific example that

$$\lim_{n \to \infty} n^{1/3} \mathrm{MMR}_l(n; T, \mathcal{G}_n) = \frac{|g'(x_0)|}{g(x_0)},$$

Tg is $n^{-1/3}$-estimable with loss function l, and denoting by g the true underlying density, no estimation procedure that performs well for all $\bar{g} \in \bar{\mathcal{G}}$, with $\mathcal{G}_n \subset \bar{\mathcal{G}}$ for all large n, can give rise to a normalized risk satisfying

$$\limsup_{n\to\infty} n^{1/3} R_l(n, t_n, g; T) < \frac{|g'(x_0)|}{g(x_0)}.$$

Of course, if for each $g \in \mathcal{G}$ the asymptotic local minimax risk is determined over a sufficiently rich subset of $\mathcal{G}$ containing g, the global minimax risk can be obtained by maximizing those local minimax risks for $g \in \mathcal{G}$.

In the next section we will give a simple method to derive a lower bound on the rate of convergence to zero of $\text{MMR}_l(n; T, \mathcal{G})$, using inequality (6.4) for an appropriate sequence of subsets $\mathcal{G}_n$ of $\mathcal{G}$. Besides the straightforward method to derive a lower bound on a minimax risk by restricting the class of densities, there is also a straightforward way to derive an upper bound for a minimax risk: for each fixed estimation procedure t_n,

$$\text{MMR}_l(n; T, \mathcal{G}_n) \le \text{MR}_l(n, t_n; T, \mathcal{G}_n) = \sup_{g\in\mathcal{G}_n} E_{g^{\otimes n}} l(|t_n(X) - Tg|).$$

We focus on lower bounds on the minimax risk, but see Example 6.1 where also the risk of a particular estimator is computed in a parametric setting.

6.2 A Minimax Lower Bound Theorem

Theorem 6.1 can be used to derive an asymptotic lower bound for the optimal rate for estimating a functional of a density that is known to belong to a class of densities $\mathcal{G}$. Before stating this theorem, we introduce the Hellinger distance between densities, a bit more general than in Section 3.1 (and also using H to denote it rather than h for obvious reasons here).

Definition 6.1 Let f and h be probability densities on a measurable space $(\Omega, \mathcal{A})$ with respect to a dominating σ-finite measure μ. The Hellinger distance $H(f, h)$ between f and h is then defined as the square root of

$$H^2(f, h) = \frac{1}{2}\int_\Omega \left(\sqrt{f(x)} - \sqrt{h(x)}\right)^2 d\mu(x) = 1 - \int_\Omega \sqrt{f(x)h(x)}\, d\mu(x).$$

The integral $\int_\Omega \sqrt{f(x)h(x)}\, d\mu(x)$ is called the Hellinger affinity.

If we use the special loss function $l(x) = |x|$, as in Theorem 6.1, we denote the corresponding minimax risk by MMR_1.

Now suppose that we want to estimate a real-valued parameter θ that can be written Tq, where q is a probability density with respect to a σ-finite measure μ on a measurable space $(\Omega, \mathcal{A})$. Let (T_n), $n \ge 1$, be a sequence of estimators of θ, based on samples of size n, that is: $T_n = t_n(X_1, \ldots, X_n)$, where $X_1, \ldots, X_n$ is a sample generated by q, and $t_n : \Omega^n \to \mathbb{R}$ is a measurable function. Furthermore, let $l\colon [0, \infty) \to [0, \infty)$ be an increasing loss function, with $l(0) = 0$. The risk of the estimator in estimating Tq, using the loss function l, is then defined by

$$E_{n,q} l\,(|T_n - Tq|)\,,$$

where $E_{n,q}$ denotes the expectation with respect to the product measure $q^{\otimes n}$, corresponding to the sample $X_1, \ldots, X_n$. Then we have the following result.

Theorem 6.1 *Let $\mathcal{G}$ be a class of probability densities on a measurable space $(\Omega, \mathcal{A})$ with respect to a σ-finite dominating measure μ, and let T be a real-valued functional on $\mathcal{G}$. Moreover, let $l\colon [0, \infty) \to \mathbb{R}$ be an increasing convex loss function, with $l(0) = 0$. Then, for any $q_1, q_2 \in \mathcal{G}$ such that the Hellinger distance $H(q_1, q_2) < 1$:*

$$\inf_{T_n} \max \left\{ E_{n,q_1} l\left(|T_n - Tq_1|\right), E_{n,q_2} l\left(|T_n - Tq_2|\right) \right\} \geq l\left(\frac{1}{4} |Tq_1 - Tq_2| \left\{ 1 - H(q_1, q_2)^2 \right\}^{2n} \right).$$

In the proof of Theorem 6.1 we need some facts concerning the Hellinger distance. A very useful relation exists between the Hellinger distance between the product densities $f^{\otimes n}(x_1, x_2, \ldots, x_n) = f(x_1)f(x_2)\ldots f(x_n)$ and $h^{\otimes n}(x_1, x_2, \ldots, x_n) = h(x_1)h(x_2)\ldots h(x_n)$ with respect to the dominating product measure $\mu^{\otimes n}$ on Ω^n, and the distance between the densities f and h on Ω:

$$1 - H^2(f^{\otimes n}, h^{\otimes n}) = \int_{\Omega^n} \sqrt{f^{\otimes n} h^{\otimes n}}\, d\mu^{\otimes n} = \tag{6.5}$$

$$= \left(\int_\Omega \sqrt{fh}\, d\mu \right)^n = \left(1 - H^2(f, h)\right)^n.$$

See also Exercise 6.2. We shall also need the inequality given in Lemma 6.1. This inequality is sometimes referred to as LeCam's inequality.

Lemma 6.1 *Let f and h be probability densities on a measurable space $(\Omega, \mathcal{A})$ with respect to a dominating measure μ. Then*

$$\left(1 - H^2(f, h)\right)^2 \leq 1 - \left(1 - \int f \wedge h\, d\mu\right)^2 \leq 2 \int f \wedge h\, d\mu.$$

Proof Writing out the square, the second inequality is trivial. The first inequality is essentially Cauchy-Schwarz:

$$\left(1 - H^2(f, h)\right)^2 + \left(1 - \int f \wedge h\, d\mu\right)^2 = \left(\int \sqrt{fh}\, d\mu\right)^2 + \left(\frac{1}{2} \int |f - h|\, d\mu\right)^2 =$$

$$= \left(\int \sqrt{fh}\, d\mu\right)^2 + \frac{1}{4} \left(\int |\sqrt{f} - \sqrt{h}| (\sqrt{f} + \sqrt{h})\, d\mu\right)^2 \leq$$

$$\leq \left(\int \sqrt{fh}\, d\mu\right)^2 + \frac{1}{4} \int (\sqrt{f} - \sqrt{h})^2\, d\mu \int (\sqrt{f} + \sqrt{h})^2\, d\mu = 1. \qquad \square$$

Proof of Theorem 6.1. By Jensen's inequality,

$$\inf_{T_n} \max \left\{ E_{n,q_1} l\left(|T_n - Tq_1|\right), E_{n,q_2} l\left(|T_n - Tq_2|\right) \right\}$$

$$\geq l\left(\inf_{T_n} \max \left\{ E_{n,q_1} l\left(|T_n - Tq_1|\right), E_{n,q_2} l\left(|T_n - Tq_2|\right) \right\} \right).$$

Hence, if we can prove the theorem for $l(x) = |x|$, we get, by the monotonicity of l:

$$\inf_{T_n} \max\{E_{n,q_1} l(|T_n - Tq_1|), E_{n,q_2} l(|T_n - Tq_2|)\} \geq l\left(\frac{1}{4}|Tq_1 - Tq_2| \left\{1 - H(q_1, q_2)^2\right\}^{2n}\right).$$

So it suffices to prove the result for $l(x) = |x|$. We get, using the triangle inequality and Lemma 6.1:

$$\max\left\{E_{n,q_1} l\,|T_n - Tq_1|, E_{n,q_2} l\,|T_n - Tq_2|\right\} \geq \frac{1}{2}\left\{E_{n,q_1}|T_n - Tq_1| + E_{n,q_2}|T_n - Tq_2|\right\}$$

$$\geq \frac{1}{2}\int \{|t_n(x_1, \ldots, x_n) - Tq_1| + |t_n(x_1, \ldots, x_n) - Tq_2|\}$$

$$\cdot \prod_{i=1}^n q_1(x_i) \wedge \prod_{i=1}^n q_2(x_i)\, d\mu(x_1) \ldots d\mu(x_n)$$

$$\geq \frac{1}{2}|Tq_2 - Tq_1| \int \prod_{i=1}^n q_1(x_i) \wedge \prod_{i=1}^n q_2(x_i)\, d\mu(x_1) \ldots d\mu(x_n)$$

$$\geq \frac{1}{4}|Tq_2 - Tq_1| \left\{1 - H\left(q_1^{\otimes n}, q_2^{\otimes n}\right)^2\right\}^2 \geq \frac{1}{4}|Tq_2 - Tq_1| \{1 - H(q_1, q_2)\}^{2n}.$$

This proves the result. □

We note that for each subclass $\mathcal{G}_n$ of $\mathcal{G}$ containing (q_n) for all n sufficiently large, the minimax risk lower bound given in Theorem 6.1 holds. In particular, the lower bound holds for all Hellinger ϵ-balls around q.

Considering the lower bound given in Theorem 6.1, and the fact that we may choose (q_n) freely as long as the condition on its Hellinger distance to q is satisfied, we can try to optimize the lower bound. This means that we should make it as large as possible. We should therefore make $|Tq_n - Tq|$ as large as possible and at the same time make $H(q_n, q)$ as small as possible. A formal way of stating this problem is to define the modulus of continuity of T over $\mathcal{G}$ locally at q, with respect to the Hellinger metric:

$$m_q(\epsilon) = \sup\{|Tq_2 - Tq| : q \in \mathcal{G} \quad \text{and} \quad H(q_2, q) \leq \epsilon\}. \tag{6.6}$$

Theorem 6.1 then leads to Corollary 6.1.

Corollary 6.1 *Let $\mathcal{G}$ be a class of densities on $\mathcal{X}$ and T a functional on $\mathcal{G}$. Fix $q \in \mathcal{G}$, and let the function m_q be defined as in (6.6). Then for each subset $\mathcal{G}_q$ of $\mathcal{G}$ containing some Hellinger ball around q,*

$$\liminf_{n\to\infty} m_q(\tau/\sqrt{n})^{-1} \mathrm{MMR}_1(n; T, \mathcal{G}_q) \geq \frac{1}{4}e^{-2\tau^2} \tag{6.7}$$

for each positive τ.

Proof Fix $\tau > 0$. For all n sufficiently large, the Hellinger ball of radius $\tau/\sqrt{n}$ around g in $\mathcal{G}$, is contained in $\mathcal{G}_q$. Now choose $(h_\epsilon) \subset \mathcal{G}$ such that $H^2(h_\epsilon, q) \leq \epsilon$ and $|Th_\epsilon - Tq| > m_q(\epsilon)(1-\epsilon)$. Then

$$\liminf_{n\to\infty} m_q(\tau/\sqrt{n})^{-1}\mathrm{MMR}_l(n; T, \mathcal{G}_q) \geq \liminf_{n\to\infty} m_q(\tau/\sqrt{n})^{-1}\mathrm{MMR}_l(n; T, \{h_{\tau/\sqrt{n}}, q\})$$

$$\geq \liminf_{n\to\infty} |Tq - Th_{\tau/\sqrt{n}}|^{-1}(1-\tau/\sqrt{n})\mathrm{MMR}_l(n; T, \{h_{\tau/\sqrt{n}}, q\}) \geq \frac{1}{4}e^{-2\tau^2}.$$

□

It is especially the behavior of the function m near zero that is important for the lower bound of the minimax risk. In many problems, of which we will see examples later, this behavior can be described by

$$m_q(\epsilon) = (c\epsilon)^r(1 + o(1)) \text{ as } \epsilon \downarrow 0 \tag{6.8}$$

for some positive parameters c and r, possibly depending on q. Then we get the following result.

Corollary 6.2 *Let $\mathcal{G}$ be a class of densities on $\mathcal{X}$ and T a functional on $\mathcal{G}$. Fix $q \in \mathcal{G}$, and let the function m_q be defined as in (6.6), allowing an asymptotic expansion as given in (6.8). Then for each subset $\mathcal{G}_q$ of $\mathcal{G}$ containing some Hellinger ball around q,*

$$\liminf_{n\to\infty} n^{r/2}\mathrm{MMR}_1(n; T, \mathcal{G}_q) \geq \frac{1}{4}\left(\frac{1}{2}c\sqrt{r}\right)^r e^{-r/2}. \tag{6.9}$$

Proof From Corollary 6.1 we have that for each $\tau > 0$

$$\liminf_{n\to\infty} n^{r/2}\mathrm{MMR}_1(n; T, \mathcal{G}_q) \geq \frac{1}{4}e^{-2\tau^2}(c\tau)^r.$$

Maximizing this lower bound with respect to τ gives the result. □

6.3 Lower Bound Based on the Van Trees Inequality

Besides the method of Theorem 6.1 there are more methods known to derive lower bounds for the rate of convergence to zero of a local minimax risk. In this section the Van Trees inequality is derived and it is shown that the modulus of continuity as defined in (6.6) also arises naturally when this inequality is used to bound the asymptotic local minimax risk from below.

Let

$$\mathcal{F} = \{f_\theta : \theta \in [0, 1]\}$$

be a family of probability densities with respect to a dominating measure μ on a measurable space $(\mathcal{Y}, \mathcal{C})$. Consider the problem of estimating a (differentiable) function $\psi(\theta)$ of θ based on a random element X having density f_θ. Let $\hat{\psi} = \hat{\psi}(X)$ be an estimator for $\psi(\theta)$ and consider its L_2-risk

$$E_\theta(\hat{\psi}(X) - \psi(\theta))^2 = \int_{\mathcal{Y}} (\hat{\psi}(x) - \psi(\theta))^2 f_\theta(x)\, d\mu(x).$$

Let ξ be any absolutely continuous probability density (with respect to Lebesgue measure) on $[0, 1]$ satisfying $\xi(0) = \xi(1) = 0$. Furthermore, define the Fisher information for θ in X by

$$I(\theta) = \int_{\mathcal{Y}} \left(\frac{\partial}{\partial\theta} \log f_\theta(x) \right)^2 f_\theta(x)\, d\mu(x), \tag{6.10}$$

and the Fisher information in the location model generated by ξ by

$$\tilde{I}(\xi) = \int_{\mathbb{R}} \left(\frac{\partial}{\partial\theta} \log \xi(x-\theta) \right)^2 \xi(x-\theta)\, dx = \int_0^1 \frac{(\frac{\partial}{\partial\theta}\xi(\theta))^2}{\xi(\theta)}\, d\theta.$$

Lemma 6.2 (Van Trees inequality) *Suppose that*

$$E_\theta \frac{\partial}{\partial\theta} \log f_\theta(X) = 0. \tag{6.11}$$

Then

$$\int_0^1 E_\theta(\hat{\psi}(X) - \psi(\theta))^2 \xi(\theta)\, d\theta \geq \frac{(\int \psi'(\theta)\xi(\theta)\, d\theta)^2}{\int I(\theta)\xi(\theta)\, d\theta + \tilde{I}(\xi)}. \tag{6.12}$$

Proof First note that

$$\int_0^1 \frac{\partial}{\partial\theta} (f_\theta(x)\xi(\theta))\, d\theta = 0 \quad \text{and} \quad \int_0^1 \psi(\theta)\frac{\partial}{\partial\theta}(f_\theta(x)\xi(\theta))\, d\theta = -\int_0^1 \psi'(\theta) f_\theta(x)\xi(\theta)\, d\theta, \tag{6.13}$$

where we use that $\xi(0) = \xi(1) = 0$ and, in the second equality, integration by parts. Using that f_θ is a probability density with respect to μ, this implies that

$$\int_{\mathcal{Y}} \int_0^1 (\hat{\psi}(x) - \psi(\theta)) \frac{\partial}{\partial\theta} f_\theta(x)\xi(\theta)\, d\theta\, d\mu(x) = \int_0^1 \psi'(\theta)\xi(\theta)\, d\theta.$$

Hence, using Cauchy-Schwarz we obtain

$$\begin{aligned}
&\int_{\mathcal{Y}} \int_0^1 (\hat{\psi}(x) - \psi(\theta))^2 \xi(\theta) f_\theta(x) d\theta\, d\mu(x) \cdot \int_{\mathcal{Y}} \int_0^1 \left(\frac{\partial}{\partial\theta} \log f_\theta(x)\xi(\theta) \right)^2 \xi(\theta) f_\theta(x) d\theta\, d\mu(x) \\
&\geq \left(\int_{\mathcal{Y}} \int_0^1 \left(\hat{\psi}(x) - \psi(\theta)\right) \left(\frac{\partial}{\partial\theta} \log f_\theta(x)\xi(\theta) \right) \xi(\theta) f_\theta(x) d\theta\, d\mu(x) \right)^2 \qquad (6.14) \\
&= \left(\int_0^1 \psi'(\theta)\xi(\theta)\, d\theta \right)^2.
\end{aligned}$$

For the second factor on the left hand side we can write

$$\int_{\mathcal{Y}}\int_0^1 \left(\frac{\partial}{\partial\theta}\log f_\theta(x)\xi(\theta)\right)^2 \xi(\theta) f_\theta(x) d\theta\, d\mu(x)$$

$$= \int_0^1 \int_{\mathcal{Y}} \left(\frac{\frac{\partial}{\partial\theta} f_\theta(x)}{f_\theta(x)} + \frac{\frac{\partial}{\partial\theta}\xi(\theta)}{\xi(\theta)}\right)^2 \xi(\theta) f_\theta(x) d\theta\, d\mu(x)$$

$$= \int_0^1 \int_{\mathcal{Y}} \left(\frac{\xi(\theta)\left(\frac{\partial}{\partial\theta} f_\theta(x)\right)^2}{f_\theta(x)} + 2\frac{\partial}{\partial\theta} f_\theta(x)\frac{\partial}{\partial\theta}\xi(\theta) + \frac{f_\theta(x)(\frac{\partial}{\partial\theta}\xi(\theta))^2}{\xi(\theta)}\right) d\theta\, d\mu(x)$$

$$= \int_0^1 I(\theta)\xi(\theta)\, d\theta + 2\int_0^1 \frac{\partial}{\partial\theta}\xi(\theta)\cdot E_\theta \frac{\partial}{\partial\theta}\log f_\theta(X)\, d\theta + \int_{\mathcal{Y}} f_\theta(x)\, d\mu(x)\tilde{I}(\xi)$$

$$= \int_0^1 I(\theta)\xi(\theta)\, d\theta + \tilde{I}(\xi),$$

using (6.11) and the fact that f_θ is a probability density with respect to μ. Hence, substituting this result in (6.14), we get Van Trees inequality (6.12). □

We will now use Lemma 6.2 to bound from below a local minimax risk of the type (6.4), with quadratic loss function $l(x) = x^2$ (making it slightly less general than the method of Theorem 6.1). To that end, first note that by Lemma 6.2,

$$\sup_{\theta\in[0,1]} E_\theta(\hat{\psi}(X) - \psi(\theta))^2 \geq \int_0^1 E_\theta(\hat{\psi}(X) - \psi(\theta))^2\xi(\theta)\, d\theta \geq \frac{(\int \psi'(\theta)\xi(\theta)\, d\theta)^2}{\int I(\theta)\xi(\theta)\, d\theta + \tilde{I}(\xi)}.$$

Since the left hand side involves $\hat{\psi}$ and the lower bound does not, this inequality implies that for any estimator $\hat{\psi}$ and absolutely continuous probability density ξ on $[0, 1]$ with $\xi(0) = \xi(1) = 0$

$$\inf_{\hat{\psi}} \sup_{\theta\in[0,1]} E_\theta(\hat{\psi}(X) - \psi(\theta))^2 \geq \frac{(\int \psi'(\theta)\xi(\theta)\, d\theta)^2}{\int I(\theta)\xi(\theta)\, d\theta + \tilde{I}(\xi)}. \tag{6.15}$$

We will restrict ourselves to the case where $\mathcal{G}$ is a convex class of densities with respect to λ on $\mathcal{X}$. Let the parameterized one-dimensional subclasses $\mathcal{G}_n$ of $\mathcal{G}$ be defined by

$$\mathcal{G}_n = \{g_\theta : \theta \in [0, 1], \text{ where } g_\theta = g + \theta(g_n - g)\}$$

where g_n is a sequence of densities satisfying the conditions of Theorem 6.1 and Lemma 6.3.

Lemma 6.3 *Let g and $(g_n)_{n=1}^\infty$ be densities on $\mathcal{X}$, such that*

$$\{x : g_n(x) > 0\} \subset \{x : g(x) > 0\}$$

for all n sufficiently large and

$$\sup_{\{x:\, g(x)>0\}} \left|\frac{g_n(x) - g(x)}{g(x)}\right| \to 0 \text{ for } n \to \infty. \tag{6.16}$$

Then there is a vanishing sequence of positive numbers c_n such that

$$(1 - c_n) \int_{\{x:\, g(x)>0\}} \frac{(g_n(x) - g(x))^2}{g(x)} \, d\lambda(x) \leq 8H^2(g_n, g) \leq$$
$$\leq (1 + c_n) \int_{\{x:\, g(x)>0\}} \frac{(g_n(x) - g(x))^2}{g(x)} \, d\lambda(x)$$

for all n.

Proof Using (6.16), the first inequality follows from

$$\int_{\{x:\, g(x)>0\}} \frac{(g_n(x) - g(x))^2}{g(x)} \, d\lambda(x) =$$
$$= \int_{\{x:\, g(x)>0\}} \left(\sqrt{g_n(x)} - \sqrt{g(x)}\right)^2 \left(1 + \sqrt{\frac{g_n(x)}{g(x)}}\right)^2 d\lambda(x) \leq$$
$$\leq 2 \sup_{\{x:\, g(x)>0\}} \left(1 + \sqrt{\frac{g_n(x)}{g(x)}}\right)^2 H^2(g_n, g) \leq (1 + \nu_n) 8H^2(g_n, g),$$

where ν_n is a vanishing sequence of positive numbers. The second inequality can be shown as follows, using that $\sqrt{1+u} \geq 1 + \frac{1}{2}u - (1 + o(1))\frac{1}{8}u^2$ for $u \to 0$.

$$H^2(g_n, g) = \frac{1}{2} \int_{\{x:\, g(x)>0\}} (\sqrt{g_n(x)} - \sqrt{g(x)})^2 \, d\lambda(x) =$$
$$= \int_{\{x:\, g(x)>0\}} \left(1 - \sqrt{1 + \frac{g_n(x) - g(x)}{g(x)}}\right) g(x) \, d\lambda(x) \leq$$
$$\leq \int_{\{x:\, g(x)>0\}} \left(-\frac{g_n(x) - g(x)}{2g(x)} + \frac{(g_n(x) - g(x))^2}{8g(x)^2}(1 + \nu_n)\right) g(x) \, d\lambda(x) =$$
$$= (1 + \nu_n) \int_{\{x:\, g(x)>0\}} \frac{(g_n(x) - g(x))^2}{8g(x)} \, d\lambda(x)$$

where $\nu_n \geq 0$ for all n and $\nu_n \to 0$. □

Now define for each n fixed, $\mathcal{Y} = \mathcal{X}^n$, $f_\theta = g_\theta^{\otimes n}$ and $\psi(\theta) = Tg_\theta$. Then we can prove the following theorem.

Theorem 6.2 *Let $\mathcal{G}$ be a convex class of densities on $\mathcal{X}$ and T a linear functional on $\mathcal{G}$. Fix $g \in \mathcal{G}$. Let (g_n) be a sequence of densities in $\mathcal{G}$ satisfying the conditions of Lemma 6.3 and*

$$\limsup_{n\to\infty} nH^2(g_n, g) \leq \tau^2 \tag{6.17}$$

for a $\tau > 0$. Then

$$\liminf_{n\to\infty} |Tg_n - Tg|^{-2}(8\tau^2 + 4\pi^2)\mathrm{MMR}_2(n; T, \mathcal{G}_n) \geq 1. \tag{6.18}$$

Proof Denoting by $\mathrm{MMR}_2(n; T, \mathcal{G}_n)$ the L_2 minimax risk, i.e., the minimax risk based on the loss function $l(x) = x^2$, we get for each n

$$\mathrm{MMR}_2(n; T, \mathcal{G}_n) = \inf_{t_n} \sup_{g \in \mathcal{G}_n} E_{g^{\otimes n}}(t_n(X) - Tg)^2 = \inf_{\hat{\psi}} \sup_{\theta \in [0,1]} E_\theta(\hat{\psi}(X) - \psi(\theta))^2$$

$$\geq \frac{\left(\int_0^1 \left(\frac{d}{d\theta} Tg_\theta\right) \xi(\theta)\, d\theta\right)^2}{\int_0^1 I(\theta)\xi(\theta)\, d\theta + \tilde{I}(\xi)}, \tag{6.19}$$

where we use (6.15).

Using linearity of T and the fact that ξ integrates to one, we get

$$\int_0^1 \left(\frac{d}{d\theta} Tg_\theta\right) \xi(\theta)\, d\theta = T(g_n - g). \tag{6.20}$$

From Exercise 6.6 it follows that the Fisher information on θ in a random vector $X \sim f_\theta = g_\theta^{\otimes n}$ ($I(\theta)$ as defined in (6.10)) is n times the information $I^{(1)}(\theta)$ on θ in $X \sim g_\theta$.

Moreover, the conditions imposed on (g_n) imply that

$$I(\theta) = nI^{(1)}(\theta) = n \int_X \left(\frac{\partial}{\partial\theta} \log g_\theta(x)\right)^2 d\lambda(x) = n \int_X \frac{(g_n(x) - g(x))^2}{g(x)\left(1 + \theta \frac{g_n(x) - g(x)}{g(x)}\right)}\, d\lambda(x)$$

$$= n \int_X \frac{(g_n(x) - g(x))^2}{g(x)}\, d\lambda(x)(1 + o(1)) \tag{6.21}$$

uniformly in $\theta \in [0, 1]$ for $n \to \infty$ (see Exercise 6.7). Therefore, by condition (6.17),

$$\limsup_{n\to\infty} n \int_0^1 I^{(1)}(\theta)\xi(\theta)\, d\theta = \limsup_{n\to\infty} 8nH^2(g_n, g) \leq 8\tau^2$$

for arbitrary ξ.

Including these observations in (6.19), for each fixed ξ, the sharpest lower bound on the L_2 minimax risk to be obtained for the fixed sequence of classes $\mathcal{G}_n$ is

$$\liminf_{n\to\infty} |Tg_n - Tg|^{-2}(8\tau^2 + \tilde{I}(\xi))\mathrm{MMR}_2(n; T, \mathcal{G}_n) \geq 1. \tag{6.22}$$

Using calculus of variations, it follows that the function

$$\xi(\theta) = 2\cos^2(\pi(\theta - 1/2))\, 1_{[0,1]}(\theta) \tag{6.23}$$

minimizes $\tilde{I}(\xi)$ over the class of permitted densities. For this ξ, we have that

$$\tilde{I}(\xi) = 4\pi^2.$$

This means that the strongest result to be obtained from (6.22) is (6.18). □

Theorem 6.1 gives, for the L_2 loss function, lower bound

$$\liminf_{n\to\infty} |Tg_n - Tg|^{-2} 16 e^{4\tau^2} \mathrm{MMR}_2(n; T, \mathcal{G}_n) \geq 1. \tag{6.24}$$

Comparing (6.18) with (6.24), we see that for each sequence (g_n) the lower bound obtained by the Van Trees inequality is sharper than the lower bound obtained by Theorem 6.1.

As in Section 6.2, we can optimize the result of Theorem 6.2 in the choice of the densities g_n.

Corollary 6.3 *Let $\mathcal{G}$ be a convex class of densities on $\mathcal{X}$ and T a linear functional on $\mathcal{G}$. Fix $g \in \mathcal{G}$, and let the function m_g be defined as in (6.6). Then for each subset $\mathcal{G}_g$ of $\mathcal{G}$ containing some Hellinger ball around g,*

$$\liminf_{n\to\infty} m_g(\tau/\sqrt{n})^{-2}\mathrm{MMR}_2(n; T, \mathcal{G}_g) \geq \frac{1}{8\tau^2 + 4\pi^2} \tag{6.25}$$

for each positive τ.

The proof is completely analogous to that of Corollary 6.1. For the important situation where the local modulus of continuity satisfies (6.8), we get the following result.

Corollary 6.4 *Let $\mathcal{G}$ be a convex class of densities on $\mathcal{X}$ and T a linear functional on $\mathcal{G}$. Fix $g \in \mathcal{G}$, and let the function m_g be defined as in (6.6), allowing an asymptotic expansion as given in (6.8). Then for each subset $\mathcal{G}_g$ of $\mathcal{G}$ containing some Hellinger ball around g,*

$$\liminf_{n\to\infty} n^r\mathrm{MMR}_2(n; T, \mathcal{G}_g) \geq \frac{1}{4}(c^2r/2)^r\left(\frac{1-r}{\pi^2}\right)^{1-r}. \tag{6.26}$$

Proof From Corollary 6.3 we have that for each $\tau > 0$

$$\liminf_{n\to\infty} n^r\mathrm{MMR}_2(n; T, \mathcal{G}_g) \geq \frac{(\tau c)^{2r}}{8\tau^2 + 4\pi^2}.$$

Maximizing this lower bound with respect to τ gives the result. □

6.4 Applications

To illustrate how Theorem 6.1 and Corollary 6.2 can be used in practical situations, we will apply the corollary to give an asymptotic lower bound for the local minimax risk in the context of some inverse problems. To start, however, we present a very familiar parametric estimation problem.

Example 6.1 (*Parameter exponential distribution*) Take for $\mathcal{G}$ the class of exponential densities on $[0, \infty)$ indexed by their scale parameters:

$$\mathcal{G} = \{g_\theta : g_\theta(x) = \theta^{-1}e^{-x/\theta}1_{[0,\infty)}(x),\ \theta > 0\}$$

We want to estimate the parameter θ ($Tg_\theta = \theta = \int xg_\theta(x)\,dx$).

From Exercise 6.10 it follows that

$$H(\theta, \nu) := H(g_\theta, g_\nu) = \sqrt{\frac{(\sqrt{\theta} - \sqrt{\nu})^2}{\theta + \nu}} = \frac{|\sqrt{\theta} - \sqrt{\nu}|}{\sqrt{\theta + \nu}}.$$

Therefore, at θ,

$$m(\epsilon) = \sup\{|\delta| : H(\theta, \theta + \delta) \leq \epsilon\} \sim 2\sqrt{2}\theta\epsilon$$

and for each class $\mathcal{G}_\theta$ in $\mathcal{G}$ containing a Hellinger ball of positive radius around g_θ,

$$\liminf_{n\to\infty} n^{1/2}\mathrm{MMR}_1(n; T, \mathcal{G}_\theta) \geq \frac{1}{4}\theta\sqrt{2/e}$$

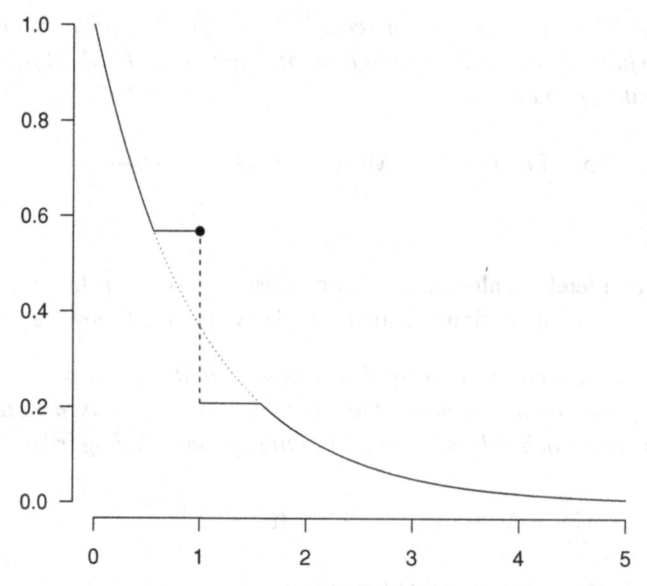

Figure 6.1 Perturbed density h_μ (solid) with decreasing density g.

What can be said about the asymptotic behavior of the estimator

$$\hat{\theta}_n = \frac{1}{n}\sum_{i=1}^{n} X_i?$$

In particular, observe that,

$$\lim_{n\to\infty} \sqrt{n} E_\theta |\hat{\theta}_n - \theta| = \theta\sqrt{2/\pi}$$

Example 6.2 (*Decreasing density*) Consider the problem of estimating a decreasing density at a fixed point. This corresponds to the problem described in Section 2.2.

Denote by $\mathcal{G}$ the class of decreasing density functions on $[0, \infty)$. Fix a decreasing density $g \in \mathcal{G}$ and a point $x_0 > 0$ such that g is differentiable at x_0 and that $g'(x_0) < 0$. We will determine the behavior of the function m for small values of ϵ, where

$$m(\epsilon) = \sup\{|h(x_0) - g(x_0)| : h \in \mathcal{G} \quad \text{and} \quad H(h, g) \le \epsilon\}.$$

This we can do by determining the function e for small values of μ, where

$$e(\mu) = \inf\{H(h, g) : h \in \mathcal{G} \quad \text{and} \quad |h(x_0) - g(x_0)| \ge \mu\},$$

since m and e are inverses of each other, in the sense that $m(\epsilon) \ge \mu \iff e(\mu) \le \epsilon$. It is easily seen that for small μ the infimum in the definition of e is attained at the function h_μ, where

$$h_\mu(x) = \begin{cases} g(x_0) + \mu & \text{for } x_0 - c_1\mu \le x \le x_0 \\ g(x_0) - c_2\mu & \text{for } x_0 < x \le x_0 + c_3\mu \\ g(x) & \text{elsewhere,} \end{cases}$$

where c_1, c_2 and c_3 are positive constants depending on μ and g such that h_μ integrates to one and is continuous at the points $x_0 - c_1\mu$ and $x_0 + c_3\mu$; see Figure 6.1. We may without loss of generality assume that $c_2 < 1$ for each small μ.

In Exercise 6.12 it is shown that

$$c_1 \to -1/g'(x_0), \quad c_2 \to 1 \quad \text{and} \quad c_3 \to -1/g'(x_0).$$

for $\mu \downarrow 0$.

Therefore, in view of Lemma 6.3, we have for $\mu \downarrow 0$,

$$e(\mu) \asymp \left(\int_0^\infty \frac{(h_\mu(x) - g(x))^2}{8g(x)} \, dx \right)^{1/2} \sim (-12g(x_0)g'(x_0))^{-1/2} \mu^{3/2},$$

giving that, for $\epsilon \downarrow 0$,

$$m(\epsilon) \sim \left((-12g(x_0)g'(x_0))^{1/2} \epsilon \right)^{2/3}.$$

Corollary 6.2 now gives that for any subclass $\mathcal{G}_g$ of $\mathcal{G}$ containing some Hellinger ball around g of positive radius,

$$\liminf_{n\to\infty} n^{1/3} \mathrm{MMR}_1(n; T, \mathcal{G}_g) \geq \frac{1}{4} \left(\frac{1}{2} \left(-12g(x_0)g'(x_0)\right)^{1/2} \sqrt{2/3} \right)^{2/3} e^{-1/3}$$
$$= \frac{1}{4} \left(-2g(x_0)g'(x_0)\right)^{1/3} e^{-1/3}$$

Example 6.3 *(Current status model)* We illustrate the use of Theorem 6.1 by deriving minimax lower bounds for the estimation in the univariate and bivariate current status models. For the univariate current status model we have the following result.

Let (F_n) be a sequence of discrete perturbations of F_0 in a neighborhood of t_0 and let G be the distribution function of the observation times. Assume that F_0 has derivative $f_0(t_0) > 0$ at t_0 and G has derivative $g(t_0) > 0$ at t_0. Then a local asymptotic minimax bound for an estimate T_n of $F_0(t_0)$ in the univariate current status model is given by

$$\liminf_{n\to\infty} n^{1/3} \max\left\{ E_{F_n} |T_n - F_0(t_0)|, E_{F_0} |T_n - F_0(t_0)| \right\} \geq \frac{1}{4} \left(\frac{f_0(t_0) F_0(t_0)\{1 - F_0(t_0)\}}{g(t_0) e} \right)^{1/3}. \tag{6.27}$$

To show this, we introduce a discrete perturbation F_n of F_0 in a neigborhood of t_0 by defining

$$F_n(t) = \begin{cases} F_0(t) & , \quad \text{if } t < t_0 - cn^{-1/3}, \\ F_0\left(t_0 - cn^{-1/3}\right), & \quad \text{if } t \in \left[t_0 - cn^{-1/3}, t_0 + cn^{-1/3}\right), \\ F_0(t) & , \quad \text{if } t \geq t_0 + cn^{-1/3}, \end{cases}$$

where $c > 0$ will be specified in the following. This means that

$$|F_n(t_0) - F_0(t_0)| = \left| F_0\left(t_0 - cn^{-1/3}\right) - F_0(t_0) \right| \sim cn^{-1/3} f_0(t_0), \; n \to \infty.$$

We now define $p_n = p_{F_n}$ by:

$$p_n(t, 1) = F_n(t)g(t), \qquad p_n(t, 0) = \{1 - F_n(t)\}\, g(t),$$

and $p_0 = p_{F_0}$ by:

$$p_0(t, 1) = F_0(t)g(t), \qquad p_0(t, 0) = \{1 - F_0(t)\}\, g(t).$$

Then $F_n(t_0) - F_0(t_0)$ has the representation

$$F_n(t_0) - F_0(t_0) = \frac{p_n(t_0, 1) - p_0(t_0, 1)}{g(t_0)}.$$

The squared Hellinger distance between p_n and p_0 is given by:

$$\frac{1}{2}\left\{\int \left\{\sqrt{F_n(t)} - \sqrt{F_0(t)}\right\}^2 dG(t) + \int \left\{\sqrt{1 - F_n(t)} - \sqrt{1 - F_0(t)}\right\}^2 dG(t)\right\}.$$

We now have:

$$\int \left\{\sqrt{F_n(t)} - \sqrt{F_0(t)}\right\}^2 dG(t) \sim \int \frac{\{F_n(t) - F_0(t)\}^2}{4F_0(t)} dG(t) \sim \frac{c^3 f_0(t_0)^2 g(t_0)}{3nF_0(t_0)},$$

and likewise

$$\int \left\{\sqrt{1 - F_n(t)} - \sqrt{1 - F_0(t)}\right\}^2 dG(t) \sim \int \frac{\{F_n(t) - F_0(t)\}^2}{4\{1 - F_0(t)\}} dG(t) \sim \frac{c^3 f_0(t_0)^2 g(t_0)}{3n\{1 - F_0(t_0)\}}.$$

So we find:

$$\frac{1}{2}\sum_{i=0}^{1} \int \left\{\sqrt{p_n(t, i)} - \sqrt{p_0(t, i)}\right\}^2 dG(t) \sim \frac{c^3 f_0(t_0)^2 g(t_0)}{6nF_0(t_0)\{1 - F_0(t_0)\}}.$$

Using (6.7) we therefore obtain as lower bound

$$\frac{1}{4} c f_0(t_0) \exp\left\{\frac{c^3 f_0(t_0)^2 g(t_0)}{3F_0(t_0)\{1 - F_0(t_0)\}}\right\}.$$

Minimizing over c yields:

$$\liminf_{n\to\infty} n^{1/3} \max\left\{E_{n,p_{F_n}} |T_n - F_0(t_0)|, E_{n,p_{F_0}} |T_n - F_0(t_0)|\right\}$$

$$\geq \frac{1}{4}\left(\frac{f_0(t_0)F_0(t_0)\{1 - F_0(t_0)\}}{g(t_0)e}\right)^{1/3}.$$

Remark Note that the lower bound is, apart from a universal constant, equal to the square root of the asymptotic variance of the maximum likelihood estimator, see Theorem 3.7.

Exercises

6.1 Argue that, whenever $\mathrm{MMR}_l(n_0, T, \mathcal{G}) < \infty$, $(\mathrm{MMR}_l(n, T, \mathcal{G}))_{n=n_0}^{\infty}$ is a decreasing sequence of positive numbers.

6.2 Verify the relation of Hellinger affinities used in (6.5), i.e., that

$$\int_{\Omega^n} \sqrt{f^{\otimes n} h^{\otimes n}} d\mu^{\otimes n} = \left(\int_{\Omega} \sqrt{fh} d\mu\right)^n.$$

6.3 The (local) modulus of continuity defined in (6.6) is a special instance of a general definition of moduli of continuity. Let $(\mathcal{X}, d_1)$ and $(\mathcal{Y}, d_2)$ be metric spaces and $T : \mathcal{X} \to \mathcal{Y}$ a mapping. Then the modulus of continuity of T, locally at $x \in \mathcal{X}$, is defined by

$$m_x(\epsilon) = \sup\{d_2(T(z), T(x)) : z \in \mathcal{X} \quad \text{and} \quad d_1(z, s) \leq \epsilon\}.$$

The (nonlocal) modulus of continuity is defined by

$$m(\epsilon) = \sup\{d_2(T(z), T(x)) : x, z \in \mathcal{X} \quad \text{and} \quad d_1(z, x) \leq \epsilon\} = \sup_{x \in \mathcal{X}} m_x(\epsilon).$$

Let $\mathcal{X} = \mathcal{Y} = \mathbb{R}$, $d_1 = d_2$ with $d_1(x, y) = |x - y|$.

a) Let $T(x) = \sin x$. Show that $m_x(\epsilon) \sim |\cos x|\epsilon$ as $\epsilon \downarrow 0$ if $\cos x \neq 0$. If $\cos x = 0$ then $m_x(\epsilon) \sim \frac{1}{2}\epsilon^2$ as $\epsilon \downarrow 0$. In this case $m(\epsilon) \sim \epsilon$ as $\epsilon \downarrow 0$.
b) Let $T(x) = x^2$. Show that for $x \neq 0$, $m_x(\epsilon) \sim |2x|\epsilon$ as $\epsilon \downarrow 0$ and observe that $m(\epsilon) = \infty$ for all $\epsilon > 0$.

6.4 Show that condition (6.11) is satisfied if the order of integration and differentiation are interchanged.

6.5 Verify (6.13).

6.6 The Fisher information for θ in X in the model $\{f_\theta : \theta \in \Theta\}$ is given by (6.10). Show that, if X denotes an n-vector of i.i.d. random elements having density $f_\theta = g_\theta^{\otimes n}$ on $\mathcal{X}^n$, the Fisher information for θ in X is n times the information for θ one of its components.

6.7 Show that (6.21) holds under the conditions imposed on g_n in Lemma 6.3.

6.8 Show that the density ξ given in (6.23) minimizes $\tilde{I}(\xi)$ over all absolutely continuous probability densities ξ on $[0, 1]$, satisfying $\xi(0) = \xi(1) = 0$.

6.9 Derive the lower bounds given in the proofs Corollaries 6.2 and 6.4 by optimizing over all values of $\tau > 0$.

6.10 Verify that in the situation of Example 6.1

$$H(\theta, \nu) := H(g_\theta, g_\nu) = \sqrt{\frac{(\sqrt{\theta} - \sqrt{\nu})^2}{\theta + \nu}} = \frac{|\sqrt{\theta} - \sqrt{\nu}|}{\sqrt{\theta + \nu}}.$$

6.11 Consider the class of uniform densities

$$\mathcal{G} = \{g_\theta : \theta > 0\} \quad \text{with} \quad g_\theta(x) = \theta^{-1} 1_{[0,\theta]}(x).$$

a) For $\theta, \nu > 0$, compute

$$H(g_\theta, g_\nu)^2 = \frac{1}{2} \int_{\mathbb{R}} \left(\sqrt{g_\theta(x)} - \sqrt{g_\nu(x)}\right)^2 dx.$$

b) For $\nu = \theta + \epsilon$, derive the asymptotic behavior (as $\epsilon \to 0$) of $H(g_\theta, g_{\theta+\epsilon})$.
c) Use (b) and the results from Section 6.2 to derive a local asymptotic minimax lower bound for estimating the parameter θ based a sample of size n from the density g_θ. Is the derived lower bound sharp?

6.12 This exercise relates to Example 6.2. Approximate g locally near x_0 linearly at x_0, and show that, for $\mu \downarrow 0$,

$$c_1 \to -1/g'(x_0), \quad c_2 \to 1 \quad \text{and} \quad c_3 \to -1/g'(x_0).$$

6.13 Derive the asymptotic behavior of $m(\epsilon)$ as $\epsilon \downarrow 0$, where

$$m(\epsilon) = \sup\{|h(x_0) - g(x_0)| : h \in \mathcal{G} \quad \text{and} \quad H(h, g) \leq \epsilon\}$$

and $\mathcal{G}$ is the class of convex decreasing densities on $[0, \infty)$ as encountered in Section 4.3. Hint: see Example 6.2 for the general approach.

Bibliographic Remarks

The problem of determining the (local) modulus of continuity of a functional over a class of functions is frequently encountered in the literature on minimax risks. See, e.g., Fan, 1993, in the context of nonparametric regression problems and Donoho and Liu, 1991, in the context of density estimation. The general form of the Van Trees inequality can be found in Van Trees, 1968, and a more recent reference on that inequality is Gill and Levit, 1995. Theorem 6.1 can also be found as Lemma 4.1 on p. 132 of Groeneboom, 1996. More information on lower bounds to the minimax risk can be found in Yu, 1997 and Chapter 2 of Tsybakov and Zaiats, 2009.

7

Algorithms and Computation

In parametric statistical models, parameter estimates can often (not always) be computed fairly simply. When the parameter space is $\mathbb{R}^p$, the computational difficulty in determining the maximum likelihood estimator is related to the structure of the score equations. When there is an explicit solution to these, there is no computational problem.

There are parametric problems, in which there are some problems in computing the estimate. For example, when the maximum likelihood estimate of the parameter $\theta > 0$ has to be computed based on a sample from the density

$$f_\theta(x) = \theta \exp\left(\theta x + 1 - e^{\theta x}\right), \quad x \geq 0, \tag{7.1}$$

the score equation does not allow for an explicit solution and one has to employ an iterative procedure to approximate the ML estimate $\hat{\theta}$. When the number of unknown parameters (dimension of θ) in a model increases, the amount of effort to compute estimators usually increases as well. Not in the least because the parameter space may be constrained.

Also in shape constrained estimation problems, the computational issues can be relatively straightforward. For example, computing the MLE of a distribution function in the current status model boils down to computing the derivative of the convex minorant of an appropriate diagram of points (see Section 2.3). In other situations, iterative schemes have to be used to obtain an approximation of the estimator. This chapter describes three approaches to such computational problems. Before that, the concept of an algorithm and issues of convergence are considered in Section 7.1. In Section 7.2, the expectation maximization (EM) algorithm is described. This algorithm was developed in the statistics community to compute ML estimators in missing data models. It will be illustrated using a parametric example of truncated exponentials and as nonparametric example the interval censoring case 2 model. Section 7.3 describes the iterative convex minorant (ICM) algorithm that can be used to maximize a concave function over the isotonic cone in $\mathbb{R}^n$. It is based on iteratively approximating the concave function by quadratic functions that have diagonal Hessian matrices. As such, the heart of this algorithm is the solution to the basic isotonic regression problem discussed in Section 2.1 as the derivative of the greatest convex minorant of a diagram of points. Yet another branch of algorithms was introduced and studied in the area of mixture models. Section 7.4 describes a general class of vertex direction algorithms that use the representation of the model as convex hull of (or convex cone generated by) a given set of densities. Finally, in Section 7.5, the algorithms considered in this chapter are applied to the problem of computing the ML estimator in the competing risk model with interval censored event times.

7.1 Algorithm: Concept and Convergence

The word algorithm is often used as a synonym of "way to compute some quantity." In this section we will make a more precise definition of what we mean by an algorithm and prove a very general convergence result. That result can then be applied to some of the algorithms to be studied in the subsequent sections.

Consider the general optimization problem

$$\text{minimize } \phi(\theta) \text{ over } \Theta \subset \mathbb{R}^k \tag{7.2}$$

where ϕ is some function defined on Θ. Of course, minimizing ϕ is equivalent to maximizing $-\phi$. Then we mean by an algorithm a triplet consisting of an initial point $\theta^{(0)} \in \Theta$, an algorithmic map A and a termination criterion. An algorithmic map is a point-to-set mapping $A\colon \Theta \to 2^{\Theta}$ assigning to each element of Θ a subset of Θ. The recipe to be followed is then

---

General Algorithm

Input:
initial point $\theta^{(0)}$
begin
  $\theta := \theta^{(0)}$;
  $k := 1$
  while termination criterion is not satisfied
  begin
    $\theta^{(k)} :=$ some element of $A(\theta^{(k-1)})$
    $k := k + 1$
  end;
end.

Consider the following simple example.

Example 7.1 The optimization problem is to minimize the function $\phi(\theta) = \theta^2$ over the set $\Theta = [1, \infty)$. Consider the algorithmic point-to-point map (where we abuse notation slightly)

$$A(\theta) = (\theta + 2)/3.$$

Then, starting from arbitrary $\theta^{(0)} \geq 1$, we see that

$$\theta^{(k)} = 3^{-k}\theta^{(0)} + 1 - 3^{-k}.$$

This sequence obviously converges to $\hat{\theta} = 1$. Note that in this situation where A is a point-to-point map, all iterates are uniquely determined by the starting value. That changes when the algorithmic map gives a set consisting of more than one point, such as

$$A(\theta) = [(\theta + 3)/4, (\theta + 2)/3]. \tag{7.3}$$

Based on this map, the sequence of iterates is not uniquely defined starting from an initial point $\theta^{(0)}$. However, in our terminology, the algorithm is uniquely defined.

Having such a general definition of an algorithm, we have to think of the properties we want the algorithm to have. Essentially, we want the sequence $(\theta^{(k)})_{k=0}^{\infty}$ to converge to the point where ϕ attains its minimum as fast as possible. In Example 7.1 there is no ambiguity in what the solution should be and convergence takes place. However, without any assumptions on the function ϕ and the set Θ, a minimizer of ϕ over Θ is not guaranteed to exist. If a minimizer exists, it is not necessarily unique and even if there is a unique minimizer, there might be lots of local minima far away from the global minimum.

A solution set of problem (7.2) is, roughly, a set of points in the parameter space we accept as limit points of an algorithm. It can, e.g., be defined as

$$\Theta_0 = \{\theta \in \Theta \colon \theta \text{ is local minimum of } \phi\}. \tag{7.4}$$

Apart from this concept, Theorem 7.1 also uses the concept of a closed mapping.

Definition 7.1 A mapping $A \colon \Theta \to 2^{\Theta}$ is closed at some point θ if for each sequence $\theta^{(k)}$ converging to θ the following holds. If $\tau^{(k)} \in A(\theta^{(k)})$ for each k and $\tau^{(k)}$ converges to some τ, then necessarily $\tau \in A(\theta)$. A mapping is closed on a subset of Θ if it is closed at each point in that subset.

Remark If A is a point-to-point mapping, continuity of A implies closedness (see Exercise 7.2). The implication in the opposite direction does not hold (see Exercise 7.3).

Theorem 7.1 *Let $\Theta \subset \mathbb{R}^k$ be closed and the nonempty set $\Theta_0 \subset \Theta$ a solution set of problem (7.2). Suppose that*

(i) Each sequence generated by the algorithmic map starting from $\theta^{(0)}$ is contained in a compact subset of Θ,
(ii) There exists a continuous function $\alpha \colon \Theta \to \mathbb{R}$ such that for each $\theta \in \Theta \setminus \Theta_0$, $\alpha(\theta') < \alpha(\theta)$ for all $\theta' \in A(\theta)$ (α is called a descent function for A),
(iii) The algorithmic map A is closed on $\Theta \setminus \Theta_0$.

Then either the algorithm stops after a finite number of steps with a point in Θ_0, or it generates an infinite sequence $(\theta^{(k)})$ such that all accumulation points of $(\theta^{(k)})$ belong to Θ_0 and $\alpha(\theta^{(k)}) \to \alpha(\theta)$ for some $\theta \in \Theta_0$.

Remark We saw that there is some arbitrariness in the choice of solution set Θ_0. In Theorem 7.1 we see that a smaller solution set imposes stronger conditions on the algorithmic map, in the sense that it should be closed over a larger set. For example, if we would define the solution set as the set of global minimizers of ϕ rather than (7.4), then condition (*ii*) in Theorem 7.1 requires the algorithmic map to "jump" out of a local minimum.

Proof Suppose that there is not an index k such that $\theta^{(k)} \in \Theta_0$. By ($i$), there is a subsequence (k_j) of (k) and a $\theta^{(\infty)} \in \Theta$ such that $\lim_{j\to\infty} \theta^{(k_j)} = \theta^{(\infty)}$. By ($ii$) this implies that $\lim_{j\to\infty} \alpha(\theta^{(k_j)}) = \alpha(\theta^{(\infty)})$. However, since $\alpha(\theta^{(k)}) \downarrow$, we have that $\lim_{k\to\infty} \alpha(\theta^{(k)}) = \alpha(\theta^{(\infty)})$. Now consider the subsequence $(k_j + 1)$ of (k). Again by (i), the sequence $\theta^{(k_j+1)}$ has a convergent subsequence, and the limit of such a subsequence is denoted by $\theta^{(\infty)\prime}$. Continuity of α then gives that $\alpha(\theta^{(\infty)\prime}) = \alpha(\theta^{(\infty)})$.

By the definition of the algorithm we know that $\theta^{(k_j+1)} \in A(\theta^{(k_j)})$. Now suppose that $\theta^{(\infty)} \notin \Theta_0$. Then, since $\theta^{(k_j)} \to \theta^{(\infty)}$ and $\theta^{(k_j+1)} \to \theta^{(\infty)\prime}$ along a subsequence, by (iii) we

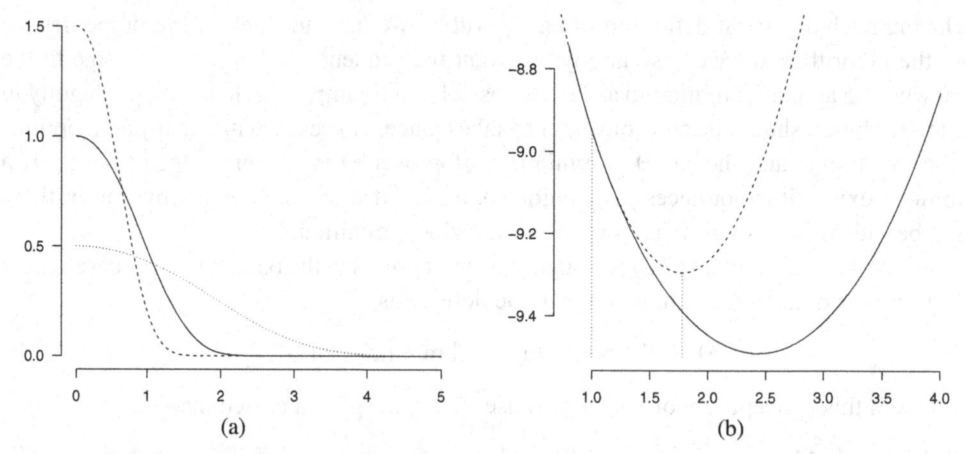

Figure 7.1 (a) The three members of the class of densities given in (7.1): $\theta = 0.5$ (dotted), $\theta = 1$ (solid) and $\theta = 1.5$ (dashed). (b) (Minus) the log likelihood function ϕ based on a sample of size 10 (solid), the current iterate $\theta^{(0)} = 1$, the quadratic approximation of ϕ at $\theta^{(0)}$ (dashed) and $\theta^{(1)} = 1.78$ as minimizer of this approximation.

have that $\theta^{(\infty)\prime} \in A(\theta^{(\infty)})$, implying (by (*ii*)) that $\alpha(\theta^{(\infty)\prime}) < \alpha(\theta^{(\infty)\prime})$. This is a contradiction and we conclude that $\theta^{(\infty)} \in \Theta_0$ necessarily. □

Example 7.2 Consider the parameter estimation problem associated with model (7.1). See Figure 7.1 for some of the densities in the model. We are given a realization $x_1, x_2, \ldots, x_n$ of an i.i.d. sample from f_θ and want to compute

$$\hat{\theta} = \operatorname{argmin}_{\theta>0}\phi(\theta)$$

where

$$\phi(\theta) = -\frac{1}{n}\sum_{i=1}^{n} \log f_\theta(x_i) = -\log\theta - \theta\bar{x} + \frac{1}{n}\sum_{i=1}^{n} e^{\theta x_i} - n.$$

The Newton algorithm to compute the $\hat{\theta}$ entails choosing an initial estimate, $\theta^{(0)}$, and defining the next iterate as minimizer of the quadratic Taylor approximation at the current iterate. See the right panel of Figure 7.1. In this example this means (see Exercise 7.7) that

$$\theta^{(k+1)} = A(\theta^{(k)}) = \operatorname{argmin}_{\theta>0}\left(\phi(\theta^{(k)}) + (\theta - \theta^{(k)})\phi'(\theta^{(k)}) + \frac{1}{2}(\theta - \theta^{(k)})^2\phi''(\theta^{(k)})\right)$$

$$= \theta^{(k)} - \frac{\phi'(\theta^{(k)})}{\phi''(\theta^{(k)})} = \theta^{(k)} + \theta^{(k)}\frac{n + n\theta^{(k)}\bar{x} - \theta^{(k)}\sum_{i=1}^{n} x_i e^{\theta^{(k)}x_i}}{n + \sum_{i=1}^{n} x_i^2 e^{\theta^{(k)}x_i}}. \quad (7.5)$$

In this example, the algorithmic point-to-point map A is continuous, hence closed.

The Newton procedure sketched in Example 7.2 is fairly general. If the function ϕ is sufficiently smooth, one can define the algorithmic map as

$$A(\theta) = \operatorname{argmin}_{\theta'\in\Theta}\phi_q(\theta'|\theta),$$

where $\phi_q(\cdot|\theta)$ is the local quadratic Taylor approximation of ϕ near θ. If $\Theta \subset \mathbb{R}^k$,

$$\phi_q(\theta'|\theta) = \phi(\theta) + (\theta' - \theta)^T \nabla\phi(\theta) + \frac{1}{2}(\theta' - \theta)^T \nabla^2\phi(\theta)(\theta' - \theta).$$

There are many situations conceivable where the Newton method does not work. If the objective function is not convex, strange things can happen (see Exercise 7.4). Moreover, even if the objective function is convex, Newton's algorithm does not necessarily converge (see Exercise 7.5). However, this specific problem (related to the absence of a descent function in general) can be circumvented by damping the Newton algorithm.

7.2 The EM Algorithm

We start with a simple parametric example to illustrate the EM algorithm. Then we will introduce the algorithm more generally and apply this to the interval censoring case 2 model. Finally, we will address some problems that are encountered when EM is defined in full generality.

(A) Description of the EM Algorithm

Example 7.3 (*Truncated Exponentials*) Consider an i.i.d. sample $X_1, X_2, \ldots, X_n$ from the exponential distribution with density

$$f_\theta(x) = \theta^{-1}e^{-x/\theta}1_{(0,\infty)}(x), \tag{7.6}$$

so that $X = (X_1, \ldots, X_n)$ has density

$$p_\theta(x) = \prod_{i=1}^n f_\theta(x_i),\ x = (x_1, \ldots, x_n) \in \mathbb{R}^n.$$

However, the vector X itself cannot be observed. Only the integer part of its components can be observed:

$$Y = T(X),\ \text{with } Y_i = \lfloor X_i \rfloor,\ 1 \le i \le n.$$

The problem is how to estimate $\theta > 0$ from Y via maximum likelihood.

If we had the complete data $x_1, \ldots, x_n$, we would write down the loglikelihood function

$$\log p_\theta(x) = \sum_{i=1}^n \log f_\theta(x_i) = -n\log\theta - \theta^{-1}\sum_{i=1}^n x_i$$

and maximize it to obtain the mean of the x_is as maximum likelihood estimate; a straightforward procedure. One step in the (iterative) EM algorithm consists of an expectation and a maximization step. During the E-step, the current iterate and the observed data are used to compute the conditional expectation of the complete loglikelihood given the data, under the distribution determined by the current iterate θ. More specifically, for the current iterate θ, the following function of θ' is determined:

$$\phi_\theta(\theta') = E_\theta\left(\log p_{\theta'}(X)|Y = y\right) = E_\theta\left(\sum_{i=1}^n \log f_{\theta'}(X_i)|Y = y\right)$$

$$= \sum_{i=1}^n E_\theta\left(\log f_{\theta'}(X_i)|Y_i = y_i\right) =: \sum_{i=1}^n \phi_\theta^{(i)}(\theta').$$

The M-step then entails maximization of $\phi_\theta(\theta')$ with respect to θ'. This maximizer is then the next iterate. Using the notation of Section 7.1, the algorithmic map is the point-to-point map

$$A(\theta) = \text{argmax}_{\theta'>0}\phi_\theta(\theta').$$

Starting from an initial value, we would like the sequence thus generated to converge to the maximum likelihood estimate.

What does this mean explicitly in our example? The function ϕ_θ can be computed once we know the conditional density of X_i given Y_i under the distribution with parameter θ. We denote this function by $k_\theta(\cdot|y)$:

$$k_\theta(x|y) = \frac{d}{dx}P_\theta(X_i \le x|Y_i = y) = \frac{e^{-x/\theta}}{\theta(e^{-y/\theta} - e^{-(y+1)/\theta})}1_{[y,y+1)}(x).$$

Therefore,

$$\phi_\theta^{(i)}(\theta') = E_\theta\left(\log f_{\theta'}(X_i)|Y_i = y_i\right) = E_\theta\left(\log(\theta'^{-1}e^{-X_i/\theta'})|Y_i = y_i\right)$$

$$= E_\theta\left(-\log\theta' - X_i/\theta'|Y_i = y_i\right) = -\log\theta' - \theta'^{-1}\int_0^\infty x k_\theta(x|y_i)\,dx =$$

$$= -\log\theta' - \theta'^{-1}\int_{y_i}^{y_i+1}\frac{xe^{-x/\theta}}{\theta(e^{-y_i/\theta} - e^{-(y_i+1)/\theta})}dx =$$

$$= -\log\theta' - \frac{y_i+\theta}{\theta'} + \frac{1}{\theta'(e^{1/\theta} - 1)}.$$

Now let $\theta > 0$ be given. As can be seen easily by differentiating $\phi_\theta(\theta') = \sum_{i=1}^n \phi_\theta^{(i)}(\theta')$ with respect to θ', ϕ_θ is maximized at

$$\theta' = \theta + \bar{y}_n - 1/(e^{1/\theta} - 1). \tag{7.7}$$

See Exercise 7.8. Therefore, the EM algorithm in this example can be described by an initial point, say $\theta^{(0)} = \bar{y} + 1/2$, the algorithmic map

$$A(\theta) = \theta + \bar{y}_n - 1/(e^{1/\theta} - 1)$$

and a suitable termination criterion, for example, "stop after 50 iterations." This algorithm was run using the following data:

$$25,\ 9,\ 4,\ 13,\ 1,\ 0,\ 5,\ 18,\ 0,\ 4.$$

The first 5 iterates thus obtained are

$$8.400000,\ 8.390082,\ 8.390070,\ 8.390070,\ 8.390070,$$

and the iterates remain unchanged till the 50th iteration.

Of course, in this simple example the EM algorithm or whatever other optimization algorithm is not actually needed. The probability mass function of Y_i is given by

$$g_\theta(y) = P([Y_i = y]) = P([X_i \in [y, y+1)]) = \int_y^{y+1} f_\theta(x)\,dx = e^{-y/\theta} - e^{-(y+1)/\theta} \quad (7.8)$$

for $y = 0, 1, 2, \ldots$. Hence, writing q_θ for the probability mass function of the vector $Y = (Y_1, \ldots, Y_n)$, we get as log likelihood function

$$\ell(\theta; y) = \log q_\theta(y) = \sum_{i=1}^n \log g_\theta(y_i) = -n\frac{\bar{y}_n + 1}{\theta} + n\log(e^{1/\theta} - 1).$$

Maximizing this with respect to θ, we get that the maximum likelihood estimator is given by

$$\hat{\theta} = \frac{1}{\log(1 + 1/\bar{y}_n)} = \frac{1}{\log(1 + 1/7.9)} \approx 8.39007. \quad (7.9)$$

Let us now turn to the general description of the EM algorithm. Our model is a measurable space $(\mathcal{X}, \mathcal{A})$ endowed with a σ-finite measure μ. A random element X takes its values in $\mathcal{X}$, and the probability induced by X is assumed to belong to a parameterized class of distributions:

$$P([X \in A]) = P_\theta(A) \text{ for } A \in \mathcal{A}.$$

Here $\theta \in \Theta \subset \mathbb{R}^m$ is the parameter. Furthermore, we assume P_θ to have a density p_θ with respect to the measure μ:

$$P_\theta(A) = \int_A p_\theta(x)\,d\mu(x), \text{ for } A \in \mathcal{A}.$$

The space $(\mathcal{X}, \mathcal{A})$ represents the hidden space, where the only partly observable element X takes its values. The actual observable random element Y is a measurable function T of X, and takes its values in what is called the observation space $(\mathcal{Y}, \mathcal{B})$. This measurable space is also endowed with a σ-finite measure, denoted by ν. The mapping T together with $\{P_\theta : \theta \in \Theta\}$ again induce probability measures $\{Q_\theta : \theta \in \Theta\}$ on $(\mathcal{Y}, \mathcal{B})$ by $Q_\theta = P_\theta T^{-1}$:

$$Q_\theta(B) = P_\theta T^{-1}(B) = P_\theta(T^{-1}(B)) \text{ for } B \in \mathcal{B}$$

We also assume Q_θ to have a density q_θ with respect to ν:

$$Q_\theta(B) = \int_B q_\theta(y)\,d\nu(y),\ B \in \mathcal{B}.$$

For the given observation y, our aim is then to compute the maximum likelihood estimator

$$\hat{\theta} = \text{argmax}_{\theta\in\Theta} q_\theta(y) = \text{argmax}_{\theta\in\Theta} \log q_\theta(y).$$

Remark In Example 7.3, μ is Lebesgue measure on hidden space $\mathcal{X} = \mathbb{R}^n$ and ν is counting measure on observation space $\mathcal{Y} = \{0, 1, 2, \ldots\}^n$. Moreover, $p_\theta = f_\theta^{\otimes n}$ and $q_\theta = g_\theta^{\otimes n}$ with f_θ and g_θ defined in (7.6) and (7.8) respectively.

Given y, the EM algorithm now proceeds as follows. Start with an initial estimate $\theta^{(0)}$ of θ. Then subsequently perform an expectation and a maximization step to obtain the next iterate $\theta^{(1)}$. Repeat these E- and M-steps until $\theta^{(m)}$ does not change in, say, the 10th decimal (or until some other criterion is met). Then take $\theta^{(m)}$ as numerical approximation of the MLE. We will explain the two basic steps that have to be taken in an iteration, denoting by θ the "current iterate."

The E-step: compute, as function of $\theta' \in \Theta$, the conditional expectation

$$\phi_\theta(\theta') = E_{P_\theta}\left\{\log p_{\theta'}(X) \mid T(X) = y\right\}. \tag{7.10}$$

If $Q_\theta(\{y\}) = P_\theta(T^{-1}(y)) > 0$ for the observed vector y, this conditional expectation can be computed by first computing the conditional density of X given $T(X) = y$, and integrating the function $x \mapsto \log p_{\theta'}(x)$ with respect to this density. If y has Q_θ-probability zero, one should be careful, but it is still possible to give meaning to definition (7.10). Further on in this section, we will show the need for the abstractly defined conditional expectation in this latter situation. For now, think of the event $T(X) = y$ as having strictly positive P_θ probability as in Example 7.3.

The M-step: find

$$\text{argmax}_{\theta' \in \Theta} \phi_\theta(\theta'). \tag{7.11}$$

As ϕ_θ is a function in θ' depending only on the observed data, this function can be computed contrary to the full log likelihood based on the data in the hidden space.

Example 7.4 (*Interval censoring case 2)* For the purpose of computing the maximum likelihood estimator of the distribution function in the interval censoring case 2 problem introduced in Example 4.7, we view the data as generated in two stages. First, the n pairs of time points $T_i < U_i$ are independently generated according to the density h, giving realized values $t_i < u_i$. These values are considered fixed in the following. The second stage is that the X_is are generated, and the random indicators Δ_i and Γ_i are produced.

We will only consider candidate estimators for F that assign all their mass to the (fixed) points v_i, where $v_1 < v_2 < \cdots < v_q$ and $V = \{v_1, \ldots, v_q\} = \{t_1, u_1, t_2, u_2, \ldots, t_n, u_n\}$. If there are no ties in the observed time points, $q = 2n$. For the purpose of maximizing the log likelihood function,

$$\ell(F) = \sum_{i=1}^{n} \delta_i \log F(t_i) + \gamma_i \log(F(u_i) - F(t_i)) + (1 - \delta_i - \gamma_i) \log(1 - F(u_i)). \tag{7.12}$$

This is no restriction, since this function only depends on F via the values F attains at the points v_i.

Now, for EM, we need a hidden space with a collection of probability measures, an observation space and a mapping T between these spaces that relates the hidden realization of interest to the observable realization. We define

$$\mathcal{X} = \{x : x = (x_1, x_2, \dots, x_n) \colon x_i \geq 0\} = \mathbb{R}_+^n$$

and endow this with its usual euclidean Borel σ-field $\mathcal{A}$. On $(\mathcal{X}, \mathcal{A})$ we define, for each

$$\theta \in \Theta = \left\{ \eta \in [0, 1]^m \colon \sum_{i=1}^{m} \eta_i = 1 \right\},$$

the probability density

$$p_\theta = [f_\theta]^{\otimes n}$$

with respect to the n-fold product of counting measure on V. The parameterization of f_θ, the density of X_i with respect to counting measure on V, is natural:

$$f_\theta(v_i) = \theta_i, \text{ for } 1 \leq i \leq m.$$

As observation space we can take $\mathcal{Y} = [\{0, 1\}^2]^n$ with its natural σ-field consisting of all subsets. The mapping T is given by

$$T(x) = \Big((1_{\{x_1 \leq t_1\}}, 1_{\{t_1 < x_1 \leq u_1\}}), \dots, (1_{\{x_n \leq t_n\}}, 1_{\{t_n < x_n \leq u_n\}}) \Big).$$

Remember that all t_is and u_is are fixed and known, so these may be "used" by the function T. Now, given some $\theta \in \Theta$ and the actually observed data y, we have (see (7.10))

$$\begin{aligned}
\phi_\theta(\theta') &= E_{P_\theta} \left\{ \log p_{\theta'}(X) \mid T(X) = y \right\} \\
&= E_{P_\theta} \left\{ \sum_{i=1}^{n} \log f_{\theta'}(X_i) \mid T(X) = y \right\} \qquad (7.13) \\
&= \sum_{i=1}^{n} E_{P_\theta} \left\{ \log f_{\theta'}(X_i) \mid (\Delta_i, \Gamma_i) = (\delta_i, \gamma_i) \right\} \\
&= \sum_{i=1}^{n} \sum_{j=1}^{m} \log \theta'_j \cdot P_\theta \left(X_i = v_j \mid (\Delta_i, \Gamma_i) = (\delta_i, \gamma_i) \right) \\
&= \sum_{j=1}^{m} c(\theta)_j \log \theta'_j,
\end{aligned}$$

where

$$c(\theta)_j = \sum_{i=1}^{n} P_\theta \left(X_i = v_j \mid (\Delta_i, \Gamma_i) = (\delta_i, \gamma_i) \right).$$

Maximizing ϕ_θ with respect to $\theta' \in \Theta$ gives (see Exercise 7.14)

$$(\text{argmax}_{\theta' \in \Theta} \phi_\theta(\theta'))_j = \frac{1}{n} c(\theta)_j = \frac{1}{n} \sum_{i=1}^{n} P_\theta \left(X_i = v_j \mid (\Delta_i, \Gamma_i) = (\delta_i, \gamma_i) \right). \qquad (7.14)$$

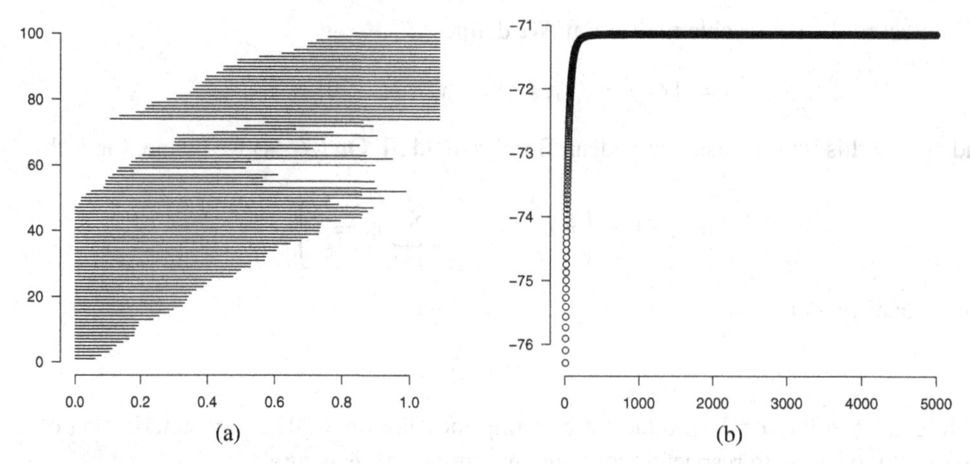

Figure 7.2 (a) Visualization of the case 2 interval censored data: every horizontal line represents the interval where the corresponding censored X_i was located according to the available information. (b) The log likelihood values of the consecutive iterates (starting from the tenth, for reasons of scaling).

To make the EM algorithm even more explicit, we need the conditional probabilities that occur in (7.14). For this, it is convinient to define for $1 \le i \le n$ the following index sets

$$J_i = \begin{cases} \{j \in \{1, 2, \ldots, q\} : v_j \le t_i\} & \text{if } \delta_i = 1, \\ \{j \in \{1, 2, \ldots, q\} : t_i < v_j \le u_i\} & \text{if } \gamma_i = 1, \\ \{j \in \{1, 2, \ldots, q\} : v_j > u_i\} & \text{if } \delta_i = \gamma_i = 0. \end{cases}$$

In fact, J_i gives the indexes of the values v_j that are possible candidates for the hidden X_i, given the associated values $(t_i, u_i, \delta_i, \gamma_i)$.

In Exercise 7.14 it is shown that

$$P_\theta\left(X_i = v_j \mid (\Delta_i, \Gamma_i) = (\delta_i, \gamma_i)\right) = \begin{cases} \theta_j / \sum_{k \in J_i} \theta_k & \text{for } j \in J_i, \\ 0 & \text{for } j \notin J_i. \end{cases}$$

Combining the E- and the M step, we obtain as general updating rule at the mth iteration:

$$\theta_j^{(m)} = \frac{\theta_j^{(m-1)}}{n} \sum_{i=1}^{n} \frac{1_{J_i}(j)}{\sum_{k \in J_i} \theta_k^{(m-1)}}. \tag{7.15}$$

Figure 7.2a visualizes a (generated) data set of size $n = 100$. The sample was generated as follows. First a sample of size $n = 100$ was generated from the distribution with distribution function F with $F(x) = \sqrt{x}$ on $[0, 1]$. For each of the resulting X_i, two observation times were sampled using the bivariate standard uniform distribution: $(T_i, U_i) = (\min\{Z_{i1}, Z_{i2}\}, \max\{Z_{i1}, Z_{i2}\})$ where $(Z_{i1}, Z_{i2}) \sim \text{Unif}\left([0, 1]^2\right)$. Here independent observations were taken for each i. Figure 7.2b shows the evolution of the log likelihood (7.12) in terms of the number of iterations obtained by using (7.15) and starting value $\theta_j^{(0)} = 1/q = 1/200$ for $1 \le j \le 200$ (note that there were no tied observations, so $q = 2n = 200$). Figure 7.3 shows two iterates obtained in this way.

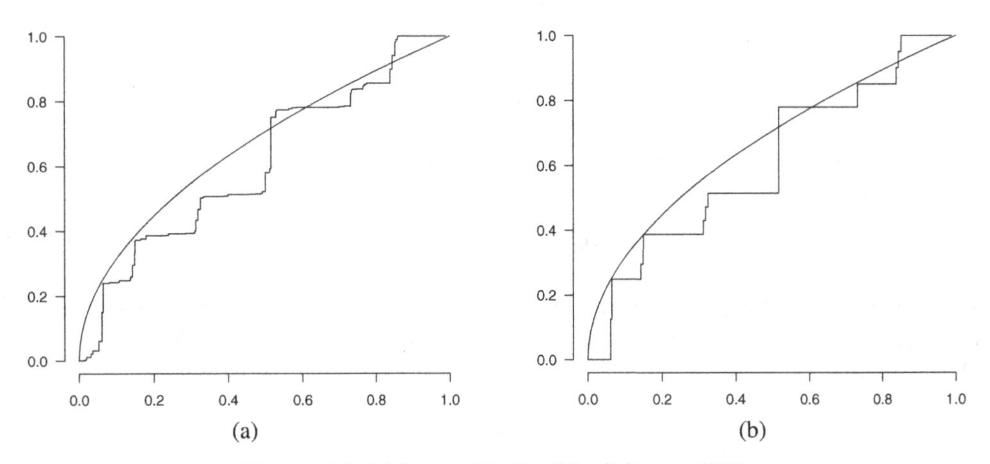

Figure 7.3 (a) Iterate 20. (b) (Final) iterate 5000.

(B) Some Heuristics for the EM Algorithm

Will the EM algorithm produce a sequence of iterates converging to a global maximum of the log likelihood? Sometimes it will and sometimes it will not! We now first give an argument explaining why the EM algorithm might work. The assumptions we impose in this informal argument also indicate issues that can cause problems for convergence of EM.

Suppose that, for the observed y, the real MLE is given by $\hat\theta$, where $\hat\theta$ is an interior point of Θ, and that the function

$$\theta \mapsto q_\theta(y),\ \theta \in \Theta$$

is differentiable on the interior of Θ. Then we must have:

$$\frac{\partial}{\partial\theta} q_\theta(y)\Big|_{\theta=\hat\theta} = 0. \tag{7.16}$$

But if the EM algorithm converges to an interior point $\theta^{(\infty)} \in \Theta$, then $\theta^{(\infty)}$ maximizes the function

$$\theta \mapsto E_{P_{\theta^{(\infty)}}}\left\{\log p_\theta(X) \mid T(X) = y\right\}. \tag{7.17}$$

This implies, assuming that (7.17) is differentiable at interior points $\theta \in \Theta$ and that we may interchange expectation and differentiation,

$$\frac{\partial}{\partial\theta} E_{P_{\theta^{(\infty)}}}\left\{\log p_\theta(X) \mid T(X) = y\right\} = E_{P_{\theta^{(\infty)}}}\left\{\frac{\partial}{\partial\theta}\log p_\theta(X) \,\Big|\, T(X) = y\right\} = 0 \tag{7.18}$$

at $\theta = \theta^{(\infty)}$.

Let the score function $\dot\ell_\theta$ in the complete data situation be defined by

$$\dot\ell_\theta(x) = \frac{\partial}{\partial\theta}\log p_\theta(x).$$

Assuming that certain interchanges of differentiation and integration (or summation) are allowed, we get from Exercise 7.15 that

$$E_{P_\theta}\left\{\dot\ell_\theta(X) \,\Big|\, T(X) = y\right\} q_\theta(y) = \frac{\partial}{\partial\theta} q_\theta(y).$$

Hence (7.18) would imply, for $\theta = \theta^{(\infty)}$,

$$\frac{\partial}{\partial\theta} q_\theta(y) = 0,$$

at a value y such that $q_\theta(y) > 0$. Or, written differently, we would have, for $\theta = \theta^{(\infty)}$,

$$\frac{\partial}{\partial\theta} \log q_\theta(y) = \frac{\frac{\partial}{\partial\theta} q_\theta(y)}{q_\theta(y)} = 0, \tag{7.19}$$

if (7.18) is satisfied and $q_\theta(y) > 0$. So (7.16) would be satisfied at $\theta = \theta^{(\infty)}$, and hence, if there is only one θ for which this score equation is zero, $\theta^{(\infty)}$ would be the MLE!

The equation

$$E_{P_\theta}\left\{\dot{\ell}_\theta(X) \,\middle|\, T(X) = y\right\} = 0, \tag{7.20}$$

that is satisfied at $\theta = \theta^{(\infty)}$, is called the self-consistency equation.

So the reason for believing that the EM algorithm might work is the fact that (7.20) implies (7.19) for $\theta = \theta^{(\infty)}$, if $q_{\theta^{(\infty)}}(y) > 0$. So, if the likelihood function $\theta \mapsto \log q_\theta(y)$ is only maximized at a value θ where the derivative with respect to θ is zero, then a stationary point $\theta^{(\infty)}$ of the EM algorithm would give the MLE.

This argument also points to potential difficulties with the EM algorithm: it might not work if the maximum is not attained at an interior point (a common situation in shape-constrained models), or if the likelihood function is not differentiable at the MLE, or if the score equation (7.19) has multiple roots, some (or all) of which do not maximize the likelihood. Indeed all these situations can occur. For example, if the EM algorithm is started for the interval censoring case 2 problem at a distribution that has a finite loglikelihood and assigns mass zero to many points v_j, the mass will remain zero at these points through all iterations; see (7.15).

(C) Monotonicity of EM

We now address the issue of monotonicity of the EM algorithm, in the sense that $\theta \mapsto -q_\theta(y)$ serves as a descent function. To start with, we assume that, for some $\theta \in \Theta$,

$$Q_\theta(\{y\}) = P_\theta(T^{-1}(y)) > 0 \tag{7.21}$$

for the observed y. The convenient consequence of this assumption is that we can write the conditional density of X given $T(X) = y$ under θ (with respect to μ on $(\mathcal{X}, \mathcal{A})$) as

$$r_\theta(x\,|\,y) = \frac{p_\theta(x)}{Q_\theta(\{y\})} 1_{T^{-1}(y)}(x) \text{ a.e. } \mu. \tag{7.22}$$

Note that $\nu(\{y\}) > 0$ by the assumption that $Q_\theta \ll \nu$ for all $\theta \in \Theta$ and (7.21). Hence, $Q_\theta(\{y\}) = q_\theta(y)\nu(\{y\})$. Without loss of generality we may assume that $\nu(\{y\}) = 1$, so that

$$q_\theta(y) = Q_\theta(\{y\}). \tag{7.23}$$

Using r_θ, we can write

$$\phi_\theta(\theta') = E_\theta\left(\log p_{\theta'}(X) \middle| T(X) = y\right) = \frac{1}{Q_\theta(\{y\})}\int_{T^{-1}(y)} p_\theta(x) \log p_{\theta'}(x)\, d\mu(x). \tag{7.24}$$

In view of the remarks made earlier, it is clear that a global convergence result for the EM algorithm is too much to hope for. What we can prove is the following monotonicity property.

Theorem 7.2 *Let $\theta^{(0)} \in \Theta$ with $q_{\theta^{(0)}}(y) = Q_{\theta^{(0)}}(\{y\}) > 0$. Denote by A the algorithmic map of the EM algorithm:*

$$A(\theta) = argmax_{\theta' \in \Theta} \phi_\theta(\theta'),$$

where ϕ_θ is given by (7.24). Then the EM algorithm is monotone, *in the sense that*

$$q_{\theta^{(m+1)}}(y) \geq q_{\theta^{(m)}}(y) \; for \; m = 0, 1, 2, \ldots$$

Proof Using (7.22) and (7.23), we have

$$\log q_\theta(y) = \log Q_\theta(\{y\}) = \log p_\theta(x) - \log r_\theta(x \mid y), \quad \text{for } x \in T^{-1}(y).$$

Hence we can write for each m and $\theta \in \Theta$ (m being the index of the mth iteration of the EM algorithm):

$$\begin{aligned}
\log q_\theta(y) &= E_{P_{\theta^{(m)}}} \{\log q_\theta(y) \mid T(X) = y\} \\
&= E_{P_{\theta^{(m)}}} \{\log p_\theta(X) - \log r_\theta(X|y) \mid T(X) = y\} \\
&= E_{P_{\theta^{(m)}}} \{\log p_\theta(X) \mid T(X) = y\} - E_{P_{\theta^{(m)}}} \{\log r_\theta(X|y) \mid T(X) = y\}. \quad (7.25)
\end{aligned}$$

Now we look separately at the two terms in the last line of (7.25), and compare the expressions we get, by replacing θ by $\theta^{(m)}$ and $\theta^{(m+1)}$, respectively. Note that we keep the distribution $P_{\theta^{(m)}}$, determining the distribution of the conditional expectation, fixed.

First of all, we get:

$$\begin{aligned}
&E_{P_{\theta^{(m)}}} \{\log p_{\theta^{(m+1)}}(X) \mid T(X) = y\} - E_{P_{\theta^{(m)}}} \{\log p_{\theta^{(m)}}(X) \mid T(X) = y\} \\
&= \phi_{\theta^{(m)}}(\theta^{(m+1)}) - \phi_{\theta^{(m)}}(\theta^{(m)}) \geq 0, \qquad (7.26)
\end{aligned}$$

since $\theta^{(m+1)}$ maximizes the conditional expectation

$$\theta \mapsto \phi_{\theta^{(m)}}(\theta) = E_{P_{\theta^{(m)}}} \{\log p_\theta(X) \mid T(X) = y\}$$

over all θ, so the value we get by plugging in $\theta^{(m+1)}$ will certainly be at least as big as the value we get by plugging in $\theta^{(m)}$.

Secondly, using Jensen's inequality, we get that

$$\begin{aligned}
&E_{P_{\theta^{(m)}}} \{\log r_{\theta^{(m+1)}}(X|y) \mid T(X) = y\} - E_{P_{\theta^{(m)}}} \{\log r_{\theta^{(m)}}(X|y) \mid T(X) = y\} \\
&= E_{P_{\theta^{(m)}}} \left\{\log \frac{r_{\theta^{(m+1)}}(X|y)}{r_{\theta^{(m)}}(X|y)} \mid T(X) = y\right\} \\
&\leq \log \left(E_{P_{\theta^{(m)}}} \left\{ \frac{r_{\theta^{(m+1)}}(X|y)}{r_{\theta^{(m)}}(X|y)} \mid T(X) = y \right\}\right) \qquad (7.27) \\
&= \log \left(\int_{T^{-1}(y)} \frac{r_{\theta^{(m+1)}}(x|y)}{r_{\theta^{(m)}}(x|y)} r_{\theta^{(m)}}(x|y)\, d\mu(x) \right) \\
&= \log \left(\int_{T^{-1}(y)} r_{\theta^{(m+1)}}(x|y)\, d\mu(x) \right) = \log 1 = 0.
\end{aligned}$$

Combining (7.26), (7.27) and (7.25), the result follows. □

Note that Example 7.3 and 7.4 both meet condition (7.21), so that Theorem 7.2 applies. Now consider the following parametric example, where one can employ the EM algorithm but where condition (7.21) is not satisfied.

Example 7.5 (*Uniform deconvolution in exponential model)* Let $Z_1, Z_2, \ldots, Z_n$ be a sample from an exponential distribution with parameter $\theta \in \Theta = (0, \infty)$ and, independent of this, a sample $U_1, U_2, \ldots, U_n$ from the uniform distribution on (0, 1). We are interested in estimating θ, but only observe the random variables $Y_1, \ldots, Y_n$ with $Y_i = Z_i + U_i$.

Denote by X the bivariate sample $((Z_1, U_1), (Z_2, U_2), \ldots, (Z_n, U_n))$, taking its value in $\mathcal{X} = [\mathbb{R}^2]^n$. Then we have, in the notation used so far, that X has a density p_θ with respect to Lebesgue measure on $\mathcal{X}$:

$$p_\theta(x) = p_\theta(z_1, u_1, \ldots, z_n, u_n) = \prod_{i=1}^{n} f_\theta(z_i) 1_{[0,1]}(u_i)$$

where $f_\theta(z) = \theta^{-1} e^{-z/\theta} 1_{[0,\infty)}(z)$. The mapping $T : [\mathbb{R}^2]^n \to \mathbb{R}^n$ is given by

$$T(x) = T(z_1, u_1, \ldots, z_n, u_n) = (z_1 + u_1, z_2 + u_2, \ldots, z_n + u_n).$$

Under P_θ, the random vector $Y = T(X)$ has a density q_θ with respect to Lebesgue measure on $\mathcal{Y} = \mathbb{R}^n$, where

$$q_\theta(y) = q_\theta(y_1, y_2, \ldots, y_n) = \prod_{i=1}^{n} g_\theta(y_i)$$

with

$$g_\theta(z) = \begin{cases} 0 & \text{if } z \le 0 \\ 1 - e^{-z/\theta} & \text{if } z \in (0, 1) \\ e^{-z/\theta}[e^{1/\theta} - 1] & \text{if } z \ge 1. \end{cases} \tag{7.28}$$

It is clear that (7.21) cannot hold for any y and θ. The convenient explicit expression of the conditional density $r_\theta(x|y)$ with respect to Lebesgue measure on $[\mathbb{R}^2]^n$ we had in the presence of (7.21) cannot be used anymore. Indeed, the conditional distribution of X given $Y = y$ is completely supported on the set

$$\{x = (z_1, u_1, \ldots, z_n, u_n) \in [\mathbb{R}^2]^n : z_i + u_i = y_i \text{ for } 1 \le i \le n\}.$$

This set has Lebesgue measure zero, so a conditional distribution of X given $Y = y$ is not absolutely continuous with respect to Lebesgue measure on $[\mathbb{R}^2]^n$.

Remark For fixed values of y with $\nu(\{y\}) = 0$, one can arbitrarily (re-) define the density $q_\theta(y)$ without changing the corresponding distribution Q_θ. If $\nu(\{y\}) = 0$ for the observed y, it is therefore not clear what $q_\theta(y)$ is; the function q_θ is only defined ν-a.s. This same fact holds true in many familiar maximum likelihood density estimation problems. It is important to fix the version of the density to be used without (or, before) using the observed data.

Let us now consider the definition and monotonicity property of the EM algorithm in the absence of condition (7.21). We will need the abstractly defined conditional expectation to define the function $\phi_\theta(\theta')$. Under condition (7.21) we could fix the observed y in advance and define the conditional expectations without considering other values that could have been observed. If (7.21) does not hold, we need to take these other possible y-values into account and carefully manage many null sets that enter the picture.

First the defining property of the conditional expectation of a measurable function $S: \mathcal{X} \to \mathbb{R}$, given $T(X) = y$. Let P be a measure on $(X, \mathcal{A})$ and $Q = PT^{-1}$ its induced measure on $(\mathcal{Y}, \mathcal{B})$. Then

$$h(y) = E_P(S(X)|T(X) = y)$$

means that $h: \mathcal{Y} \to \mathbb{R}$ is a $\mathcal{B}$–Borel measurable function satisfying

$$\int_B h(y)\, dQ(y) = \int_{T^{-1}(B)} S(x)\, dP(x) \text{ for all } B \in \mathcal{B}.$$

This function h exists and is Q-a.e. uniquely determined. Now consider one EM step, starting at $\theta^{(m)}$. Then the function $\phi_{\theta^{(m)}}$ is to be computed at all values θ'. In fact, this is a function of y as well. Now, for our fixed $\theta^{(m)}$, we can for each $\theta' \in \Theta$ define the function

$$\phi_\theta(\theta'; y) = E_{P_{\theta(m)}}(\log p_{\theta'}(X)|T(X) = y).$$

In fact, for each θ' we select a version of this conditional expectation, which is $Q_{\theta^{(m)}}$-a.e. unique.

Second, we get:

$$\begin{aligned}
&E_{P_{\theta(m)}} \{\log k_{\theta^{(m+1)}}(X|y) \mid T(X) = y\} - E_{P_{\theta(m)}} \{\log k_{\theta^{(m)}}(X|y) \mid T(X) = y\} \\
&E_{P_{\theta(m)}} \left\{\log \frac{k_{\theta^{(m+1)}}(X|y)}{k_{\theta^{(m)}}(X|y)} \mid T(X) = y\right\} \\
&\le \log E_{P_{\theta(m)}} \left\{ \frac{k_{\theta^{(m+1)}}(X|y)}{k_{\theta^{(m)}}(X|y)} \mid T(X) = y\right\},
\end{aligned} \tag{7.29}$$

where (the conditional form of) Jensen's inequality is used in the last step. But we have:

$$E_{P_{\theta(m)}} \left\{ \frac{k_{\theta^{(m+1)}}(X|y)}{k_{\theta^{(m)}}(X|y)} \mid T(X) = y\right\} = 1, \text{ a.e. } [Q_{\theta^{(m)}}]. \tag{7.30}$$

This is seen in the following way. Let the function $g: \mathcal{Y} \to \mathbb{R}$ represent the left hand side of (7.30):

$$g(y) = E_{P_{\theta(m)}} \left\{ \frac{k_{\theta^{(m+1)}}(X|y)}{k_{\theta^{(m)}}(X|y)} \mid T(X) = y\right\}. \tag{7.31}$$

Then g is a $\mathcal{B}$-measurable function that is defined by the following relation:

$$\int_B g(y)\, dQ_{\theta^{(m)}}(y) = \int_{T^{-1}(B)} \frac{k_{\theta^{(m+1)}}(x|T(x))}{k_{\theta^{(m)}}(x|T(x))}\, dP_{\theta^{(m)}}(x), \ \forall B \in \mathcal{B} \tag{7.32}$$

(using the general definition of conditional expectations). But since we can write

$$\frac{k_{\theta^{(m+1)}}(x|T(x))}{k_{\theta^{(m)}}(x|T(x))} = \frac{p_{\theta^{(m+1)}}(x)}{q_{\theta^{(m+1)}}(T(x))} \cdot \frac{q_{\theta^{(m)}}(T(x))}{p_{\theta^{(m)}}(x)},$$

the right hand side of (7.32) can be written:

$$\int_{T^{-1}(B)} \frac{p_{\theta^{(m+1)}}(x)}{q_{\theta^{(m+1)}}(T(x))} \cdot q_{\theta^{(m)}}(T(x))\,d\mu(x) = \int_{T^{-1}(B)} \frac{q_{\theta^{(m)}}(T(x))}{q_{\theta^{(m+1)}}(T(x))}\,dP_{\theta^{(m+1)}}(x)$$

$$= \int_B \frac{q_{\theta^{(m)}}(y)}{q_{\theta^{(m+1)}}(y)}\,dQ_{\theta^{(m+1)}}(y) = \int_B q_{\theta^{(m)}}(y)\,d\nu(y) = \int_B 1\,dQ_{\theta^{(m)}}(y), \tag{7.33}$$

implying, using (7.32) and (7.33),

$$g(y) = E_{P_\theta^{(m)}}\left\{ \frac{k_{\theta^{(m+1)}}(X|y)}{k_{\theta^{(m)}}(X|y)} \mid T(X) = y \right\} = 1 \text{ a.e. } [Q_{\theta^{(m)}}].$$

So (neglecting things happening on sets of $Q_{\theta^{(m)}}$-measure zero) we get that the last expression in (7.29) is equal to zero!

Now, by combining (7.25), (7.26) and (7.29), we get

$$q_{\theta^{(m+1)}}(y) - q_{\theta^{(m)}}(y) \geq 0,$$

i.e., the likelihood for the parameter θ in the observation space is increased (at least not decreased) at each step of the EM algorithm.

For the general theory on change of variables in integrals with respect to measures, see, e.g., Billingsley, 1995, third edition (the first edition of this book contained an incorrect result of this type).

7.3 The Iterative Convex Minorant Algorithm

In Section 2.1, we considered the problem of minimizing the quadratic function

$$r \mapsto Q(r) = \frac{1}{2}\sum_{i=1}^{n}(r_i - y_i)^2 w_i \tag{7.34}$$

over the closed convex cone $C = \{r \in \mathbb{R}^n : r_1 \leq r_2 \leq \cdots \leq r_n\}$. Here $y \in \mathbb{R}^n$ is a fixed vector to be projected and the weight vector $w \in \mathbb{R}^n$ satisfies $w_i > 0$ for all i. In Lemma 2.1 it was derived that the solution to this optimization problem can be constructed using the convex minorant of a cumulative sum diagram of points.

Often, one has to minimize a more complicated convex function ϕ over the set C. If the function ϕ is smooth, the Newton approach to this problem would be to replace this minimization problem by a sequence of quadratic minimization problems where those problems are chosen such that the objective function ϕ and its (local) quadratic approximation coincide at the subsequent iterates up till second order. Such a quadratic minimization over C is usually complicated in itself. In view of Lemma 2.1, it is therefore natural to replace the full quadratic approximation of a function ϕ by another one, of the form (7.34). If, at some iterate, we can approximate the function ϕ by a weighted sum of squares of this form, we can solve the iteration-optimization by taking the left derivative of the convex minorant of a collection of points. This is the basic idea behind the iterative convex minorant (ICM) algorithm. Let us now state a smoothness condition on C and more formally describe the algorithm.

Condition 7.1 (i) ϕ is convex, continuous and attains its minimum over C at a unique point $\hat{\beta}$,
(ii) ϕ is continuously differentiable on the set $\{\beta \in \mathbb{R}^n : \phi(\beta) < \infty\}$.

Write $\nabla\phi$ for the gradient of ϕ and $(\cdot, \cdot)$ for the usual inner product in $\mathbb{R}^n$. It can be proved analogously to Lemma 7.2 (see Exercise 7.21) that

$$\hat{\beta} = \text{argmin}_{\beta \in C} \phi(\beta)$$

if and only if $\hat{\beta} \in C$ satisfies the optimality conditions

$$(\hat{\beta}, \nabla\phi(\hat{\beta})) = 0 \text{ and } (\beta, \nabla\phi(\hat{\beta})) \geq 0 \ \ \forall \beta \in C. \tag{7.35}$$

The first thing to do is to find an appropriate approximation for ϕ at a certain point $\gamma \in C$ with $\phi(\gamma) < \infty$. By the assumption on ϕ, we can write for each positive diagonal $n \times n$ matrix D (which may depend on γ):

$$\begin{aligned} \phi(\beta) - \phi(\gamma) &= (\beta - \gamma)^T \nabla\phi(\gamma) + \frac{1}{2}(\beta - \gamma)^T D(\beta - \gamma) + o(\|\beta - \gamma\|) \\ &= c_\gamma + \frac{1}{2}(\beta - \gamma + D^{-1}\nabla\phi(\gamma))^T D(\beta - \gamma + D^{-1}\nabla\phi(\gamma)) + o(\|\beta - \gamma\|) \end{aligned}$$

as $\beta \to \gamma$. Here $c_\gamma = \frac{1}{2}\nabla\phi(\gamma)^T D^{-1} \nabla\phi(\gamma)$ does not depend on β. In the terminology in Section 7.1, the most natural algorithmic map to consider is the point-to-point map

$$B(\gamma) = \text{argmin}_{\beta \in C} \frac{1}{2}(\beta - \gamma + D^{-1}\nabla\phi(\gamma))^T D(\beta - \gamma + D^{-1}\nabla\phi(\gamma)). \tag{7.36}$$

Described as generally as this, the algorithm does not converge. See Exercise 7.5 and 7.10, where examples are given where the algorithm oscillates between two iterates.

In other examples it may happen that the value of ϕ at some iterate is infinite, so that the algorithm is not even well defined. However, we can modify the map B to get a globally convergent algorithm. The key to this modification is entailed in Lemma 7.1. It states that when moving from β to $B(\beta)$ via the line segment connecting these points, initially the value of ϕ decreases. In other words, the mapping B generates a descent direction for ϕ at each $\beta \in C \setminus \{\hat{\beta}\}$ such that $\phi(\beta) < \infty$.

Lemma 7.1 *Let ϕ satisfy conditions 7.1 and $\beta \in C \setminus \{\hat{\beta}\}$ satisfy $\phi(\beta) < \infty$. Then*

$$\phi(\beta + \lambda(B(\beta) - \beta)) < \phi(\beta)$$

for all $\lambda > 0$ sufficiently small.

Proof Fix $\beta \in C \setminus \{\hat{\beta}\}$ with $\phi(\beta) < \infty$ and define the function ψ on $[0, 1]$ as follows:

$$\psi(\lambda) = \phi(\beta + \lambda(B(\beta) - \beta)).$$

It suffices to show that the right derivative of ψ at zero,

$$\psi'(0) = (B(\beta) - \beta)^T \nabla\phi(\beta),$$

is strictly negative. From the definition of $B(\beta)$ and the fact that $\beta \in C$, it follows that

$$B(\beta)^T \left(D(\beta)(B(\beta) - \beta) + \nabla\phi(\beta) \right) = 0 \tag{7.37}$$

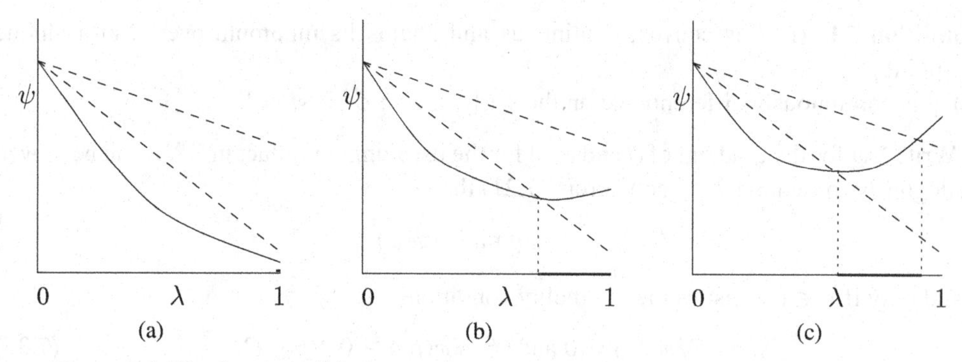

Figure 7.4 The three possible forms of the set returned by the algorithmic map C in the parametrization $\psi(\lambda) = \phi(\beta + \lambda(B(\beta) - \beta))$.

and

$$\beta^T\Big(D(\beta)(B(\beta) - \beta) + \nabla\phi(\beta)\Big) \geq 0. \tag{7.38}$$

Subtracting (7.38) from (7.37) we see that

$$\Big(B(\beta) - \beta\Big)^T D(\beta)\Big(B(\beta) - \beta\Big) + \psi'(0) \leq 0. \tag{7.39}$$

Note that the assumption $\beta \neq \hat{\beta}$ implies that $\beta \neq B(\beta)$. Therefore, since $D(\beta)$ is positive definite, the first term at the left hand side of (7.39) is strictly positive, so that $\psi'(0) < 0$. □

Using Lemma 7.1 we can construct an algorithm that converges to $\hat{\beta}$. The idea behind this (modified) iterative convex minorant algorithm is to select a point $\beta^{(k+1)}$ from the segment

$$\mathrm{seg}(\beta^{(k)}, B(\beta^{(k)})) = \Big\{\beta^{(k)} + \lambda(B(\beta^{(k)}) - \beta^{(k)}) \colon \lambda \in [0, 1]\Big\}$$

such that the value of ϕ decreases sufficiently when moving from $\beta^{(k)}$ to $\beta^{(k+1)}$. This modification is called damping and is also often applied in general Newton algorithms. One way to formalize this idea is to define the algorithmic map C

$$C(\beta) = \begin{cases} \{B(\beta)\} & \text{if } \phi(B(\beta)) < \phi(\beta) + (1-\epsilon)\nabla\phi(\beta)^T(B(\beta) - \beta) \\ \{y \in \mathrm{seg}(\beta, B(\beta)) \colon (1-\epsilon)\nabla\phi(\beta)^T(y - \beta) \leq \phi(y) - \phi(\beta) \leq \\ \qquad\qquad \epsilon\nabla\phi(\beta)^T(y - \beta)\} \text{ elsewhere,} \end{cases} \tag{7.40}$$

where $\epsilon \in (0, 1/2)$ is fixed. Writing $\psi(\lambda) = \phi(\beta + \lambda(B(\beta) - \beta))$ for $\lambda \in [0, 1]$, Figure 7.4 illustrates the idea behind the definition of C. Note that $\psi'(0) = \nabla\phi(\beta)^T(B(\beta) - \beta) < 0$ by Lemma 7.1. If $\psi(1) < \psi(0) + (1-\epsilon)\psi'(0)$, $C(\beta) = B(\beta)$. If $\psi(1) \geq \psi(0) + (1-\epsilon)\psi'(0)$, $C(\beta)$ is the set of vectors y in the segment connecting β and $B(\beta)$ corresponding to λ with $\psi(0) + (1-\epsilon)\lambda\psi'(0) \leq \psi(\lambda) \leq \psi(0) + \epsilon\lambda\psi'(0)$.

Before proving that this algorithm converges under general conditions, some comments on the practical implementation. To completely specify the algorithm for practical implementation, we should fix an initial point, a rule to determine $\beta^{(k+1)}$ from $C(\beta^{(k)})$ and a termination criterion. As an initial point we take any $\beta^{(0)} \in C$ with $\phi(\beta^{(0)}) < \infty$. As a rule to choose $\beta^{(k+1)}$ from $C(\beta^{(k)})$ we propose to choose $\beta^{(k+1)} = B(\beta^{(k)})$ whenever it belongs to $C(\beta^{(k)})$, and, otherwise, perform a binary search for an element of $C(\beta^{(k)})$ in the segment

$\text{seg}(\beta^{(k)}, B(\beta^{(k)}))$. See the pseudocode that follows for an exact description of this binary search, which can easily be seen to terminate after a finite number of steps. Finally, we base our stopping criterion on (7.35), where we use that the inequality part of (7.35) is equivalent to the conditions

$$\sum_{i=j}^{n} \frac{\partial}{\partial \beta_i} \phi(\hat{\beta}) \begin{cases} \geq 0 & \text{for } 1 \leq j \leq n \\ = 0 & \text{for } j = 1. \end{cases}$$

Modified iterative convex minorant algorithm

Input:
$\eta > 0$: accuracy parameter;
$\epsilon \in (0, 1/2)$: line search parameter;
$\beta^{(0)} \in C$: initial point satisfying $\phi(\beta^{(0)}) < \infty$;

begin
  $\beta := \beta^{(0)}$;
  while $|\sum_{i=1}^{n} \beta_i \frac{\partial}{\partial \beta_i} \phi(\beta)| > \eta$ **or** $|\sum_{i=1}^{n} \frac{\partial}{\partial \beta_i} \phi(\beta)| > \eta$ **or** $\min_{1 \leq j \leq n} \sum_{i=j}^{n} \frac{\partial}{\partial \beta_i} \phi(\beta) < -\eta$ **do**
  begin
    $\tilde{y} := \text{argmin}_{y \in C} (y - \beta + D(\beta)^{-1} \nabla \phi(\beta))^T D(\beta)(y - \beta + D(\beta)^{-1} \nabla \phi(\beta))$;
    if $\phi(\tilde{y}) < \phi(\beta) + \epsilon \nabla \phi(\beta)^T (\tilde{y} - \beta)$ **then**
      $\beta := \tilde{y}$
    else
      begin
        $\lambda := 1; s := 1/2; z := \tilde{y}$;
        while $\phi(z) < \phi(\beta) + (1 - \epsilon) \nabla \phi(\beta)^T (z - \beta)$ (I) **or**
              $\phi(z) > \phi(\beta) + \epsilon \nabla \phi(\beta)^T (z - \beta)$ (II) **do**
        begin
          if (I) **then** $\lambda := \lambda + s$;
          if (II) **then** $\lambda := \lambda - s$;
          $z := \beta + \lambda(\tilde{y} - \beta)$;
          $s := s/2$;
        end;
        $\beta := z$;
      end;
  end;
end.

To prove convergence of the modified ICM algorithm to the point $\hat{\beta}$, we use Theorem 7.1.

Theorem 7.3 *Let the function $\phi: \mathbb{R}^n \to (-\infty, \infty]$ satisfy Condition 7.1 and $\beta^{(0)} \in C$ satisfy $\phi(\beta^{(0)}) < \infty$. Let the mapping $\beta \mapsto D(\beta)$ take values in the set of positive definite $(n \times n)$ diagonal matrices such that $\beta \mapsto D(\beta)$ is continuous on the set*

$$K = \{\beta \in C : \phi(\beta) \leq \phi(\beta^{(0)})\}. \tag{7.41}$$

Then an algorithm generated by the mapping C, as defined in (7.40), converges to $\hat{\beta}$.

Proof From Lemma 7.1 it follows that the mapping C is well defined and has ϕ as a descent function. From this observation it follows that

$$\{\beta^{(k)}: k \geq 0\} \subset K,$$

where K is as defined in (7.41). From Condition 7.1 and the fact that $\phi(\beta^{(0)}) < \infty$, it follows that K is compact. Therefore, in view of Theorem 7.1, closedness of C at each $\beta \in K \setminus \{\hat{\beta}\}$ would imply global convergence of the algorithm.

Fix $\beta \in K \setminus \{\hat{\beta}\}$ and a sequence $(\beta^{(k)})$ in K such that $\beta^{(k)} \to \beta$. Let $\gamma^{(k)} \in C(\beta^{(k)})$ with $\gamma^{(k)} \to \gamma$ for some $\gamma \in K$. To prove closedness of C, we have to prove that $\gamma \in C(\beta)$.

First note that continuity of the mapping $\beta \mapsto D(\beta)$ on K and Condition 7.1 yield that

$$\nabla\phi(\beta^{(k)}) \to \nabla\phi(\beta) \text{ and } B(\beta^{(k)}) \to B(\beta) \tag{7.42}$$

as $k \to \infty$. From this it follows that $\gamma \in \text{seg}(\beta, B(\beta))$ necessarily. Now consider the two different situations that can occur.

The first situation is that

$$\phi(B(\beta^{(k)})) \leq \phi(\beta^{(k)}) + (1-\epsilon)\nabla\phi(\beta^{(k)})^T(B(\beta^{(k)}) - \beta^{(k)})$$

for infinitely many values of k. Letting k tend to infinity along a subsequence k_j where this inequality holds, we get from (7.42) that

$$\phi(B(\beta)) \leq \phi(\beta) + (1-\epsilon)\nabla\phi(\beta)^T(B(\beta) - \beta)$$

so that $C(\beta) = \{B(\beta)\}$. Moreover, along the same subsequence it follows from the definition of C that $\gamma^{(k_j)} = B(\beta^{(k_j)})$. Therefore, for $j \to \infty$, $\gamma^{(k_j)} \to B(\beta)$ by the continuity of B. This shows that $\gamma = B(\beta) \in C(\beta)$, as was to be proved.

The other possibility is that for all k sufficiently large

$$\phi(B(\beta^{(k)})) > \phi(\beta^{(k)}) + (1-\epsilon)\nabla\phi(\beta^{(k)})^T(B(\beta^{(k)}) - \beta^{(k)}).$$

Letting $k \to \infty$ and using (7.42), it then follows that

$$\phi(B(\beta)) \geq \phi(\beta) + (1-\epsilon)\nabla\phi(\beta)^T(B(\beta) - \beta).$$

Therefore, according to the definition of C and the fact that $\gamma \in \text{seg}(\beta, B(\beta))$, $\gamma \in C(\beta)$ whenever

$$\phi(\gamma) - \phi(\beta) \in [(1-\epsilon)\nabla\phi(\beta)^T(\gamma - \beta), \epsilon\nabla\phi(\beta)^T(\gamma - \beta)].$$

This, however, immediately follows from the fact that for all k sufficiently large

$$\phi(\gamma^{(k)}) - \phi(\beta^{(k)}) \in [(1-\epsilon)\nabla\phi(\beta^{(k)})^T(\gamma^{(k)} - \beta^{(k)}), \epsilon\nabla\phi(\beta^{(k)})^T(\gamma^{(k)} - \beta^{(k)})],$$

$\beta^{(k)} \to \beta$, $\gamma^{(k)} \to \gamma$ and $\nabla\phi(\beta^{(k)}) \to \nabla\phi(\beta)$. □

To illustrate the algorithm, we apply it to the interval censoring problem as in Example 7.4.

Example 7.6 (*Interval censoring case 2)* Recall from Example 7.4 that we denote by $v_1 < v_2 < \cdots < v_q$ the ordered observed times (so the t_is as well as the u_is); an ML estimator can be found that does not put mass outside this set of v_js. If there are no ties (we will assume this in this example), $q = 2n$. However, in view of log likelihood function (7.12), the number of potential support points (points at which there is mass) of the ML estimator can be reduced further. Indeed, suppose i identifies a subject with $\delta_i = 0$. Then the corresponding $F(u_i)$ does not occur in the log likelihood; only $F(t_i)$ occurs. Similarly, $F(t_i)$ does not occur in the log likelihood if i is such that $\gamma_i = \delta_i = 0$. Let us therefore perform a further reduction of the parameter space, removing all v_js that correspond to u_is with $\delta_i = 1$ and t_is with $\gamma_i = \delta_i = 0$. Abusing notation slightly, we represent the remaining points by $v_1 < v_2 < \cdots < v_p$ (where $p < q$ whenever the reduction step makes a difference).

Writing, for $1 \leq j \leq p$, $\beta_j = \sum_{k=1}^{j} \theta_k$ for the distribution function F evaluated at v_j, we obtain the natural parametrization for the ICM algorithm. Now define the following partition of $\{1, \ldots, p\}$ three disjoint sets of indices I_1, I_{2a}, I_{2b} and I_3:

$$
\begin{aligned}
I_1 &= \{1 \leq j \leq p \colon v_j = t_i \text{ for some } i \text{ with } \delta_i = 1\} \\
I_{2a} &= \{1 \leq j \leq p \colon v_j = t_i \text{ for some } i \text{ with } \gamma_i = 1\} \\
I_{2b} &= \{1 \leq j \leq p \colon v_j = u_i \text{ for some } i \text{ with } \gamma_i = 1\} \text{ and} \\
I_3 &= \{1 \leq j \leq p \colon v_j = u_i \text{ for some } i \text{ with } \gamma_i = \delta_i = 0\}.
\end{aligned}
$$

Moreover, define the function $k \colon I_{2a} \to I_{2b}$ as follows:

$$
k(j) = m \text{ if } v_j = t_i \text{ with } \gamma_i = 1 \text{ and } v_m = u_i \text{ for this } i.
$$

In other words: the v_j corresponding to $j \in I_{2a}$ equals by definition a t_i for some $i \in \{1, \ldots, n\}$ with $\gamma_i = 1$. The u_i for this subject i also appears in the set of v_js: exactly as v_m, where $m \in I_{2b}$. Of course, $k(j) > j$ automatically. For example, if 4 successive realizations of (T, U, Δ, Γ) are given by

$$
(4, 12, 0, 1), \ (3, 7, 1, 0), \ (6, 11, 0, 0) \text{ and } (9, 14, 1, 0), \tag{7.43}
$$

then the ordered and pooled observed time points are given by 3, 4, 6, 7, 9, 11, 12, 14. This set reduces to the following v_js: 3, 4, 9, 11, 12 (e.g., the 6 vanishes because it is t_3 with corresponding $\delta_i = \gamma_i = 0$). Moreover, $I_1 = \{1, 3\}$ since $v_1 = 3$ corresponds to t_2, which has associated $\delta_2 = 1$, and similarly $v_3 = 9$ corresponds to t_4 with associated $\delta_4 = 1$. Also, $I_{2a} = \{2\}$ since $v_2 = 4 = t_1$ with $\gamma_1 = 1$, and $I_{2b} = \{5\}$ since $v_5 = 12 = u_1$ with $\gamma_1 = 1$. The function k is simply $k(2) = 5$. Using this notation, we can rewrite minus the log likelihood defined in (7.12) as

$$
\phi(\beta) = -\left(\sum_{j \in I_1} \log \beta_j + \sum_{j \in I_{2a}} \log(\beta_{k(j)} - \beta_j) + \sum_{j \in I_3} \log(1 - \beta_j) \right).
$$

Note that in this representation we sum over the indexes of the points $v_1, \ldots, v_p$ rather than over the n observation points.

The aim is to minimize this function over the p-dimensional cone C. In order to implement the ICM algorithm, we need the first and second partial derivatives of ϕ with respect to the parameters β_j for $1 \leq j \leq p$. These first derivatives are given by

$$
r_j(\beta) = \frac{\partial \phi}{\partial \beta_j}(\beta) = \begin{cases} -\beta_j^{-1} & \text{for } j \in I_1 \\ (\beta_{k(j)} - \beta_j)^{-1} & \text{for } j \in I_{2a} \\ -(\beta_j - \beta_{k^{-1}(j)})^{-1} & \text{for } j \in I_{2b} \\ (1 - \beta_j)^{-1} & \text{for } j \in I_3. \end{cases}
$$

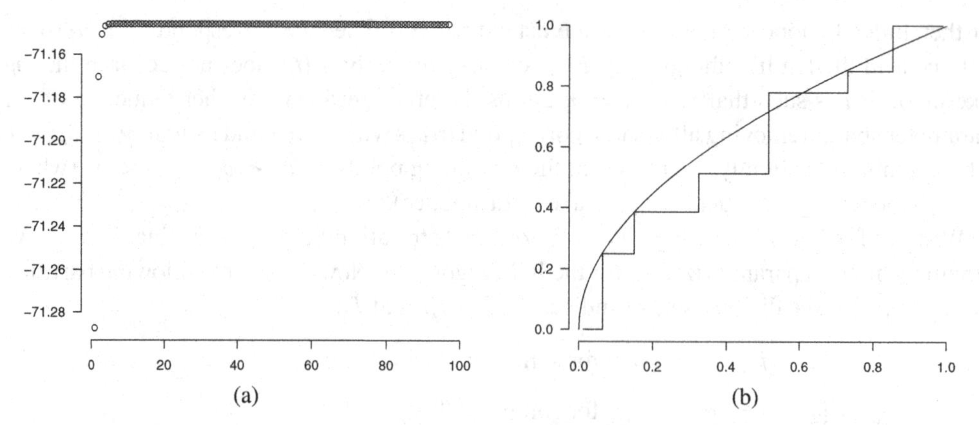

Figure 7.5 (a) The log likelihood values of the consecutive iterates (starting from the fourth, for reasons of scaling) of the ICM algorithm. (b) The iterate after 100 iterations together with the underlying distribution function.

The diagonal entries of the Hessian matrix of ℓ are given by

$$d_j(\beta) = \frac{\partial^2 \phi}{\partial \beta_j^2}(\beta) = \begin{cases} \beta_j^{-2} & \text{for } j \in I_1 \\ (\beta_{k(j)} - \beta_j)^{-2} & \text{for } j \in I_{2a} \\ (\beta_j - \beta_{k^{-1}(j)})^{-2} & \text{for } j \in I_{2b} \\ (1 - \beta_j)^{-2} & \text{for } j \in I_3. \end{cases}$$

Note that the Hessian matrix is reasonably sparse, in the sense that many off-diagonal elements $\partial^2\phi/\partial\beta_j\partial\beta_i$ for $i \neq j$ are zero (see also Exercise 7.23).

Now, at a given iterate satisfying $\phi(\beta) < \infty$, say $\beta^{(0)}$, a local quadratic approximation of ϕ with diagonal Hessian (apart from an additive constant not depending on β) is given by

$$\sum_{i=1}^{p} \left(\beta_i - \beta_i^{(0)} + r_j(\beta^{(0)})/d_j(\beta^{(0)})\right)^2 d_i(\beta^{(0)}).$$

The minimizer of this function over C can be computed using the remark following Lemma 2.1. Using the diagram of points consisting of $P_0 = (0, 0)$ and for $1 \le j \le p$

$$P_j = \left(\sum_{i=1}^{j} d_i(\beta^{(0)}), \sum_{i=1}^{j} \beta_i^{(0)} d_i(\beta^{(0)}) - r_i(\beta^{(0)})\right),$$

the jth component ($1 \le j \le p$) of the minimizer is given by the left derivative of the convex minorant of this diagram of points, evaluated at P_j.

Using the same data as in Example 7.4 and starting from the uniform distribution on $v_1, \ldots, v_p$ (so $\beta_j^{(0)} = j/p$ for $1 \le j \le p$), Figure 7.5a shows the values of the log likelihood at the successive iterations. Figure 7.5b shows the resulting approximation after 100 iterations. It is interesting that the value of the log likelihood attained by the EM algorithm after 5000 iterations (-71.14584) is attained by the ICM algorithm after only 11 (fast) iterations.

7.4 Vertex Direction Algorithms

Consider the normal (or Gaussian) deconvolution problem introduced in Example 4.1. The sampling density of Z is then given by a (location) mixture of unit variance Gaussian densities. It can be written as

$$g(z) = \int_{-\infty}^{\infty} \frac{1}{\sqrt{2\pi}} \exp\left(-\tfrac{1}{2}(z-\theta)^2\right) d\mu(\theta) = \int_{-\infty}^{\infty} g_\theta(z)\, d\mu(\theta)$$

for some mixing probability measure μ. In other words, the density g belongs to the convex hull of the set of basis functions

$$\mathcal{G} = \{g_\theta : \theta \in \mathbb{R}\}, \text{ where } g_\theta(z) = \frac{1}{\sqrt{2\pi}} \exp\left(-\tfrac{1}{2}(z-\theta)^2\right).$$

All deconvolution models are mixture models. Also other classes of densities can be identified as mixtures, e.g., observation densities in interval censoring models, the class of decreasing densities and the class of convex decreasing densities (see Exercise 7.25). In this section we consider M-estimators in such models. These are defined as solution to a (usually infinite dimensional) minimization problem over the set of mixtures.

A whole class of algorithms, called vertex direction algorithms, has been developed that can be applied to solve these optimization problems. The iterations in algorithms of this type consist of two steps. First, starting from a current iterate, one or more profitable directions for the objective function are determined. This step entails minimizing a function over a low dimensional set Θ, parameterizing the set of basis functions (or vertices) of the model. In the Gaussian deconvolution problem $\Theta = \mathbb{R}$. Subsequently, given the current iterate, the basis functions thus determined are used to decrease the objective function. In many examples this latter step boils down to a moderate-dimensional minimization problem. Before explaining these algorithms in more detail, we first state the general optimization problem.

Consider the following type of optimization problem

$$\text{minimize } \phi(g) \quad \text{for} \quad g \in C \tag{7.44}$$

where ϕ is a convex function defined on (a superset of) a convex set of functions C. We assume throughout that ϕ has a unique minimizer over C.

Assumption 7.1 ϕ is a convex function on C such that for each $f, g \in C$ where ϕ is finite, the function $t \mapsto \phi(g + t(f-g))$ is continuously differentiable for $t \in (0, 1)$.

Now define, for each $g \in C$ and h, a function such that for some $\epsilon > 0$, $g + \epsilon h \in C$,

$$D_\phi(h; g) = \lim_{\epsilon \downarrow 0} \epsilon^{-1} (\phi(g + \epsilon h) - \phi(g)).$$

Note that this quantity exists (possibly equal to ∞) by convexity of ϕ. As we will see, a choice often made for h is $h = f - g$ for some arbitrary $f \in C$. In that case we have

$$D_\phi(f - g; g) = \lim_{\epsilon \downarrow 0} \epsilon^{-1} (\phi(g + \epsilon(f - g)) - \phi(g)).$$

The following simple but important result gives necessary and sufficient conditions for $\hat{g}$ to be the solution of (7.44).

Lemma 7.2 *Suppose that ϕ satisfies Assumption 7.1. Then*

$$\hat{g} = \text{argmin}_{g \in C} \phi(g) \quad \textit{if and only if} \quad D_\phi(f - \hat{g}; \hat{g}) \geq 0 \quad \textit{for all} \;\; f \in C.$$

Proof First we prove $\Rightarrow$. Suppose $\hat{g} = \text{argmin}_{g \in C} \phi(g)$ and choose $f \in C$ arbitrarily. Then, for $\epsilon \downarrow 0$

$$0 \leq \epsilon^{-1}(\phi(\hat{g} + \epsilon(f - \hat{g})) - \phi(\hat{g})) \downarrow D_\phi(f - \hat{g}; \hat{g}).$$

Now $\Leftarrow$. For arbitrary $f \in C$, write τ for the convex function $\epsilon \mapsto \phi(\hat{g} + \epsilon(f - \hat{g}))$ and note that

$$\phi(f) - \phi(\hat{g}) = \tau(1) - \tau(0) \geq \tau'(0+) = D_\phi(f - \hat{g}; \hat{g}) \geq 0.$$

□

An interesting special case emerges when C is the convex hull of a class of functions, as in the Gaussian deconvolution model

$$\mathcal{G} = \{g_\theta : \theta \in \Theta \subset \mathbb{R}^k\}, \tag{7.45}$$

in the sense that

$$C = \text{conv}(\mathcal{G}) = \left\{ g = \int_\Theta g_\theta \, d\mu(\theta) : \mu \text{ probability measure on } \Theta \right\}. \tag{7.46}$$

Example 7.7 (*Convex decreasing density)* The class of convex decreasing densities on $[0, \infty)$ has representation (7.46) with

$$g_\theta(x) = \frac{2(\theta - x)}{\theta^2} 1_{(0,\theta)}(x), \quad \theta > 0.$$

See Exercise (7.25)

Example 7.8 (*Mixture of unit variance normals)* The Gaussian deconvolution problem entails estimation of a density (and associated mixing distribution) that belongs to the convex hull of the class of normal densities with unit variance:

$$g_\theta(x) = \frac{1}{\sqrt{2\pi}} e^{-\frac{1}{2}(x-\theta)^2}.$$

In the examples just considered, usually one has a sample $Z_1, Z_2, \ldots, Z_n$ from the unknown density $g \in C$, and wants to estimate the underlying density based on that sample. We now consider two types of nonparametric shape constrained density estimators: least squares (LS) estimators and maximum likelihood (ML) estimators.

We define a LS estimate of the density in C as minimizer of the function

$$\phi(g) = \frac{1}{2} \int_0^\infty g(z)^2 \, dt - \int_0^\infty g(z) \, d\,\mathbb{G}_n(z) \tag{7.47}$$

over the class C. Here $\mathbb{G}_n$ is the empirical distribution function of the sample. The reason for calling this estimator a LS estimator is that if the empirical distribution function $\mathbb{G}_n$ had a density g_n with respect to Lebesgue measure,

$$\frac{1}{2}\int (g(t)-g_n(t))^2\,dt = \frac{1}{2}\int g(t)^2\,dt - \int g(t)g_n(t)\,dt + \frac{1}{2}\int g_n(t)^2\,dt$$

would be the natural function to minimize for a least squares estimator. The last term in this expression does not depend on g. The first two terms correspond to (7.47). See Exercise 1.1 and (2.27) for a related reasoning in the context of density estimation and estimating an increasing hazard function. Note that for objective function (7.47)

$$D_\phi(h;g) = \lim_{\epsilon\downarrow 0}\epsilon^{-1}\left(\phi(g+\epsilon h)-\phi(g)\right) = \int h(x)g(x)\,dx - \int h(x)\,d\,\mathbb{G}_n(x).$$

As maximum likelihood estimate we define the minimizer of the function

$$\phi(g) = -\int \log g(x)\,d\,\mathbb{G}_n(x)$$

over the class of densities C. Note that for this function

$$D_\phi(h;g) = \lim_{\epsilon\downarrow 0}\epsilon^{-1}\left(\phi(g+\epsilon h)-\phi(g)\right) = -\int \frac{h(x)}{g(x)}\,d\mathbb{G}_n(x).$$

For both objective functions ϕ, the function D_ϕ has the linearity property stated as follows.

Assumption 7.2 The function ϕ has the property that for each $g \in C$ and $f = \int_\Theta g_\theta\,d\mu_f(\theta) \in C$,

$$D_\phi(f-g;g) = \int_\Theta D_\phi(g_\theta - g; g)\,d\mu_f(\theta). \tag{7.48}$$

Under this additional assumption, the nonnegativity condition in Lemma 7.2 that has to hold for each $f \in C$ may be restricted to functions $f \in \mathcal{G}$.

Lemma 7.3 *Suppose that C =conv($\mathcal{G}$) with $\mathcal{G}$ as in (7.45) and that ϕ satisfies Assumptions 7.1 and 7.2. Then*

$$\hat{g} = \text{argmin}_{g\in C}\phi(g) \quad \textit{if and only if} \quad D_\phi(g_\theta - \hat{g};\hat{g}) \geq 0 \;\; \textit{for all} \;\; \theta \in \Theta.$$

Proof This follows immediately from Lemma 7.2, the fact that $g_\theta \in C$ and (7.48) □

For the situation of Lemma 7.3, there is a variety of algorithms to solve (7.44) that can be called "of vertex direction (VD) type." A common feature of VD algorithms is that they consist of two basic steps. Given a current iterate g, find a value of $\theta \in \Theta$ such that $D_\phi(g_\theta - g; g)$ is negative. (If such a value cannot be found, the current iterate is optimal!) This means that traveling from the current iterate in the direction of g_θ would (initially) decrease the value of the function ϕ.

Having found such a feasible profitable direction from the current iterate, the next step is to solve some low-dimensional optimization problems to get to the next iterate.

For the original VD algorithm, the first step is naturally implemented as follows. Given the current g, find $\hat\theta$ corresponding to the minimizer of $D_\phi(g_\theta - g; g)$ over Θ. The second step is then to choose the function

$$g^{(new)} = (1-\hat\epsilon)g + \hat\epsilon g_{\hat\theta}$$

where $\hat{\epsilon}$ is given by

$$\hat{\epsilon} = \text{argmin}_{\epsilon\in[0,1]}\phi((1-\epsilon)g + \epsilon g_{\hat{\theta}}).$$

In other words, the next iterate is the optimal convex combination of the current iterate and the most promising vertex in terms of the directional derivative. It is clear that usually the next iterate has one more support point than the current iterate.

Another algorithm of VD type is the vertex exchange algorithm, which not only uses the parameter $\hat{\theta}$ corresponding to the minimizer of $D_\phi(g_\theta - g; g)$, but also the maximizer $\check{\theta}$ of $D_\phi(g_\theta - g; g)$ restricted to the basis functions currently represented in the current iterate to get a direction. The set of parameters corresponding to basis functions represented in a mixture g is called the support of g. Denote by $\mu_g(\{\check{\theta}\})$ the mass assigned to $\check{\theta}$ by the mixing distribution corresponding to g. Then the direction given by the algorithm is $g + \mu_g(\{\check{\theta}\})(g_{\hat{\theta}} - g_{\check{\theta}})$. The new iterate becomes

$$g + \hat{\epsilon}\mu_g(\{\check{\theta}\})(g_{\hat{\theta}} - g_{\check{\theta}})$$

where

$$\hat{\epsilon} = \text{argmin}_{\epsilon\in[0,1]}\phi(g + \epsilon\mu_g(\{\check{\theta}\})(g_{\hat{\theta}} - g_{\check{\theta}})).$$

If $\hat{\epsilon} = 1$, the point $\check{\theta}$ is eliminated from the support of the current iterate, and the mass assigned to $\check{\theta}$ by the old mixing distribution is moved to the new point $\hat{\theta}$. It is clear that in this algorithm the number of support points of the iterate can increase by one, remain the same, or also decrease by one during one iteration (if $\hat{\epsilon} = 1$ and $\hat{\theta}$ already belongs to the support). In specific examples, the number of support points of the solution $\hat{g}$ is known to be smaller than a constant N which only depends on the data (and is known in advance).

Another variation on the theme is called the intra simplex direction algorithm. The set of all local minima $\{\theta_1, \dots, \theta_m\}$ of $D_\phi(g_\theta - g; g)$, where D_ϕ is negative, is determined and the optimal convex combination of the current iterate and all vertices $g_{\theta_1}, \dots, g_{\theta_m}$ is the new iterate. This final step is to minimize a convex function in the variables $\epsilon_1, \dots, \epsilon_m \in [0, 1]$ under the constraint $0 \le \sum_{i=1}^m \epsilon_i \le 1$.

Another variation on the original VD algorithm is the Simar algorithm, originally applied to solve a Poisson mixture problem. It sticks to the original idea of picking one θ corresponding to a profitable direction. The second step differs from those indicated earlier in the sense that the the new iterate is chosen to minimize the objective function over a finite dimensional subset of the mixture class rather than taking a convex combination of the current iterate with some other functions. Denote by S_g the set of support points of the mixing measure corresponding to a function $g \in C$. Then, given $\hat{\theta}$, the next iterate is given by

$$g^{(new)} = \text{argmin}_{h\in C(g)}\phi(h), \quad \text{where} \quad C(g) = \{h \in C\colon S_h \subset S_g \cup \{\hat{\theta}\}\}.$$

It is to be noted that support points can (and usually do) vanish during this second step.

We now illustrate the support reduction algorithm to compute the least squares estimator in convex regression.

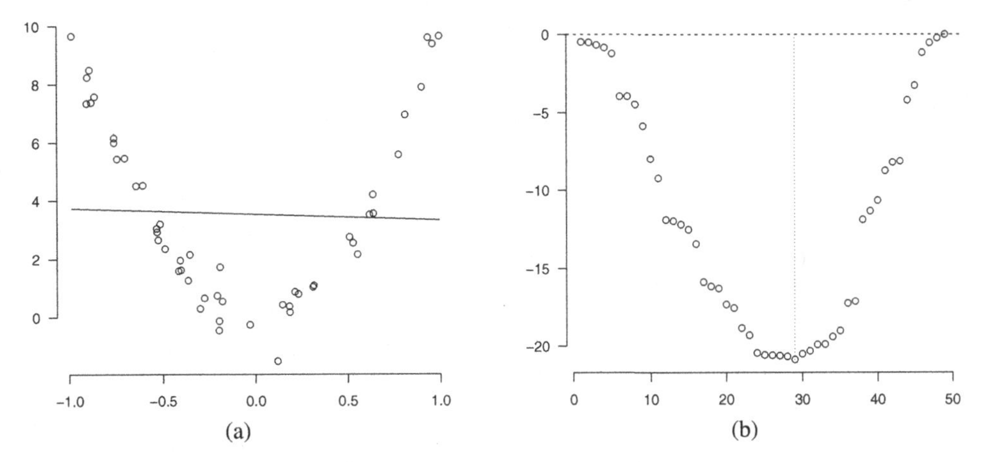

Figure 7.6 Scatter plot of the dataset of size $n = 50$ from the convex regression model (a) and the associated function D defined in (7.49) (b).

Example 7.9 (*Least squares estimation of a convex regression function*) Consider the convex regression problem introduced in Section 4.3. Data $(x_1, y_1), \ldots, (x_n, y_n)$ are given, and these are assumed to be generated by the following model:

$$Y_i = r_0(x_i) + \epsilon_i$$

where r_0 is a convex function on $\mathbb{R}$ and the ϵ_is are independent centered random variables with finite second moment. Assume (if there are no ties without loss of generality) that $x_1 < x_2 < \cdots < x_n$. The least squares estimate for r_0 is defined as the minimizer of the function

$$\phi(r) = \frac{1}{2} \sum_{i=1}^{n} (y_i - r(x_i))^2$$

over the class of all convex regression functions r on $\mathbb{R}$. As seen in Section 4.3, we may (and will) restrict ourselves to estimating r_0 by a piecewise linear function on the range of the x_is with knots restricted to the x_is. Existence and uniqueness of this estimator were established in Lemma 4.1. Now note that all convex functions of this type can be represented as

$$r(x) = \sum_{i=1}^{n} \alpha_i r_i(x), \quad x \in [x_1, x_n]$$

where $r_i(x) = (x_{i+1} - x)_+$ for $1 \le i \le n-1$, $r_n(x) \equiv 1$, $\alpha_{n-1}, \alpha_n \in \mathbb{R}$; therefore the class of functions can be represented as a (closed) convex cone generated by the functions $r_1, r_2, \ldots, , \pm r_{n-1}, \pm r_n$ and the optimization problem fits within the setting of vertex direction methods.

To explain the algorithm, consider the data set of size $n = 50$ showed in Figure 7.6a. The line in the picture is the usual least squares linear regression. This is a feasible solution (the function is convex), but clearly not the optimal solution in the convex regression problem at hand. Figure 7.6b shows the directional derivative function evaluated at the current iterate,

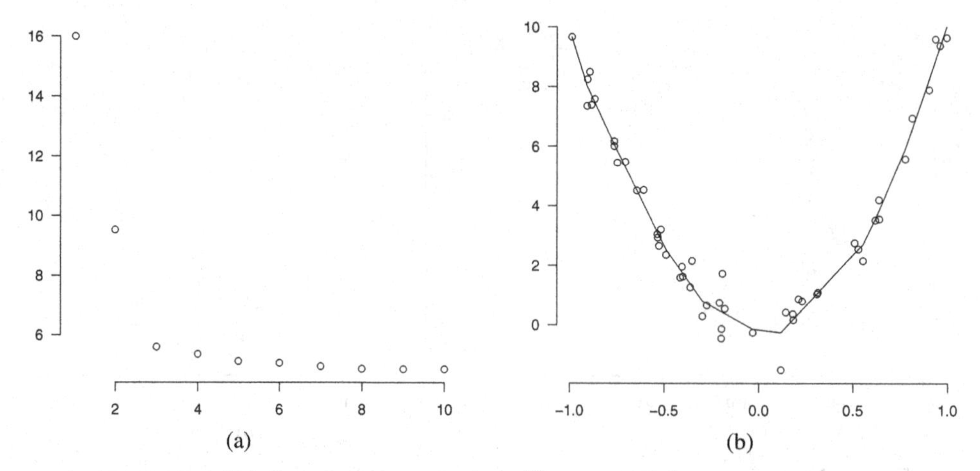

Figure 7.7 (a) The successive values $\phi(r^{(k)})$ for the iterates produced by the Support Reduction algorithm (except for the first for reasons of scaling). (b) The scatter plot with the (final) iterate produced by the algorithm after 11 iterations.

in the direction of generators $r_1, \ldots, r_{n-1}$:

$$D(j) = \lim_{\epsilon \downarrow 0} \epsilon^{-1}\left(\phi(r^{(0)} + \epsilon r_j) - \phi(r^{(0)})\right) = \sum_{i=1}^{j}(y_i - r^{(0)}(x_i))(x_{j+1} - x_i)$$
$$= \sum_{k=1}^{j}(x_{k+1} - x_k)\sum_{i=1}^{k}(y_i - r^{(0)}(x_i)). \tag{7.49}$$

By definition, starting from the linear regression, the derivatives in the directions r_{n-1} and r_n are zero (a property that will be forced at the subsequent iterates as well; see the following). If $D(j)$ would be nonnegative for all j, the current iterate would be the optimal solution. As can be seen in the graph of D, this is certainly not the case. There are various js for which the directional derivatives are still negative. In the next step, based on D, we choose the j such that $D(j)$ is most negative. In this case, it is index 29, as indicated in the graph of D. Now a minimization of ϕ is performed over all piecewise linear functions having knots at x_1, x_{28} and x_{50} and satisfying

$$D(n-1) = \sum_{i=1}^{n}(y_i - r(x_i))x_i = 0 \text{ and } D(n) = \sum_{i=1}^{n}(y_i - r(x_i)) = 0.$$

If this function is feasible (i.e., convex in this example), this is the new iterate. If not, a support reduction step is performed. This means that starting from the old iterate, the new (unrestricted) iterate is approached via the line segment connecting the two points in $\mathbb{R}^n$. Before reaching the new iterate, at some point the convex combination of the two will become infeasible, in the sense that the convexity restriction will be violated if one would go closer to the suggested unrestricted minimizer. At this point, the point where this violation takes place is removed from the set of support points and a new unrestricted iterate is computed. This process is continued until a new iterate is found that is feasible. It can be proved that this algorithm leads to a new feasible iterate having ϕ value strictly smaller than the old value. Moreover, the resulting algorithm can be shown to converge. Figure 7.7 shows the values of ϕ evaluated at the subsequent iterates as well as the iterate emerging after 11 iterations.

We end this section with with the bivariate interval censoring problem introduced in Section 5.2.

Example 7.10 *(Bivariate interval censoring case 2)* In this model (see Section 5.2), the log likelihood function is given by (5.19).

For computational purposes, it is convenient to introduce rectangles to which the unobservable observations are known to belong (see Figure 5.7), where we represent the (one-sided) unbounded rectangles by finite rectangles with upper or lower bounds outside the range of the observed data. In this set up we have to maximize

$$\sum f_i \log H_i' p, \tag{7.50}$$

where $p = (p_1, \ldots, p_m)'$ is a vector of probability masses at possible points of mass (x_j, y_j) and H_i is a vector of length m, consisting of ones and zeros, where the component H_{ij} is equal to 1 if the point (x_j, y_j) is contained in the rectangle

$$[L_{i1}, R_{i1}] \times [L_{i2}, R_{i2}],$$

and is zero, otherwise. The masses p_j should be nonnegative and sum to 1. This optimization can easily be handled by using iterative quadratic minimization and the support reduction algorithm.

An `R` package, called `MLEcens`, written by Marloes Maathuis, is available for computing the MLE. The algorithm determines rectangles where the MLE has mass via a preliminary reduction algorithm, and next computes the mass of the MLE in these rectangles, using the support reduction algorithm. The reduction algorithm is described in Maathuis, 2005. The `R` package uses as an example a data set, studied in Betensky and Finkelstein, 1999.

We computed the MLE by a *C* program based on the support reduction algorithm with iterative quadratic minimization of Groeneboom, Jongbloed and Wellner, 2008, which is also used in the `R` package `MLEcens`. There is an extensive discussion on where to put the mass once one has determined rectangles that can have positive mass, see, e.g., Sun, 2006, Section 7.3, Gentleman and Vandal, 2002, Bogaerts and Lesaffre, 2004, and Maathuis, 2005. The algorithm for computing these rectangles, proposed in Maathuis, 2005, seems at present to be the fastest.

We propose a method that avoids computation of these canonical rectangles, also avoids discussion of whether one should place the mass of the MLE at the left lower bound or the right upper bound of the rectangles. We specify in advance a set of points where one allows mass to be placed. In this way we obtain an MLE on a sieve, where the sieve consists of distributions having discrete mass at these points. The bottleneck in the computation of the MLE for the bivariate interval censoring problem is not the determination of the canonical rectangles, but the computation of the mass the MLE puts on these canonical rectangles, since there usually are very many.

The latter phenomenon shows up in particular in simulations. As an example, simulating data from the distribution with density $f(x, y) = x + y$ on the unit square, with a uniform observation distribution, we got for sample size $n = 5000$ about $5 \cdot 10^5$ possible rectangles where the masses could be placed, which is (at present) an almost prohibitive number if one wants to do simulations for the limit behavior of the MLE on an ordinary table computer. Ultimately, the discussion on these matters should be determined by insights into the distribution theory of the MLE or the MLE on the chosen sieve. But unfortunately, if the underlying distributions are not assumed to be purely discrete, at present nothing is known about this, in contrast with the situation in dimension 1.

We found the following method for choosing points of possible mass to work well in practice. For each observation rectangle $[t_{i1}, u_{i1}] \times [t_{i2}, u_{i2}]$ we generate uniformly a point x_i in $[t_{i1}, u_{i1}]$ and a point y_i in $[t_{i2}, u_{i2}]$. This yields n points (x_i, y_i) of possible positive mass. This presupposes that the rectangles have finite bounds, but this is the usual set-up for computing the MLE: the infinite bounds are replaced by bounds just outside the rectangle, containing the observations. Natural choices for the right and upper bounds in the Betensky-Finkelstein data would be 24 and 30, since the intervals have right bounds at multiples of 3, and since the largest observation on the first axis is 21 and on the second axis 27.

To illustrate the results of this method on simulated data, we show the estimate of the bivariate distribution function, which we call MLE again, since it is the MLE on a sieve, in Figure 5.8 (in Section 5.2) together with a picture of the real underlying distribution function for a sample of size $n = 5000$. Although there are 5000 possible points of mass initially, the support reduction algorithm only selects 169 of these as points of positive mass. Figure 5.9 and Figure 5.10 are also based on this simulation study.

The data of Betensky and Finkelstein, 1999, are given in Table 7.1, where the rectangles to which the hidden observations are known to belong are denoted by $[L_{i1}, R_{i1}] \times [L_{i2}, R_{i2}]$, $i = 1, \ldots, n$. The frequencies of the hidden observations belonging to these rectangles are given in the 5th and 10th column. There are 87 observation rectangles and the total sample size, taking the frequencies into account, is 204. The table is also given in Sun, 2006, table 7.1, p. 165, but there the rectangles are slightly changed from the data in Betensky and Finkelstein, 1999, by lowering the left bounds by 1 if they are larger than zero. Since we do not see a pressing reason for doing that, we just give the data here as they were given by Betensky and Finkelstein, 1999. If the upper bound R_{ij} is unknown, we put $R_{ij} = \infty$ and if the lower bound L_{ij} is unknown, we put $L_{ij} = -\infty$.

The maximal intersection rectangles where the MLE will put its mass are given in Table 7.2a. They can be computed, for example, by applying the reduction algorithm, used in the `R` package `MLEcens`.

To facilitate the comparison with the existing literature, we will only discuss the MLE, based on the preliminary reduction to rectangles which can have mass, and not follow the procedure we used for computing the MLE on a sieve in the simulation from the density $f(x, y) = x + y$ on $[0, 1]^2$. We will use the convention of putting the mass of the MLE in the left lower corner of these rectangles, except that we do not allow values less than zero (so we replace $-\infty$ by 0), and compute the MLE by the support reduction algorithm of Groeneboom, Jongbloed and Wellner, 2008. The result is shown in Table 7.2, where the masses of the MLE are given. It is seen that this is in close correspondence with Table 7.2 on p. 166 of Sun, 2006, apart from the slightly different definition of the rectangles. The `R` package `MLEcens` also gives this result (in all 9 decimals).

As remarked earlier, we have some freedom in choosing the input to the support reduction algorithm. We can first compute the canonical rectangles, and allow as possible points of mass the left lower corners of these rectangles, or we could just allow points of mass at, e.g., the points $(3i, 3j)$, $i, j = 0, 10$. In both cases the support reduction algorithm arrives at

Table 7.1 *The Betensky-Finkelstein Data*

L_{i1}	R_{i1}	L_{i2}	R_{i2}	Frequency	L_{i1}	R_{i1}	L_{i2}	R_{i2}	Frequency
0	3	0	–	3	6	–	6	–	3
0	3	3	–	1	6	–	9	–	1
0	3	6	–	3	9	–	0	–	2
0	6	6	–	1	9	–	9	–	3
0	3	9	–	1	9	–	12	–	1
0	3	12	–	5	12	–	0	–	5
0	3	15	–	5	12	–	6	–	1
0	6	15	–	1	12	–	9	–	4
3	3	3	–	1	12	–	12	–	10
3	3	6	–	1	15	–	0	–	3
3	3	9	–	3	15	–	3	–	1
3	6	9	–	2	15	–	6	–	1
3	6	12	–	3	15	–	9	–	2
3	3	15	–	2	15	–	12	–	8
3	6	15	–	2	15	–	15	–	9
3	6	18	–	1	18	–	0	–	1
3	3	21	–	1	18	–	6	–	1
6	6	0	–	2	18	–	9	–	1
6	9	0	–	1	18	–	12	–	1
6	9	9	–	1	18	–	15	–	3
6	6	12	–	1	18	–	18	–	6
6	9	12	–	2	21	–	15	–	1
6	6	15	–	1	–	0	0	–	9
6	9	15	–	1	–	0	3	–	3
6	6	18	–	1	–	0	6	–	10
6	9	18	–	2	–	0	9	–	6
9	9	0	–	1	–	0	12	–	8
9	12	0	–	2	–	0	15	–	5
9	9	9	–	2	–	0	18	–	4
9	12	9	–	1	–	0	21	–	1
9	12	12	–	3	0	–	0	3	1
9	9	15	–	1	6	–	0	6	1
9	12	24	–	1	6	–	6	6	1
9	9	27	–	1	12	–	0	3	1
12	12	0	–	1	12	–	0	6	1
12	15	0	–	1	15	–	0	3	1
12	15	6	–	1	21	–	15	15	1
12	15	15	–	1	3	–	–	0	1
12	15	21	–	1	9	–	–	0	1
0	–	0	–	6	12	–	–	0	1
3	–	0	–	2	0	3	0	6	1
6	–	0	–	1	3	6	6	12	1
6	–	3	–	2	9	9	9	9	1
–	0	–	0	1					

Table 7.2 *Maximal Intersection Rectangles and Masses of MLE*

(a) Canonical Rectangles				(b) Masses of MLE		
L_{j1}	R_{j1}	L_{j2}	R_{j2}	L_{j1}	L_{j2}	Mass MLE
0	0	0	0	0	0	0.013676984
0	0	21	–	0	21	0.307533525
3	3	21	–	3	21	0.087051863
6	6	6	6	6	6	0.014940282
6	6	18	–	6	18	0.062521573
9	9	9	9	9	9	0.010009349
9	9	27	–	9	27	0.071073995
12	12	0	0	12	0	0.004836043
12	12	24	–	12	24	0.053334241
15	15	0	0	15	0	0.042456241
15	15	21	–	15	21	0.021573343
21	–	15	15	21	15	0.044427509
21	–	18	–	21	18	0.266565054

exactly the same result, given in Table 7.2b. Note that the points $(3i, 30)$ and $(30, 3j)$ allow extra mass outside the region of the observation points, since the largest observation points on the first and second coordinates are 21 and 27, respectively. We need these extra points, since the indicators $\Delta_{i1}^{(3)}$ and $\Delta_{i2}^{(3)}$ can give the information that there are hidden observations outside the region formed by the observation points.

A picture of the MLE, based on the marginal data for the first coordinate, together with the first marginal df of the MLE for the bivariate distribution function, is shown in Figure 7.8. It is seen that the two estimates are very similar, and only start differing a bit on the last interval.

A picture of the bivariate MLE for the Betensky-Finkelstein data, restricted to the rectangle $[0, 18] \times [0, 24]$, is shown in Figure 7.9. It can be seen from this picture that the steep increase of the first marginal df of the MLE at zero is due to the "ridge" for the larger values of the second coordinate. For the meaning of the codings CMV and MAC, see Betensky and Finkelstein, 1999, or Sun, 2006, Section 7.3.

7.5 The MLE in the Competing Risks Model with Interval Censoring

In this section we compare the performance of the algorithms, discussed in the preceding sections for the MLE in the competing risk model with interval censored data. Note that this also covers the so-called mixed case interval censoring where there can be more than two observations T_i per unobservable hidden variable X_i, since only the surrounding interval matters for the computation of the MLE. We recall that the log likelihood, divided by n, has the form

$$\ell(F) = \frac{1}{n}\sum_{i=1}^{N}\sum_{k=1}^{K}\{\Delta_{ik,1}\log F_k(T_i) + \Delta_{ik,2}\log\{F_k(U_i) - F_k(T_i)\} + \Delta_{i,K+1}\log\{1 - F_+(U_i)\}\} \tag{7.51}$$

where $T_i < U_i$.

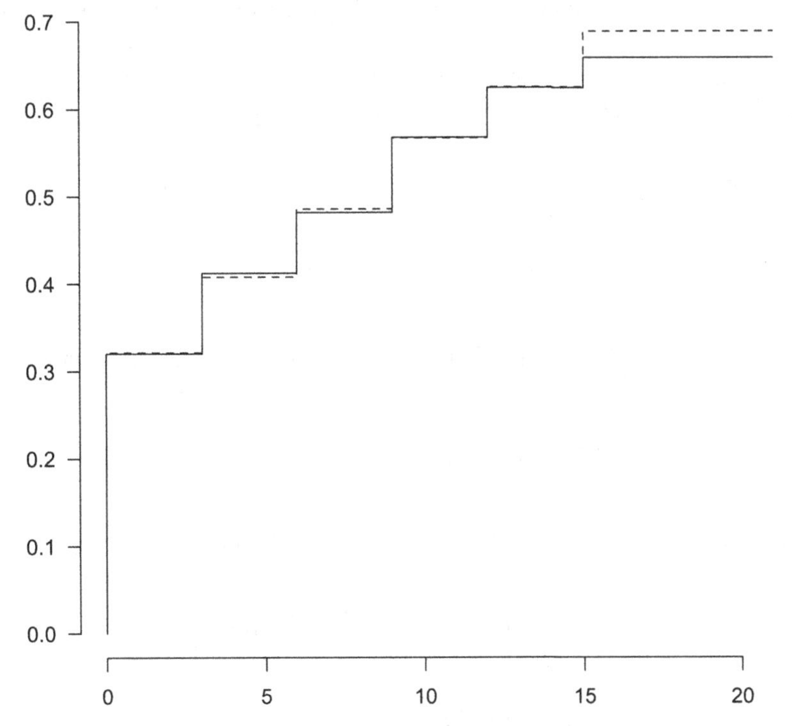

Figure 7.8 The first marginal distribution function of the data set in Betensky and Finkelstein, 1999, computed on the interval [0, 21) (the largest observation point on the first coordinate is 21). The solid line gives the MLE of the first marginal distribution function, based on the marginal data, while the dashed line gives the first marginal distribution function of the MLE of the bivariate MLE.

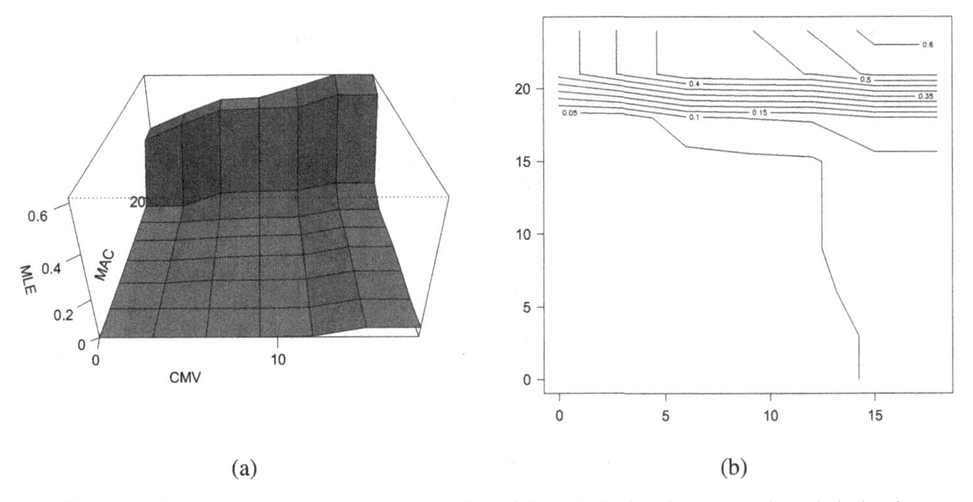

Figure 7.9 Surface (a) and level plots (b) of the MLE for the Betensky-Finkelstein data, restricted to the interval $[0, 18] \times [0, 24]$.

For the EM algorithm, we parameterize via the point masses p_{ik} of F_k at the points V_i, where V_i belongs to the set $T_1, \dots, T_n, U_1, \dots, U_n$, allowing extra mass $p_{n_k+1,k}$ for mass for F_k at infinity, if there are n_k points which are relevant for F_k. Taking the derivative of the criterion function with respect to p_{ik}, we get the iteration steps:

$$p_{ik}^{(m+1)} = n^{-1}\sum_{j=1}^{n}\left\{\frac{\Delta_{jk,1}}{F_k(T_j)}1_{\{T_j\ge V_i\}} + \frac{\Delta_{jk,2}}{F_k(U_j)-F_k(T_j)}1_{\{V_i\in(T_j,U_j]\}} + \frac{\Delta_{j,K+1}}{1-F_+(U_j)}1_{\{U_j<V_i\}}\right\}p_{ik}^{(m)},$$

for $i = 1, \dots, n_k + 1$. This leads to the following iterations for the distribution functions:

$$\begin{aligned}F_k^{(m+1)}(V_{(i)}) &= n^{-1}\sum_{j=1}^{n}\sum_{\ell=1}^{i}\left\{\frac{\Delta_{jk,1}}{F_k(T_j)}1_{\{T_j\ge V_{(\ell)}\}} + \frac{\Delta_{jk,2}}{F_k(U_j)-F_k(T_j)}1_{\{V_{(\ell)}\in(T_j,U_j]\}} + \frac{\Delta_{j,K+1}}{1-F_+(U_j)}1_{\{U_j<V_{(\ell)}\}}\right\}p_{(\ell)k}^{(m)}\\ &= \int\left\{\frac{\delta_{k1}F_k^{(m)}(V_{(i)}\wedge t)}{F_k^{(m)}(t)} + \frac{\delta_{k2}\{F_k^{(m)}(V_{(i)}\wedge u)-F_k^{(m)}(V_{(i)}\wedge t)\}}{F_k^{(m)}(u)-F_k^{(m)}(t)} + \frac{\delta_{K+1}\{F_k^{(m)}(V_{(i)})-F_k^{(m)}(V_{(i)}\wedge u)\}}{1-F_+^{(m)}(u)}\right\}d\mathbb{P}_n(t,u,\delta).\end{aligned}$$

Note that the right-hand side has the interpretation as a conditional expectation under the distribution $F^{(m)}$ at the mth iteration step:

$$E_{F^{(m)}}\left\{\mathbb{F}_{nk}(V_{(i)})\,\Big|\,(T_j, U_j, \Delta_{jk,1}, \Delta_{jk,2}, \Delta_{j,K+1}),\ j = 1, \dots, n\right\},$$

where $\mathbb{F}_{nk}(V_{(i)})$ is the empirical distribution function of the kth subvariable, assuming that the mass of $F^{(m)}$ is concentrated in the observation points.

For the iterative convex minorant algorithm we use the characterization

$$\begin{aligned}&\int_{t'\ge t}\left\{\frac{\delta_{k1}}{\hat F_{nk}(t')} - \frac{\delta_{k2}}{\hat F_{nk}(u)-\hat F_{nk}(t)}\right\}d\mathbb{P}_n(t,u,\delta)\\ &\quad+\int_{u\ge t}\left\{\frac{\delta_{k2}}{\hat F_{nk}(u)-\hat F_{nk}(t)} - \frac{\delta_{K+1}}{1-\hat F_{n+}(u)(t)}\right\}d\mathbb{P}_n(t,u,\delta)\\ &\le \lambda_n,\end{aligned} \tag{7.52}$$

for each $t \ge 0$, where λ_n is defined by

$$\lambda_n = 1 - \int\frac{\delta_{K+1}}{1-\hat F_{n+}(u)}\,d\mathbb{P}_n(t,u,\delta). \tag{7.53}$$

Note that this characterization is an extension of the characterization for the current status model in Lemma 5.2 of Section 5.1, where the inequalities are reversed since we consider the tail regions $[t, \infty)$ in the integrals.

The solution can be computed by the iterative convex minorant algorithm in the following way. Let us call, for each k, an observation time T_i such that $\Delta_{ik,1} = 1$ be an observation of type 1 (for k) and an observation time U_i such that $\Delta_{ik,2} = 1$ be an observation of type 2 (for k). Moreover, let $V_{1k} < V_{2k} < \cdots < V_{n_k,k}$ be the strictly ordered observation points of both types for the kth component function. We now form a cusum diagrams with weights $w_{jk} = w_{jk}(F_k)$ for F_k, defined by:

$$w_{jk} = \int_{V_{jk} \le t < V_{j+1,k}} \left\{ \frac{\delta_{k1}}{F_k(t)^2} + \frac{\delta_{k2}}{\{F_k(u) - F_k(t)\}^2} \right\} d\mathbb{P}_n + \int_{V_{jk} \le u < V_{j+1,k}} \left\{ \frac{\delta_{k2}}{\{F_k(u) - F_k(t)\}^2} + \frac{\delta_{K+1}}{\{1 - F_+(u)\}^2} \right\} d\mathbb{P}_n,$$

for $j = 1, \ldots, n_k$, where $V_{n_k+1,k} = \infty$. The y-coordinates of these cusum diagrams are the cumulative sums of $F(V_{jk})w_{jk} + v_{jk}$, where

$$v_{jk} = \int_{V_{jk} \le t < V_{j+1,k}} \left\{ \frac{\delta_{k1}}{F_k(t)} - \frac{\delta_{k2}}{F_k(u) - F_k(t)} \right\} d\mathbb{P}_n + \int_{V_{jk} \le u < V_{j+1,k}} \left\{ \frac{\delta_{k2}}{F_k(u) - F_k(t)} - \frac{\delta_{K+1}}{1 - F_+(u)} \right\} d\mathbb{P}_n.$$

If the largest order statistic of the observations is an observation with corresponding $\Delta_{i,K+1} = 1$ (meaning that this observation is a censoring time for all K risks, indicating there is survival beyond this point), the cusum diagram is of the form

$$\left(\sum_{j=1}^{i} w_{jk}, \sum_{j=1}^{i} w_{jk} F_k(V_j) + v_{jk} \right), \quad i = 1, \ldots, n_k, \tag{7.54}$$

for each k, otherwise it is of the form

$$\left(\sum_{j=1}^{i} w_{jk}, \sum_{j=1}^{i} w_{jk} F_k(V_j) + v_{jk} - \lambda_n 1_{\{i=n_k\}} \right), \quad i = 1, \ldots, n_k, \tag{7.55}$$

where λ_n is defined by (7.53). As usual, the point $(0, 0)$ is added to the cusum diagrams. The MLE is now computed by iteratively computing the greatest convex minorants of the cusum diagrams and taking its left derivative $\tilde{F}_k$. Note that, using this approach, the left derivative $\tilde{F}_k^{(m+1)}$ minimizes at the $(m + 1)$th step the sum of squares

$$\sum_{i=1}^{n_k-1} \left\{ F_k^{(m)}(V_{ik}) + \frac{v_{jk}^{(m)}}{w_{jk}} - F_k(V_{i,k}) \right\}^2 w_{ik} + \left\{ F_k^{(m)}(V_{n_k,k}) + \frac{v_{j,n_k}^{(m)}}{w_{j,n_k}} - \lambda^{(m)} - F_k(V_{n_k,k}) \right\}^2 w_{n_k,k},$$

over K-tuples $F = (F_1, \ldots, F_K)$ of distribution functions, where

$$\lambda^{(m)} = 1 - \int \frac{\delta_{K+1}}{1 - F_+^{(m)}(u)} \, d\mathbb{P}_n(t, u, \delta).$$

After computing the left continuous slopes of the K cusum diagrams, line search is performed which approximately minimizes over $\alpha \in (0, 1]$:

$$-\ell\left(F^{(m)} + \alpha\big(\tilde{F}^{(m+1)} - F^{(m)}\big)\right) + \alpha\lambda^{(m)}\left(\tilde{F}_+^{(m+1)}(V_{max}) - F_+^{(m)}(V_{max})\right),$$

where $\ell(F)$ is defined by (7.51), $\tilde{F}^{(m+1)} = (\tilde{F}_1^{(m+1)}, \ldots, \tilde{F}_K^{(m+1)})$ consists of the left continuous slopes of the cusum diagrams (7.54) or (7.55) and

$$V_{max} = \max_{k=1,\ldots,K} V_{n_k}.$$

We choose as the input for the next iteration

$$F^{(m+1)} = F^{(m)} + \alpha\big(\tilde{F}^{(m+1)} - F^{(m)}\big),$$

where $\alpha \in (0, 1]$ is the value, determined by the line search.

Finally, the `R` package `MLEcens` computes the MLE using the vertex direction method. This means that at each iteration quadratic minimization is used, combined with Armijo's rule for line search, to go from one quadratic minimization iteration step to the next quadratic minimization step, where the weights of the quadratic minimization are adjusted. The procedure is in fact completely analogous to the method used for bivariate interval censoring, case 2. The prototype of this method is given in Groeneboom, Jongbloed and Wellner, 2008, in the treatment of the Aspect experiment in quantum statistics.

It is of interest to compare the performance of the three methods for computing the MLE. To this end, we consider the competing risk model with current status data (so the middle term in the log likelihood (7.51) drops out), where $K = 2$. We consider the model

$$\mathbb{P}\{X \le t | Y = k\} = 1 - e^{-kt},\ t \ge 0, \qquad \mathbb{P}\{Y = 1\} = \frac{1}{3}, \qquad \mathbb{P}\{Y = 2\} = \frac{2}{3},$$

leading to:

$$F_1(t) = \frac{1}{3}\left\{1 - e^{-t}\right\}, \qquad F_2(t) = \frac{2}{3}\left\{1 - e^{-2t}\right\},\ t \ge 0.$$

The observation distribution is (independently) uniform on the interval $[0, 3]$. A picture of these subdistribution functions and their maximum likelihood estimators for the competing risk model with current status data is given in Figure 7.10.

The performance of the three algorithms to compute the MLE is given in Table 7.3 for sample sizes $n = 100, 1000, 10000$ and 25000. The time was measured in seconds, using the C procedure `clock()`. For the vertex direction method the original C program, which was the basis for the `R` routine in `MLEcens`, was used to give a fair comparison. The values given here are more or less typical for the performance. The algorithm was considered to have converged if the Fenchel conditions were satisfied at accuracy 10^{-9}, that is:

$$\left|\sum_{k=1}^{K} \int F_k^{(m)}(t) \left\{\frac{\delta_{k1}}{F_k^{(m)}(t)} - \frac{\delta_{K+1}}{1 - F_+^{(m)}(t)} - \lambda^{(m)} 1_{\{t = V_{n_k,k}\}}\right\} d\mathbb{P}_n(t, \delta)\right| < 10^{-9} \tag{7.56}$$

Table 7.3 *Comparison of the Performance of Methods to Compute the MLE*

n	EM	ICM	Vertex Direction
100	0.63	0.002	0.004
1000	–	0.06	0.66
10,000	–	10.6	153.9
25,000	–	35.6	1630.8

and

$$\int_{t\le V_{jk}} \left\{ \frac{\delta_{k1}}{F_k^{(m)}(t)} - \frac{\delta_{K+1}}{1-F_+^{(m)}(t)} - \lambda^{(m)} 1_{\{t=V_{n_k,k}\}} \right\} d\mathbb{P}_n(t,\delta) > -10^{-9},$$

$$j = 1,\dots,n_k,\ k = 1,\dots,K. \tag{7.57}$$

It is seen that the iterative convex minorant algorithm is always fastest and that the difference in computing time increases for increasing sample size. EM did not converge within 100,000 iterations for $n = 1000, 10000$ and 25000; for $n = 1000$ and after 1446 seconds the inner product (7.56) was still larger than 10^{-7} (although it possibly would have converged for 10^6 iterations). On the other hand, EM needs around 4000 iterations for sample

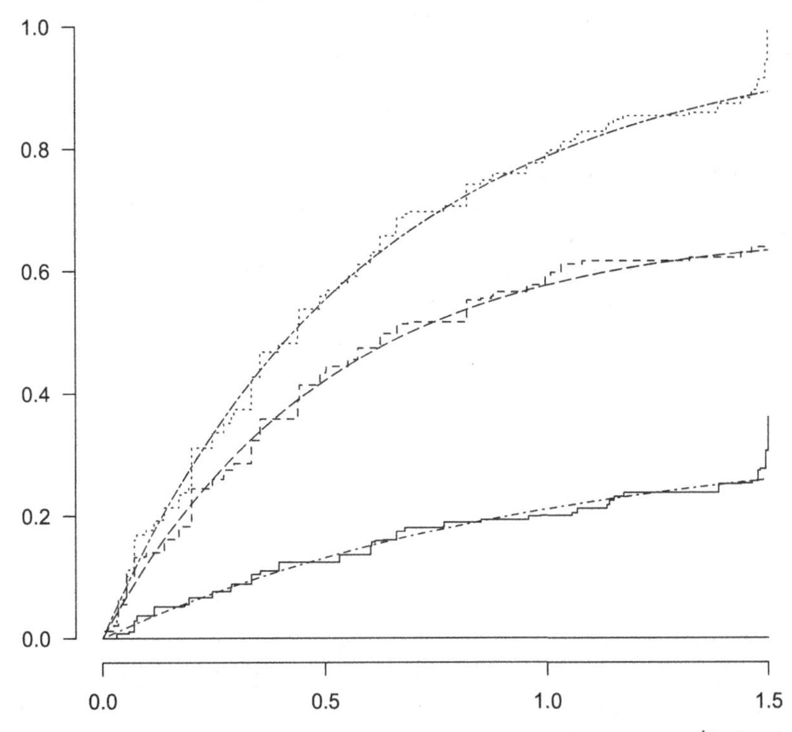

Figure 7.10 The two distribution functions $F_k(t) = (k/3)\{1 - e^{-kt}\}$ for the competing risk model with current status data and their MLEs for a sample of size $n = 25000$. The upper curves give the sum function F_+ and its estimator $\hat{F}_{n+}$.

size 100. The iterative convex minorant algorithm only needs between 40 and 150 iterations for all cases. Of course we only have given the numbers for one model (and one laptop computer) here, and further research will be needed to give an overall picture of the behavior. It seems, though, that the iterative convex minorant algorithm and the vertex direction method are far superior to the EM algorithm for this problem.

Exercises

7.1 Show that, given starting value $\theta^{(0)} > 1$, a sequence of iterates using the algorithmic map (7.3) can be any sequence $(\theta^{(k)})_{k=0}^{\infty}$ such that

$$\theta^{(k)} \in \left[1 + 4^{-k}(\theta^{(0)} - 1), 1 + 3^{-k}(\theta^{(0)} - 1)\right], \quad k = 0, 1, 2, \ldots$$

7.2 Show that a continuous point-to-point map is closed.

7.3 Show that closedness of a point-to-point map does not imply continuity. Think, e.g., of the function $f : \mathbb{R} \to \mathbb{R}$ with $f(x) = 1/x$ if $x \neq 0$ and $f(x) = 0$ if $x = 0$.

7.4 In this exercise it will be seen that for nonconvex functions with local minima, the limit point of Newton's algorithm depends on the starting value. It is not guaranteed that the limiting value corresponds to the global minimum of the function. Consider the function

$$f(x) = \frac{1}{4}x^4 - \frac{1}{3}x^3 - x^2$$

on $\mathbb{R}$.

a) Verify that this function is not convex but that it does attain its minimum value.

b) The basic Newton steps for minimizing this function,

$$x_{k+1} = x_k - \frac{f'(x_k)}{f''(x_k)},$$

need not be well defined (second derivative could be zero). Nevertheless, implement this algorithm (using your favorite package), where you can generate sequential iterates using a starting value x_0 of your choice.

c) Run the algorithm using starting value $x_0 = -2$. See Figure 7.11a.

d) The same as (c), but now with starting value $x_0 = 3$.

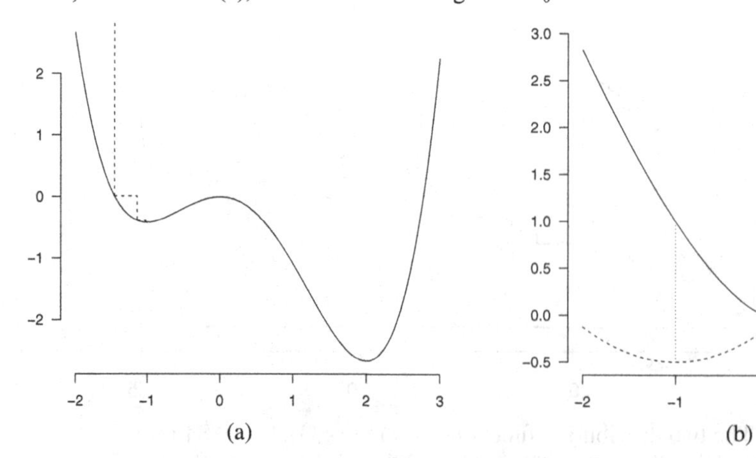

Figure 7.11 (a) The function f defined in Exercise 7.4, with Newton iterates starting from $x_0 = -2$. (b) The function defined in Exercise 7.5, with the local quadratic approximation at $x_0 = 1$ with its associated minimum (x_1)

7.5 In this exercise it will be seen that Newton's algorithm applied to a convex function does not necessarily converge. Consider the convex function $f(x) = |x|^{3/2}$ on $\mathbb{R}$ as shown in Figure 7.11b.

a) Make the formal Newton steps for minimizing f over $\mathbb{R}$ (as given in Exercise 7.4(b)) explicit for this function.

b) Verify that whatever starting value is chosen, the algorithm does not have a descent function.

c) Observe that the direction proposed by the algorithm is a descent direction and that the algorithm can be made to converge by proper damping.

7.6 Consider the situation of Example 7.2, based on the model defined in (7.1).

a) Show that a random variable with density f_θ given in (7.1) can be generated as follows, based on a random variable U that is uniformly distributed on $[0, 1]$:

$$X = \frac{1}{\theta} \log(1 - \log U).$$

b) Choose a parameter $\theta_0 > 0$ and sample size $n \in \mathbb{N}$. Generate a sample of size n from the corresponding density f_{θ_0} and implement the Newton procedure described in Example 7.2. Does the algorithm converge? Experiment with various values of the parameter θ_0 and sample size n.

7.7 Verify (7.5).

7.8 Show that (7.7) corresponds to the solution of the M-step in the truncated exponential model.

7.9 Show that $\theta = A(\theta) \iff \theta = \hat{\theta}$ in Example 7.3.

7.10 To show the algorithm based on the algorithmic map B of (7.36), take $n = 2$ and the function

$$\phi(\beta) = \beta_2 - \beta_1 + \frac{1}{2}\beta_1\beta_2 + \frac{3}{4}(\beta_1^2 + \beta_2^2).$$

Moreover, take $\beta^{(0)} = (1, 1)^T$ and $D \equiv I$, the identity matrix. Show that $\beta^{(k)} = (1, 1)^T$ for k even and $\beta^{(k)} = (-1, -1)^T$ for k odd in this example.

7.11 Show in the context of Example 7.3 that $E_{\hat{\theta}}(\bar{X}|Y = y) = A(\hat{\theta})$, which by Exercise 7.9 equals $\hat{\theta}$.

7.12 Prove that the EM algorithm described in Example 7.3 converges to $\hat{\theta}$.

7.13 This exercise relates to Example 7.4. Show that maximizing ϕ_θ with respect to $\theta' \in \Theta$ gives

$$\theta'_j = \frac{1}{n} c(\theta)_j = \frac{1}{n} \sum_{i=1}^{n} P_\theta\left(X_i = v_j \;\middle|\; (\Delta_i, \Gamma_i) = (\delta_i, \gamma_i)\right).$$

Hint: relax the constraint $\sum_{j=1}^{m} \theta_j = 1$ using a Lagrange multiplier.

7.14 In the setting of Example 7.4 show that

$$P_{\theta^{(0)}}\left(X_i = v_j \;\middle|\; (\Delta_i, \Gamma_i) = (\delta_i, \gamma_i)\right) = \begin{cases} \theta_j^{(0)} / \sum_{k \in J_i} \theta_k^{(0)} & \text{for } j \in J_i, \\ 0 & \text{for } j \notin J_i. \end{cases}$$

7.15 Use properties of conditional expectations, and assume that certain interchanges of differentiation and integration (or summation) are allowed to show that:

$$E_{P_\theta}\left\{\dot{\ell}_\theta(X) \;\middle|\; T(X) = y\right\} q_\theta(y) = \frac{\partial}{\partial \theta} q_\theta(y).$$

7.16 Show that whenever the iterate $\theta^{(m-1)}$ in (7.15) satisfies $\sum_j \theta_j^{(m-1)} = 1$, the same holds true for $\theta^{(m)}$. This guarantees that starting the iterations given in (7.15) with a genuine probability vector $\theta^{(0)}$ leads to probability vectors $\theta^{(m)}$ for $m = 1, 2, \ldots$.

7.17 This exercise relates to Example 7.3 and (7.20). From Example 7.3 we see that

$$\dot{\ell}_\theta(x) = \frac{\partial}{\partial\theta}\log p_\theta(x) = \sum_{i=1}^{n}\frac{\partial}{\partial\theta}\log f_\theta(x_i) = n\theta^{-2}(-\theta + \bar{x}_n).$$

Show that

$$E_{P_\theta}\left\{\dot{\ell}_\theta(X)\,\middle|\, T(X) = y\right\} = 0$$

if and only if $\theta = \hat{\theta}$, where $\hat{\theta}$ is given in (7.9). Hence, the self-consistency equation holds in the truncated exponential example. Moreover, there is only one θ satisfying the self-consistency equation.

7.18 Show that the convolution of the exponential density with parameter θ and standard uniform density is given by (7.28).

7.19 In Section 4.6 the ML estimator in deconvolution models is introduced. As illustration, the estimator was computed using the EM algorithm. Write $z_1 < z_2 < \cdots < z_n$ for the ordered realizations of a sample from $g = k * dF$ where k is a decreasing density on $[0,\infty)$ and F a distribution function supported on $[0,\infty)$.

a) Show that in this case the ML estimator for F is a discrete distribution function with mass concentrated on the observed data points.

b) For a discrete distribution function F supported on $\{z_1, z_2, \ldots, z_n\}$ define the probability vector p with components $p_i = F(z_i) - F(z_i-)$ for $1 \le i \le n$. Given a probability vector $p^{(\ell)} = (p_1^{(\ell)}, \ldots, p_n^{(\ell)})$, show that the conditional expectation (under $p^{(\ell)}$) of the full log likelihood given the observed data is given by

$$Q(p|p^{(\ell)}) = \sum_{j=1}^{n}\left(\sum_{i=j}^{n}\frac{k(z_i - z_j)}{g^{(\ell)}(z_i)}\right)p_j^{(\ell)}\log p_j$$

where $g^{(\ell)}(z) = \sum p_m^{(\ell)}k(z - z_m)$ is the convolution density using probability vector $p^{(\ell)}$.

c) Starting from a probability vector $p^{(0)} = (p_1^{(0)}, \ldots, p_n^{(0)})$, show that the explicit iteration step in the EM algorithm is given by

$$p_j^{(\ell+1)} = \frac{\tau_j^{(\ell)}}{\sum_{m=1}^{n}\tau_m^{(\ell)}} \text{ with } \tau_j^{(\ell)} = p_j^{(\ell)}\sum_{i=j}^{n}\frac{k(z_i - z_j)}{g^{(\ell)}(z_i)}.$$

d) Generate a sample $X_1, \ldots, X_n$ using a distribution function F on $[0,\infty)$ and a sample $Y_1, \ldots, Y_n$ from a decreasing density k on $[0,\infty)$. For example, follow Section 4.6 and take

$$F(x) = \int_0^x f(y)\,dy \text{ with } f(y) = \frac{2}{9}y(1-y)1_{[0,3]}(y) \text{ and } k(x) = \sqrt{\frac{2}{\pi}}e^{-x^2/2}1_{[0,\infty)}(x).$$

Then define $Z_i = X_i + Y_i$. Use the uniform distribution on $\{z_1, z_2, \ldots, z_n\}$ as starting distribution and implement the iterative scheme of (b). Run a number of iterations and observe that the log likelihood at the successive iterates increases in line with Theorem 7.2.

7.20 In Section 2.3 an explicit construction is given for the ML estimator in the current status model. The EM algorithm can also be used to compute this estimator. Follow the steps of Example 7.4 to derive an explicit iteration scheme in the spirit of (7.15) to compute the ML estimator using EM.

7.21 Prove (7.35).

7.22 Show that the index set $I_3 = \{4\}$ in the context of the data given in (7.43).

7.23 Verify in the context of Example 7.6 that

$$\frac{\partial^2 \phi}{\partial \beta_i \partial \beta_j} = 0 \text{ if } i \in I_1 \text{ and } j \in I_3.$$

Identify more zero entries of the Hessian matrix of ϕ and observe that the Hessian matrix is sparse in general.

7.24 Suppose $X_1, \ldots, X_n$ are i.i.d. according to the density

$$f_{p,\mu_1,\mu_2}(x) = \frac{1}{\sqrt{2\pi}} \left(pe^{-(x-\mu_1)^2/2} + (1-p)e^{-(x-\mu_2)^2/2} \right)$$

on $\mathbb{R}$ with $-\infty < \mu_1 < \mu_2 < \infty$ and $p \in [0, 1]$. Write down the log likelihood for (p, μ_1, μ_2) and design (and implement) a vertex direction method to compute the maximum likelihood estimator of this parameter vector.

7.25 Consider $g_\theta(x) = 2(\theta - x)/\theta^2 1_{[0,\theta]}(x)$ for $\theta > 0$ as introduced in Example 7.7. Show that the class of mixture densities corresponding to the basic class of densities $\{g_\theta : \theta > 0\}$ is exactly the class of convex decreasing densities on $[0, \infty)$.

7.26 Define the class uniform densities by $\mathcal{G} = \{g_\theta : \theta > 0\}$ with $g_\theta(x) = \theta^{-1} 1_{[0,\theta]}(x)$. Show that the class of mixture densities corresponding to $\mathcal{G}$ is exactly the class of decreasing densities on $[0, \infty)$.

7.27 Verify (7.49).

Bibliographic Remarks

The description of algorithms used here is in line with that used in Bazaraa et al., 2006. The EM algorithm was first thoroughly studied in Dempster et al., 1977. The book by McLachlan and Krishnan, 2007, is fully dedicated to the EM algorithm, and there it is mentioned that the EM algorithm was already introduced in a specific model in Newcomb, 1886. Convergence of the algorithm was rigorously proved in Wu, 1983, using the method described in Section 7.1. The ICM algorithm was introduced in Groeneboom and Wellner, 1992, and convergence of its modified version established in Jongbloed, 1998b. Also combined versions (hybrid algorithms) consisting of alternating steps of the EM and the ICM algorithm have been proposed and used, e.g., in Wellner and Zhan, 1997, and Jongbloed, 2001. The original VD algorithm was introduced in Wynn, 1970, and Fedorov, 1971, in the context of optimal experimental design. An extension of the basic algorithm (Simar's algorithm) was applied to the problem of Poisson demixing in Simar, 1976, and further studied in Böhning, 1982. The vertex exchange algorithm was introduced and studied in Böhning, 1986; the intra simplex direction method by Lesperance and Kalbfleisch, 1992. The support reduction algorithm was defined and proved to converge in Groeneboom, Jongbloed and Wellner, 2008. There it was also applied to a dataset from the Aspect experiment from quantum statistics. The MLE for the Betensky-Finkelstein dataset is also discussed in Section 7.3.3 of Sun, 2006, who also refers to Gentleman and Vandal, 2002, and Bogaerts and Lesaffre, 2004, for discussions of the computation of the MLE for this data set. The computation of the rectangles where the MLE puts mass is also discussed in Song, 2001. Another type of algorithm that can be used to solve optimization problems as discussed in this chapter are interior point methods. For an overview of those methods, see Wright, 1997.

8

Shape and Smoothness

As seen in the examples discussed so far, shape-restricted estimators often satisfy the required shape constraint with minimal smoothness properties. The Grenander density estimator is decreasing, but discontinuous (see Figure 2.4). The least squares estimator for a convex decreasing density is convex and decreasing, but its derivative is discontinuous (see Figure 4.9). Similar observations can be made for other models. Sometimes, there are reasons to assume that an underlying distribution function is smooth. In other situations (as will be encountered in Chapter 9), smoothness of an estimated model is needed in a proof that a bootstrap method works.

In this chapter, the problem of estimating a smooth shape-constrained function is considered. The estimation of smooth functions without shape constraints has received quite some attention since the 1950s. Methods such as kernel smoothing and spline fitting have been widely applied and studied thoroughly. In order to obtain smooth shape-constrained estimators, various approaches are possible. A first is to smooth the nonsmooth shape-constrained estimator. In Section 8.1 this approach is illustrated using the maximum likelihood estimator (MLE) in the current status model. A related method interchanges the order of smoothing and maximizing in this procedure. In Section 8.2 it is first illustrated using the problem of estimating a decreasing density on $[0, \infty)$ as introduced in Section 2.2. Instead of using the empirical distribution function in the definition of the log likelihood, a smooth estimator for the observation distribution function is used and then the corresponding smoothed (log) likelihood maximized to obtain an estimator. This method is also very natural if only binned observations are available. This will be seen in the context of Wicksell's problem as introduced in Section 4.1. Another method is to first estimate the distribution without using the shape constraint and process this estimator in such a way that it satisfies the shape constraint without losing its smoothness. Yet another method that can be used for M-estimators (such as maximum likelihood or least squares) is to restrict the maximization over a class of smooth shape constrained functions or penalize for roughness. This method is illustrated in Section 8.3, using the problem of estimating an increasing hazard rate, as introduced in Section 2.6. In Section 8.4 an example of this approach (monotonic rearrangements) is illustrated in the context of estimating a nondecreasing regression function and a distribution function in the deconvolution model.

Sections 8.5, 8.6 and 8.7 illustrate the methods of Sections 8.1 and 8.2 in some other models. The maximum smoothed likelihood estimator for the current status model is studied in Section 8.5. In Section 8.6 the smoothed maximum likelihood estimator in the interval censoring case 2 problem is defined and studied. Finally, the smoothed maximum likelihood estimator is applied in a two-dimensional problem, the bivariate interval censoring problem, in Section 8.7.

8.1 Smoothing a Shape-Constrained Estimator

Having a nonparametric nonsmooth shape constrained estimator, a smooth estimator can be obtained by smoothing that basic estimator. If the smoothing does not disturb the shape constraint, the smoothed estimator will be a smooth shape restricted estimator. In the case where the basic estimator is the maximum likelihood estimator, the resulting estimator is called a smoothed maximum likelihood estimator (SMLE).

In this section, we illustrate the method using the current status model introduced in Section 2.3. The function of interest is the distribution function F (and the corresponding density f with respect to Lebesgue measure, if it exists) of the event times $X_1, \ldots, X_n$. However, only information on the current status of subject i at time T_i is available, so the data consist of

$$(T_1, \Delta_1), \ldots, (T_n, \Delta_n), \quad \text{where} \quad \Delta_i = 1_{\{X_i \le T_i\}}.$$

Based on these data, the MLE $\hat{F}_n$ can be explicitly constructed by taking the derivative of the greatest convex minorant of a diagram of points; see Section 2.3. For the smoothing part, we use kernel estimators. Given a symmetric smooth probability density K on $[-1, 1]$, the kernel, with corresponding distribution function $\mathbb{K}$ and derivative K', and bandwidth $h > 0$, define scaled versions of $\mathbb{K}$, K and K' by

$$\mathbb{K}_h(u) = \mathbb{K}(u/h), \; K_h(u) = \frac{1}{h} K(u/h), \quad \text{and} \quad K_h'(u) = \frac{1}{h^2} K'(u/h).$$

We will use the triweight kernel, defined by

$$K(x) = \frac{35}{32}\left(1 - x^2\right)^3 1_{[-1,1]}(x).$$

Define the SMLE $\tilde{F}_{nh}$ for F by

$$\tilde{F}_{nh}(t) = \int \mathbb{K}_h(t-u)\, d\hat{F}_n(u) = \int_{t-h}^{t+h} \hat{F}_n(u) K_h(t-u)\, du, \tag{8.1}$$

where $\hat{F}_n$ is the MLE. Clearly, as $\hat{F}_n$ is a piecewise constant step function, estimating the density f associated with F cannot be done by simple differentiation. However, the SMLE can be used like that. Indeed, the SMLE $\tilde{f}_{nh}$ for f can be defined by

$$\tilde{f}_{nh}(t) = \int K_h(t-u)\, d\hat{F}_n(u).$$

Now note that monotonicity of $\hat{F}_n$ implies monotonicity of $\tilde{F}_{nh}$, since the latter merely is a mixture of translated (monotone) functions K_h (see Exercise 8.1). Consistency of $\tilde{F}_{nh}(t)$, for $0 < h = h_n \to 0$ follows easily from uniform consistency of $\hat{F}_n$ and (8.1), implying that

$$\hat{F}_n(t-h) \le \tilde{F}_{nh}(t) \le \hat{F}_n(t+h). \tag{8.2}$$

See also Exercise 8.2. Derivation of the asymptotic distribution theory of the SMLE is more involved. In Section 11.3 asymptotic normality of $\tilde{F}_n(t)$ will be derived in Theorem 11.4.

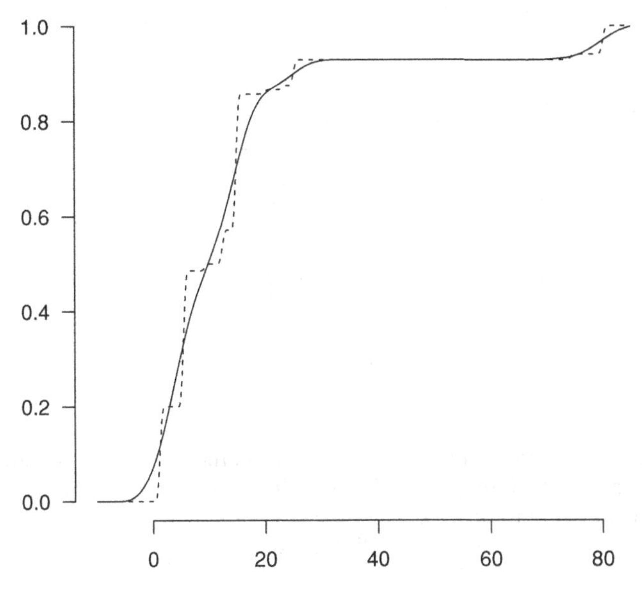

Figure 8.1 The smoothed MLE based on the Rubella data with $h = 8$ (solid) and $h = 1$ (dashed) based on the triweight kernel.

Figure 8.1 shows the estimator based on the rubella data described in Section 2.3 for two choices for the bandwith: $h = 8$ (solid) and $h = 1$ (dashed).

As is clear from Figure 8.1, a disturbing property of $\tilde{F}_n$ is that in case the first point of jump of MLE $\hat{F}_n$ is smaller than h, the smoothed MLE assigns positive mass to the negative half line. To overcome this problem, some boundary correction method is needed. See Exercise 8.14 for a simple possibility of boundary correction. In Sections 8.5, 9.2 and 11.3 more sophisticated methods with better performance will be used.

Now consider the problem of bandwidth selection. This problem does not occur with the plain MLE, but is the price to be paid for exploiting smoothness. The asymptotically mean squared error optimal rate of convergence for the bandwidth $h = h_n$ turns out to be $n^{-1/5}$; see Theorem 11.4. Since the expression for the asymptotically optimal constant involves the underlying (unknown) density f of X_i, this asymptotically optimal constant cannot be used directly. One option would be to estimate the optimal constant. A natural way of doing that would be based on kernel smoothing, using a pilot bandwidth typically larger than $n^{-1/5}$. We will describe another heuristic data dependent way to choose such a (local) bandwidth, based on the bootstrap.

Choose $x > 0$. In order to find an approximately MSE optimal bandwidth to estimate $F(x)$, we wish to minimize the function

$$h \mapsto \text{MSE}\,(h; n|T_1, \ldots, T_n) = E_F\left[\left(\tilde{F}_n(x; h) - F(x)\right)^2 |T_1, \ldots, T_n\right]$$

as function of h. Here we make the dependence of $\tilde{F}_n$ on h explicit. The expectation is taken with respect to the Δ_is, for fixed observation points $T_1, \ldots, T_n$, where the Δ_is are conditionally independent (given the T_is) and Bernoulli($F(T_i)$) distributed. This function, however, cannot be computed since the underlying F is unknown. The function

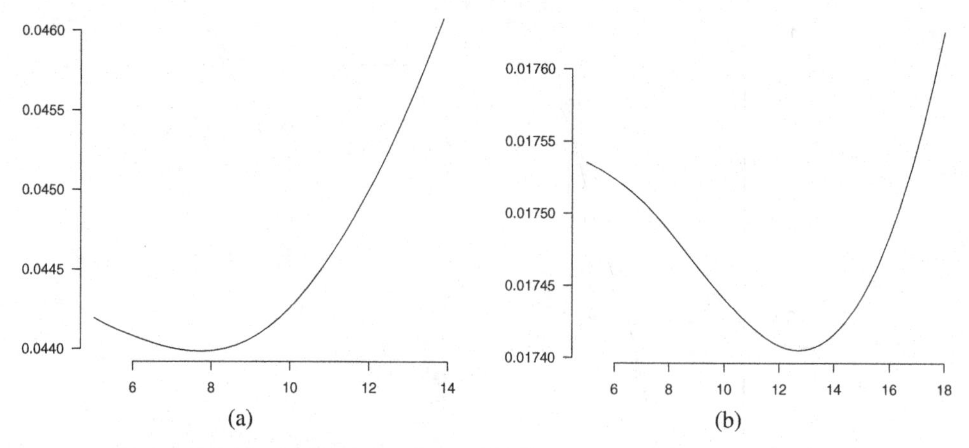

Figure 8.2 The approximated conditional (on the T_is) MSE on a grid of bandwidths. (a) The picture for $x = 20$, resulting in an optimal h of approximately 7.7. (b) the function for $x = 35$, yielding $h \approx 12.7$. The pictures are based on $B = 10000$ repetitions.

can be estimated by

$$h \mapsto \widehat{\text{MSE}}\,(h; n|T_1, \ldots, T_n) = E_{\tilde{F}_{n;hp}}\left[\left(\tilde{F}_{n;h}(x) - \tilde{F}_{n;h_p}(x)\right)^2 |T_1, \ldots, T_n\right]$$

where $\tilde{F}_{n;h_p}$ is an initial smooth estimator for F, say the smoothed maximum likelihood estimator with pilot bandwidth h_p.

The procedure is now to iteratively draw the $\Delta_i^* \sim$Bernoulli$(\tilde{F}_{n;h_p}(T_i))$, for $1 \le i \le n$, independently and compute the resulting $(\tilde{F}_{n;h}(x) - \tilde{F}_{n;h_p}(x))^2$ for values of h on a fine grid. Having determined, say, $B = 10000$ of these vector of squared losses, $(\tilde{F}_{n;h}^{[j]}(x) - \tilde{F}_{n;h_p}(x))^2$, $1 \le j \le B$ the Monte Carlo approximation of the function function $\widehat{\text{MSE}}$,

$$h \mapsto \frac{1}{B}\sum_{j=1}^{B}\left(\tilde{F}_{n;h}^{[j]}(x) - \tilde{F}_{n;h_p}(x)\right)^2,$$

can be determined on the grid of h-values chosen. For the point $x = 20$, this function based on $B = 10000$ realizations using $\tilde{F}_{n;h_p}$ with bandwidth $h_p = 10$ is given in the left panel of Figure 8.2. The right panel shows the picture for $x = 35$.

In Sections 8.6 and 8.7, SMLEs are studied in the more complicated univariate and bivariate interval censoring case 2 model.

8.2 Maximizing a Smoothed Objective Function

The maximum likelihood estimator for a shape constrained density maximizes the log likelihood

$$\ell(g) = \sum_{i=1} \log g(z_i) = n \int \log g(z)\, d\mathbb{G}_n(z) \qquad (8.3)$$

over the shape-restricted class of observation densities. Here $\mathbb{G}_n$ is the empirical distribution function based on the data $z_1, \ldots, z_n$. In case of estimating a decreasing density on $[0, \infty)$, the maximization can be restricted to decreasing densities that are constant on the intervals $[0, z_{(1)}]$, $(z_{(1)}, z_{(2)}]$ up to $(z_{(n-1)}, z_{(n)}]$ (see Section 2.2). Writing

$$g_n(z) = \sum_{i=1}^{n} \frac{1_{[z_{(i-1)}, z_{(i)})}(z)}{n(z_{(i)} - z_{(i-1)})},$$

where $z_{(0)} = 0$, this allows the following rewrite of the log likelihood:

$$\ell(g) = n \int \log g(z)\, g_n(z)\, dz = -n \int g_n(z) \log \frac{g_n(z)}{g(z)}\, dz - n \int g_n(z) \log g_n(z)\, dz, \quad (8.4)$$

where the first integral at the right hand side is minus (n times) the Kullback Leibler divergence of g with respect to g_n. The second term does not depend on g. This amounts to saying that the maximum likelihood (Grenander) estimator minimizes the Kullback Leibler divergence of densities in the class of decreasing densities on $[0, \infty)$ with respect to the wiggly density estimate g_n; see also the end of Section 2.2.

From the construction of the Grenander estimator it is clear that it inherits its discontinuous behavior from the discontinuity of g_n. A natural idea to obtain an estimator with more smoothness is therefore to replace the density estimate g_n in (8.4) by a smoother one or, equivalently, replace the empirical distribution in (8.3) by a smoother estimate of the distribution function. For this reason, the resulting estimator is called the maximum smoothed likelihood estimator (MSLE). Replacing the empirical distribution by a smoother estimator in the log likelihood (or in another objective function) is the basic idea behind the general method that is illustrated using the decreasing density problem in this section.

For the model at hand, the method will be illustrated with a kernel estimator as basic (generally not decreasing) density estimator. A first result is that the estimator can be computed in the same spirit as the Grenander estimator itself, where the integrated density estimator $\tilde{G}_n$ replaces the empirical distribution function $\mathbb{G}_n$.

Lemma 8.1 *The MSLE for a decreasing density, maximizing the function*

$$\tilde{\ell}(g) = \int \log g(z)\, d\tilde{G}_n(z)$$

over all decreasing densities on $[0, \infty)$, *is given by the right derivative of the concave majorant of the function* $x \mapsto \tilde{G}_n(z)$ *over* $[0, \infty)$.

From this lemma, it is immediately clear that whenever the initial function $\tilde{G}_n$ is already concave on $[0, \infty)$, the MSLE will actually be $\tilde{G}_n$ itself (see Exercise 8.5). The asymptotics of the MSLE will therefore be straightforward in case $\tilde{G}_n$ will become concave with probability tending to one. The uniform convergence result in the following lemma (which follows from Theorem C in Silverman, 1978) shows that for a range of bandwidth choices the derivative of the kernel estimator will converge to the derivative of the underlying density uniformly.

Lemma 8.2 *Let* $\tilde{g}_{n,h}$ *be a kernel estimator based on a sample of size n from a decreasing density g on* $(0, \infty)$. *Assume the kernel function K continuously differentiable and that the conditions of Theorem C in Silverman, 1978, are satisfied for K and* $r = 1$. *Moreover,*

Table 8.1 *Frequencies of Observed Diameters in Cells* $(i + 0.5, i + 1.5]$, $i = 1, 2, \ldots 14$

i	1	2	3	4	5	6	7	8	9	10	11	12	13	14
Freq	52	146	197	210	184	143	95	57	31	15	7	4	2	1

Note: The unit of measurement is mm.

suppose $h \to 0$ *and* $(nh^3)^{-1}\log(1/h) \to 0$. *Then for each* $\epsilon > 0$,

$$\sup_{x \geq \epsilon} |\tilde{g}'_{n,h}(x) - g'(x)| \to 0$$

almost surely.

It is well known that the optimal rate for choosing the bandwidth in the standard kernel estimation problem is $n^{-1/5}$ (see also Exercise 8.4). If the underlying g satisfies the conditions of Lemma 8.2 and $[a, b]$ is an interval where this g has a derivative strictly bounded away from zero, Lemma 8.2 shows that by taking this rate, the estimator $\tilde{g}_{n,h}$ will with probability one become monotone on $[a, b]$. Using some additional localization argument will show that on (a, b) the MSLE will with probability tending to one (for n tending to infinity) coincide with $\tilde{g}_{n,h}$ and hence its (pointwise) asymptotic distribution for $x \in (a, b)$ will be the same as that of $\tilde{g}_{n,h}(x)$.

Another interesting application of the idea of maximizing a smoothed criterion function occurs when there are binned observations. A nice example of this is the dataset originally studied by Wicksell in 1925 (see Section 4.1). Wicksell worked with diameters in mm (rather than squared radii) and used cells of width 1 mm to describe the data. The original measured diameters are given in Table 8.1. One could rewrite the equations relating the observation distribution and the distribution of interest in terms of diameters (or look at the equations Wicksell used), but we will instead use the binned observations of the squared circle radii, together with the equations derived in Section 4.1. Figure 8.3 shows the resulting (variable bin width) histogram of squared circle radii.

The maximum likelihood estimator in Wicksell's problem is not well defined; see Exercise 4.8. In the current situation, we do not have the precise data, since these are binned. This makes it straightforward to define a maximum smoothed likelihood estimator. This estimator is obtained by using the histogram estimate $\tilde{g}$ in the definition of the log likelihood (which is available) rather than the empirical distribution function of the data (which is not available). As previously mentioned, this estimator can be viewed as Kullback Leibler projection of the histogram-type estimator on the class of sampling densities that can occur in Wicksell's problem. Another approach would be to first apply the inverse Wicksell transformation to density estimate $\tilde{g}$, obtaining a function that is not monotone, and projecting this on the class of monotone functions. We will take a closer look at this type of estimator.

To start with, define the inverse estimator that is obtained by using relation (4.1) expressing F in terms of g, plugging in the histogram density estimate $\tilde{g}$ of Figure 8.3. Denote by t_i $(i = 1, 2, \ldots, m)$ the break points of the histogram estimator $\tilde{g}$ and by $\tilde{g}_i$ the value of $\tilde{g}$ on the interval $[t_i, t_{i+1})$. By convention take $t_0 = 0$ and $t_{m+1} = \infty$. Note that $\tilde{g}_0 = \tilde{g}_m = 0$.

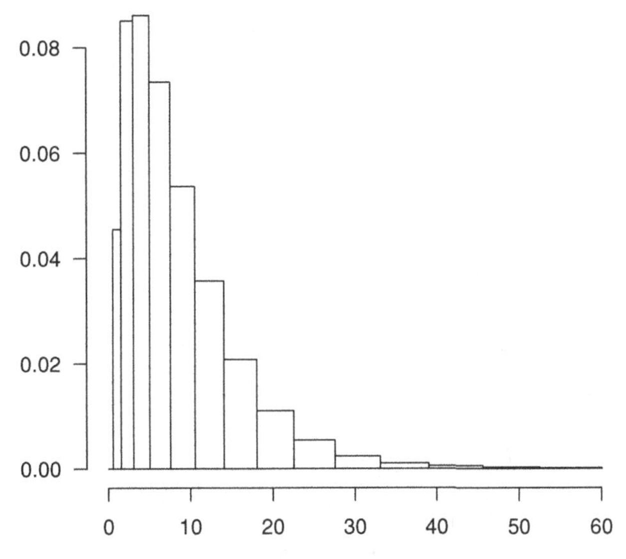

Figure 8.3 The probability (variable binwidth) histogram based on the diameters of Table 8.1, transformed to squared circle radii.

Then the histogram density can be represented by

$$\tilde{g}(x) = \sum_{i=1}^{m-1} \tilde{g}_i 1_{[t_i, t_{i+1})}(x), \quad x \geq 0.$$

Following the (nonsmoothed) approach in Section 4.1, we define

$$V_n^{naive}(z) = \int_z^{\infty} (y - z)^{-1/2} \tilde{g}(y)\, dy.$$

The plug-in estimate thus obtained is, for $z \in [t_i, t_{i+1})$, $0 \leq i \leq m$, given by

$$1 - F^{naive}(z) = \frac{V_n^{naive}(z)}{V_n^{naive}(0)} = \frac{\tilde{g}_i \sqrt{t_{i+1} - z} + \sum_{j=i+1}^{m-1} \tilde{g}_j \left(\sqrt{t_{j+1} - z} - \sqrt{t_j - z}\right)}{\sum_{i=1}^{m-1} \tilde{g}_i (\sqrt{t_{i+1}} - \sqrt{t_i})}.$$

See Figure 8.4 for a picture of this estimate. It is clear that the estimate does not satisfy the conditions needed for a distribution function. It takes values below zero and it is also initially decreasing.

Nevertheless, this estimator can be used as an ingredient to obtain a monotone estimator via isotonization. The resulting estimator can be interpreted as projection estimator obtained by applying the inverse Wicksell transformation to a smooth estimator of the observation density (at least, more smooth than the empirical distribution function). More concretely, the estimator will be given by $1 - V_n / V_n(0)$ where V_n minimizes the following function over all decreasing functions on $[0, \infty)$:

$$V \mapsto \frac{1}{2} \int_0^{\infty} V(x)^2\, dx - \int V(x) V_n^{naive}(x)\, dx,$$

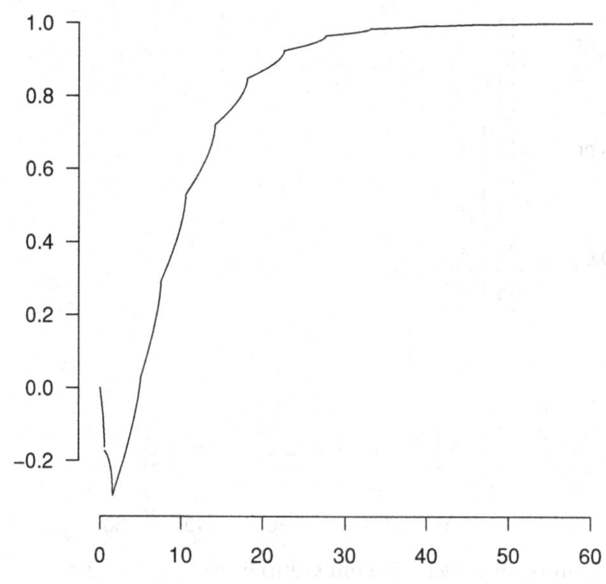

Figure 8.4 The plug-in estimate of the distribution function F based on density estimate $\tilde{g}$ of Figure 8.3.

which is (as also seen in Section 2.6) a least squares type estimator. In fact, its solution is obtained by taking the right derivative of the least concave majorant of the function $\tilde{U}_n$ given by

$$\tilde{U}_n(z) = \int_0^z V_n^{naive}(y)\,dy = \frac{4}{3}\sum_{j=1}^{m-1} \tilde{g}_j \left(t_{j+1}^{3/2} - t_j^{3/2}\right) - \frac{4}{3}\tilde{g}_i \left(t_{j+1} - z\right)^{3/2}$$

$$-\frac{4}{3}\sum_{j=i+1}^{m-1} \tilde{g}_j \left(\left(t_{j+1} - z\right)^{3/2} - \left(t_j - z\right)^{3/2}\right)$$

for $z \in [t_i, t_{i+1})$, $0 \le i \le m$.

See Figure 8.5 for $\hat{U}_n$ and its least concave majorant and Figure 8.6 for the resulting estimator of F given by $1 - \hat{U}_n'/\hat{U}_n'(0)$. This example shows that there is a variety of possible estimation procedures based on optimizing an objective function based on a smoothed version of the empirical distribution function. In Section 8.5, the MSLE in the current status model will be studied, where a problem that is often present near the boundary of the domain (e.g., near zero) is also dealt with.

8.3 Penalized M-Estimation

Another method to obtain smooth shape constrained estimators is via penalization. Not only can smoothness be forced via penalization, but also consistency problems near the boundary (as observed in Exercises 2.7 and 4.11) can be prevented. In this section both problems need to be solved in the model used to illustrate the approach: estimating a smooth increasing hazard rate on the interval $[0, a]$ for some $a > 0$. Recall from Section 2.6 that $Z_1, \ldots, Z_n$

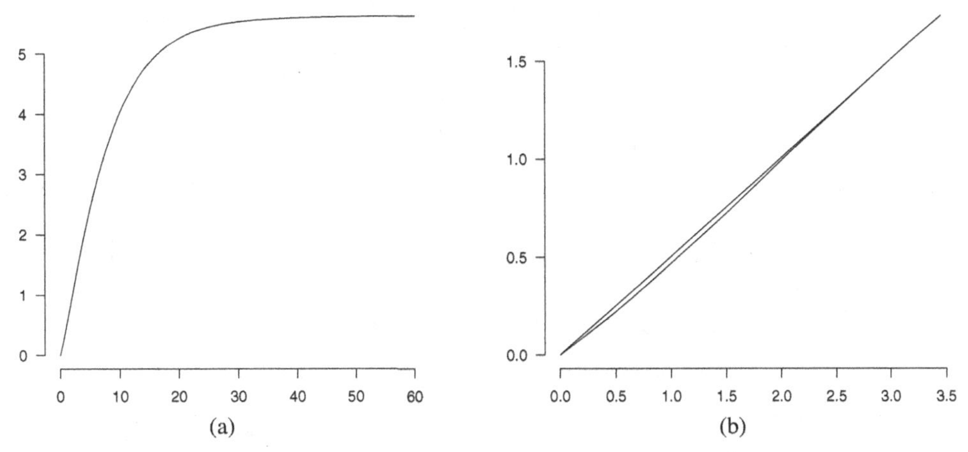

Figure 8.5 (a) The functions $\tilde{U}_n$ and its least concave majorant $\hat{U}_n$. (b) The picture locally; the functions $\tilde{U}_n$ and $\hat{U}_n$ only differ on a small interval near zero.

are i.i.d. from a distribution with nondecreasing hazard rate h on $[0, a]$. The problem is to estimate h based on $Z_1, \ldots, Z_n$. A possible estimator is the least squares estimator $\hat{h}_n$ minimizing (2.27),

$$h \mapsto \frac{1}{2}\int_0^a h(x)^2\,dx - \int_{[0,a]} h(x)\,d\mathbb{H}_n(x)$$

over all increasing hazard rates on $[0, a]$. Here $\mathbb{H}_n$ is the empirical cumulative hazard function based on $Z_1, \ldots, Z_n$. This estimator allows for an explicit representation as left derivative of the greatest convex minorant of cusum diagram (2.29). According to Exercise 2.23, $\hat{h}_n$ is inconsistent at $z = 0$ and $z = a$, whereas it is uniformly consistent on closed subintervals

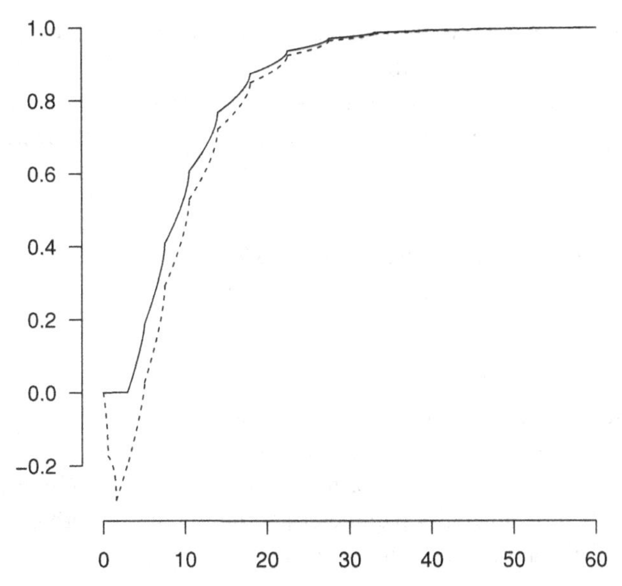

Figure 8.6 The estimate of the distribution function F based on the function $\hat{U}_n$. The dashed curve is the plug-in estimate.

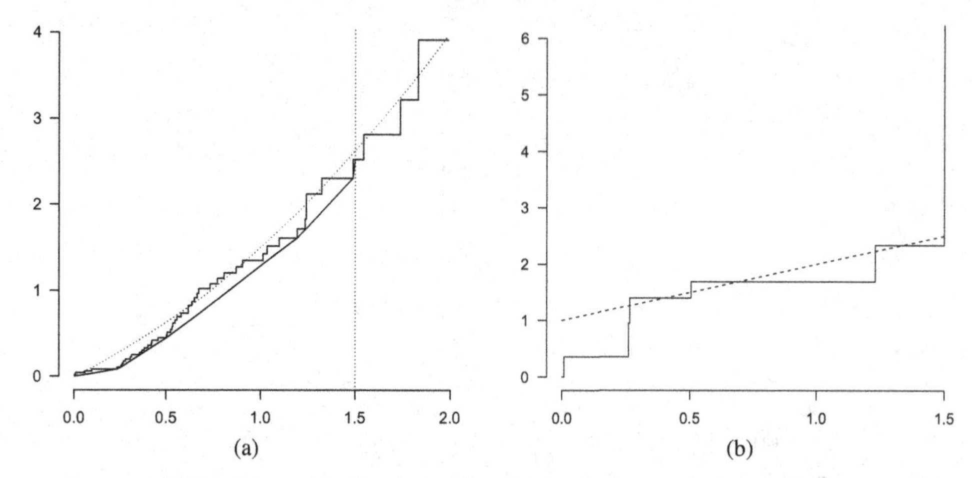

Figure 8.7 (a) The nonpenalized cumulative sum diagram with its convex minorant on the interval $[0, 1.5]$ based on a sample of size $n = 50$ from the distribution with hazard rate $h(x) = 1 + x$ on $[0, \infty)$. (b) The corresponding estimate of the hazard rate on the interval $[0, 1.5]$.

of $(0, a)$. We now introduce a penalized least squares criterion and show that it leads to a uniformly consistent estimator for h_0. For nonnegative parameters α_n and β_n, define

$$h \mapsto \frac{1}{2}\int_0^a h(x)^2\,dx - \int_{[0,a]} h(x)\,d\mathbb{H}_n(x) - \alpha_n h(0) + \beta_n h(a). \tag{8.5}$$

The minimizer of this function over all nondecreasing hazard rates on $[0, a]$ is given by the left derivative of the greatest convex minorant of the cusum diagram given by

$$\left(Z_{(i)}, \mathbb{H}_n(Z_{(i)}-) + \alpha_n\right),\ Z_{(i)} < a, \qquad \left(a, \mathbb{H}_n(a-) + \alpha_n - \beta_n\right). \tag{8.6}$$

See Exercise 8.7. Figure 8.7 shows the cumulative sum diagram and resulting estimate for h when $\alpha_n = \beta_n = 0$, so in the nonpenalized case.

For vanishing α_n and β_n, uniform consistency of the resulting estimator on intervals of the type $[\delta, a - \delta]$ follows from Lemma 3.1. Uniform consistency of the penalized estimator on the interval $[0, a]$ is given in Lemma 8.3, which is Corollary 2.1 in Groeneboom and Jongbloed, 2013a.

Lemma 8.3 *Let h_0 be continuous and nondecreasing on $[0, a]$ and have continuous (one sided) derivatives at 0 and a. Take $\alpha_n, \beta_n \sim n^{-2/3}$. Then:*

$$\sup_{[0,a]} |\hat{h}_n(x) - h_0(x)| = O_P\left(n^{-1/4}\right). \tag{8.7}$$

See Figure 8.8 for this penalized estimator based on the same data as Figure 8.7. It is clear from the picture that the hazard estimate near the boundary improves by penalization.

Having a uniformly consistent estimator for h_0, the next step is to impose smoothness by penalization. Let $\lambda \geq 0$ be a penalty parameter and define the smooth penalized local least squares estimator of h_0 on $[0, a]$ as minimizer of

$$\Phi_\lambda(h) = \int_0^a \left(h(x) - \hat{h}_n(x)\right)^2 dx + \lambda \int_0^a h'(x)^2\,dx \tag{8.8}$$

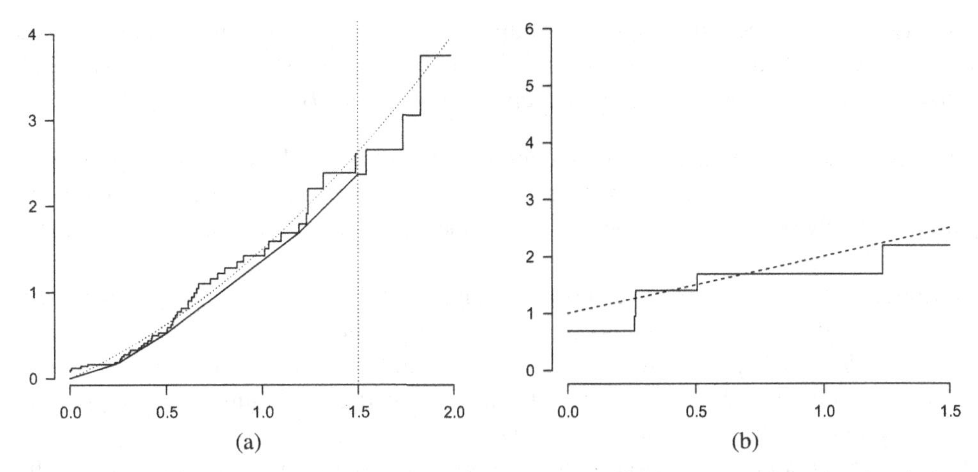

Figure 8.8 The penalized cumulative sum diagram with its convex minorant on the interval [0, 1.5] (a) and the resulting estimate for h based the same data as Figure 8.7(b). The penalization parameters are chosen $\alpha = 0.08$ and $\beta = 0.16$.

over the set of differentiable functions h on $[0, a]$, where $\hat{h}_n$ is the monotone (on $[0, a]$) piecewise constant estimate that minimizes (8.5). Our first lemma gives the minimizer of Φ_λ over the class of smooth functions on $[0, a]$ under boundary constraints at 0 and a. Note that the estimator is not required to be monotone at this stage.

Lemma 8.4 *Let $\kappa_1, \kappa_2 > 0$. Then the unique minimizer of Φ_λ over all smooth functions on $[0, a]$ such that $h(0) = \kappa_1$ and $h(a) = \kappa_2$ exists and is given by*

$$h(x) = h_1(x) + c_1 e^{-x/\sqrt{\lambda}} + c_2 e^{-(a-x)/\sqrt{\lambda}} \tag{8.9}$$

where

$$h_1(x) = \frac{1}{2}\lambda^{-1/2} \int_0^a e^{-|y-x|/\sqrt{\lambda}} \hat{h}_n(y)\, dy, \tag{8.10}$$

and c_1 and c_2 are chosen such that h satisfies the imposed boundary constraints.

Proof Writing

$$I(h) = \int_0^a G(x, h, h')\, dx = \int_0^a \left(h(x) - \hat{h}_n(x)\right)^2 dx + \lambda \int_0^a h'(x)^2\, dx, \tag{8.11}$$

we get Euler's differential equation from calculus of variations:

$$G_h - \frac{d}{dx} G_{h'} = 0.$$

This equation is to be solved under under the boundary conditions $h(0) = \kappa_1$ and $h(a) = \kappa_2$. This results in the second order integral equation

$$h''(x) = \lambda^{-1} \left(h(x) - \hat{h}_n(x)\right) \tag{8.12}$$

with boundary constraints; see also Exercise 8.8.

A particular solution to (8.12) is given by (8.10). Note that h_1 is a smoothed version of $\hat{h}_n$ in the sense of kernel smoothing, where $\sqrt{\lambda}$ is the bandwidth and the kernel is the Laplace density (which has unbounded support). Adding the solutions to the homogeneous equation multiplied by constants c_1 and c_2 respectively, the unique solution to the boundary value problem is obtained by choosing c_1 and c_2 appropriately in (8.9). □

Now note that the requirement of monotonicity was not included in the optimization problem. In case $a = \infty$ (a situation actually not covered), this would mean that the function h_1 would actually be monotone since $h_1(x)$ is a local average of the monotone function $\hat{h}_n$ near x (see Exercise 8.1).

Various choices can be made to determine c_1 and c_2. One is to fix the value of the solution at the value of the initial (consistent) estimator $\hat{h}_n$. This choice would lead to a monotone estimator of h on $[0, a]$; see Exercise 8.9. Another choice would be to minimize Φ_λ over all nondecreasing hazard rates h on $[0, a]$. This can be done by minimizing Φ_λ over the functions given in (8.9) as function of c_1 and c_2.

One can easily think of variations on the two-stage penalization method described in this section. One could, for example, also start off with a criterion function including penalty terms for both the boundary behavior and smoothness properties. Alternatively, one could interchange the order and first construct a smooth estimator followed by defining criterion functions including penalization terms for the behavior of the estimator near the boundaries.

8.4 Monotonic Rearrangements of Smooth Estimators

A simple method that can be used to convert a nonmonotone function into a monotone function is monotonic rearrangement. To understand the idea, consider a (measurable) real valued function f on $[0, 1]$, say. Then, by definition, the sets $L_v(f)$, where

$$L_v(f) = \{x \in [0, 1] : f(x) \le v\}, \quad \text{for } v \in \mathbb{R},$$

are nested in the sense that $L_u(f) \subset L_v(f)$ whenever $u \le v$. Therefore, the function m_f, defined by

$$m_f(v) = \lambda\,(L_v(f)), \quad v \in \mathbb{R}$$

where λ denotes Lebesgue measure, is nondecreasing in v. The monotonic (consider now increasing) rearrangement of f is given by

$$\tilde{f}(x) = m_f^{-1}(x).$$

It is clear that monotonicity of m_f implies monotonicity of $\tilde{f}$. A way of thinking of $\tilde{f}$ is that it represents the quantile function of the random variable $f(U)$, where U has the uniform distribution on $[0, 1]$. Indeed,

$$f(U) \le v \iff U \in L_v(f), \text{ so } P(f(U) \le v) = m_f(v).$$

Another way of looking at the monotonic rearrangement is by discretization. Suppose f is continuous, let $0 = x_1 < x_1 < \cdots < x_m = 1$ be a dense equidistant grid in $[0, 1]$ and define $y_i = f(x_i)$ for $0 \le i \le m$. Then on this grid the monotonic rearrangement of f can be approximated by

$$\tilde{f}(x_i) = y_{(i)}, \tag{8.13}$$

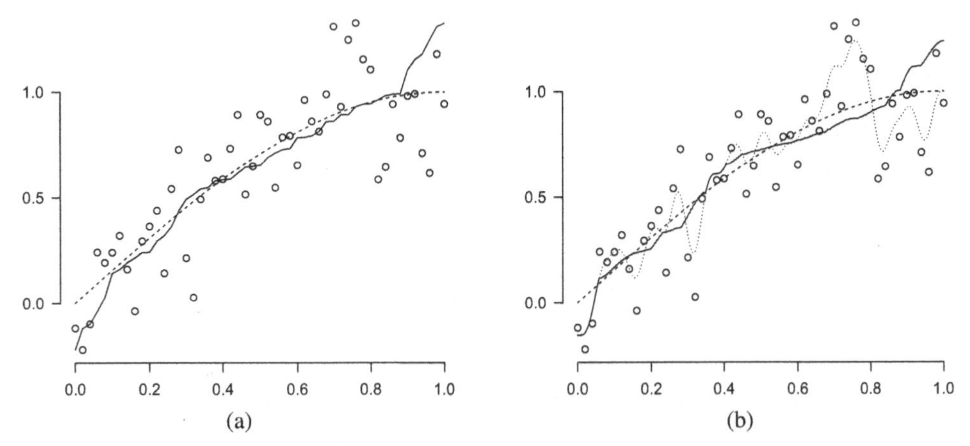

Figure 8.9 Monotonic rearrangement estimates of the monotone regression function based on a sample of size 50 according to model (8.14) with $\mu(x) = \sin(\pi x/2)$ (dashed lines in the plots), $\sigma = 0.2$. (a) The discrete rearrangement. (b) A rearrangement based on the Nadaraya Watson estimator (dotted line in the plot) based on the data, with normal kernel and bandwidth 0.05.

where $y_{(i)}$ denotes the ith order statistic of the y_is. Natural extensions of this approach can be easily constructed, e.g., by choosing a different measure λ to start with, taking bigger intervals, and so on.

Let us now see where the approach leads in the setting of monotone regression. We observe pairs (x_i, y_i) where $x_i = i/n$ and y_i is a realized value of

$$Y_i = \mu(x_i) + \epsilon_i \tag{8.14}$$

with $\epsilon_1, \ldots, \epsilon_n$ i.i.d. normally distributed with zero expectation and variance σ^2 and μ nondecreasing. A first naive approach could be to estimate μ only at the observed design points by $\hat{\mu}(x_i) = y_{(i)}$, the ith order statistic of the observed responses. See the left picture in Figure 8.9. To show that also smooth estimators can be obtained, we use the Nadaraya-Watson regression estimator for μ as basic estimator and construct its monotonic rearrangement. To this end we use the discrete sorting approximation of (8.13).

The method of monotonic rearrangement can of course also be used based on a nonsmooth initial estimate. The result will then in general also be nonsmooth, but still monotone. This can be illustrated using the deconvolution problem introduced in Section 2.4. Figure 8.10 shows the naive (nonmonotone) estimator of the distribution function F also shown in Figure 2.10. Note that the naive estimator does not satisfy the range constraint that holds for distribution functions (values in $[0, 1]$). However, since this estimator is by definition equal to one to the right of the highest observation $z_{(n)}$, the monotonic rearrangement estimator will satisfy the range constraint in this example. See also Exercise 8.12. If monotonicity were only imposed on a finite interval, this would not be true.

Exercise 8.13 shows how the method of monotonic rearrangements can be used in the current status problem. There the method does not give sensible results if it is applied without any smoothing, but with smoothing (constructing a reasonable initial estimator), it will lead to a well behaved estimator.

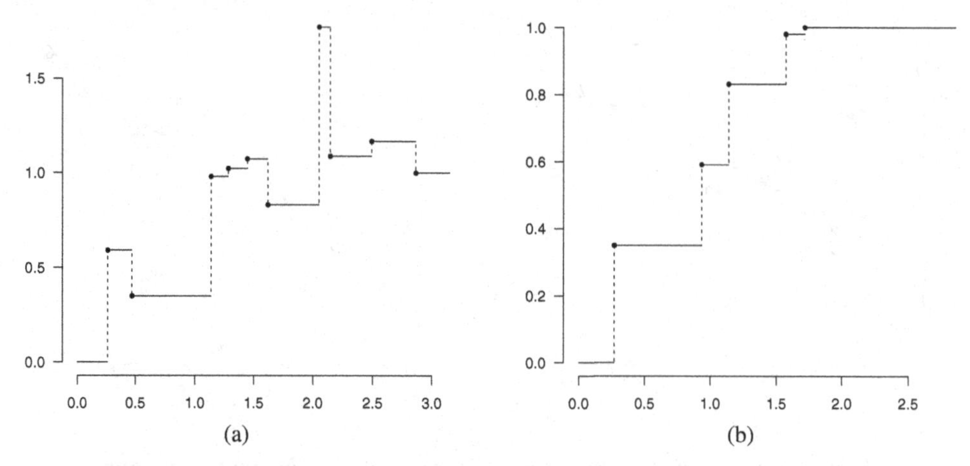

Figure 8.10 Naive estimator of F based on linear interpolation of (the left continuous version of) U_n in Figure 2.9 (a) and the monotonic rearrangement estimator based on this (b).

8.5 Maximum Smoothed Likelihood Estimation in the Current Status Model

The MLE for current status is characterized in Lemma 2.3. The characterizing (in-) equalities can be represented as

$$\int_{[t,\infty)} \{\delta - \hat{F}_n(t)\}\, d\mathbb{P}_n(t,\delta) \geq 0,\ t \in \mathbb{R}, \tag{8.15}$$

with equality if t is a point of mass of $\hat{F}_n$. This characterization leads to a convex minorant algorithm for the computation of the MLE and it was crucial in the derivation of the asymptotic distribution of the MLE in Section 3.8.

If one is willing to assume smoothness of the underlying distribution function, it is possible to construct a much better estimator than the MLE, in the sense that it converges to the underlying distribution at a faster rate. In Section 8.1, this was done by smoothing the MLE, yielding the SMLE. The theory for the SMLE will be treated in Section 11.3. Alternatively, the approach of Section 8.2 can be taken, by first smoothing the estimates of the underlying distributions and next maximizing over the distribution function one wants to estimate. The maximizer is then the MSLE. In this section we shall consider the MSLE.

We assume again that the support of the distribution to be estimated is an interval $[a,b]$. In order to define a smoothed objective function, we define the kernel estimate g_{nh} of the observation density g by

$$g_{nh}(t) = \int K_h(t-x)\, d\mathbb{G}_n(x),\ \ t \in [a+h, b-h] \tag{8.16}$$

and, for $t \in [a, a+h]$, we use a boundary corrected expression

$$g_{nh}(t) = \alpha((t-a)/h) \int K_h(t-x)\, g_n(x)\, dx + \beta((t-a)/h) \int \frac{t-x}{h} K_h(t-x)\, dx.$$

Here the functions α and β are chosen in such a way that

$$\alpha(u)\int_{x=-1}^{u} K(x)\,dx + \beta(u)\int_{x=-1}^{u} x\,K(x)\,dx = 1,\ u \in [0,1], \tag{8.17}$$

and

$$\alpha(u)\int_{x=-1}^{u} xK(x)\,dx + \beta(u)\int_{x=-1}^{u} x^2\,K(x)\,dx = 0,\ u \in [0,1]. \tag{8.18}$$

For $t \in [b-h, b]$, we define

$$g_{nh}(t) = \alpha((b-t)/h)\int K_h(t-x)\,g_n(x)\,dx - \beta((b-t)/h)\int \frac{t-x}{h}\,K_h(t-x)\,dx.$$

Similarly, we define the kernel estimator of the subdensity g^1 of the observation times corresponding to $\delta = 1$ by

$$g_{nh}^1(t) = \int \delta K_h(t-x)\,d\mathbb{P}_n(x,\delta), \quad t \in [a+h, b-h] \tag{8.19}$$

and for $t \in [a, a+h]$

$$\begin{aligned} g_{nh}^1(t) &= \alpha((t-a)/h)\int \delta\,K_h(t-x)\,d\mathbb{P}_n(x,\delta)\\ &\quad + \beta((t-a)/h)\int \delta\,\frac{t-x}{h}\,K_h(t-x)\,d\mathbb{P}_n(x,\delta), \end{aligned}$$

where the functions α and β satisfy (8.17) and (8.18). Finally we define for $t \in [b-h, b]$:

$$\begin{aligned} g_{nh}^1(t) &= \alpha((b-t)/h)\int \delta\,K_h(t-x)\,d\mathbb{P}_n(x,\delta)\\ &\quad - \beta((b-t)/h)\int \delta\,\frac{t-x}{h}\,K_h(t-x)\,d\mathbb{P}_n(x,\delta), \end{aligned}$$

The MSLE $\hat F_{nh}^{MSL}$ now maximizes the objective function

$$\tilde\ell(F) = \int_a^b \log F(u)\,g_{nh}^1(u)\,du + \int_a^b \log\{1-F(u)\}\left\{g_{nh}(u) - g_{nh}^1(u)\right\}\,du. \tag{8.20}$$

The key to the analysis of the MSLE is the continuous cumulative sum diagram

$$\left\{\left(G_{nh}(t), G_{nh}^1(t)\right),\ t \in [a,b]\right\},\quad G_{nh}(t) = \int_a^t g_{nh}(u)\,du,\quad G_{nh}^1(t) = \int_a^t g_{nh}^1(u)\,du, \tag{8.21}$$

replacing the ordinary cumulative sum diagram

$$\left\{\left(\mathbb{G}_n(t), \int_{u\in[a,t]} \delta\,\mathbb{P}_n(u,\delta)\right),\quad t \in [a,b]\right\},$$

which can be used to compute the MLE. It is proved in Groeneboom et al., 2010, Theorem 3.1 on p. 356, that the MSLE is defined by the right continuous slope of the greatest convex minorant of the cusum diagram (8.21). That is, the MSLE $\hat F_{nh}^{MSL}$, maximizing (8.20), is given at the point t by $S_n(t)$, where $S_n(t)$ is the right continuous slope of the function C_n, evaluated

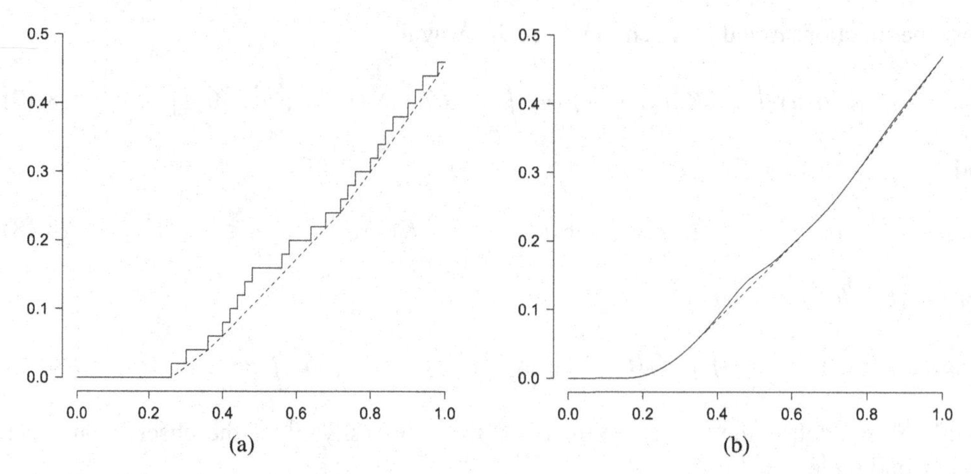

Figure 8.11 Unsmoothed (a) and smoothed (b) cusum diagram for a sample of size $n = 50$ for the current status model, where the observation distribution and the distribution of the hidden variable are both Uniform(0, 1). Bandwidth $h = n^{-1/2} \approx 0.1414$.

at the point $G_{nh}(t)$ in the time scale G_{nh}, and where

$$\{(G_{nh}(t), C_n(t)),\ t \in [a, b]\}$$

is the lower convex hull of (8.21). Figure 8.11 shows the cusum diagram for the ordinary MLE and the MSLE for a sample of size $n = 50$, where the bandwidth is taken $h = n^{-1/2} \approx 0.1414$, which is obviously too small, but still shows a discrepancy between the cusum diagram and its greatest convex minorant. Figure 8.12 shows the corresponding MLE, SMLE (to be discussed in Section 11.3) and MSLE.

Figure 8.13 shows the cusum diagram for the ordinary MLE and the MSLE for a sample of size $n = 100$, where the bandwidth is taken $h = n^{-1/5} \approx 0.3981$, which is the right order, and which shows a cusum diagram which is convex itself. Figure 8.14 shows the corresponding MLE, SMLE and MSLE.

For the rest of this section we assume that the interval $[a, b]$ is given by $[0, M]$. The following result is proved in Groeneboom et al., 2010 (Corollary 3.4 on p. 359).

Lemma 8.5 *Let the conditions of Theorem 11.4 be satisfied. Furthermore, let* $h = h_n \sim cn^{-\alpha}$, *where* $\alpha \in (0, 1/3)$, *and let the naive estimator* $\hat{F}_{nh}^{\text{naive}}$ *of* F_0 *be defined by*

$$\hat{F}_{nh}^{\text{naive}}(t) = \frac{g_{nh}^1(u)}{g_{nh}(t)},$$

where g_{nh} *and* g_{nh}^1 *are defined by (8.16) and (8.19). Then, if* $0 < m < M' < M$, *we have:*

$$\mathbb{P}\left\{\hat{F}_{nh}^{\text{naive}} = \hat{F}_{nh}^{MSL} \text{ on } [m, M']\right\} \longrightarrow 1,$$

as $n \to \infty$. *Consequently, for all* $t \in (0, M)$ *the asymptotic distributions of (the rescaled)* $\hat{F}_{nh}^{MSL}(t)$ *and* $\hat{F}_{nh}^{\text{naive}}(t)$ *are the same.*

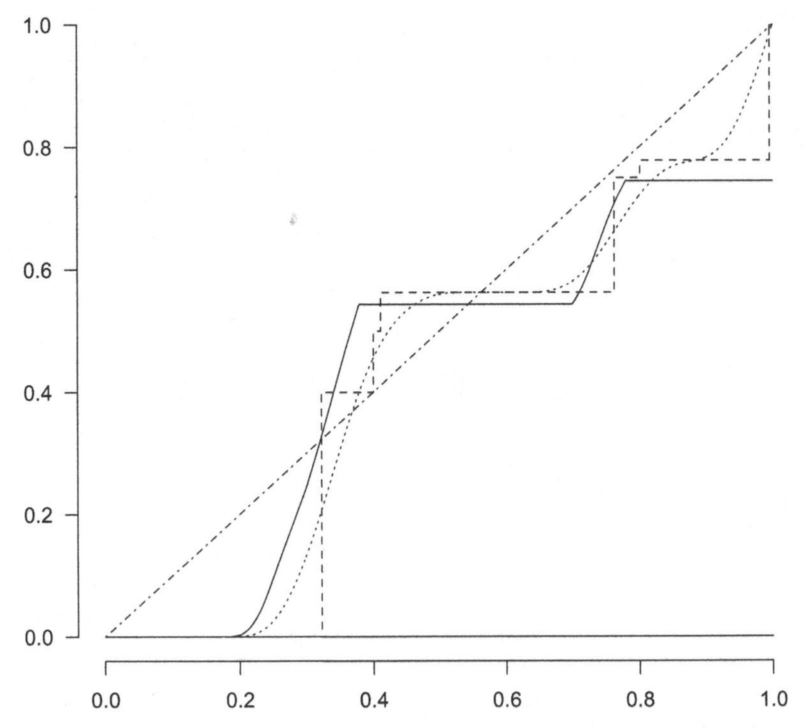

Figure 8.12 MSLE (solid), SMLE (dotted), MLE (dashed) and the uniform underlying distribution function (dashed-dotted) for the data of Figure 8.11.

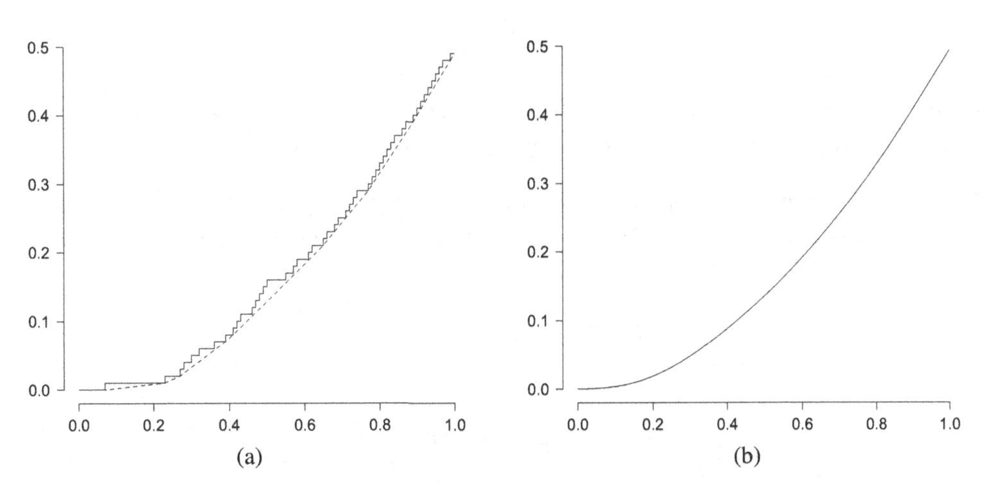

Figure 8.13 Unsmoothed (a) and smoothed (b) cusum diagram for a sample of size $n = 100$ for the current status model, where the observation distribution and the distribution of the hidden variable are both Uniform(0, 1). Bandwidth $h = n^{-1/5} \approx 0.3981$.

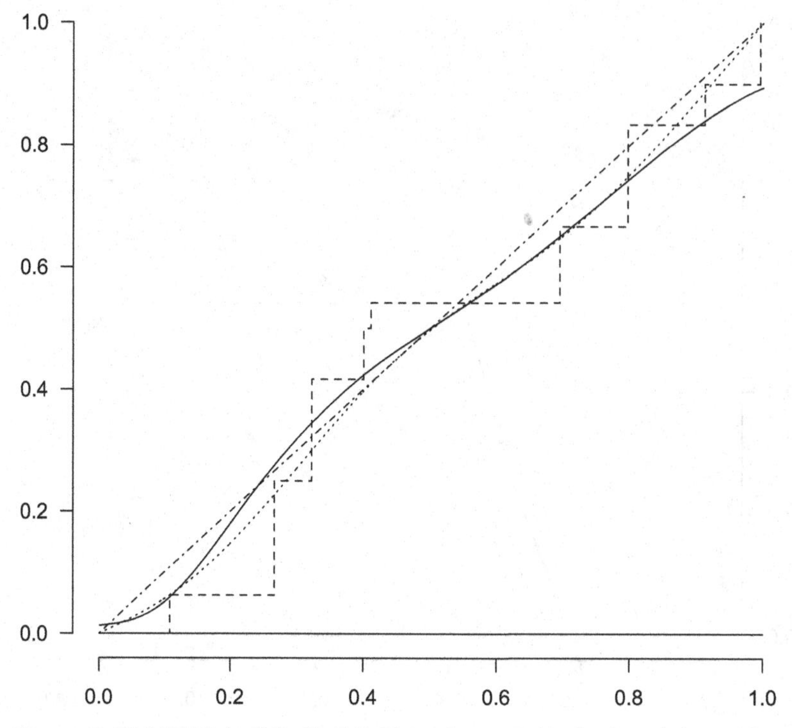

Figure 8.14 MSLE (solid), SMLE (dotted), MLE (dashed) and the uniform underlying distribution function (dashed-dotted) for the data of Figure 8.13.

In this way the asymptotic study of the MSLE is reduced to the study of the ratio of kernel estimators

$$\frac{g_{nh}^1(u)}{g_{nh}(u)},$$

for which the monotonicity constraints are no longer active. This can be analyzed by standard methods and we get the following result (Theorem 3.5 on p. 360 of Groeneboom et al., 2010).

Theorem 8.1 *Let the conditions of Theorem 11.4 be satisfied. Fix $t \in (0, M)$ so that f_0'' and g'' exist and are continuous at t. Let $h = h_n \sim cn^{-1/5}$ $(c > 0)$ be the bandwidth used in the definition of g_{nh} and g_{nh}^1. Then*

$$n^{2/5}\{\hat{F}_{nh}^{MSL} - F_0(t)\} \xrightarrow{\mathcal{D}} N(\mu, \sigma^2),$$

e where

$$\mu = \frac{1}{2}c^2\left\{f_0'(t) + \frac{2f_0(t)g'(t)}{g(t)}\right\}\int u^2K(u)\,du,$$

and

$$\sigma^2 = \frac{F_0(t)(1 - F_0(t))}{cg(t)}\int K(u)^2\,du.$$

It is seen from the formulation of Theorems 11.4 and 8.1 that the asymptotic variance of the SMLE and the MSLE is the same, and that the difference between the estimators shows up in the bias. One cannot say that one estimator is uniformly better than the other. Theorem 3.5 on p. 360 of Groeneboom et al., 2010, has the condition that $g(t)f_0'(t) + 2f_0(t)g'(t) \neq 0$, but this not necessary for the validity of the theorem; it is only for the conclusion that choice of bandwidth $cn^{-1/5}$ leads to the optimal rate. See also the remark following Theorem 11.4.

Results, similar to Theorems 11.4 and 8.1, can be derived for the corresponding estimates of the density f_0. In this case the rate of convergence drops to $n^{-2/7}$, which is the usual rate for the estimation of derivatives of densities (the inverse nature of the estimation problem makes the estimation of the distribution function F_0 comparable to the estimation of a density and estimation of the density f_0 comparable to the estimation of the derivative of a density). For details on this we refer to Groeneboom et al., 2010.

8.6 Smooth Estimation for Interval Censoring Case 2

In this section, we illustrate the smoothed maximum likelihood method introduced in Section 8.1 as well as the maximum smoothed likelihood estimator of Section 8.5 for the interval censoring case 2 problem. This problem is introduced in Section 4.7 and studied from an algorithmic point of view in Example 7.4. Basically, there is an (unobservable) i.i.d. sample $X_1, \ldots, X_n$ having distribution function F on $[0, \infty)$ and independent of this an observable i.i.d. sample of random vectors (T_i, U_i), $1 \le i \le n$ with (joint) density g such that $T_i < U_i$ with probability one. The observed data contain for each i the information that of the intervals $[0, T_i]$, $(T_i, U_i]$ or (U_i, ∞) contains X_i. The data are therefore given by

$$(T_i, U_i, \Delta_{i1}, \Delta_{i2}) \text{ where } \Delta_{i1} = 1_{\{X_i \le T_i\}} \quad \text{and} \quad \Delta_{i2} = 1_{\{T_i < X_i \le U_i\}}.$$

As running data example in this section, we use the data set, studied in Betensky and Finkelstein, 1999; see also Table 7.1. In view of the MLE described in Example 7.4, the smoothed MLE is rather straightforward to compute. First compute the MLE, using, e.g., the ICM algorithm of Section 7.3, and then smooth the resulting distribution function using kernel smoothing. This estimator is automatically monotone and smooth (see Exercise 8.1).

We recall the setting for this problem, also allowing for ties in the data (as present in the example considered here). The log likelihood to be maximized is given by

$$\sum_{i=1}^{n} f_i \left\{\Delta_{i1} \log F(T_i) + \Delta_{i2} \log\{F(U_i) - F(T_i)\} + \Delta_{i3} \log\{1 - F(U_i)\}\right\}, \tag{8.22}$$

where f_i denotes the number of ties at the ith observation pair (T_i, U_i) (also called the multiplicity), and where

$$\Delta_{i3} = 1 - \Delta_{i1} - \Delta_{i2}.$$

Denoting the MLE, obtained in this way, by $\hat{F}_n$, an SMLE $\tilde{F}_{n,h}$ can be simply obtained by computing

$$\tilde{F}_{n,h}(x) = \int \mathbb{K}\left(\frac{x-y}{h}\right) d\hat{F}_n(y), \tag{8.23}$$

Table 8.2 *First Marginal of the Betensky-Finkelstein Data*

T_i	U_i	k in Δ_{ik}	Frequency
0	–	1	47
0	3	2	19
0	6	2	2
3	3	2	8
3	6	2	9
6	6	2	5
6	9	2	7
9	9	2	6
9	12	2	7
12	12	2	1
12	15	2	4
–	0	3	7
–	3	3	3
–	6	3	9
–	9	3	7
–	12	3	23
–	15	3	25
–	18	3	13
–	21	3	2
			204

where h is the chosen bandwidth and where $\mathbb{K}$ is an integrated positive probability kernel:

$$\mathbb{K}(u) = \int_{-\infty}^{u} K(y)\,dy, \tag{8.24}$$

choosing for K for example the triweight kernel, given by

$$K(u) = \frac{35}{32}\left(1 - u^2\right)^3 1_{[-1,1]}(u).$$

See also Section 8.5.

The data for the first marginal of the observation set in Betensky and Finkelstein, 1999, are displayed in Table 8.2. It is seen that this table exhibits some peculiarities; for example, there are some degenerate intervals (8 observations are in the interval [3, 3], 5 observations in the interval [6, 6], etc.).

We now reduce the optimization problem to the problem of maximizing

$$\sum_{i=1}^{n} f_i \log\left\{\sum_{j\in J_i} p_j\right\},$$

where the p_j are in this case the probability masses at the points $0, 3, 6, \ldots, 21$, where we add an extra point larger than 21, say 22, to allow for extra mass to the right of all observations. Furthermore, J_i is the set of indices consistent with the ith observation. In the present case

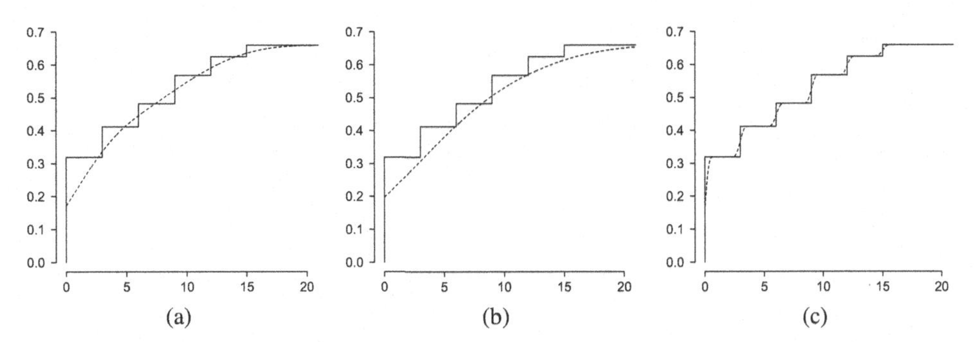

Figure 8.15 The MLE (solid) and SMLE (dashed) for the data of Table 8.2. (a) The bandwidth is taken h; (b) $2h$; and (c) $0.2h$, where $h = 21 \cdot N^{-1/5}$ and N is the total number of observations.

we get from Table 8.2:

$$47 \log p_1 + 19 \log (p_1 + p_2) + 2 \log (p_1 + p_2 + p_3) + \ldots,$$

where we number the p_j in a obvious way, using the ordering of the observation points $0, 3, 6, \ldots,$ augmented with the point to the right of all observation points. Note that some of the p_j can be zero. In fact, the MLE puts zero masses p_j at the points 18 and 21 in this example.

In Figure 8.15 the MLE is shown, together with an MSLE, using the triweight kernel and bandwidth

$$h = 21 \cdot N^{-1/5}, \quad \text{where} \quad N = \sum_{i=1}^{n} f_i$$

is the number of observations. The factor 21 corresponds to the length of the interval where the estimation is done. This choice of h is a common and simple, although not very sophisticated, way of picking a bandwidth in density estimation. The effect of the choice of bandwidth is shown varying via the choices of the bandwidths h, $2h$ and $0.2h$, respectively. If one takes a very small bandwidth, as in Figure 8.15c, the SMLE is of course very close to the MLE.

The MLE maximizes objective function (8.22), which, divided by the total number of observations, can be written in the form

$$\int \log F(t)\, d\mathbb{P}_n(t, u, 1, 0) + \int \log\{F(u) - F(t)\}\, d\mathbb{P}_n(t, u, 0, 1) \\ + \int \log\{1 - F(u)\}\, d\mathbb{P}_n(t, u, 0, 0), \qquad (8.25)$$

where $\mathbb{P}_n$ is the empirical measure of the observations $(T_i, U_i, \Delta_{i1}, \Delta_{i2})$. Writing G_{n1} for the empirical subdistribution function of the T_is with $\Delta_{i1} = 1$, G_{n3} for the empirical subdistribution function of the U_is with $\Delta_{i1} = \Delta_{i2} = 0$ and G_{n2} for the bivariate empirical

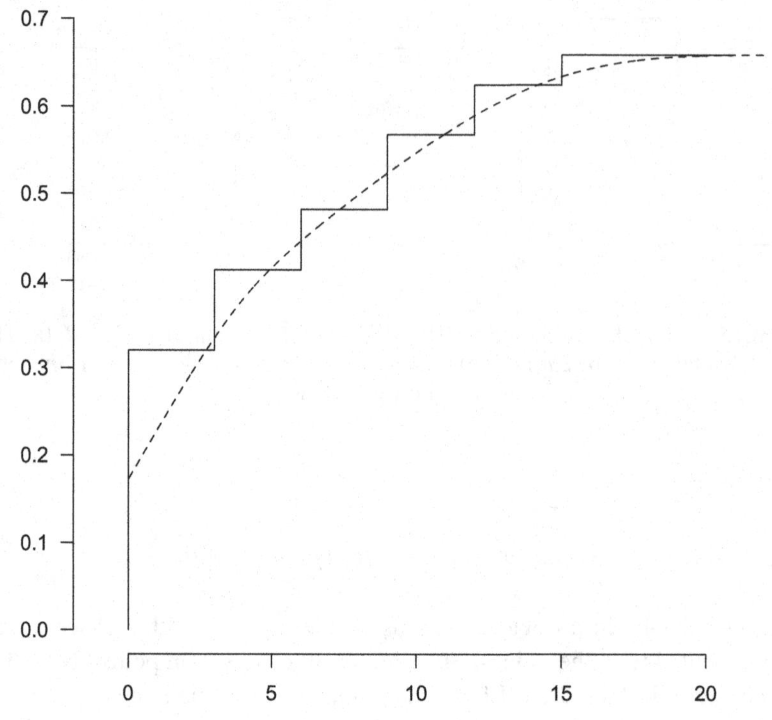

Figure 8.16 MLE (solid) and the SMLE (dashed) for the data of Table 8.2.

subdistribution function of the pairs (T_i, U_i) with $\Delta_{i2} = 1$, (8.25) can be rewritten as

$$\int \log F(t)\, dG_{n1}(t) + \int \log\{F(u) - F(t)\}\, dG_{n2}(t,u) + \int \log\{1 - F(u)\}\, dG_{n3}(u). \tag{8.26}$$

Replacing the subdistribution functions G_{nj} in (8.26) by smoothed versions $\tilde{G}_{nj}$ one gets the smoothed log likelihood:

$$\int \log F(t)\, d\tilde{G}_{n1}(t) + \int \log\{F(u) - F(t)\}\, d\tilde{G}_{n2}(t,u) + \int \log\{1 - F(u)\}\, d\tilde{G}_{n3}(u). \tag{8.27}$$

Maximizing this function over distribution functions F, yields the maximum smoothed likelihood estimator (MSLE) of F.

A picture of the SMLE, together with the MLE, is shown in Figure 8.16, where the same bandwidth is used as in Figure 8.15. We also used the triweight kernel, just as in Figure 8.15. For the estimation of the density, corresponding to the two-dimensional distribution function $\tilde{G}_{n2}$, a product of two triweight kernels was used, and the same bandwidth as used before was used for both coordinates.

If we have information that the support of the distribution is a known finite interval (information on this matter was less clear in the earlier example, and we therefore did not use a boundary correction), it is often advisable to introduce a boundary correction to avoid bias at the upper end of the interval. For example, if the support is known to be equal to $[a, b]$, we replace (8.23) by

$$\tilde{F}_{n,h}(x) = \int \{\mathbb{K}_h(t - x) + \mathbb{K}_h(t + x - 2a) - \mathbb{K}_h(2b - t - x)\}\, d\hat{F}_n(x), \qquad (8.28)$$

which is similar to the boundary correction method to be used for the SMLE in Section 11.3.

8.7 Smooth Estimation in the Bivariate Interval Censoring Model

In Section 5.2, the bivariate interval censoring problem is introduced. Analogously to the approach in Section 8.1 and 8.6, the SMLE for this bivariate interval censoring model is (at interior points) easily defined. If $\hat{F}_n$ is the MLE of the bivariate distribution function, the SMLE of the density is defined by

$$\tilde{f}_{n,h_1,h_2}(x, y) = \int K_{h_1}(x - u) K_{h_2}(y - v)\, d\hat{F}_n(u, v), \qquad (8.29)$$

where, as used before, for $h > 0$,

$$K_h(u) = h^{-1} K(u/h),$$

and K is a kernel of the usual kind, such as the triweight kernel. We do not necessarily have to use the product of two one-dimensional kernels for the estimate $\tilde{f}_{nh}$, but this seems the simplest choice. The estimate of the two-dimensional distribution function is now obtained by integrating $\tilde{f}_{n,h_1,h_2}$ over both coordinates. Alternatively, we can define as estimate of the bivariate distribution function

$$\tilde{F}_{n,h_1,h_2}(x, y) = \int \mathbb{K}((x - u)/h_1) \mathbb{K}((y - v)/h_2)\, d\hat{F}_n(u, v), \qquad (8.30)$$

where the integrated kernel $\mathbb{K}$ is defined by

$$\mathbb{K}(x) = \int_{-\infty}^{x} K(u)\, du, \ x \in \mathbb{R}.$$

For a sample of size $n = 1000$, generated by the density $f(x, y) = x + y$ on $[0, 1]^2$, bivariate current status data were obtained using uniformly distributed observation times on $[0, 1]^2$. The SMLE is shown in Figure 8.17, where the bandwidth $h = n^{-1/6}$ is used on both coordinates. We also use the boundary correction, to be discussed in Section 11.3, for the univariate current status problem, so the actual definition of $\tilde{F}_{n,h_1,h_2}$, used on the whole domain, is

$$\tilde{F}_{n,h_1,h_2}(x, y) = \int \left\{\mathbb{K}\left(\frac{x - u}{h_1}\right) - \mathbb{K}\left(\frac{2 - x - u}{h_1}\right) + \mathbb{K}\left(\frac{x + u}{h_1}\right)\right\}$$
$$\cdot \left\{\mathbb{K}\left(\frac{y - v}{h_2}\right) - \mathbb{K}\left(\frac{2 - y - v}{h_2}\right) + \mathbb{K}\left(\frac{y + v}{h_2}\right)\right\} d\hat{F}_n(u, v), \qquad (8.31)$$

if the domain is $[0, 1]^2$ (see (8.28)), with obvious extensions to more general rectangles. The level curves for the SMLE are shown in Figure 8.18.

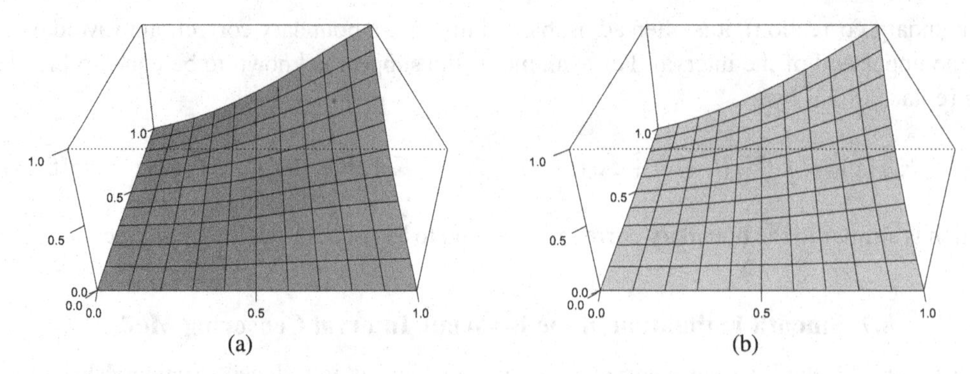

(a) (b)

Figure 8.17 The SMLE (a) and the underlying distribution function $F_0(x,y) = \frac{1}{2}xy(x+y)$ (b) for a sample of size $n = 1000$ from the density $f_0(x,y) = x + y$ on $[0,1]^2$. The observation density is uniform. The bandwidth for the SMLE is $n^{-1/6}$ in both directions.

A picture of the MLE and the SMLE for the Betensky-Finkelstein data is shown in Figure 8.19 and the picture of the level curves in Figure 8.20. It seems that the SMLE might be the more sensible estimate.

It is also straightforward to define the MSLE for smooth estimation of the distribution in these models. It was argued in Section 5.2 that the MLE for bivariate interval censoring is a maximizer of the log likelihood (divided by n) given by (5.20). The MSLE is

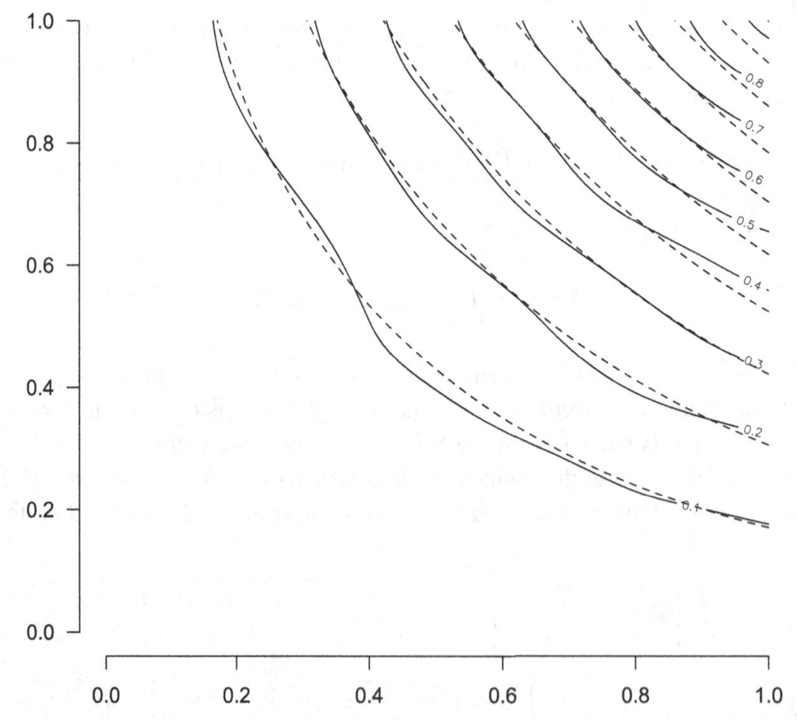

Figure 8.18 The level plots of the SMLE and the underlying distribution function (dashed), for the same sample as in Figure 8.17. The dashed curves are the corresponding levels of the underlying distribution function.

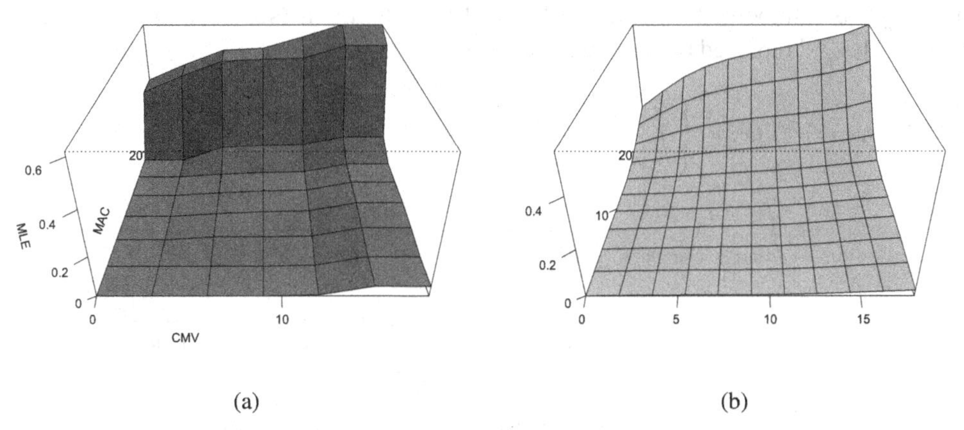

Figure 8.19 The MLE (a) and SMLE (b) for the Betensky-Finkelstein data, restricted to the interval $[0, 18] \times [0, 24]$.

therefore the maximizer of

$$
\begin{aligned}
\ell(F) = &\int \log F(t, v)\, d\tilde{V}_n^{(11)} + \int_{t \le u,\, v \le w} \log\{F(u, w) - F(t, w) - F(u, v) + F(t, v)\}\, d\tilde{V}_n^{(22)} \\
&+ \int_{u \ge t} \log\{F(u, v) - F(t, v)\}\, d\tilde{V}_n^{(21)} + \int_{w \ge v} \log\{F(t, w) - F(t, v)\}\, d\tilde{V}_n^{(12)} \\
&+ \int_{u \ge t} \log\{F_1(u) - F_1(t) - F(u, w) + F(t, w)\}\, d\tilde{V}_n^{(23)} \\
&+ \int_{w \ge v} \log\{F_2(w) - F_2(v) - F(u, w) + F(u, v)\}\, d\tilde{V}_n^{(32)} \\
&+ \int_{w \ge v} \log\{F_1(t) - F(t, w)\}\, d\tilde{V}_n^{(13)} + \int_{u \ge t} \log\{F_2(v) - F(u, v)\}\, d\tilde{V}_n^{(31)} \\
&+ \int \log\{1 - F_1(u) - F_2(w) + F(u, w)\}\, d\tilde{V}_n^{(33)},
\end{aligned}
\tag{8.32}
$$

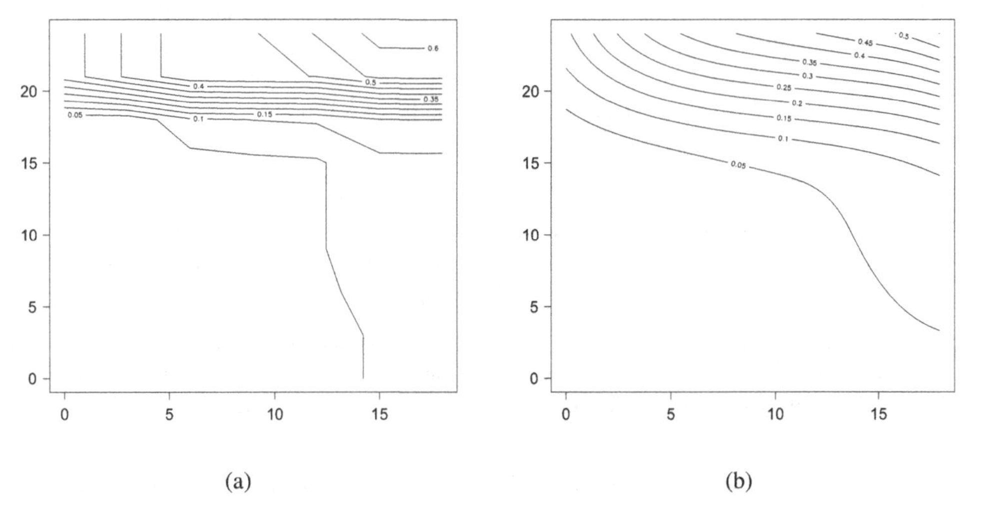

Figure 8.20 Contour plot of the MLE (a) and SMLE (b) for the Betensky-Finkelstein data, restricted to the interval $[0, 18] \times [0, 24]$.

where $\tilde{V}_n^{(ij)}$ is a smoothed version of $V_n^{(ij)}$. EM-type iterations (see Section 7.2), based on (5.21), with $V_n^{(ij)}$ replaced by $\tilde{V}_n^{(ij)}$, take the form:

$$
\begin{aligned}
f^{(m+1)}(x,y) = f^{(m)}(x,y)\Bigg\{&\int_{t\ge x,\,v\ge y}\frac{d\tilde{V}_n^{(11)}}{F^{(m)}(t,v)}\\
&+\int_{t<x\le u,\,v<y\le w}\frac{d\tilde{V}_n^{(22)}}{F^{(m)}(u,w)-F^{(m)}(t,w)-F^{(m)}(u,v)+F^{(m)}(t,v)}\\
&+\int_{t<x\le u,\,v\ge y}\frac{d\tilde{V}_n^{(21)}}{F^{(m)}(u,v)-F^{(m)}(t,v)}+\int_{t\ge x,\,v<y\le w}\frac{d\tilde{V}_n^{(12)}}{F^{(m)}(t,w)-F^{(m)}(t,v)}\\
&+\int_{t<x\le u,\,w\ge y}\frac{d\tilde{V}_n^{(23)}}{F_1^{(m)}(u)-F_1^{(m)}(t)-F^{(m)}(u,w)+F^{(m)}(t,w)}\\
&+\int_{u\ge x,\,v<y\le w}\frac{d\tilde{V}_n^{(32)}}{F_2^{(m)}(w)-F_2^{(m)}(v)-F^{(m)}(u,w)+F^{(m)}(u,v)}\\
&+\int_{t<x\le u,\,v\ge y}\frac{d\tilde{V}_n^{(13)}}{F_1^{(m)}(t)-F^{(m)}(t,w)}+\int_{t\ge x,\,v<y\le w}\frac{d\tilde{V}_n^{(31)}}{F_2^{(m)}(v)-F^{(m)}(u,v)}\\
&+\int_{u<x,\,w<y}\frac{d\tilde{V}_n^{(33)}}{1-F_1^{(m)}(u)-F_2^{(m)}(w)+F^{(m)}(u,w)}\Bigg\}. \qquad (8.33)
\end{aligned}
$$

Once $f^{(m+1)}$ is determined, one can also compute $F^{(m+1)}$, and this can be given as input to the next iterations. But it is clear that in practice one has to discretize, and perform these iterations on a grid of points. The computational burden is considerable and we still do not have experience with this method for the present model.

Exercises

8.1 Let G be a nondecreasing function on $\mathbb{R}$ and K a probability density on $\mathbb{R}$. Show that, for all $h > 0$, the function

$$\tilde{G}(x) = \frac{1}{h}\int G(y)K\left(\frac{x-y}{h}\right)dy$$

is also nondecreasing on $\mathbb{R}$.

8.2 Use uniform strong consistency of the MLE for the distribution function in the current status model to show that the SMLE for this distribution function as defined in Section 8.1 is consistent. Identify conditions needed on the bandwidth $h = h_n$.

8.3 In the setting of Section 8.2, define the smoothed maximum likelihood estimator of a decreasing density. Show that on the interval $[h, \infty)$ it is monotone (where $h > 0$ is the bandwidth chosen and the kernel function K has support $[-1, 1]$). Also think of a method of sampling from this estimator.

8.4 Consider a sequence of i.i.d. random variables with density g on $\mathbb{R}$. Suppose $x \in \mathbb{R}$ is such that g is twice continuously differentiable at x and let K be a kernel density as described in

Section 8.1. Define the kernel estimator of $g(x)$ based on $X_1, \ldots, X_n$ with bandwidth $h > 0$ by

$$g_{n,h}(x) = \frac{1}{nh}\sum_{i=1}^{n} K\left(\frac{x - X_i}{h}\right). \tag{8.34}$$

Show that, for $n \to \infty$ and $h \to 0$ such that $nh \to \infty$,

$$E g_{n,h}(x) = g(x) + \frac{1}{2}h^2 g''(x)\int_{-1}^{1} y^2 K(y)\,dy + o(h^2)$$

and

$$\mathrm{Var}(g_{n,h}(x)) = \frac{g(x)}{nh}\int_{-1}^{1} k(y)^2\,dy + o\left(\frac{1}{nh}\right).$$

Conclude that the asymptotically mean squared error (MSE) optimal for the bandwidth is given by $h = h_n = cn^{-1/5}$ for some $c > 0$. Also determine the (asymptotically) MSE optimal value for c.

8.5 Let g be an arbitrary probability density on $\mathbb{R}$. Show that for all other probability densities h on $\mathbb{R}$,

$$\int \log g(z)\,g(z)\,dz \geq \int \log h(z)\,g(z)\,dz.$$

8.6 Define and compute the maximum smoothed likelihood estimator for the distribution function F in Wicksell's problem, based on the binned data from Table 8.1. This entails defining the smoothed (log) likelihood and using techniques from Chapter 7 to compute its maximizer.

8.7 Show that indeed the minimizer of (8.5) over all nondecreasing hazards on $[0, a]$ is given by the left derivative of the greatest convex minorant of (8.6).

8.8 Consider the problem of minimizing the functional I given in (8.11),

$$I(h) = \int_0^a \left(h(x) - \hat{h}_n(x)\right)^2 dx + \lambda \int_0^a h'(x)^2\,dx$$

over all continuously differentiable functions h. Fix a continuously differentiable hazard function h on $[0, a]$.

a) Let $\nu\colon [0, a] \to \mathbb{R}$ be a continuously differentiable function. Show that

$$\lim_{\delta\to 0} \delta^{-1}\left(I(h + \delta\nu) - I(h)\right) = 2\int_0^a \nu(x)\left(h(x) - \hat{h}_n(x)\right) + 2\lambda h'(x)\nu'(x)\,dx.$$

b) Now suppose ν satisfies $\nu'(0) = \nu'(a) = 0$. Then show that

$$\lim_{\delta\to 0} \delta^{-1}\left(I(h + \delta\nu) - I(h)\right) = 2\int_0^a \left(h(x) - \hat{h}_n(x) - \lambda h''(x)\right)\nu(x)\,dx.$$

c) Argue from (b) that a necessary condition for h to be optimal is that it satisfies the second order differential equation given in (8.12),

$$h''(x) = \lambda^{-1}\left(h(x) - \hat{h}_n(x)\right).$$

Hint: use (b) for an appropriate sequence of test functions ν.

8.9 Let $\hat{h}_n$ be a bounded nondecreasing hazard function on $[0, a]$. Consider the problem of minimizing (8.8), with $\lambda > 0$, over all hazard functions h on $[0, a]$ such that $h(0) = \hat{h}_n(0)$ and

$h(a) = \hat{h}_n(a)$. Argue that the resulting minimizer will also be a nondecreasing hazard rate on $[0, a]$.

8.10 Apply the penalization approach of Section 8.3 to the problem of estimating a smooth decreasing density.

8.11 a) Consider the function $f(x) = 1 - x$ on the interval $[0, 1]$. Construct the (increasing) monotonic rearrangement of this function. Also construct the decreasing rearrangement of this function.

b) Do the same for the function $f(x) = -8x^3 + 12x^2 - 5x + 1$; see Figure 8.21.

c) Now generate data from an exponential density, compute a (not decreasing) kernel estimator for the density and approximate its monotonic rearrangement numerically. Note that it makes sense to work with the class of sets $L_v = \{x : f(x) \geq v\}$.

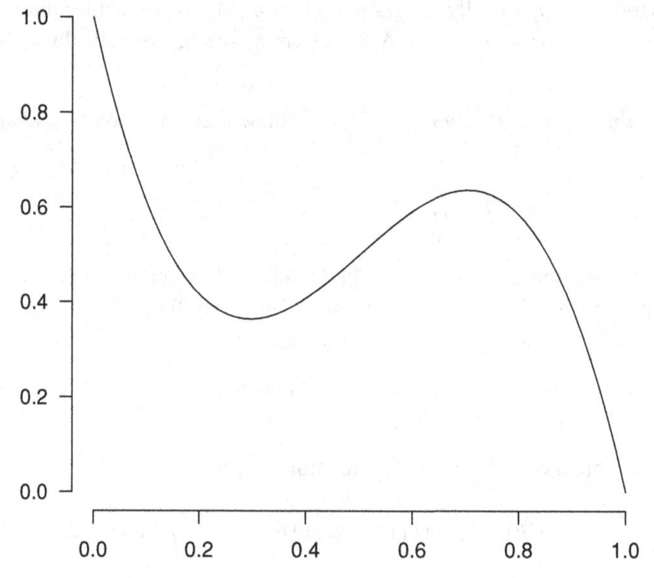

Figure 8.21 The nonmonotone function $f(x) = -8x^3 + 12x^2 - 5x + 1$ of Exercise 8.11 (b).

8.12 Verify that the monotonic rearrangement estimator for the distribution function in the deconvolution problem as discussed in Section 8.4 indeed takes values in the interval $[0, 1]$.

8.13 We consider current status data given distinct inspection times $t_1 < \cdots < t_n$. The observable indicators are independent Bernoulli random variables with $\Delta_i \sim \text{Bin}(1, F(t_i))$.

a) Construct an estimator for F based on monotonic rearrangements and the basic (unrestricted) MLE of F at the points t_i.

b) Describe how to construct a monotonic rearrangement estimator for F based on the Nadaraya Watson kernel regression estimator of F. What will be the behavior of this estimator for (too) small values of the bandwidth and (too) large values of the bandwidth?

8.14 Generate a sample of size 100 from the standard exponential density g. Denote the data by $x_1, \ldots, x_{100}$.

a) Plot the kernel density estimator g_{nh} of the sampling density based on the data obtained; see (8.34). Use the triweight kernel function K. Observe that the value of the estimate is too small in a right neighborhood of zero.

b) Show that for the situation in (a),

$$Eg_{nh}(0) = \int_0^1 K(v)e^{-hv}\,dv,$$

and argue from this that the kernel density estimator is inconsistent at zero. Define a new data set of size 200, consisting of $\pm x_1, \pm x_2, \ldots, \pm x_{100}$. Also plot the kernel density estimate based on these data. Denoting this estimator by f_{nh}.

c) Argue that the function f_{nh} is symmetric around zero.

d) Define $\tilde{g}_{nh}(x) = 2f_{nh}(x)1_{[0,\infty)}(x)$, and plot this function. Observe that it satisfies the shoulder condition $\tilde{g}'_{nh}(0+) = 0$.

e) Argue that $\tilde{g}_{nh}(0)$ is consistent for $g(0)$ and, in view of (d), argue that $\tilde{g}'_{nh}(0+)$ is not consistent for $g'(0)$.

Bibliographic Remarks

Kernel estimators for estimating smooth functions were introduced in Rosenblatt, 1956b. Introductory books to the subject include Silverman, 1986, and Wand and Jones, 1995. Early references on combining shape (order) constraints and smoothness are Mammen, 1991, and Ramsay, 1998. Density estimation based on smoothing the MLE of a distribution function is studied in the uniform deconvolution model in Groeneboom and Jongbloed, 2003. In the monotone density model, Eggermont and LaRiccia, 2000, and Van der Vaart and Van der Laan, 2003, study smooth monotone estimators. For the current status model, smooth estimation of the distribution function, density and hazard rate was studied in Groeneboom et al., 2010. Nonparametric density estimation with a roughness penalty is introduced in Good and Gaskins, 1971. Many results on maximum penalized likelihood estimation can be found in Eggermont and LaRiccia, 2001a,b. Penalization approaches to avoid inconsistency of estimators at boundary points are found in Woodroofe and Sun, 1993, for estimating a monotone density, Pal, 2008, for estimating a monotone regression function, and Groeneboom and Jongbloed, 2013b for estimating a monotone hazard rate. Monotonic rearrangements of functions are introduced in Bennett and Sharpley, 1988, used to define a unimodal density estimator in Fougères, 1997, and used in the context of estimating a monotone regression function in Dette et al., 2006. There is quite some literature on boundary corrections for density estimators. Albers, 2012, gives an overview of a number of approaches to this.

9

Testing and Confidence Intervals

The usual setting for nonparametric tests for shape restrictions on, for example, a hazard function is that one tests that a hazard is constant against the alternative that the hazard is strictly increasing or decreasing. An interesting new angle is taken in Gijbels and Heckman, 2004, and Hall and Van Keilegom, 2005, where it is tested that a hazard is increasing or decreasing against the alternative that there are local disturbances of the monotonicity. We discuss this problem and various approaches that can be adopted in this situation in Section 9.1.

Another type of testing problems with shape constraints are two- or k-sample problems. In Section 9.2 we discuss various procedures that can be used to test whether two samples from distributions with decreasing densities on $[0, \infty)$ actually originate from the same (decreasing) density.

Methods for k-sample test problems in the situation of right censored data have been around for a long time; in the interval censoring setting, such tests have only been developed rather recently. The reason for this is probably that the treatment of right-censored data is much closer to classical theory, where one has $\sqrt{n}$-convergence, asymptotic normality, and so on. In Section 9.3 we describe various procedures for the two-sample tests with current status observations and in Section 9.4 we further explore how the two-sample test can be formulated for the more general interval censoring model.

In the testing problems discussed in this chapter, bootstrap methods for determining critical values for the test statistics play a crucial role. To show that the bootstrap works, the nature of the original estimator of the model parameters (under the null hypothesis) from which one takes the bootstrap samples becomes important. It turns out that smooth monotone estimators as discussed in Chapter 8 can be used to show that a bootstrap technique to approximate critical values works in practice.

The classical bootstrap is used again in Section 9.5 on the construction of confidence intervals. In this case resampling from a smooth estimate of the distribution creates too much bias and it seems better to stay close to the original data by just resampling with replacement from them. By using the bootstrap, the problem of having to estimate parameters in the construction of the confidence intervals is avoided. We base the confidence intervals on the SMLE and on density estimates, obtained by smoothing the MLE. The confidence intervals for values of the distribution function are compared with confidence intervals directly based on the MLE itself, as presented in Banerjee and Wellner, 2005.

9.1 Testing for a Monotone Hazard

The problem of estimating a monotone hazard rate based on a sample from its corresponding distribution was introduced in Section 2.6. Also smooth monotone estimators were studied in Section 8.3. We now consider testing problems in the setting of monotone hazard rates. To that end, consider a sequence of i.i.d. random variables $X_1, X_2, \ldots$ with density function g_0 on $[0, \infty)$. Denote the distribution function, hazard function and cumulative hazard function associated with g_0 by G_0, h_0 and H_0, respectively, and recall the relations between these functions:

$$h_0(x) = \frac{g_0(x)}{1 - G_0(x)}, \quad H_0(x) = -\log(1 - G_0(x)), \quad G_0(x) = 1 - e^{-H_0(x)}. \tag{9.1}$$

Already in the 1960s, procedures were developed to test the null hypothesis of a constant hazard rate (corresponding to an exponential distribution) against the alternative of an increasing hazard (presence of aging). One popular test statistic in this context is the test statistic of Proschan and Pyke, 1967.

Example 9.1 *(Proschan-Pyke test)* Fix the sample size n and denote by $X_{(1)} < X_{(2)} < \cdots < X_{(n)}$ the order statistics of the sample. Taking $X_{(0)} = 0$, define the normalized spacings as

$$S_i = (n - i + 1)\left(X_{(i)} - X_{(i-1)}\right), \quad 1 \le i \le n. \tag{9.2}$$

Then, if g_0 were the exponential density with rate λ (so $h_0(x) = \lambda$ on $[0, \infty)$), the S_is are independent and exponentially distributed random variables with rate λ. See Exercise 9.2. If the data are not exponentially distributed, but still have a convex cumulative hazard function, it can be shown (see Exercise 9.4) that

$$S_1 \gtrsim S_2 \gtrsim \cdots \gtrsim S_n,$$

where $\gtrsim$ denotes stochastic ordering. Hence, if for a given sample of X_is resulting S_i values tend to be bigger for smaller values of i, this points in the direction of an increasing hazard rate. In Proschan and Pyke, 1967, this is formalized by defining indicators

$$V_{ij} = 1_{S_i > S_j}, \quad 1 \le i, j \le n$$

and as test statistic

$$V_n = \sum_{i=1}^{n} \sum_{j=i}^{n} V_{ij}.$$

Large values of V_n indicate that in relatively many pairs (i, j) with $i < j$ satisfy $S_i > S_j$. Hence large values of V_n should lead to rejection of the null hypothesis that the hazard rate is constant in favor of the alternative that the hazard rate is increasing.

A convenient feature of the statistic V_n is that it is scale invariant, so its distribution under the null hypothesis does not depend on the (further unspecified) rate of the underlying exponential distribution. This allows for efficient computation of critical values using Monte Carlo simulation.

Another problem that was more recently raised in Gijbels and Heckman, 2004, and Hall and Van Keilegom, 2005, concerns testing (local) monotonicity of h_0. More precisely, given an interval $I \subset [0, \infty)$, one could test the hypothesis that h_0 is increasing (or decreasing) on the interval I against the alternative that it is not. We restrict ourselves here to intervals of the type $[0, a]$ and increasing hazard rates and consider the null hypothesis for fixed $a > 0$

$$\text{Hyp}_0: \forall x, y \in [0, a] \text{ with } x \leq y,\ h_0(x) \leq h_0(y)$$

against the alternative that this monotonicity does not hold. Various procedures have been proposed to address this problem.

One, adopted by Gijbels and Heckman, 2004, is directly related to the test of Example 9.1. Actually, the test statistic is a multiscale version of the Proschan-Pyke statistic, in the sense that local versions of the statistic V_n are combined into a statistic by taking the maximum of these local versions, including various scales of interval lengths. For details, see Gijbels and Heckman, 2004.

In Hall and Van Keilegom, 2005, another approach is taken. The idea is that convexity of H_0 on $[0, a]$ can be stated as

$$H_0(x+y) - 2H_0(x) + H_0(x-y) \geq 0 \text{ for all } x \in (0, a) \text{ and } y \in \mathbb{R} \text{ such that } x \pm y \in [0, a].$$

See Exercise 9.1. Estimating the cumulative hazard by the empirical cumulative hazard

$$\mathbb{H}_n(x) = \begin{cases} -\log\{1 - \mathbb{G}_n(x)\}, & x \in [0, X_{(n)}), \\ \infty, & x \geq X_{(n)}, \end{cases} \tag{9.3}$$

they define the following test statistic:

$$U_n = \int_{x=0}^{a} \int_{y=-x\vee(x-a)}^{x\wedge(a-x)} (\max\{0, 2\mathbb{H}_n(x) - \mathbb{H}_n(x+y) - \mathbb{H}_n(x-y)\})^r\, w(x, y) w(x, y)\, dy dx, \tag{9.4}$$

where w is a nonnegative weight function and $r > 0$. The null hypothesis is rejected for values of U_n that are too large. Critical values for the test statistic are obtained by a bootstrap procedure. The bootstrap data are generated using a kernel density estimate of the sampling density. The bandwidth is chosen in a particular way. First a pilot bandwidth is chosen, usually leading to a nonconvex corresponding cumulative hazard estimator. Then the bandwidth of the density estimator is increased until the corresponding cumulative hazard is convex on $[0, a]$. Using the density estimate thus obtained, bootstrap realizations of the statistic U_n are generated, and the $(1 - \alpha)$-quantile of the realized bootstrap values is taken as critical value. For an application of this test statistic (without formal testing) in paleobiology, see Steinsaltz and Orzack, 2011. There the hypothesis is formulated as follows: "What is the probability that samples from a given nonconvex survivorship curve can mislead us into thinking it was convex?"

Another type of test statistic for this problem is a distance between two estimators for the cumulative hazard function: one estimator that approximates H_0 well only under Hyp_0 and one that works well also if Hyp_0 does not hold. This approach is adopted in Durot, 2008, and Groeneboom and Jongbloed, 2013c. An estimator for H_0 without assuming Hyp_0 is just the empirical cumulative hazard function given in (9.3). An estimator of h_0 under Hyp_0 is the least squares estimator defined as minimizer of (2.27) in Section 2.6. On $[0, a]$ it is given by the left-continuous derivative of the greatest convex minorant (GCM) of the empirical cumulative hazard function given by (9.3), restricted to $[0, a]$. The estimator $\hat{H}_n$ of H_0 under

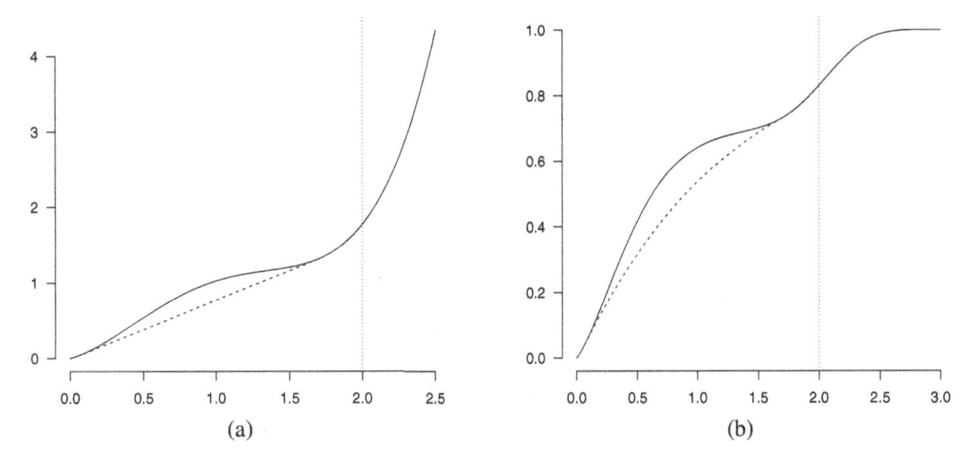

Figure 9.1 (a) A nonconvex cumulative hazard H_0 and its greatest convex minorant on [0, 2] (dashed). (b) The corresponding distribution functions.

the null hypothesis Hyp_0 is therefore defined by

$$\hat{H}_n(x) = \begin{cases} \mathbb{H}_n(x) & x > a \\ \text{GCM}(y \mapsto \mathbb{H}_n(y)\colon 0 \le y \le a)(x) & x \in [0, a]. \end{cases} \tag{9.5}$$

Note that this estimator is continuous at a, if $a \neq X_i$ for all i.

A possible test statistic for testing Hyp_0 is an empirical L_1 distance between the two estimators for H_0 just considered:

$$T_n = \int_{[0,a]} \{\mathbb{G}_n(x-) - \hat{G}_n(x)\}\, d\mathbb{G}_n(x) \tag{9.6}$$

where $\hat{G}_n$ is the distribution function corresponding to $\hat{H}_n$: $\hat{G}_n(x) = 1 - \exp\{-\hat{H}_n(x)\}$. Note that $T_n \geq 0$, since $\hat{H}_n$ is the greatest convex minorant (hence a minorant) of $\mathbb{H}_n$ on $[0, a]$. Also note that under the alternative hypothesis, T_n will tend to be higher than under the null hypothesis. See Figure 9.1 for a cumulative hazard function that does not satisfy Hyp_0 together with its greatest convex minorant on the interval [0, 2]. When the sample size tends to infinity, $\mathbb{H}_n$ will converge to this underlying cumulative hazard H_0 on [0, 2] whereas $\hat{H}_n$ will converge to the convex minorant of H_0. Figure 9.1b shows the distribution functions corresponding H_0 and its greatest convex minorant on [0, 2]. Other measures of distance between $\mathbb{G}_n$ and $\hat{G}_n$ may also be used to define a test statistic. Examples are

$$S_n = \int_{[0,a]} \{\mathbb{H}_n(x-) - \hat{H}_n(x)\}\, d\mathbb{G}_n(x) \text{ and } D_n = \sup_{[0,a]} \left(\mathbb{H}_n(x) - \hat{H}_n(x)\right). \tag{9.7}$$

The latter statistic is studied in Durot, 2008.

(A) Approximation of Critical Values

In order to obtain critical values for statistic T_n defined in (9.6) (or any of the other statistics defined), there are various possible approaches. The first is to use that its distribution under any H that obeys Hyp_0 is stochastically bounded by its distribution under the distribution

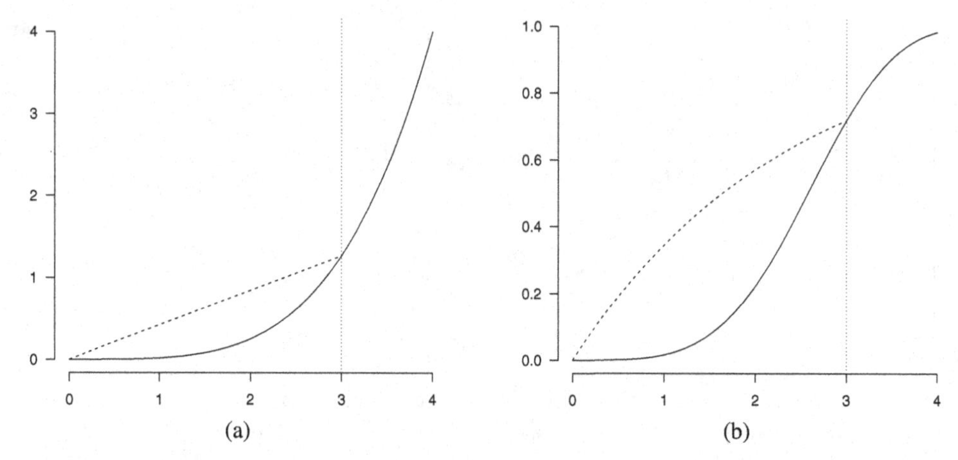

Figure 9.2 (a) A convex cumulative hazard H_0 and its corresponding cumulative hazard H_a for $a = 3$ defined by (9.8). (b) The corresponding distribution functions.

function with the cumulative hazard function that is obtained by linear interpolation on the interval $[0, a)$:

$$H_a(x) = \frac{H(a)}{a} x 1_{[0,a)}(x) + H(x)1_{[a,\infty)}(x). \tag{9.8}$$

See Figure 9.2.

Lemma 9.1 *For each cumulative hazard function H that is convex on* $[0, a]$,

$$P_H(T_n \geq t) \leq P_{H_a}(T_n \geq t) \tag{9.9}$$

for all $t \geq 0$.

Proof Let $E_1, E_2, \ldots, E_n$ be an i.i.d. sequence of standard exponential random variables. Define

$$X_i = H^{-1}(E_i), \quad Y_i = H_a^{-1}(E_i) \text{ for } 1 \leq i \leq n.$$

Then the X_is and the Y_is are samples from the distributions with cumulative hazard H and H_a respectively (see Exercise 9.3). Denote by U_n the test statistic (9.6) based on the Y_is and by V_n the statistic based on the X_is. Furthermore, define the function $\phi\colon [0, a] \to [0, a]$ by

$$\phi(x) = H_a^{-1}(H(x)). \tag{9.10}$$

Note that ϕ is convex and increasing on $[0, a]$ and that $Y_i = \phi(X_i) \leq X_i$ for all i) (see Exercise 9.10). Moreover, $\mathbb{G}_n^X(x) = \mathbb{G}_n^Y(\phi(x))$, since

$$\{i : X_i \leq x\} = \{i : \phi(X_i) \leq \phi(x)\} = \{i : Y_i \leq \phi(x)\}.$$

Consequently, also $\mathbb{H}_n^X(x) = \mathbb{H}_n^Y(\phi(x))$, where these functions refer to the empirical cumulative hazards based on the samples of X_is and Y_is respectively. Now define $\bar{H}_n(x) = \hat{H}_n^Y(\phi(x))$, where the latter denotes the greatest convex minorant of the empirical hazard function based on the Y_is, evaluated at $\phi(x)$. Then $\bar{H}_n$ is *a minorant* of $\mathbb{H}_n^X$, i.e.,

$$\bar{H}_n(x) = \hat{H}_n^Y(\phi(x)) \leq \mathbb{H}_n^Y(\phi(x)) = \mathbb{H}_n^X(x).$$

Moreover, it is also convex. Indeed, using monotonicity and convexity of $\hat{H}_n^Y$ and convexity of ϕ we have for $\alpha \in (0, 1)$ and $x, y \in [0, a]$

$$\bar{H}_n(\alpha x + (1-\alpha)y) = \hat{H}_n^Y(\phi(\alpha x + (1-\alpha)y)) \le \hat{H}_n^Y(\alpha\phi(x) + (1-\alpha)\phi(y))$$
$$\le \alpha \hat{H}_n^Y(\phi(x)) + (1-\alpha)\hat{H}_n^Y(\phi(y)) = \alpha \bar{H}_n(x) + (1-\alpha)\bar{H}_n(y).$$

Hence, the convex minorant $\bar{H}_n$ of $\mathbb{H}_n^X$ is smaller than or equal to the greatest convex minorant $\hat{H}_n^X$ of $\mathbb{H}_n^X$:

$$\bar{H}_n(x) \le \hat{H}_n^X(x) \le \mathbb{H}_n^X(x) \Rightarrow \mathbb{G}_n^X(x) - \hat{G}_n^X(x) \le \mathbb{G}_n^X(x) - \bar{G}_n(x),$$

where we use the obvious notation relating cumulative hazards to distribution functions. This implies that

$$U_n := \int_{[0,a]} \left(\mathbb{G}_n^Y(x-) - \hat{G}_n^Y(x)\right) d\mathbb{G}_n^Y(x) = \int_{[0,a]} \left(\mathbb{G}_n^Y(\phi(x)-) - \hat{G}_n^Y(\phi(x))\right) d\mathbb{G}_n^Y(\phi(x))$$
$$= \int_{[0,a]} (\mathbb{G}_n^X(x-) - \bar{G}_n(x))\, d\mathbb{G}_n^X(x) \ge \int_{[0,a]} \left(\mathbb{G}_n^X(x-) - \hat{G}_n^X(x)\right) d\mathbb{G}_n^X(x) =: V_n.$$

Noting that $P_H(T_n \ge t) = P(V_n \ge t)$ and $P_{H_a}(T_n \ge t) = P(U_n \ge t)$, the result follows. □

If $H(a)$ were known, the distribution of T_n under H_a could be approximated efficiently using Monte Carlo simulation. In practice, however, $H(a)$ is unknown. In order to really use the approximation, an estimate for H_0 at a is needed. This estimation, combined with the stochastic domination of Lemma 9.1, can be called a bootstrap procedure. A natural and consistent estimator for $H_0(a)$ is of course given by $\mathbb{H}_n(a)$.

It is clear that if the function H_0 is strictly convex on $[0, a]$, the lower bound of Lemma 9.1 may be quite rough. In that case, the convex minorant of its empirical version will tend to wrap tightly around this function whereas in case the cumulative hazard is linear on $[a, b]$ (as in the exponential case), this difference will tend to be bigger. This is related to the discrepancy between the limit processes as described in Section 3.9 and Section 3.10. The following theorem, proved in Groeneboom and Jongbloed, 2013a, describes the asymptotic behavior of T_n, if the underlying hazard is strictly increasing on $[0, a]$.

Theorem 9.1 *Let h_0 be strictly increasing and positive on the interval $I = [0, a]$, with a bounded continuous derivative, staying away from zero on I. Moreover, let $\zeta(t)$ be the distance at t of the process $x \mapsto W(x) + x^2$ on $\mathbb{R}$ to its greatest convex minorant, where W is two-sided Brownian motion, originating from zero, and let ET_n satisfy*

$$ET_n = n^{-2/3}\mu_{G_0} + o\left(n^{-5/6}\right), \tag{9.11}$$

where

$$\mu_{G_0} = E\zeta(0)\int_0^a \left(\frac{2h_0(t)g_0(t)}{h_0'(t)}\right)^{1/3} dG_0(t), \tag{9.12}$$

and T_n is defined by (9.6). Then

$$n^{5/6}\left\{T_n - n^{-2/3}\mu_{G_0}\right\} \xrightarrow{\mathcal{D}} N\left(0, \sigma_{G_0}^2\right)$$

where $N\left(0,\sigma_{G_0}^2\right)$ *is a normal distribution with mean zero and variance* $\sigma_{G_0}^2$, *defined by*

$$\sigma_{G_0}^2 = 2\int_0^\infty \operatorname{covar}(\zeta(0),\zeta(s))\,ds\int_0^a\left(\frac{2h_0(t)g_0(t)}{h_0'(t)}\right)^{4/3}\,dG_0(t). \tag{9.13}$$

It is shown in Groeneboom, 2011, that

$$E\zeta(0) = \tfrac{2}{3}E\max_{t\in\mathbb{R}}\left\{W(t)-t^2\right\}\approx 0.52712.$$

To test the hypothesis that the hazard rate is strictly increasing on $[0,a]$, one could try to estimate the parameters μ_{G_0} and $\sigma_{G_0}^2$ of Theorem 9.1 and use the limiting normal distribution for the critical values. In case the data point toward the alternative hypothesis, reflected in the shape-constrained estimator of h_0' approaching zero, it is not straightforward how to deal with this. For the bootstrap procedure, which we will now describe, it is clear how to apply it in this case.

The proposed method runs as follows. First estimate the cumulative hazard function under the null hypothesis by a smooth estimator, having the property that the corresponding hazard satisfies the null hypothesis. Then draw samples of size n from this estimate B times and compute B times the bootstrap version of the test statistic: $T_{n,i}^*$, $1\le i\le B$. Finally, approximate the distribution of T_n under the true cumulative hazard function H_0 (assumed to belong to $H_{[a,b]}$) by the empirical distribution of these bootstrap values and its critical value at (for example) level 10% by the 90th percentile of this generated set of bootstrap values.

In order to prevent inconsistency of the estimator for h_0 at the endpoints, we use a penalized version of $\hat h_n$, $\hat h_n^{[p]}$ as defined in Section 8.3. This estimator is the derivative of the penalized cusum diagram consisting of the points

$$(0,0),\qquad \left(X_{(i)},\mathbb{H}_n(X_{(i)}-)+2n^{-2/3}\right),\ X_{(i)}<a,\qquad (a,\mathbb{H}_n(a-)). \tag{9.14}$$

The left derivative of the present cusum diagram minimizes the criterion (8.5) with $\alpha_n=\beta_n=2n^{-2/3}$, over all nondecreasing functions h on $[0,a]$. For the reason for choosing a penalty of order $cn^{-2/3}$, see Lemma 8.3.

For $x\in[0,a]$, we estimate the hazard by kernel smoothing of $\hat h_n^{[p]}$. Let K be the triweight kernel

$$K(u)=\frac{35}{32}\left(1-u^2\right)^3 1_{[-1,1]}(u),\ u\in\mathbb{R}. \tag{9.15}$$

This is a mean zero probability density with second moment $1/9$. Then, define for bandwidth $b_n>0$

$$\tilde h_n(x)=\int K_{b_n}(x-y)\,d\hat H_n^{[p]}(y)=\int K_{b_n}(x-y)\,\hat h_n^{[p]}(y)\,dy, \tag{9.16}$$

where $K_{b_n}(u)=K(u/b_n)/b_n$. The corresponding estimate of h_0' and H_0 are then given by

$$\tilde h_n'(x)=\int K_{b_n}(x-y)\,d\hat h_n^{[p]}(y)\text{ and }\tilde H_n(x)=\int_0^x \tilde h_n(u)\,du. \tag{9.17}$$

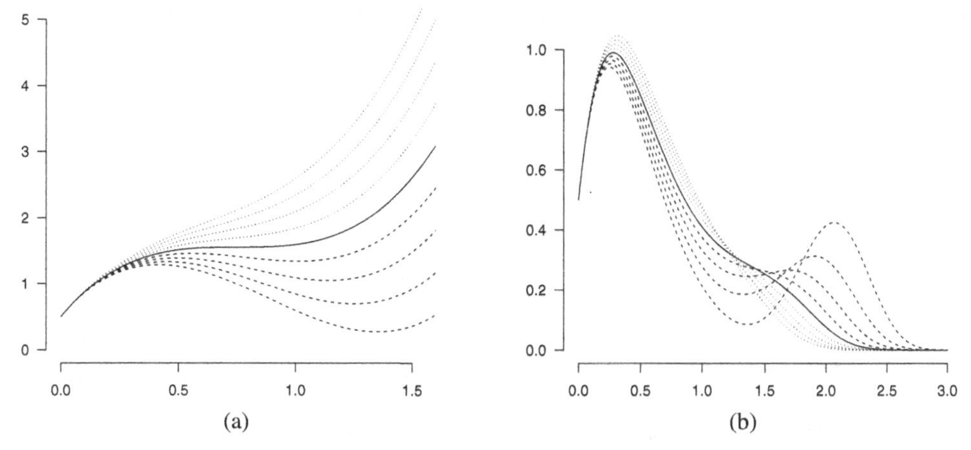

Figure 9.3 (a) The hazard functions $h^{(d)}$ for $d = -1, -0.75, -0.50, -0.25$ (dashed), $d = 0$ (full curve) and $d = 0.25, 0.50, 0.75, 1$ (dotted) corresponding to distribution functions (9.19). (b) The corresponding density functions.

To illustrate the behavior of the various tests we introduce the family of hazards $\{h^{(d)} : d \in [-1, 1]\}$, also considered in Hall and Van Keilegom, 2005:

$$h^{(d)}(x) = \frac{1}{2} + \frac{5}{2}\left\{\left(x - \frac{3}{4}\right)^3 + \left(\frac{3}{4}\right)^3\right\} + dx^2,\ x \ge 0. \tag{9.18}$$

The corresponding distribution functions on $[0, \infty)$ are given by:

$$F^{(d)}(x) = 1 - \exp\left\{-\frac{1}{2}x - \frac{5}{2}\left\{\frac{1}{4}\left(x - \frac{3}{4}\right)^4 + \left(\frac{3}{4}\right)^3 x\right\} - \frac{1}{3}dx^3 + \frac{5}{8}\left(\frac{3}{4}\right)^4\right\}. \tag{9.19}$$

If $d > 0$ we get a strictly increasing hazard; if $d < 0$, the hazard is decreasing between $\frac{3}{4} - \frac{2}{15}d \pm \frac{2}{15}\sqrt{d^2 - 45d/4}$ and if $d = 0$ the hazard has a stationary point at $x = \frac{3}{4}$. See Figure 9.3a for some hazards in this family and Figure 9.4 for the projection estimate and penalized projection estimate based on a sample of size $n = 100$ from the distribution with hazard rate $h^{(1)}$.

The rather different nature of the isotonic projection of the hazard rate and the projection of Hall and Van Keilegom, 2005 is illustrated in Figure 9.5a, where $d = -1$. The hazard estimate of Hall and Van Keilegom, 2005, given by the dashed-dotted curve in Figure 9.5, extends (with positive values) to the left of zero and has a slower increase to the right of 2.0 than the actual hazard, which is given by the black curve (which is clearly not monotone). The isotonic projection, on the other hand, only lives on $[0, \infty)$, and follows the steep increase of the real hazard to the right of 2.0, whereas it only locally corrects for the nonmonotonicity. The interval on which the hazard was estimated (and made monotone) was $[0, F^{-1}(0.95)] \approx [0, 2.31165]$, where $F = F^{(d)}$ in (9.19) with $d = -1$. On the other hand, for distribution functions at the other end of the family $\{F^{(d)} : d \in [-1, 1]\}$ at $d = 1$, and therefore deep inside the null hypothesis region, the starting bandwidth for the calibration of the Hall and Van Keilegom, 2005, method immediately gives an increasing hazard on the

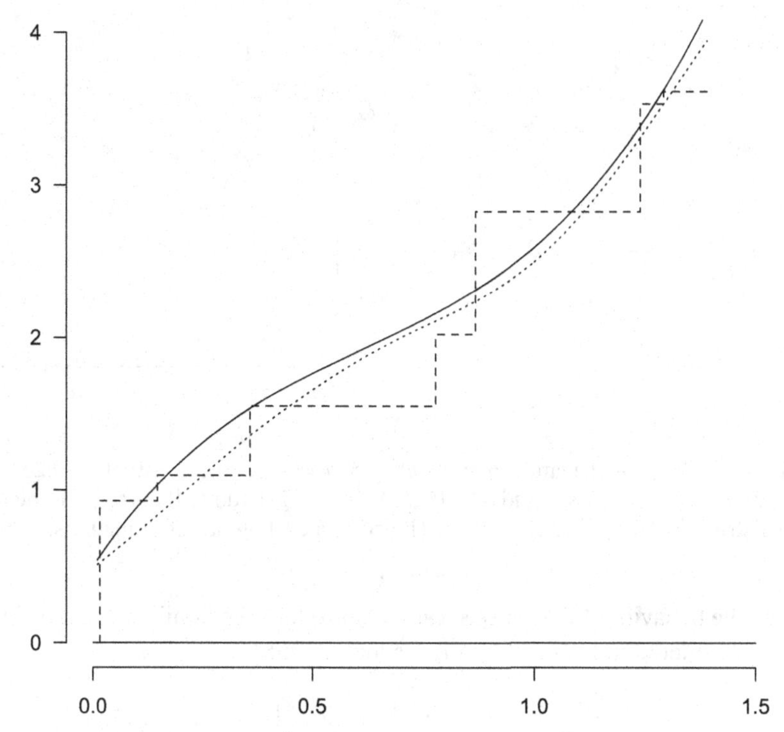

Figure 9.4 The estimate $\tilde{h}_n$ (dotted) of the hazard $h^{(1)}$ (solid) of the family $\{h^{(d)}: d \in [-1, 1]\}$ for a sample of size $n = 100$, together with the (penalized with $2n^{-2/3} \approx 0.093$) isotonic estimate $\hat{h}_n$ (dashed) on the 95% percentile interval $[0, \left(F^{(1)}\right)^{-1}(0.95)]$. Bandwidth $b_n = n^{-1/5} \approx 0.398$.

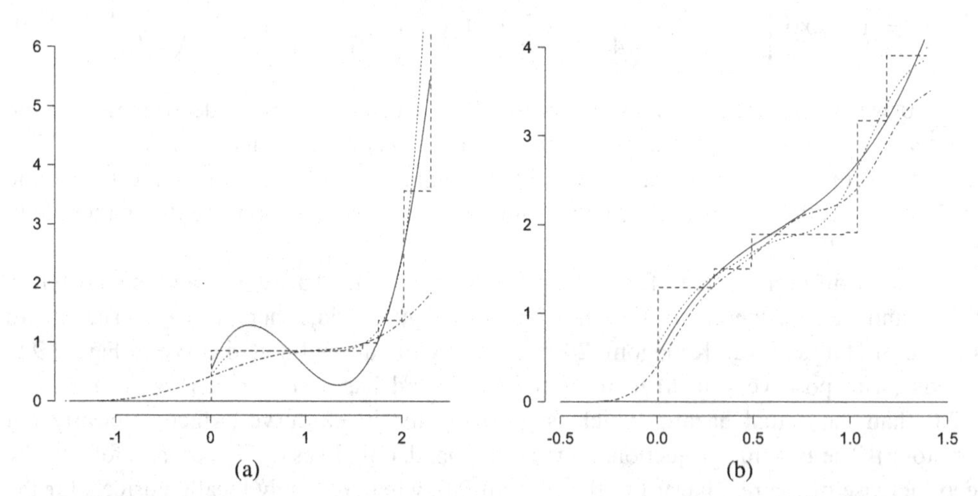

Figure 9.5 The real hazard function $h^{(d)}$ (solid), the isotonic estimate $\hat{h}_n^{(p)}$ of the hazard (dashed), the Hall and Van Keilegom estimate (dashed-dotted) of the hazard (after calibration) and the smoothed isotonic estimate $\tilde{h}_n$ (dotted), using bandwidth $b_n = n^{-1/5}$, for a sample of size $n = 100$ from the distribution function $F^{(d)}$. (a) Corresponds to $d = -1$; (b) to $d = 1$.

interval $[0, F^{-1}(0.95)] \approx [0, 1.39778]$, where $F = F^{(d)}$ with $d = 1$, and the projections of the two methods are less different; see Figure 9.5b.

In justifying this method for approximating critical values for T_n, we use the following bootstrap version of Theorem 9.1, proved in Groeneboom and Jongbloed, 2012.

Theorem 9.2 *Let the conditions of Theorem 9.1 be satisfied and let, in addition, h_0 have a bounded second derivative on $[a, b]$. Let $\tilde{H}_n$ be the estimate of the cumulative hazard function under the null hypothesis, defined by (9.17), and based on a sample $X_1, \ldots, X_n$ from G_0, where we take a bandwidth b_n, satisfying $b_n \asymp n^{-1/5}$. Let $X_1^*, \ldots, X_n^*$ be a bootstrap sample generated by $\tilde{H}_n$ and let $\mathbb{G}_n^*$ and $\hat{G}_n^*$ be the (bootstrap) empirical distribution function and corresponding estimate $\hat{G}_n^*$, based on the greatest convex minorant of the function $x \mapsto -\log(1 - \mathbb{G}_n^*(x-))$, respectively. Finally, let T_n^* be defined by*

$$T_n^* = \int_{[a,b]} \{\mathbb{G}_n^*(x-) - \hat{G}_n^*(x)\}\, d\mathbb{G}_n^*(x).$$

Then we have, in probability, as $n \to \infty$,

$$n^{5/6}\left\{T_n^* - n^{-2/3}\mu_{G_0} \mid X_1, \ldots, X_n\right\} \xrightarrow{\mathcal{D}} N\left(0, \sigma_{G_0}^2\right),$$

where μ_{G_0} and $\sigma_{G_0}^2$ are given in Theorem 9.1. Hence T_n^* and T_n have the same asymptotic critical values under the hypothesis of the theorem, and we can use the bootstrap procedure, sampling from $\tilde{H}_n$, to estimate these critical values.

(B) A Simulation Study

We now compare the power behavior of the test based on test statistic T_n, defined by (9.6), with other test statistics for the family of distributions with hazard rate (9.18) and distribution function (9.19). See Figure 9.3 for these hazard rates and corresponding densities. The bootstrap resampling for T_n was done by taking $B = 2000$ samples from the estimate $\tilde{H}_n$ defined in (9.17) with bandwidth $b_n = n^{-1/5}$. For the estimator $\hat{h}_n^{[p]}$ on which $\tilde{H}_n$ is based (see (9.14)), the penalty was taken equal to $2n^{-2/3}$. The sample was generated by first generating a standard exponential sample $E_1, \ldots, E_n$, producing the bootstrap sample via $X_i^* = \tilde{H}_n^{-1}(E_i)$, $1 \le i \le n$ (see Exercise 9.3). In this way, B values T_n^* were obtained. The critical value is taken to be the 90th percentile of these values of T_n^*.

We make comparisons with tests based on the statistics U_n defined in (9.4) and D_n given in (9.7). For determining a critical value for D_n, again $B = 2000$ random standard exponential samples were generated, and the value

$$D_n^* = \sup_{x \in [0, \mathbb{H}_n(a)]} \{\mathbb{H}_n^*(x) - \hat{H}_n^*(x)\} \tag{9.20}$$

was determined for each such bootstrap sample (taking the interval $[0, \mathbb{H}_n(a)]$ as interval of convexity). The critical value was then taken to be the 90th percentile of the so obtained values of D_n^*. Note that this procedure is equivalent to the procedure that first estimates the (constant) hazard rate on $[0, a]$ by $\mathbb{H}_n(a)/a$, then takes bootstrap samples from the exponential distribution with this hazard rate and finally determines the supremum distance between the two resulting estimators on the interval $[0, a]$. This method is the same as

Table 9.1 *Estimated Powers* ($\times 10^3$) *for Model (9.19), Where* $\alpha = 0.1$, $n = 50$, *and* $d = -1, -0.9, \ldots, -0.1$

	d									
1	−1	−0.9	−0.8	−0.7	−0.6	−0.5	−0.4	−0.3	−0.2	−0.1
T_n	888	726	583	429	345	261	213	190	147	145
U_n	833	636	467	361	297	234	200	183	152	151
D_n	42	29	24	21	16	18	15	18	15	17
T_n'	258	162	111	57	40	28	22	15	9	4

Note: The estimation interval is $[0, F_0^{-1}(0.95)]$.

described in Lemma 9.1; see Exercise 9.8. The critical value for U_n was determined as described directly after (9.4).

In Table 9.1, four tests are compared: the test based on T_n with bootstrap approximation of the critical value, the tests based on U_n and D_n and the test based on T_n using the method of Lemma 9.1 to approximate the critical value (this is denoted by T_n'). For U_n and D_n the critical values are determined as just described. In this table the tests are compared on the fixed interval $[0, a] = [0, F_0^{-1}(0.95)]$. In all cases we generated 2,000 samples, and also $B = 2000$ bootstrap samples from each original sample.

The simulations for U_n took rather long, since repeated density estimation is needed at each step in view of the needed calibration of the bandwidth to create a nondecreasing hazard in the original sample. Also, one has to compute an estimator of the distribution function, the density, and the derivative of the density to check whether one gets a nondecreasing hazard on the chosen interval at the critical bandwidth. The estimation of the density and its derivative was speeded up by using fast Fourier transform, and the distribution function was computed by numerically integrating the density estimate.

It is seen from Table 9.1 that the bootstrap test based on T_n is more powerful for the alternatives $F^{(d)}$ for $d \in [-1, -0.3]$ than the test based on U_n. Table 9.2 shows that the test based on U_n is rather anticonservative (or liberal). This seems to suggest that the high power in the region $d \in [-0.3, 0]$ is at least partly due to the anticonservative behavior of this test. The test based on D_n is very conservative for this interval, as is to be expected,

Table 9.2 *Estimated Rejection Probabilities* ($\times 10^3$) *for Model (9.19) under the Null Hypothesis, Where* $\alpha = 0.1$, $n = 50$, *and* $d = 0, 0.1, \ldots, 1$

	d										
	0	0.1	0.2	0.3	0.4	0.5	0.6	0.7	0.8	0.9	1.0
T_n	129	116	120	112	99	99	89	89	90	86	82
U_n	146	138	132	130	124	122	110	103	102	99	110
D_n	21	19	18	13	15	18	21	24	15	18	21
T_n'	3	3	4	1	2	3	2	1	1	1	0

Note: The estimation interval is $[0, F_0^{-1}(0.95))$.

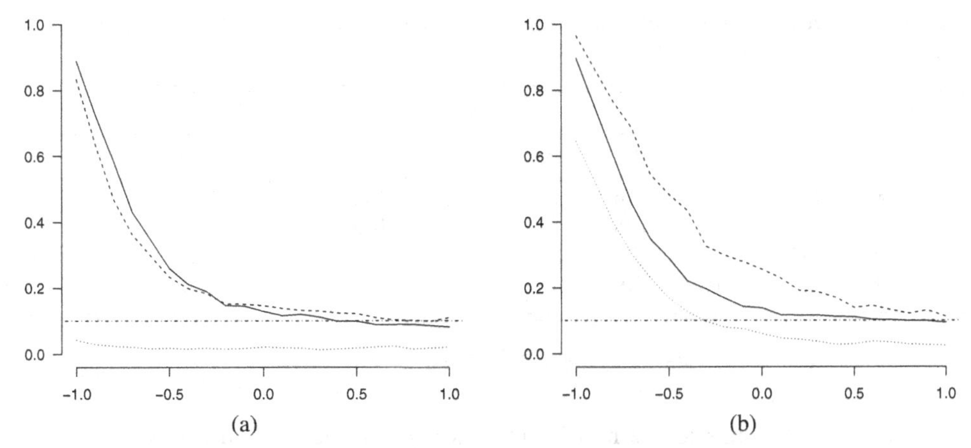

Figure 9.6 The estimated power curves for the family $\{F^{(d)}: d \in [-1, 1]\}$, for the T_n-procedure (solid), the U_n-procedure (dashed) and the D_n-procedure (dotted). The sample size $n = 50$ and the testing interval is $[0, a]$ with $a = F_0^{-1}(0.95)$ (a) and $a = F_0^{-1}(0.80)$ (b).

since the estimated critical value is based on the exponential (worst case) distribution. The test based on T_n has a middle position: it is more conservative than the test based on U_n but less conservative than the test based on D_n. A graphical comparison of the power functions is given in the left panel of Figure 9.6.

Interestingly, the power of the test based on D_n increases considerably if we take a smaller interval $[0, F_0^{-1}(0.80)]$. In fact, this test is derived under the assumption that not all order statistics belong to the interval $[0, a]$. But this often happens if we take $[0, a] = [0, F_0^{-1}(0.95)]$, in particular for the bootstrap samples.

Another reason for the higher power of the test based on D_n in this situation is the fact that the isotonic projection of the hazard $h^{(d)}$, for $d \in [-1, 0]$ is almost constant on the interval $[0, F_0^{-1}(0.8))$, since we miss the steeply increasing part of the hazard from $F_0^{-1}(0.8)$ to $F_0^{-1}(0.95)$, so sampling from the isotonic projection is almost the same as sampling (locally) from an exponential distribution in this case.

For $a = F_0^{-1}(0.8)$ the results are shown in Tables 9.3 and 9.4, and Figure 9.6b. The test based on U_n is very powerful, but also very anticonservative in this situation. For example,

Table 9.3 *Estimated Powers ($\times 10^3$) for Model (9.19), where* $\alpha = 0.1$, $n = 50$, *and* $d = -1, -0.9, \ldots, -0.1$

	d									
	-1	-0.9	-0.8	-0.7	-0.6	-0.5	-0.4	-0.3	-0.2	-0.1
T_n	897	750	604	456	349	290	221	197	168	142
U_n	965	864	766	686	544	483	434	326	299	279
D_n	645	524	399	303	231	168	127	97	80	75
T_n'	742	569	395	253	181	121	67	63	38	30

Note: The hypothesized interval of monotonicity is $[0, F_0^{-1}(0.8)]$.

Table 9.4 *Estimated Rejection Probabilities* ($\times 10^3$) *for Model (9.19) under the Null Hypothesis, where* $\alpha = 0.1$, $n = 50$, *and* $d = 0, 0.1, \ldots, 1$

	d										
	0	0.1	0.2	0.3	0.4	0.5	0.6	0.7	0.8	0.9	1.0
T_n	138	117	116	116	112	111	103	102	99	99	94
U_n	256	229	192	188	170	139	145	132	121	131	112
D_n	60	47	43	37	27	29	37	34	28	26	25
T_n'	24	19	16	9	13	9	5	6	6	4	4

Note: The hypothesized interval of monotonicity is $[0, F_0^{-1}(0.8)]$.

for $d = 0$ (which belongs to the null hypothesis region) one gets an estimated rejection probability of more than 25% instead of the desired 10%.

It is also of interest to compare the powers of the procedure based on bootstrapping from a penalized and smoothed version of the hazard, with the powers obtained by just bootstrapping from the isotonic piecewise constant hazard estimate $\hat{h}_n$ without any smoothing or penalizing. This is done in Figure 9.7, showing that the difference is not very large for this family (and this sample size). The general trend is that bootstrapping from $\hat{h}_n$ gives slightly more conservative critical values. Note that Theorem 9.2 applies to the procedure based on $\tilde{h}_n$, not to that based on $\hat{h}_n$.

9.2 k-Sample Tests for Decreasing Densities

In Section 2.2 the problem of estimating a monotone density is introduced. The nonparametric maximum likelihood (Grenander) estimator in this setting is studied from an asymptotic point

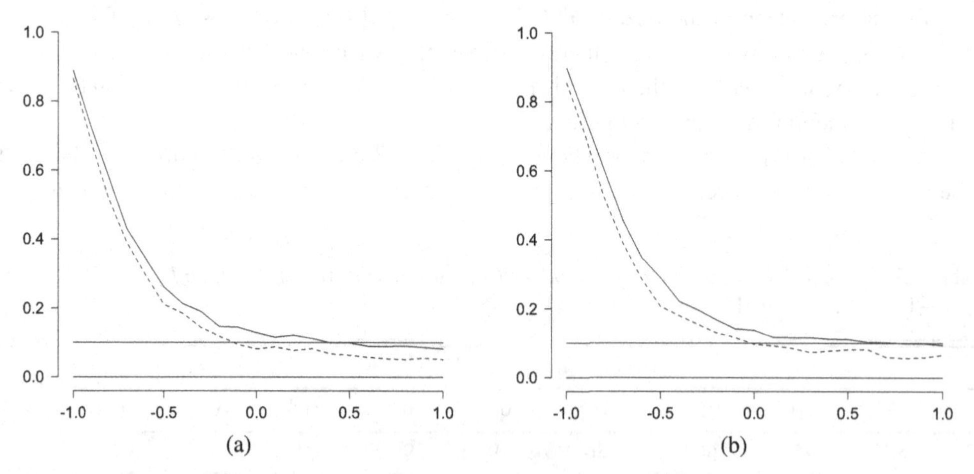

Figure 9.7 The estimated power curves for the family $\{F^{(d)} : d \in [-1, 1]\}$ of the isotonic test statistic T_n, defined by (9.6), for critical values estimated by bootstrapping from a penalized and smoothed isotonic estimate (solid) and for critical values estimated by bootstrapping from the isotonic estimate itself (dotted). The sample size $n = 50$ and the estimation interval is $[a, b] = [0, F_0^{-1}(0.95)]$ in (a); $[a, b] = [0, F_0^{-1}(0.8)]$ in (b).

of view in Section 3.2 and 3.6. In this section we study a k-sample problem for decreasing densities. More specifically, consider independent samples $X_{i1}, X_{i2}, \ldots, X_{i,n_i}$ of size n_i from a decreasing densities f_i are taken, where $i = 1, \ldots, k$. We wish to test the hypothesis

$$\text{Hyp}_0: f_1 = \cdots = f_k.$$

We consider various test statistics that can be used to perform this test. The first is rather straightforward: the likelihood ratio test. Under the null hypothesis, the complete sample can be viewed as sample of size $n = \sum_{i=1}^{k} n_i$ from a single decreasing density f and the maximum likelihood estimator is the Grenander estimator introduced in Section 2.2. In case of the alternative hypothesis, the maximum likelihood estimator for f_i is the Grenander estimator based on the observed X_{ij} values for $1 \le j \le n_i$. Then the (log) likelihood ratio statistic for testing Hyp_0 against the alternative that the densities are not the same is given by

$$T_n = \sum_{i=1}^{k} \sum_{j=1}^{n_i} \log\left(\frac{\hat f_{ni}(X_{ij})}{\hat f_n(X_{ij})}\right), \qquad \sum_{i=1}^{k} n_i = n, \tag{9.21}$$

where $\hat f_{ni}$ is the Grenander estimate based on the ith sample and $\hat f_n$ is the Grenander estimate based on the combined samples. The hypothesis can then be rejected for large values of this T_n.

Alternative test statistics are based on other measures of distance. Durot et al., 2013, for example, consider L_1-type tests based on

$$S_{n1} = \sum_{i<j} \int_a^b \left|\hat f_{ni}(t) - \hat f_{nj}(t)\right| \, dt \tag{9.22}$$

and

$$S_{n2} = \sum_{i=1}^{k} \int_a^b \left|\hat f_{ni}(t) - \hat f_n(t)\right| \, dt, \tag{9.23}$$

where the $\hat f_{ni}$ are the Grenander estimates, defined earlier, for the k samples, and $\hat f_n$ is the Grenander estimate based on the combined samples.

Statistics of this type are shown to be asymptotically normal under Hyp_0 in Durot et al., 2013, under the following conditions:

(A0) The estimators $F_{n1}, F_{n2}, \ldots, F_{nk}$ are independent estimators of the distribution functions $F_1, \ldots, F_k$ of the k samples, respectively, and for every $j = 1, 2, \ldots, k$, the estimator $F_{nj}: [a, b] \to \mathbb{R}$ is a cadlag step process. For example, the F_{nj} could be the empirical distribution functions of the jth sample.

(A1) For each $j = 1, 2, \ldots, k$, the function $f_j: [a, b] \mapsto \mathbb{R}$ is decreasing and continuously differentiable, such that $0 < \inf_{t\in[a,b]} |f_j'(t)| \le \sup_{t\in[a,b]} |f_j'(t)| < \infty$ (in our case f_j is the density in the jth sample).

(A2) For each $j = 1, 2, \ldots, k$, there exists a constant $C_j > 0$, such that for all $x \ge 0$ and $t = a, b$, the process $M_{nj} = F_{nj} - F_j$ satisfies

$$\mathbb{E}\left[\sup_{u\in[a,b],\, x/2\le|t-u|\le x} \left(M_{nj}(t) - M_{nj}(u)\right)^2\right] \le \frac{C_j x}{n_j}.$$

Furthermore, it is assumed that there exists an embedding into Brownian bridges (or, for tests in the regression model, in Brownian motion):

(A3) For each $j = 1, 2, \dots, k$, there exists a Brownian bridge B_{nj}, an increasing function $L_j \colon [a, b] \mapsto \mathbb{R}$ with $\inf_{t\in[a,b]} L'_j(t) > 0$, and constants $q > 6$ and $C > 0$, such that for all $x \in (0, n_j]$:

$$\mathbb{P}\left\{ n_j^{1-1/q} \sup_{t\in[a,b]} \left| M_{nj}(t) - n_j^{-1/2} B_{nj} \circ L_j(t) \right| > x \right\} \le C x^{-q}.$$

Since the F_{nj} are assumed to be independent, we can assume without loss of generality that the B_{nj} are independent. Note that, for $j = 1, 2, \dots, k$, we can write

$$B_{nj}(t) = W_{nj}(t) - \xi_{nj} t, \qquad \text{for } t \in [a, b], \tag{9.24}$$

where the W_{nj} are independent Brownian motions and $\xi_{nj} \equiv 0$, if B_{nj} is Brownian motion, and $\xi_{nj} \sim N(0, 1)$ independent of B_{nj}, if B_{nj} is Brownian bridge. Finally, the following smoothness assumption is required.

(A4) There exist a $\theta \in (3/4, 1]$ and $C > 0$, such that for all $x, y \in [a, b]$ and $j = 1, 2, \dots, k$,

$$|f'_j(x) - f'_j(y)| \le C|x - y|^\theta \text{ and } |L''_j(x) - L''_j(y)| \le C|x - y|^\theta.$$

In order to formulate the first limit theorem, we introduce the random variables

$$\zeta_j(c) = \operatorname{argmax}_{u\in\mathbb{R}} \left\{ W_j(u + c) - u^2 \right\}, \qquad \text{for } c \in \mathbb{R} \text{ and } j = 1, 2, \dots, k, \tag{9.25}$$

where the argmax function is the supremum of the times at which the maximum is attained, $W_1, W_2, \dots, W_k$ are independent standard two-sided Brownian motions. The following theorem (Theorem 2.1 in Durot et al., 2013) establishes asymptotic normality for test statistic S_{n1}.

Theorem 9.3 *Assume (A0), (A1), (A2), (A3), (A4) and let S_{n1} be defined by (9.22). Let ζ_j be defined in* (9.25), *for $j = 1, 2, \dots, k$, with independent standard Brownian motions $W_1, W_2, \dots, W_k$. If $f_0 = f_1 = \dots = f_k$, then*

$$n^{1/6}\left(n^{1/3} S_{n1} - m_1\right) \xrightarrow{\mathcal{D}} N\left(0, \sigma_1^2\right), \qquad n \to \infty$$

where $N(0, \sigma_1^2)$ is a normal distribution with variance

$$\sigma_1^2 = 8 \sum_{i<j} \sum_{l<m} \int_a^b \int_0^\infty \operatorname{cov}\left(|Y_{si}(t) - Y_{sj}(t)|, |Y_{sl}(0) - Y_{sm}(0)|\right) \mathrm{d}t \, \mathrm{d}s,$$

and where

$$m_1 = \sum_{i<j} \int_a^b |4 f'_0(s)|^{1/3} \mathbb{E}\left|Y_{si}(0) - Y_{sj}(0)\right| ds,$$

with

$$Y_{sj}(t) = \frac{L'_j(s)^{1/3}}{c_j^{1/3}} \zeta_j \left(\frac{c_j^{1/3} t}{L'_j(s)^{1/3}} \right), \quad \text{for } j = 1, 2, \dots, k, \quad c_j = \lim_{n\to\infty} \frac{n_j}{n}. \tag{9.26}$$

The expressions for μ_1 and σ_1^2 can be simplified somewhat, since in our case $L_j = F_j$, the distribution function of the jth sample, see p. 943 of Durot et al., 2013. This implies that σ_1^2 does not depend on f_0 under Hyp_0. This phenomenon was probably first observed in Groeneboom, 1985, in connection with the asymptotics of the L_1 distance of the Grenander estimator to the real (decreasing) density (see also Groeneboom et al., 1999).

To formulate a similar result for the test statistic S_{n2}, we need one more piece of notation. First, for each fixed $t \in [a, b]$, define

$$\begin{aligned} \tilde{\zeta}_{t0}(c) &= \text{argmax}_{u \in \mathbb{R}} \left\{ \tilde{W}_{t0}(u+c) - u^2 \right\}, \\ \hat{\zeta}_{t0}(c) &= \text{argmax}_{u \in \mathbb{R}} \left\{ \hat{W}_{t0}(u+c) - u^2 \right\}, \end{aligned} \tag{9.27}$$

where

$$\begin{aligned} \tilde{W}_{t0}(u) &= \sum_{j=1}^{k} \left(\frac{c_j L_j'(t)}{L_0'(t)} \right)^{1/2} W_j(u) \\ \hat{W}_{t0}(u) &= \sum_{j=1}^{k} c_j^{1/2} W_j \left(n^{1/3} \left\{ L_j \circ L_0^{-1}(L_0(t) + n^{-1/3}u) - L_j(t) \right\} \right), \end{aligned} \tag{9.28}$$

with $W_1, W_2, \ldots, W_k$ being the independent standard Brownian motions used to define (9.25) and

$$L_0 = F_0 = \sum_{i=1}^{k} c_j F_j, \qquad c_j = \lim_{n \to \infty} \frac{n_j}{n} \tag{9.29}$$

where F_j is the distribution functions of the jth sample. Note that for $t \in [a, b]$ fixed, due to (9.29), the processes $\tilde{W}_{t0}$ and $\hat{W}_{t0}$ are distributed as standard Brownian motion, which means that $\tilde{\zeta}_{t0}(c)$ and $\hat{\zeta}_{t0}(c)$ have the same distribution as $\zeta_j(c)$. We are now in the position to formulate the second main theorem of this section.

Theorem 9.4 *Assume (A0), (A1), (A2), (A3), (A4) and let S_{n2} be defined by (9.23). Let ζ_j, $\tilde{\zeta}_{t0}$ and $\hat{\zeta}_{t0}$ be defined in (9.25) and (9.27), respectively, with independent standard Brownian motions $W_1, W_2, \ldots, W_k$. If $f_0 = f_1 = \cdots = f_k$, then*

$$n^{1/6} \left(n^{1/3} S_{n2} - m_2 \right) \xrightarrow{\mathcal{D}} N\left(0, \sigma_2^2\right), \qquad n \to \infty$$

where $N(0, \sigma_2^2)$ is a normal distribution with variance

$$\sigma_2^2 = 8 \sum_{i=1}^{k} \sum_{j=1}^{k} \int_a^b \int_0^\infty \text{cov}\left(|Y_{si}(t) - Y_{s0}(t)|, |Y_{sj}(0) - Y_{s0}(0)| \right) \,dt\,ds,$$

with Y_{sj} defined in (9.26) and

$$Y_{s0}(t) = L_0'(s)^{1/3} \tilde{\zeta}_{t0} \left(\frac{t}{L_0'(s)^{1/3}} \right).$$

Furthermore, m_2 may depend on n and is defined by

$$m_2 = \sum_{j=1}^{k} \int_a^b |4 f_0'(t)|^{1/3} \mathbb{E} \left| L_0'(t)^{1/3} \hat{\zeta}_{t0}(0) - \frac{L_j'(t)^{1/3}}{c_j^{1/3}} \zeta_j(0) \right| dt.$$

If, in addition, $L_j = a_j L$, for all $j = 1, 2, \ldots, k$, for a given function $L : [a, b] \to \mathbb{R}$ and given real numbers a_j, then $\hat{\zeta}_{t0} = \tilde{\zeta}_{t0}$ and m_2 no longer depends on n.

Since the normalizing constants are rather intractable for the purpose of a statistical test, it is suggested in Durot et al., 2013, to estimate the critical values of a test by using a bootstrap method.

(A) Finding the Critical Value by a Bootstrap Procedure

It is known that the standard bootstrap typically does not work for Grenander-type estimators, see, e.g., Kosorok, 2008a, Sen et al., 2010. These authors propose a smooth bootstrap based on generating from a kernel smoothed density estimate. Here we also consider a smoothed bootstrap. This will require the use of a smooth estimator $\tilde{f}_n$ which, under the null hypothesis $f_1 = \cdots = f_k = f_0$, satisfies bootstrap versions of assumptions (A0)–(A4). It is proved in Durot et al., 2013 that the following general property will be sufficient for the bootstrap to work.

(A*) The estimator $\tilde{f}_n$ is continuously differentiable on $[a, b]$. Furthermore, there exists events A_n and real numbers $\theta \in (3/4, 1]$ and $\varepsilon_n > 0$, such that $\mathbb{P}(A_n) \to 1$ and $n^\gamma \varepsilon_n \to \infty$ for any $\gamma > 0$, as $n \to \infty$, and such that the following three properties hold on A_n:

$$\sup_{t \in [a,b]} |\tilde{f}_n(t) - f_0(t)| = o(n^{-1/3}), \tag{9.30}$$

$$\sup_{t \in [a,b]} |\tilde{f}_n'(t) - f_0'(t)| = o(n^{-1/6}), \tag{9.31}$$

and for all $x, y \in [a, b]$,

$$|\tilde{f}_n'(x) - \tilde{f}_n'(y)| \le |x - y|^\theta / \varepsilon_n. \tag{9.32}$$

By means of the estimator $\tilde{f}_n$, one builds bootstrap versions $S_{nk}^\star$ of test statistics S_{nk}, for $k = 1, 2$, in such a way that under the null hypothesis and conditionally on the original observations, $n^{1/6}(n^{1/3} S_{nk}^\star - m_k)$ converges in distribution to the Gaussian law with mean zero and variance σ_k^2, in probability, i.e.,

$$\sup_{t \in \mathbb{R}} \left| \mathbb{P}^\star \left\{ \frac{n^{1/6}(n^{1/3} S_{nk}^\star - m_k)}{\sigma_k} \le t \right\} - \Phi(t) \right| \to 0, \quad \text{in probability, as } n \to \infty, \tag{9.33}$$

where m_k and σ_k^2 are the limit bias and variance given in Theorems 9.3 and 9.4, Φ denotes the distribution function of the standard normal distribution and $\mathbb{P}^\star$ is the conditional probability given the original observations. In this case, for a fixed level $\alpha \in (0, 1)$, one can estimate the α-upper percentile point $q_{nk}^\star(\alpha)$ of the conditional distribution of $S_{nk}^\star$ and consider the critical region

$$\left\{ S_{nk} > q_{nk}^\star(\alpha) \right\}. \tag{9.34}$$

If assumptions (A0)–(A4) are fulfilled, then Theorems 9.3 and 9.4 together with (9.33) ensure that the test with critical region (9.34) has asymptotic level α.

For $t \in [a + h_n, b - h_n]$, we define

$$\tilde{f}_n(t) = \frac{1}{h_n} \int_{\mathbb{R}} K\left(\frac{t - x}{h_n}\right) \mathrm{d}F_{n0}(x), \tag{9.35}$$

where K is a twice continuously differentiable kernel, with a bounded third derivative, symmetric around zero, and having support $[-1, 1]$. Near the boundary, we correct the bias by means of boundary kernels. That is, we construct linear combinations of $K(u)$ and $uK(u)$ with coefficients depending on the value near the boundary. For $t \in [a, a + h_n] \cup [b - h_n, b]$, define

$$\tilde{f}_n(t) = \int \frac{1}{h_n} K_{B,t}\left(\frac{t - x}{h_n}\right) \mathrm{d}F_{n0}(x), \tag{9.36}$$

with, taking $u = (t - x)/h$,

$$K_{B,t}(u) = \begin{cases} \alpha\left(\dfrac{t - a}{h_n}\right) K(u) + \beta\left(\dfrac{t - a}{h_n}\right) uK(u), & t \in [a, a + h_n], \\[2ex] \alpha\left(\dfrac{b - t}{h_n}\right) K(u) - \beta\left(\dfrac{b - t}{h_n}\right) uK(u), & t \in [b - h_n, b], \end{cases} \tag{9.37}$$

where for $s \in [-1, 1]$, the coefficients $\alpha(s)$ and $\beta(s)$ are determined by

$$\begin{aligned} \alpha(s) \int_{-1}^{s} K(u)\,\mathrm{d}u + \beta(s) \int_{-1}^{s} uK(u)\,\mathrm{d}u &= 1, \\ \alpha(s) \int_{-1}^{s} uK(u)\,\mathrm{d}u + \beta(s) \int_{-1}^{s} u^2 K(u)\,\mathrm{d}u &= 0. \end{aligned} \tag{9.38}$$

The following lemma guarantees that $\tilde{f}_n$ satisfies condition (A*). For the proof, see Durot et al., 2013; see also Section 8.5.

Lemma 9.2 *Let $\tilde{f}_n(t)$ be defined by* (9.35) *for all $t \in [a + h_n, b - h_n]$ and by* (9.36) *on $[a, a + h_n)$ and $(b - h_n, b_n]$. Assume $h_n = R_n n^{-\gamma}$, where $0 < R_n + R_n^{-1} = O_P(1)$ and $\gamma \in (1/6, 1/5]$. If $f_0 = f_1 = \cdots = f_k$ is twice continuously differentiable on $[a, b]$ and (A3) holds with $\sup_{t \in [a,b]} L'_j(t) < \infty$, for each $j = 1, 2, \ldots, k$, then $\tilde{f}_n$ satisfies (A*).*

As noted in Durot et al., 2013, it is tempting to consider bootstrapping from the Grenander estimator $\tilde{f}_n = \hat{f}_{n0}$ itself, where $\hat{f}_{n0}$ is based on the pooled samples, since $\hat{f}_{n0}$ does not depend on any tuning parameter. From the results in Kosorok, 2008a, and Sen et al., 2010, it appears that bootstrapping from the Grenander does not work for simulating the local behavior of the statistic. However, in our situation we are bootstrapping statistics that are integrals of the difference of two Grenander estimators, so it is not clear whether the results by Kosorok, 2008a, and Sen et al., 2010, will be relevant. We take a look at bootstrapping from $\hat{f}_{n0}$ in the simulation study that follows, where we follow Durot et al., 2013.

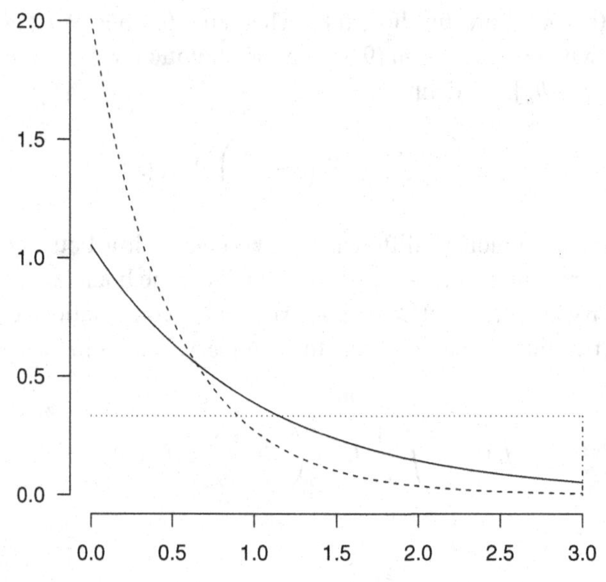

Figure 9.8 Densities of the type (9.39) for $\lambda = 0$ (dotted), $\lambda = 1$ (solid) and $\lambda = 2$ (dashed).

(B) A Simulation Study

We consider a three sample test in the monotone density model. The three densities f_1, f_2 and f_3 are chosen from the family of exponential densities truncated to the interval $[0, 3]$:

$$f(x,\lambda) = \begin{cases} \lambda e^{-\lambda x}(1 - e^{-3\lambda})^{-1}, & \lambda > 0; \\ 1/3, & \lambda = 0, \end{cases} \tag{9.39}$$

for $x \in [0, 3]$ and $f(x, \lambda) = 0$ otherwise. See Figure 9.8.

Under the null hypothesis $f_1 = f_2 = f_3 = f_0$ the bootstrap samples are generated from a pooled estimate for f_0 based on the pooled sample of size $n = n_1 + n_2 + n_3$. We correct the estimator at the boundaries by (9.36). According to Lemma 9.2 the bootstrap should work in this situation. We first investigate how the bandwidth can be chosen data-adaptively.

(C) Choice of Bandwidth

In the experiments, the least squares cross-validation function, as a function of the bandwidth h, is given by

$$\mathrm{LSCV}(h) = \int \tilde{f}_{n,h}(x)^2\,dx - \frac{2n}{n-1}\int \tilde{f}_{n,h}(x)\,\mathrm{d}F_{n0}(x) + \frac{2K(0)}{(n-1)h}, \tag{9.40}$$

where $\tilde{f}_{n,h}$ is the (smooth) estimate of the density, based on the pooled samples, with bandwidth h and F_{n0} is the empirical distribution function of the pooled samples. Note that if $\tilde{f}_{n,h}$ is the ordinary kernel estimator determined with the empirical distribution function F_{n0}, then LSCV(h) is an unbiased estimator of the mean integrated squared error minus the

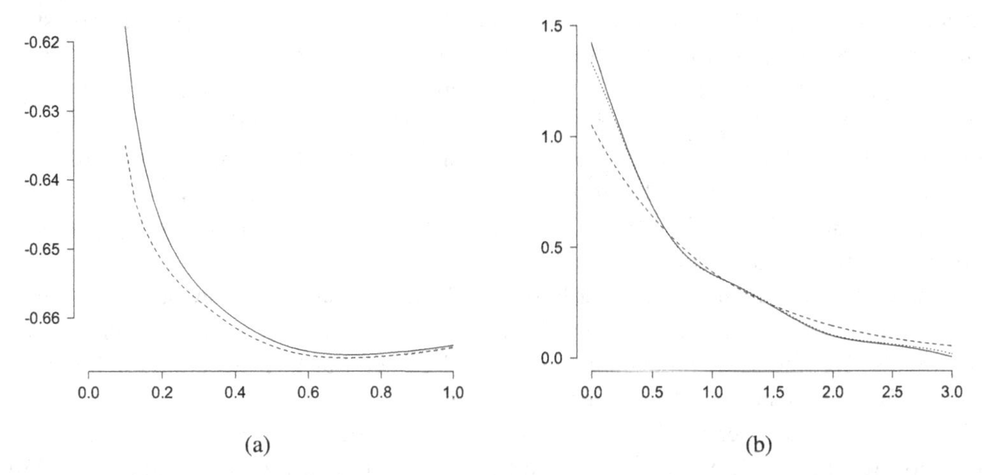

Figure 9.9 Cross-validation functions LSCV(h) for the smoothed Grenander (solid) and the ordinary kernel estimator with boundary correction (9.36), dashed in (a). The resulting density estimates are given in (b): smoothed Grenander (solid), ordinary kernel estimate (dotted) and the true underlying density (dashed).

squared L_2-norm of the density f_0 (which does not depend on h):

$$E\int(\tilde{f}_{n,h}(t)-f_0(t))^2\,dt-\int f_0^2(t)\,dt.$$

The boundary correction method (9.36) generally leads to a convex cross-validation curve with a clear minimum, as shown in Figure 9.9a. The cross-validation curve for the smoothed Grenander (solid) with the boundary correction (9.36) attains its minimum for a value of h close to 0.725, and the resulting kernel estimate (solid) is shown in Figure 9.9b. The cross-validation curve in Figure 9.9a, based on a sample of size $n = 300$ from a truncated exponential distribution on $[0, 3]$, of the ordinary kernel estimate (dashed) with the boundary correction (9.36) and the kernel estimate based on the Grenander are rather close, with a minimum at approximately the same bandwidth. But this kernel estimate will not necessarily be decreasing and we actually prefer a decreasing density such as the smoothed Grenander, which belongs to the allowed class of densities, for generating the bootstrap samples.

The overall performance of the smoothed Grenander estimator with boundary correction (9.36) seems to be the best. Therefore, for $t \in [h, 3-h]$ the smooth estimate for generating the bootstrap samples is defined as

$$\tilde{f}_{n,h}(t)=\int K_h(t-x)\,\mathrm{d}\hat{F}_{n0}(x), \tag{9.41}$$

where K is a symmetric kernel with support $[-1, 1]$ and $\hat{F}_{n0}$ is the least concave majorant of the empirical distribution function F_{n0}. We correct the kernel density estimate at the boundaries of $[0, 3]$ by means of (9.36) with $\hat{F}_{n0}$ instead of F_{n0}.

(D) Simulating the Level and Power under Alternatives

To investigate the finite sample power at a given combination $(\lambda_1, \lambda_2, \lambda_3)$, three samples of sizes n_j from $f_j = f(\cdot, \lambda_j)$, for $j = 1, 2, 3$, are generated, and the value of the test statistics

S_{n1} and S_{n2} is computed, as defined in (9.22) and (9.23). Next, 10,000 times three bootstrap samples of sizes n_1, n_2 and n_3 from the pooled estimate $\tilde{f}_{n,h}$ were generated, and the values $S^\star_{n1}$ and $S^\star_{n2}$ of both test statistics and their 5th upper percentiles $q^\star_{nk}(0.05)$ are determined, for $k = 1, 2$. This whole procedure is repeated $B = 10000$ times and the number of times is counted that the values of the test statistics S_{n1} and S_{n2} exceed the corresponding 5th upper percentiles $q^\star_{n1}(0.05)$ and $q^\star_{n2}(0.05)$, respectively. By dividing this number by B, this provides an approximation of the finite sample power of both test statistics at underlying truncated exponentials with parameters λ_1, λ_2 and λ_3.

In view of the comments made after Lemma 9.2, we compare the behavior of the smooth bootstrap procedure with bootstrapping from the pooled Grenander estimate itself. To this end, we also run the same procedure as described earlier, but then generate the bootstrap samples from $\hat{f}_{n0}$ instead of the smooth estimate $\tilde{f}_{n,h}$.

To investigate the performance under the null hypothesis, $\lambda_1 = \lambda_2 = \lambda_3$ were taken equal to the values $0.1, 0.5, 1, 2, \ldots, 6$ and equal sample sizes $n_j = 100$ and $n_j = 250$, for $j = 1, 2, 3$. The simulated levels are determined by means of $B = 10000$ repetitions. The simulations to investigate the finite sample power at alternatives are done with sample sizes $n_1 = n_2 = n_3 = 100$ and alternatives for which $\lambda_1 = \lambda_2 = 1$ and λ_3 varies between 0 and 3.5 by steps of 0.1.

(E) Comparison with True Power

Finally, in order to calibrate the finite sample power obtained from bootstrapping, the true finite sample power for a given choice $(\lambda_1, \lambda_2, \lambda_3)$ was also (approximately) determined. To this end, 10,000 samples of size $n = n_1 + n_2 + n_3$ were generated from the mixture density

$$f_0(x) = \left(\frac{n_1}{n} \frac{\lambda_1 e^{-\lambda_1 x}}{1 - e^{-3\lambda_1}} + \frac{n_2}{n} \frac{\lambda_2 e^{-\lambda_2 x}}{1 - e^{-3\lambda_2}} + \frac{n_3}{n} \frac{\lambda_3 e^{-\lambda_3 x}}{1 - e^{-3\lambda_3}} \right) 1_{[0,3]}(x). \tag{9.42}$$

This mixture density is considered to be the least favorable density among all densities under the null hypothesis, in the case of three truncated exponentials with parameters λ_1, λ_2 and λ_3. For each of the samples the values of the test statistics S_{n1} and S_{n2} were computed and used to determine the 5th upper percentiles $q_{nk}(0.05)$, $k = 1, 2$, for both test statistics. Next, another 10,000 times three samples of sizes n_j from $f_j = f(\cdot, \lambda_j)$ were generated, both test statistics were computed and the number of times it exceeds the corresponding 5th percentile $q_{nk}(0.05)$ was counted. Dividing these numbers by 10,000 provides an approximation of the true finite sample power for a given choice $(\lambda_1, \lambda_2, \lambda_3)$. Note that such a calibration is not implementable in practice since it requires knowledge of f_1, f_2 and f_3, but it may serve as a benchmark for the power obtained from bootstrapping, in the simulations.

(F) Implementation

We now give some more detail on how bootstrapping from the smoothed Grenander has been implemented. First consider the estimate defined in (9.41) for $t \in [h, 3 - h]$. One possibility to implement this estimate would be to use numerical integration of $\hat{f}_{n0}$. However, one can avoid this by a summation by parts procedure.

Let $p_1, \ldots, p_m$ be the jump sizes of the Grenander estimator at the points of jump $\tau_1 < \cdots < \tau_m \in (0, 3)$, where τ_m is the largest order statistic. Note that $\hat{f}_{n0}$ is left-continuous and that $\hat{f}_{n0}$ always has a jump down to zero at the last order statistic. We now define

$$\mathbb{K}_h(x) = \int_{x/h}^{\infty} K(u)\,du, \qquad x \in \mathbb{R}. \tag{9.43}$$

In the simulations the triweight kernel

$$K(u) = \frac{35}{32}\left(1 - u^2\right)^3 1_{[-1,1]}(u)$$

was taken. Then, defining $\tau_0 = 0$, for $t \in [h, 3-h]$, we can write

$$\begin{aligned}\tilde{f}_{n,h}(t) &= \sum_{i=1}^{m}\left\{\sum_{j=i}^{m} p_j\right\}\int_{\tau_{i-1}}^{\tau_i} K_h(t-x)\,\mathrm{d}x = \sum_{j=1}^{m} p_j \int_0^{\tau_j} K_h(t-x)\,\mathrm{d}x\\ &= \sum_{j=1}^{m} p_j \int_{(t-\tau_j)/h}^{t/h} K(u)\,\mathrm{d}u = \sum_{j=1}^{m} p_j \int_{(t-\tau_j)/h}^{1\wedge(t/h)} K(u)\,\mathrm{d}u = \sum_{j=1}^{m} p_j \mathbb{K}_h(t-\tau_j),\end{aligned}$$

so that for $t \in [h, 3-h]$, the estimate $\tilde{f}_{n,h}(t)$ can now be computed as a finite sum over the jumps p_i of the Grenander estimator $\hat{f}_{n0}$. We then still have to define $\tilde{f}_{n,h}(t)$ for $t \in [0, h) \cup (3-h, 3]$. To this end, for $j = 0, 1, 2$, let

$$\mathbb{K}_h^{(j)}(x) = \int_{-\infty}^{x/h} u^j K(u)\,\mathrm{d}u.$$

Note that $\mathbb{K}_h^{(0)}(t) = 1 - \mathbb{K}_h(t)$, where $\mathbb{K}_h$ is defined in (9.43). As before, we get for $t < h$,

$$\begin{aligned}\tilde{f}_{n,h}(t) &= \int \left\{\alpha\left(\frac{t}{h}\right) K_h(t-x) + \beta\left(\frac{t}{h}\right)\frac{t-x}{h} K_h(t-x)\right\} \hat{f}_{n0}(x)\,dx\\ &= \alpha\left(\frac{t}{h}\right)\sum_{j=1}^{m} p_j \int_{(t-\tau_j)/h}^{t/h} K(u)\,du + \beta\left(\frac{t}{h}\right)\sum_{j=1}^{m} p_j \int_{(t-\tau_j)/h}^{t/h} uK(u)\,du\\ &= \alpha\left(\frac{t}{h}\right)\sum_{j=1}^{m} p_j\left\{\mathbb{K}_h^{(0)}(t) - \mathbb{K}_h^{(0)}(t-\tau_j)\right\} + \beta\left(\frac{t}{h}\right)\sum_{j=1}^{m} p_j\left\{\mathbb{K}_h^{(1)}(t) - \mathbb{K}_h^{(1)}(t-\tau_j)\right\},\end{aligned}$$

where ϕ and ψ are defined by (9.38), and similarly for $t > 3-h$,

$$\tilde{f}_{n,h}(t) = \alpha\left(\frac{3-t}{h}\right)\sum_{j=1}^{m} p_j\left\{1 - \mathbb{K}_h^{(0)}(t-\tau_j)\right\} + \beta\left(\frac{3-t}{h}\right)\sum_{j=1}^{m} p_j \mathbb{K}_h^{(1)}(t-\tau_j).$$

This means that also near the boundaries of $[0, 3]$, the estimator $\tilde{f}_{n,h}(t)$ can be computed in terms of finite sums over the jumps of the Grenander estimator $\hat{f}_{n0}$.

(G) Results

First the level of the tests under the null hypothesis of all λs equal to some λ_0 is investigated, where λ_0 varied over $0.1, 0.5, 1, 2, \ldots, 6$. The significance level $\alpha = 0.05$ was taken and the

Table 9.5 *Simulated Levels of S_{n1} and S_{n2} under the Null Hypothesis*

	$n_1 = n_2 = n_3 = 100$				$n_1 = n_2 = n_3 = 250$			
	Genander		Smooth		Grenander		Smooth	
λ_0	S_{n1}	S_{n2}	S_{n1}	S_{n2}	S_{n1}	S_{n2}	S_{n1}	S_{n2}
0.1	.0122	.0215	.0109	.0195	.0173	.0256	.0166	.0241
0.5	.0337	.0352	.0381	.0424	.0393	.0402	.0407	.0500
1	.0417	.0405	.0474	.0477	.0422	.0424	.0479	.0499
2	.0462	.0436	.0560	.0592	.0466	.0449	.0546	.0504
3	.0473	.0443	.0582	.0591	.0496	.0474	.0554	.0545
4	.0474	.0444	.0545	.0560	.0498	.0478	.0559	.0563
5	.0455	.0486	.0593	.0577	.0530	.0541	.0487	.0508
6	.0441	.0459	.0600	.0597	.0553	.0496	.0541	.0526

Note: The target value is $\alpha = 0.05$.

bootstrap experiments were performed with $n_1 = n_2 = n_3 = 100$ and $n_1 = n_2 = n_3 = 250$. The results are listed in Table 9.5. It can be seen that close to $\lambda_0 = 0$, which corresponds to the uniform distribution, the attained level is much too small. For large λ_0 the attained levels tend to be somewhat too large. Note that the simulated levels obtained from bootstrapping from the Grenander itself are comparable to the ones obtained from the smooth bootstrap.

Next, the power under alternatives of the form $f_1 = f_2 = f(\cdot, 1)$ and $f_3 = f(\cdot, \lambda)$ was investigated, with $n_1 = n_2 = n_3 = 100$. A picture of the power estimates of the smoothed Grenander, using cross-validation for the bandwidth choice, is shown in Figure 9.10 together with the estimates obtained by bootstrapping from the Grenander estimator. Figure 9.10a displays the powers simulated by generating bootstrap samples from the ordinary Grenander

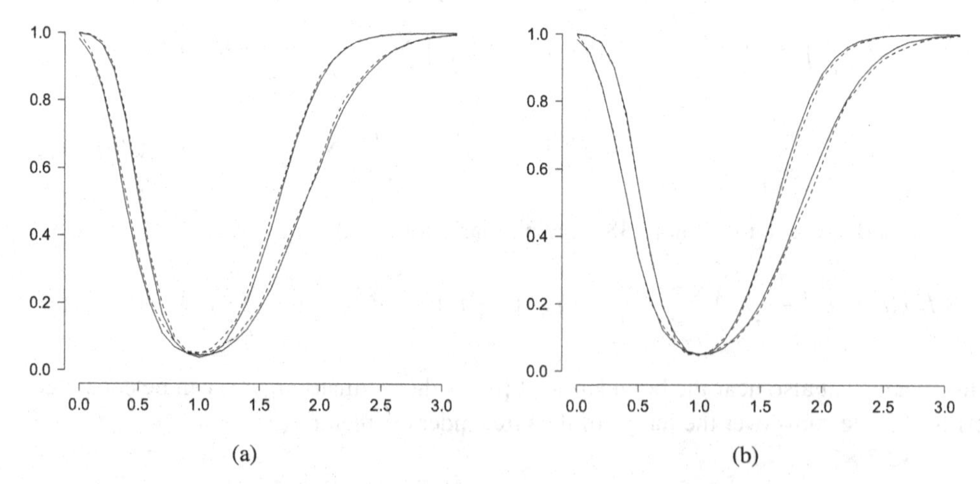

Figure 9.10 Simulated powers (solid) from bootstrapping (in (a) from the Grenander estimator and in (b) from the smoothed Grenander estimator) and estimated true powers (dashed) of S_{n1} (lower curves) and S_{n2} (upper curves), for $\lambda = 0, 0.1, 0.2, \ldots, 3.5$. The level of the test is taken to be 0.05.

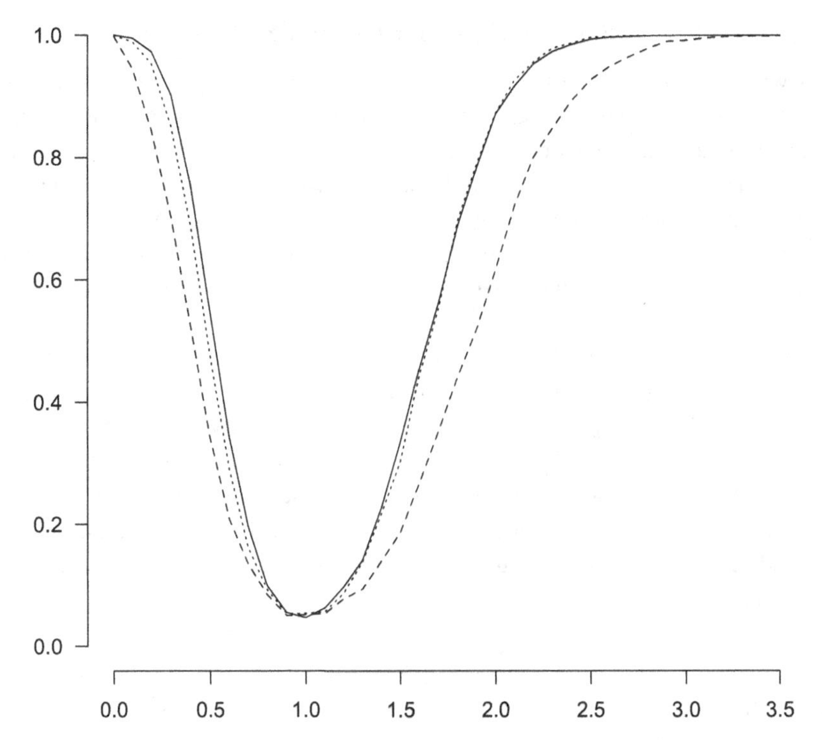

Figure 9.11 Estimated true powers of the LR test (9.21) (dotted) and the test based on S_{n1} (dashed) and S_{n2} (solid), for $\lambda = 0, 0.1, 0.2, \ldots, 3.5$. The level of the test is taken to be 0.05.

estimator (solid curves) and the direct estimates of the true power (dashed curves). The top solid and dashed curves correspond to test statistic S_{n2}. This test statistic seems to be uniformly more powerful (over the alternatives considered here) than test statistic S_{n1}, which corresponds to the bottom solid and dashed curves. Figure 9.10b displays the powers simulated by generating bootstrap samples from the smoothed Grenander estimator (solid curves) and the same direct estimates of the true power (dashed curves). Again the top solid and dashed curves correspond to test statistic S_{n2}.

The simulated powers in Figure 9.10a, based on bootstrapping from the ordinary Grenander, tend to be conservative. The simulated powers in Figure 9.10b, based on bootstrapping from the smoothed Grenander, tend to be slightly anticonservative. Note that, similar to the simulated levels in Table 9.5, there is hardly any difference between the results when using the smooth bootstrap or when bootstrapping from the ordinary Grenander. This seems to be an interesting point for further research.

We finally make a comparison with the LR test statistic, defined by (9.21) (this comparison was not made in Durot et al., 2013). To this end, we recomputed the powers of the two tests, generating the bootstrap samples from the least favorable mixture density (9.42), and also computed the powers of the LR test in this way. It is seen from Figure 9.11 that the powers of the LR test are very close to the powers of the test based on S_{n2} in this example. Whether this means that these tests have some asymptotic equivalence property is still an open question at this point.

9.3 Two-Sample Tests for Current Status Data

In contrast with the number of two-sample tests available for right censored data, there is very little for interval censored data. Permutation tests for the two-sample problem with interval censored data have been considered in Peto and Peto, 1972. Since they rely on the permutation distribution, such tests can only be used when the censoring mechanism is the same in both samples. For right censored data, permutation tests, considered as conditional tests, have been shown to be asymptotically independent of the censoring distributions in the two sample problem in Neuhaus, 1993, but it is doubtful that this property persists for interval censored data.

We consider here a likelihood ratio based test for testing that two samples come from the same distribution, if current status censoring is present. To be more specific, we have a sample $T_1, T_2, \dots, T_m$ from a distribution with density g_1 along with indicators $(\Delta_1, \Delta_2, \dots, \Delta_m)$ where $\Delta_i = 1_{\{X_i \le T_i\}}$ and the X_is are independent of the T_js and distributed according to a distribution function F_1 (see also Section 2.3). We also have a sample $(T_{m+1}, \Delta_{m+1}), \dots (T_N, \Delta_N)$ of size n $(N = m + n)$, independent of the first, where the T_is are distributed according to a density g_2 and the X_is according to a distribution function F_2. Again $\Delta_i = 1_{\{X_i \le T_i\}}$. The hypotheses to be tested are

$$\text{Hyp}_0 : F_1 = F_2 \text{ against } F_1 \neq F_2.$$

We first discuss the likelihood ratio test. Under the null hypothesis of equality of F_1 and F_2 we have to maximize

$$\sum_{i=1}^{N} \{\Delta_i \log F(T_i) + (1 - \Delta_i) \log (1 - F(T_i))\}, \quad N = m + n,$$

over all distribution functions F. Without the restriction of the null hypothesis we have to maximize

$$\ell(F_1, F_2) = \sum_{i=1}^{m} \{\Delta_i \log F_1(T_i) + (1 - \Delta_i) \log (1 - F_1(T_i))\}$$
$$+ \sum_{i=m+1}^{N} \{\Delta_i \log F_2(T_i) + (1 - \Delta_i) \log (1 - F_2(T_i))\}$$

over all pairs of distribution functions (F_1, F_2).

This means that under the null hypothesis the MLE $\hat{F}_N$ evaluated at $T_{(k)}$ is given by the left hand derivative of the greatest convex minorant of the cusum diagram of the points $(0, 0)$ and the points

$$\left(i, \sum_{j \le i} \Delta_{(j)}\right), \quad i = 1, \dots, N \tag{9.44}$$

evaluated at $T_{(k)}$. Here $\Delta_{(j)}$ denotes the indicator corresponding to the jth order statistic $T_{(j)}$ in $\{T_1, \dots, T_N\}$ (see Section 2.3). Denote by $T_{(k1)}$ the kth order statistic of the T_is corresponding to the first sample $\{T_1, \dots, T_m\}$. Under the alternative hypothesis, the MLE $\hat{F}_{N1}$ of F_1 at $T_{(k1)}$ is then given by the left continuous slope of the greatest convex minorant of the cumulative sum diagram consisting of the points (0, 0) and the points

$$\left(i, \sum_{j \le i} \Delta_{(j1)}\right), \quad i = 1, \dots, m, \tag{9.45}$$

evaluated at $T_{(k1)}$, where $\Delta_{(j1)}$ is the indicator corresponding to $T_{(j1)}$. Similarly the MLE $\hat{F}_{N2}$ of F_2 at T_{k2} is given by the left hand derivative of the greatest convex minorant of the cumulative sum diagram of the points (0, 0) and the points

$$\left(i, \sum_{j \le i} \Delta_{(j2)}\right), \quad i = 1, \dots, n, \tag{9.46}$$

where $\Delta_{(j2)}$ is the indicator corresponding to jth order statistic $T_{(j2)}$ of the second sample $\{T_{m+1}, \dots, T_{m+n}\}$.

Then the log likelihood ratio test statistic is given by:

$$\begin{aligned}\tilde{V}_N = \ell(\hat{F}_{N1}, \hat{F}_{N2}) - \ell(\hat{F}_N, \hat{F}_N) &= \sum_{i=1}^{m} \left\{\Delta_i \log \frac{\hat{F}_{N1}(T_i)}{\hat{F}_N(T_i)} + (1 - \Delta_i) \log \frac{1 - \hat{F}_{N1}(T_i)}{1 - \hat{F}_N(T_i)}\right\} \\ &+ \sum_{i=m+1}^{N} \left\{\Delta_i \log \frac{\hat{F}_{N2}(T_i)}{\hat{F}_N(T_i)} + (1 - \Delta_i) \log \frac{1 - \hat{F}_{N2}(T_i)}{1 - \hat{F}_N(T_i)}\right\},\end{aligned} \tag{9.47}$$

where the terms with coefficients Δ_i and $1 - \Delta_i$ are defined to be zero if Δ_i and $1 - \Delta_i$ are zero, respectively.

The properties of this LR test are still largely unknown. This is different for an LR-type test based on maximum smoothed likelihood estimators introduced in Section 8.5. We will now discuss this latter test in more detail.

(A) A Likelihood Ratio Test, Based on Maximum Smoothed Likelihood Estimators

As mentioned in Section 8.5, smoothing the empirical distribution function based on non-negative data usually gives problems near zero in the sense that the resulting estimator might (and probably will) assign a positive probability to the negative half axis. In order to avoid such problems, we restrict the domain on which the test statistic is computed to an interval $[a, b] \subset (0, M)$, where $[0, M]$ is the support of the underlying densities, corresponding to the distribution functions F_1 and F_2 of the hidden variables. Inspired by the LR statistic given in (9.47), we consider the statistic V_N given by

$$\begin{aligned}V_N = &\frac{2m}{N} \int_{t \in [a,b]} \left\{\tilde{h}_{N1}(t) \log \frac{\tilde{F}_{N1}(t)}{\tilde{F}_N(t)} + \{\tilde{g}_{N1}(t) - \tilde{h}_{N1}(t)\} \log \frac{1 - \tilde{F}_{N1}(t)}{1 - \tilde{F}_N(t)}\right\} dt \\ &+ \frac{2n}{N} \int_{t \in [a,b]} \left\{\tilde{h}_{N2}(t) \log \frac{\tilde{F}_{N2}(t)}{\tilde{F}_N(t)} + \{\tilde{g}_{N2}(t) - \tilde{h}_{N2}(t)\} \log \frac{1 - \tilde{F}_{N2}(t)}{1 - \tilde{F}_N(t)}\right\} dt\end{aligned} \tag{9.48}$$

where $\tilde{F}_{N1}$, $\tilde{F}_{N2}$ and $\tilde{F}_N$ are the maximum smoothed likelihood estimators (MSLEs) for the first, second and combined sample, respectively, and $\tilde{g}_{Ni}$ and $\tilde{h}_{Ni}$ are kernel estimates of the relevant observation densities, defined in the following. As explained in Section 8.5, the MSLEs for the combined samples and the first and second sample are computed by replacing the cumulative sum diagrams (9.44), (9.45) and (9.46) by the continuous cumulative sum diagrams

$$\left(\tilde{G}_N(t), \tilde{H}_N(t)\right), \quad t \in [0, M], \tag{9.49}$$

$$\left(\tilde{G}_{N1}(t), \tilde{H}_{N1}(t)\right), \quad t \in [0, M], \tag{9.50}$$

and

$$\left(\tilde{G}_{N2}(t), \tilde{H}_{N2}(t)\right), \quad t \in [0, M], \tag{9.51}$$

respectively, where $\tilde{G}_N$, $\tilde{G}_{Ni}$, $\tilde{H}_N$, $\tilde{H}_{Ni}$ and their derivatives are defined in the following way.

The densities $\tilde{g}_{Ni}$ and $\tilde{h}_{Ni}$ are kernel estimators with bandwidth b_N, defined by

$$\tilde{g}_{Ni}(t) = \int K_{b_N}(t-u)\, d\mathbb{G}_{Ni}(u), \qquad \tilde{h}_{Ni}(t) = \int K_{b_N}(t-u)\,\delta\, d\mathbb{P}_{Ni}(u, \delta). \tag{9.52}$$

Here $\mathbb{G}_{N1}$ is the empirical distribution of the observations $T_1, \ldots, T_m$ of the first sample and $\mathbb{P}_{N1}$ is the empirical distribution of the observations $(T_1, \Delta_1), \ldots, (T_m, \Delta_m)$ of the first sample, with the analogous definitions of $\mathbb{G}_{N2}$ and $\mathbb{P}_{N2}$ for the second sample. The densities $\tilde{g}_N$ and $\tilde{h}_N$ are defined by

$$\tilde{g}_N = \alpha_N \tilde{g}_{N1} + \beta_N \tilde{g}_{N2}, \qquad \tilde{h}_N = \alpha_N \tilde{h}_{N1} + \beta_N \tilde{h}_{N2}, \qquad \alpha_N = \frac{m}{N}, \qquad \beta_N = 1 - \alpha_N.$$

The kernel K_b is defined in the usual way by

$$K_b(u) = \frac{1}{b} K\left(\frac{u}{b}\right),$$

for a bandwidth $b > 0$, where K is a symmetric positive kernel with compact support. For example the triweight kernel

$$K(u) = \frac{35}{32}\left(1 - u^2\right)^3 1_{[-1,1]}(u), \tag{9.53}$$

which is the kernel used in the simulation study reported in the following.

For $t \in [0, b_N]$ and $t \in [M - b_N, M]$ a boundary kernel is used, defined by a linear combination of $K(u)$ and $uK(u)$. Other ways of bias correction at the boundary are also possible (see Exercise 8.14), but it seems absolutely necessary to use such a correction in order to obtain a reasonable behavior at the boundary. Using boundary kernels, the simple property that the distribution function of the estimator can be obtained by just integrating the kernel is lost, and indeed the estimates of the distribution functions were obtained by numerically integrating the estimates of the densities (and not by integrating the kernels). Using these conventions, we define

$$\tilde{G}_{Ni}(t) = \int_0^t \tilde{g}_{Ni}(u)\, du, \qquad \tilde{H}_{Ni}(t) = \int_0^t \tilde{h}_{Ni}(u)\, du,$$

$$\tilde{G}_N = \alpha_N \tilde{G}_{N1} + \beta_N \tilde{G}_{N2}, \qquad \tilde{H}_N = \alpha_N \tilde{H}_{N1} + \beta_N \tilde{H}_{N2},$$

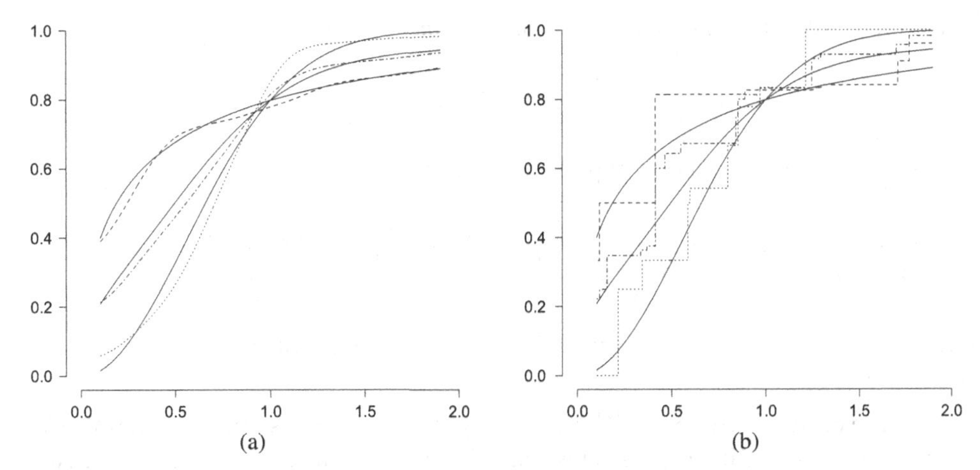

Figure 9.12 MSLEs and MLEs on $[a, b]$ for samples of size $m = n = 250$ from the Weibull densities (9.54). G_1 and G_2 are uniform on $[0, 2]$, and the interval $[a, b] = [0.1, 1.9]$. (a) The MSLE estimates; (b) the MLEs. The dashed curves give the estimates for the first sample ($\alpha_1 = 0.5$), the dotted curves the estimates for the second sample ($\alpha_2 = 2$), and the dashed-dotted curves the estimates for the combined samples. The solid curves give the corresponding actual distribution functions for these three situations. The bandwidth for the computation of the MSLEs was $b_N = 2N^{-1/5} \approx 0.57708$, where $N = m + n = 500$.

and use the corresponding numerical integrals in the continuous cusum diagrams (9.49) to (9.51). A picture of the MSLE estimators and the MLE estimators for samples of size 250 from two different Weibull distributions with densities

$$\alpha_1 \lambda x^{\alpha_1 - 1} e^{-\lambda x^{\alpha_1}}, \quad \alpha_2 \lambda x^{\alpha_2 - 1} e^{-\lambda x^{\alpha_2}}, \quad x > 0, \quad \alpha_1 = 0.5, \quad \alpha_2 = 2, \; \lambda = 1.6, \tag{9.54}$$

respectively, where $\alpha_1 = 0.5$ holds for the first sample and $\alpha_2 = 2$ for the second sample, is shown in Figure 9.12.

The following result shows that the test statistic V_N is, for a suitable choice of the bandwidth, an asymptotic pivot under the null hypothesis of equality of the two distribution functions F_1 and F_2 of the hidden variables in the two samples.

Theorem 9.5 *Let the test statistic V_N be defined by (9.48), using a bandwidth b_N such that $b_N \asymp N^{-\alpha}$, where $2/9 < \alpha < 1/3$. Furthermore, let F stay away from 0 and 1 on $[a, b]$ and have a bounded continuous second derivative f' on an interval (a', b') containing $[a, b]$, and let g_1 and g_2 be continuous densities that stay away from 0 on $[a, b]$, with continuous bounded second derivatives on the interval (a', b'). Let the log likelihood ratio statistic V_N, based on the MSLEs, be defined by (9.48). Then we have in probability, if the distribution functions of the hidden variables in the two samples are both equal to F and $m/N \to \alpha \in (0, 1)$, as $N \to \infty$,*

$$N\sqrt{\frac{b_N}{b-a}} \left\{ (V_N | T_1, \ldots, T_N) - \frac{b-a}{N b_N} \int K(u)^2 \, du \right\} \xrightarrow{\mathcal{D}} N(0, \sigma_K^2), \tag{9.55}$$

where $N(0, \sigma_K^2)$ denotes a normal distribution with mean zero and variance

$$\sigma_K^2 = 2\int\left\{\int K(u+v)K(u)\,du\right\}^2 dv.$$

Remark To say that (9.55) holds in probability means that

$$\mathbb{P}\left\{N\sqrt{\frac{b_N}{b-a}}\left\{V_N - \frac{b-a}{Nb_N}\int K(u)^2\,du\right\} \le x \;\middle|\; T_1,\dots,T_N\right\} \xrightarrow{p} \Phi(x/\sigma_K),$$

for each $x \in \mathbb{R}$, where Φ is the standard normal distribution function and $\xrightarrow{p}$ denotes convergence in probability.

Remark The condition $b_N \gg N^{-1/3}$ is necessary for having the asymptotic equivalence of the MSLEs to ratios of kernel estimators (see Corollary 3.4 in Groeneboom et al., 2010), and $b_N \ll N^{-2/9}$ prevents the bias from entering, which causes the asymptotic distribution of V_N to become dependent on the observation densities g_1 and g_2. The bias term drops out if the observation densities g_1 and g_2 are the same in the two samples.

However, we prefer to work with a larger bandwidth, at the cost of introducing a bias term, depending on the underlying distributions, as shown in Theorem 9.6. It turns out that this bias term does not cause problems, if critical values are computed using the bootstrap procedure to be discussed later in this section. The key to this is that the bias term is estimated automatically in the bootstrap resampling from a smooth estimate of F and that, by Theorem 9.2, the difference between this estimate of the bias and the deterministic bias is sufficiently small, so that we can replace it by the deterministic bias in the central limit theorem for the bootstrap test statistic.

Theorem 9.6 *Let the test statistic V_N be defined by (9.48), using a bandwidth b_N such that $b_N \asymp N^{-\alpha}$, where $1/5 < \alpha \le 2/9$. Furthermore, let F stay away from 0 and 1 on $[a, b]$ and have a bounded continuous second derivative f' on an interval (a', b') containing $[a, b]$, and let g_1 and g_2 be continuous densities that stay away from 0 on $[a, b]$, with continuous bounded second derivatives on the interval (a', b'). Let the log likelihood ratio statistic V_N, based on the MSLEs, be defined by (9.48). Then we have in probability, if the distribution functions of the hidden variables in the two samples are both equal to F and $m/N \to \alpha \in (0, 1)$, as $N \to \infty$,*

$$N\sqrt{\frac{b_N}{b-a}}\left\{(V_N|T_1,\dots,T_N) - \frac{b-a}{Nb_N}\int K(u)^2\,du\right.$$
$$\left. - \alpha_N\beta_N\int_{t=a}^{b}\frac{f(t)^2\{g_1'(t)g_2(t) - g_2'(t)g_1(t)\}^2}{F(t)\{1-F(t)\}\bar{g}_N(t)g_1(t)g_2(t)}\,dt\left\{\int u^2K(u)\,du\right\}^2 b_N^4\right\}$$
$$\xrightarrow{\mathcal{D}} N(0, \sigma_K^2),$$

where $\bar{g}_N$ is defined by:

$$\bar{g}_N(t) = \alpha_N g_1(t) + \beta_N g_2(t)$$

and $N(0,\sigma_K^2)$ denotes a normal distribution with mean zero and variance σ_K^2 defined as in Theorem 9.5.

Remark If $b_N \asymp N^{-1/5}$ the situation becomes even more complicated. If the observation densities g_1 and g_2 are the same, one still gets the asymptotic normality result, as shown in the following theorem. But if the densities g_1 and g_2 are different, extra nonnegligible random terms enter because of the presence of the bias term. We will not discuss this further here.

Theorem 9.7 *Let the test statistic V_N be defined by (9.48), using a bandwidth b_N such that $b_N \asymp N^{-\alpha}$, where $1/5 \le \alpha < 1/3$. Furthermore, let F stay away from 0 and 1 on $[a,b]$ and have a bounded continuous second derivative f' on an interval (a',b') containing $[a,b]$, and let $g_1 = g_2$ be a continuous density that stays away from 0 on $[a,b]$, with a continuous bounded second derivative on the interval (a',b'). Then we have in probability, if the distribution functions of the hidden variables in the two samples are both equal to F and $m/N \to \alpha \in (0,1)$, as $N \to \infty$,*

$$N\sqrt{\frac{b_N}{b-a}}\left\{\left(V_N|T_1,\ldots,T_N\right) - \frac{b-a}{Nb_N}\int K(u)^2\,du\right\} \xrightarrow{\mathcal{D}} N(0,\sigma_K^2),$$

where $N(0,\sigma_K^2)$ denotes a normal distribution with mean zero and variance σ_K^2 defined as in Theorem 9.5.

Remark We used a conditional formulation, since conditional tests are used in the bootstrap approach, but the convergence in distribution will also hold in Theorems 9.5 to 9.7, if we do not condition on $T_1,\ldots,T_N$.

(B) A Bootstrap Method for Determining the Critical Value

Having defined a test statistic and knowing its asymptotic behavior under the null hypothesis, we still need to take one extra step to make the method work in practice. We will now describe a bootstrap approach to determine critical values in practice.

First compute the MSLE $\tilde F_{N,\tilde b_N}$ for the combined sample, using a bandwidth $\tilde b_N \asymp N^{-1/5}$. Then, using the observations $T_1,\ldots,T_m$ and $T_{m+1},\ldots,T_N$ of the two samples, generate corresponding bootstrap values $\Delta_1^*,\ldots,\Delta_m^*$ and $\Delta_{m+1}^*,\ldots,\Delta_N^*$ by letting the Δ_i^* be independent Bernoulli $(\tilde F_{N,\tilde b_N}(T_i))$ random variables. So in practice we generate independent Uniform(0, 1) variables U_i^*, and let Δ_i^* be equal to 1 if $U_i^* < \tilde F_{N,\tilde b_N}(T_i)$ and 0 otherwise. If the observation distributions, generating $T_1,\ldots,T_m$ and $T_{m+1},\ldots,T_N$, respectively, are different, this structure is preserved in this procedure; in the computation of the MSLEs $\tilde F_{Nj}^*$ in the bootstrap samples, the estimates $\tilde g_{Nj}$ of g_j in the original samples are used, for $j = 1,2$. Repeating this procedure B times yields B bootstrap values $V_{N,i}^*$, $1 \le i \le B$, of the test statistic. The distribution of V_N under the null hypothesis is now approximated by the empirical distribution of these bootstrap values and the critical value at (for example) level 5% by the 95th percentile of this set of bootstrap values $V_{N,i}^*$.

In justifying this method for the test statistic V_N, the following theorem in Groeneboom, 2012b, can be used.

Theorem 9.8 *Let, under either of the conditions of Theorems 9.5 to 9.7, $\tilde{F}_{N,\tilde{b}_N}$ be the MSLE of F under the null hypothesis, defined by the slope of the cusum diagram (9.49), where the bandwidth $\tilde{b}_N$ satisfies $\tilde{b}_N \asymp N^{-1/5}$. Let V_N^* be defined by*

$$V_N^* = \frac{2m}{N}\int_{t\in[a,b]}\left\{\tilde{h}_{N1}^*(t)\log\frac{\tilde{F}_{N1}^*(t)}{\tilde{F}_N^*(t)} + \{\tilde{g}_{N1}(t) - \tilde{h}_{N1}^*(t)\}\log\frac{1-\tilde{F}_{N1}^*(t)}{1-\tilde{F}_N^*(t)}\right\}dt$$

$$+\frac{2n}{N}\int_{t\in[a,b]}\left\{\tilde{h}_{N2}^*(t)\log\frac{\tilde{F}_{N2}^*(t)}{\tilde{F}_N^*(t)} + \{\tilde{g}_{N2}(t) - \tilde{h}_{N2}^*(t)\}\log\frac{1-\tilde{F}_{N2}^*(t)}{1-\tilde{F}_N^*(t)}\right\}dt \quad (9.56)$$

where $\tilde{F}_N^*$, $\tilde{F}_{N1}^*$ and $\tilde{F}_{N2}^*$ are the MSLEs, computed for the samples $(T_1, \Delta_1^*), \ldots, (T_m, \Delta_m^*)$ and $(T_{m+1}, \Delta_{m+1}^*), \ldots, (T_N, \Delta_N^*)$, and where the Δ_i^* are Bernoulli $(\tilde{F}_{N,b_N}(T_i))$ random variables, generated in the way described before the statement of this theorem; $\tilde{g}_{Ni}$ and $\tilde{h}_{Ni}^*$ are kernel estimates of the relevant observation densities, just as in (9.52), where

$$\tilde{h}_{N1}^*(t) = m^{-1}\sum_{i=1}^m \Delta_i^* k_{b_N}(t - T_i), \qquad \tilde{h}_{N2}^*(t) = n^{-1}\sum_{i=m+1}^N \Delta_i^* k_{b_N}(t - T_i)$$

with the same bandwidth b_N as taken in the original samples, and where the densities $\tilde{g}_{N1}$ and $\tilde{g}_{N2}$ are the same as in the original samples.

Then, under Hyp$_0$ *that the conditional distribution function of V_N^*, given $(T_1, \Delta_1), \ldots, (T_N, \Delta_N)$, rescaled in the same way as in Theorems 9.5 to 9.7 (depending on the choice of bandwidth b_N and presence or absence of the condition $g_1 = g_2$), converges at each $x \in \mathbb{R}$ in probability to $\Phi(x/\sigma_K)$, where Φ is the standard normal distribution function.*

If the null hypothesis does not hold, the same scheme is followed. The critical value is again determined by first computing the Δ_i^*, using the MSLE $\tilde{F}_{N,\tilde{b}_N}$, based on the combined sample.

The method of Theorem 9.8 is illustrated in Figure 9.13, where the empirical distribution function of 1,000 samples of the standardized test statistic V_N, for sample sizes $m = n = 250(= N/2)$, is compared with the empirical distribution function of 1,000 bootstrap samples of the standardized test statistic V_N^*, drawn using the method of Theorem 9.8 from one original sample. Note that the bootstrap simulation reproduces the original distribution quite well if the samples are drawn from the same distribution, but that one needs much larger sample sizes for the normal approximation to become accurate; at these sample sizes the distribution is still rather skewed. This is also noted in Groeneboom, 2012b, where the bootstrap method of computing critical values is advised for precisely this reason.

(C) Other Nonparametric Tests

Most tests that have been proposed for this problem are based on a comparison of simple functionals of the Δ_i. Under the assumption that the observation times T_i have the same

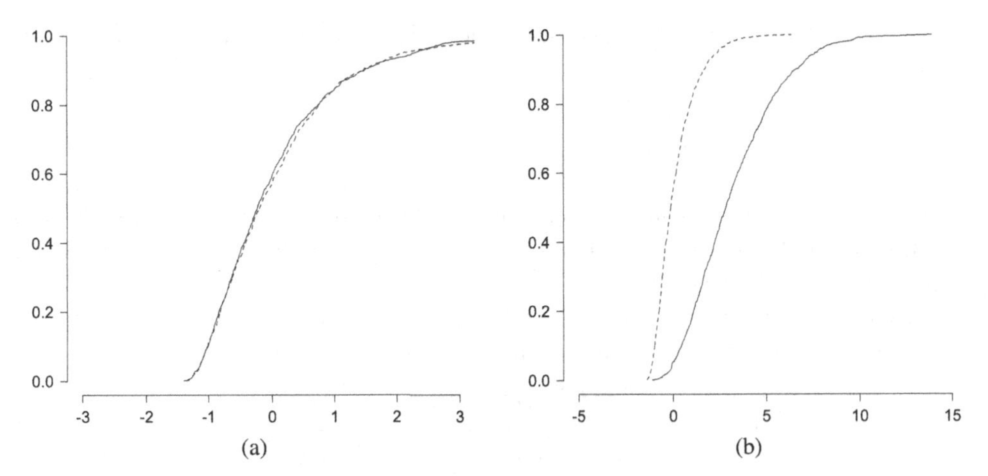

Figure 9.13 (a) The empirical distribution function (solid) of V_N, standardized as in Theorem 9.7, for 1,000 samples of sample sizes $m = n = 250$, where the samples are drawn from a Weibull distribution with parameters $\alpha = 0.5$ and $\lambda = 1.6$, and the empirical distribution function of 1,000 bootstrap samples of V_N^* (dashed), again standardized as in Theorem 9.7, where the bootstrap samples are drawn using the method of Theorem 9.8 from one original sample. (b) The same, but now the first sample is drawn from a Weibull distribution with parameters $\alpha = 0.5$ and $\lambda = 1.6$, and the second sample from a Weibull distribution with parameters $\alpha = 1$ and $\lambda = 1.6$. The dashed curve for the bootstrap samples of V_N^* are based on drawing from the MSLE of the distribution, based on the combined sample.

distribution in the two samples, the following test statistic is proposed in Sun, 2006:

$$\beta_N \sum_{i=1}^{m} \Delta_i - \alpha_N \sum_{i=m+1}^{N} \Delta_i . \tag{9.57}$$

The variance of $N^{-1/2}$ times (9.57) is given by:

$$\alpha_N \beta_N \int F(t)\, dG(t) \left\{1 - \int F(t)\, dG(t)\right\}, \tag{9.58}$$

if $G_1 = G_2 = G$; see Exercise 9.12. Now consider as (rescaled) test statistic

$$U_N = \frac{\tilde{U}_N}{\hat{\sigma}_N} \tag{9.59}$$

where $\tilde{U}_n$ is defined by

$$\tilde{U}_N = N^{-1/2} \left\{ \beta_N \sum_{i=1}^{m} \Delta_i - \alpha_N \sum_{i=m+1}^{N} \Delta_i , \right\}$$

and where

$$\hat{\sigma}_N^2 = \alpha_N \beta_N \int \hat{F}_N(t)\, d\mathbb{G}_N(t) \left\{1 - \int \hat{F}_N(t)\, d\mathbb{G}_N(t)\right\}. \tag{9.60}$$

Note that if $G_1 = G_2$, $\tilde{U}_N$ has expectation zero and variance (9.58) under the null hypothesis. Moreover, for $N \to \infty$,

$$\frac{1}{N}\sum_{i=1}^{N} \Delta_i \xrightarrow{p} \int F(t)\, dG(t), \tag{9.61}$$

where F is the limit (mixture) distribution of the combined samples (which is the underlying distribution under Hyp_0). Hence U_N tends to a standard normal distribution under the null hypothesis, if $G_1 = G_2 = G$.

Andersen and Rønn, 1995, consider a test based on an L_2-type distance between the two ML estimators based on the two samples at hand. In particular, they define

$$W_N = \frac{\sqrt{N}\int_0^a \{\hat{F}_{N1}(t)^2 - \hat{F}_{N2}(t)^2\}\, d\mathbb{G}_N(t)}{\sqrt{\frac{4}{\alpha_N \beta_N}\int_0^a \hat{F}_N(t)^3\{1 - \hat{F}_N(t)\}\, d\mathbb{G}_N(t)}}. \tag{9.62}$$

Under the conditions of Theorem 9.5, and assuming $G_1 = G_2$, they show that under Hyp_0,

$$W_N \xrightarrow{\mathcal{D}} N(0, 1), \tag{9.63}$$

where $N(0, 1)$ is the standard normal distribution.

(D) A Simulation Study

We now compare the LR test based on the MSLEs, V_N, with the methods based on U_N and W_N defined in (9.59) and (9.62) and with the original likelihood ratio test, based on $\tilde{V}_N$, defined by (9.47). For V_N, the critical values are determined by (Bernoulli) bootstrapping the Δ_i, using the MSLE $\tilde{F}_{N,\tilde{b}_N}(T_i)$ for the combined samples at the observations T_i, by taking 1,000 bootstrap samples and determining the 95th percentile of the bootstrap test statistics so obtained. As bandwidth for the SMLE $b_N = 2N^{-1/5}$ was used, except for the bootstrap resampling for the MLE, where the bandwidth $b_N = N^{-1/5}$ was used (in this case taking $b_N = 2N^{-1/5}$ made the test based on the MLEs very conservative). Moreover, the kernel (9.53) was used in computing $\tilde{F}_N$, as described before. For U_N and W_N we just took 1.96 as our critical value for the absolute value of the test statistic, since the convergence to the standard normal distribution is reasonably fast for these test statistics under the null hypothesis. In this way one can rather quickly compute tables with estimated rejection probabilities for these test statistics. The results are shown in Tables 9.6 to 9.13.

As mentioned earlier, the critical values for the original likelihood ratio test, based on $\tilde{V}_N$, were also determined by (Bernoulli) resampling from the MSLE under the null hypothesis $\tilde{F}_N$, based on the combined samples. This method still has no theoretical justification, but it is interesting to try it out. We also used this method in the example on the data in Hoel and Walburg, 1972, using the SMLE instead of the MSLE. An interesting question is whether the bootstrapping from the MLE $\hat{F}_N$ works for these statistics, in view of negative results in this direction on bootstrapping of the pointwise behavior from the Grenander estimator. These negative findings seem to be countered somewhat by positive findings on bootstrapping from the MLE for global statistics, as for example reported in Section 9.2, although under the null hypothesis a slightly anticonservative behavior was observed here. This was also tried out in Groeneboom, 2012b, where again anticonservative behavior under the null hypothesis was

Table 9.6 *Estimated Levels*

$g_1 = g_2$	λ, α_i			Under H_0		
$m = n = 50$	1.6, 0.5	1.6, 1.0	1.6, 2.0	0.58, 0.5	0.58, 1.0	0.58, 2.0
V_N	0.041	0.058	0.045	0.049	0.049	0.059
$\tilde{V}_N$	0.045	0.051	0.041	0.052	0.046	0.055
U_N	0.050	0.060	0.047	0.054	0.058	0.052
W_N	0.055	0.066	0.087	0.061	0.061	0.072

Note: The estimation interval is [0.1, 1.9], and $m = n = 50$; $g_1(t) \equiv \frac{1}{2}$, $g_2(t) \equiv \frac{1}{2}$, $\alpha_1 = \alpha_2$. The intended level is $\alpha = 0.05$.

Table 9.7 *Estimated Levels*

$g_2(t) = \frac{1}{4}(2-t)^3$	λ, α_i			Under H_0		
$m = n = 50$	1.6, 0.5	1.6, 1.0	1.6, 2.0	0.58, 0.5	0.58, 1.0	0.58, 2.0
V_N	0.049	0.051	0.045	0.049	0.049	0.059
$\tilde{V}_N$	0.051	0.055	0.049	0.044	0.050	0.056
U_N	0.422	0.745	0.950	0.262	0.540	0.885
W_N	0.122	0.108	0.130	0.326	0.302	0.276

Note: The estimation interval is [0.1, 1.9], and $m = n = 50$; $g_1(t) \equiv \frac{1}{2}$, $g_2(t) = \frac{1}{4}(2-t)^3$, $\alpha_1 = \alpha_2$. The intended level is $\alpha = 0.05$.

Table 9.8 *Estimated Levels*

$g_1 = g_2$,	λ, α_i			Under H_0		
$m = n = 250$	1.6, 0.5	1.6, 1.0	1.6, 2.0	0.58, 0.5	0.58, 1.0	0.58, 2.0
V_N	0.051	0.049	0.052	0.053	0.032	0.040
$\tilde{V}_N$	0.048	0.049	0.059	0.053	0.045	0.054
U_N	0.050	0.060	0.047	0.054	0.058	0.052
W_N	0.055	0.066	0.087	0.061	0.061	0.072

Note: The estimation interval is [0.1, 1.9], and $m = n = 250$; $g_1(t) \equiv \frac{1}{2}$, $g_2(t) \equiv \frac{1}{2}$, $\alpha_1 = \alpha_2$. The intended level is $\alpha = 0.05$.

Table 9.9 *Estimated Levels*

$g_2(t) = \frac{1}{4}(2-t)^3$	λ, α_i			Under H_0		
$m = n = 250$	1.6, 0.5	1.6, 1.0	1.6, 2.0	0.58, 0.5	0.58, 1.0	0.58, 2.0
V_N	0.044	0.050	0.051	0.049	0.044	0.051
$\tilde{V}_N$	0.045	0.051	0.041	0.052	0.054	0.058
U_N	0.970	1.000	1.000	0.840	0.996	1.000
W_N	0.181	0.135	0.102	0.513	0.491	0.410

Note: The estimation interval is [0.1, 1.9], and $m = n = 250$. The intended level is $\alpha = 0.05$; $g_1(t) = \frac{1}{2}$, $g_2(t) = \frac{1}{4}(2-t)^3$, $\alpha_1 = \alpha_2$.

Table 9.10 *Powers for Different Shapes, If* $m = n = 50$

$g_1 = g_2$	$\lambda, \alpha_1, \alpha_2$		Different Shapes	
$m = n = 50$	1.6, 0.5, 1.0	1.6, 0.5, 2.0	0.58, 0.5, 2.0	0.58, 1.0, 2.0
V_N	0.174	0.675	0.470	0.207
$\tilde{V}_N$	0.125	0.533	0.364	0.173
U_N	0.061	0.069	0.045	0.053
W_N	0.062	0.110	0.179	0.146

Note: The estimation interval is [0.1, 1.9].

Table 9.11 *Powers for Different Shapes, If* $m = n = 250$

$g_1 = g_2$	$\lambda, \alpha_1, \alpha_2$		Different Shapes	
$m = n = 250$	1.6, 0.5, 1.0	1.6, 0.5, 2.0	0.58, 0.5, 2.0	0.58, 1.0, 2.0
V_N	0.606	1.000	0.990	0.787
$\tilde{V}_N$	0.440	1.000	0.974	0.610
U_N	0.076	0.132	0.062	0.076
W_N	0.088	0.112	0.583	0.406

Note: The estimation interval is [0.1, 1.9].

Table 9.12 *Powers for Different Baseline Hazards, Same Shape, If* $m = n = 50$

$g_1 = g_2$	$\lambda, \alpha_i, \theta$			Different Baseline Hazards		
$m = n = 50$	1.6, 0.5, 1.25	1.6, 0.5, 1.5	1.6, 0.5, 2	0.58, 2, 1.25	0.58, 2, 1.5	0.58, 2, 2
V_N	0.138	0.283	0.632	0.091	0.208	0.480
$\tilde{V}_N$	0.097	0.218	0.498	0.082	0.171	0.342
U_N	0.108	0.198	0.441	0.100	0.151	0.333
W_N	0.147	0.352	1.000	0.103	0.293	0.681

Note: The estimation interval is [0.1, 1.9]. The parameters α_i are either both 0.5 or both 2 and $\lambda = 1.6$ or 0.58; $\theta = 1.25$, 1.5 or 2.

Table 9.13 *Powers for Different Baseline Hazards, Same Shape, If* $m = n = 250$

$g_1 = g_2$	$\lambda, \alpha_i, \theta$			Different Baseline Hazards		
$m = n = 250$	1.6, 0.5, 1.25	1.6, 0.5, 1.5	1.6, 0.5, 2	0.58, 2, 1.25	0.58, 2, 1.5	0.58, 2, 2
V_N	0.377	0.873	1.000	0.227	0.689	0.995
$\tilde{V}_N$	0.246	0.728	0.996	0.171	0.505	0.964
U_N	0.324	0.721	0.971	0.200	0.495	0.921
W_N	0.473	0.912	1.000	0.337	0.835	1.000

Note: The estimation interval is [0.1, 1.9]. The parameters α_i are either both 0.5 or both 2 and $\lambda = 1.6$ or 0.58; $\theta = 1.25$, 1.5 or 2.

observed. For this reason bootstrapping from the MSLE was preferred, also for bootstrapping the test, based on the MLEs. We report these findings too in the second line of the tables, which are taken from Groeneboom, 2012b. The question on whether bootstrapping from the MLE itself works in this case is still not answered.

In the first simulations, the observation densities g_1 and g_2 for the observation times T_i are uniform on $[0, 2]$. Because $g_1 = g_2$, Theorem 9.7 applies. This allows us to resample from the MSLE $\tilde{F}_N$, which was also used in the computation of the test statistic for the original samples. The samples of X-values were taken from

$$\alpha_1 \lambda x^{\alpha_1 - 1} e^{-\lambda x^{\alpha_1}}, \ x > 0, \tag{9.64}$$

and the second sample from the density

$$\alpha_2 \lambda \theta x^{\alpha_2 - 1} e^{-\lambda \theta x^{\alpha_2}}, \ x > 0, \tag{9.65}$$

where $\lambda = 1.6$ or $\lambda = 0.58$, and $\alpha_i = 0.5, 1.0$ or 2.0. The value of θ is 1, 1.25 or 2.

The powers and levels computed in the following for the test statistics V_N (MSLEs) are determined by taking 1,000 samples from the original distributions and taking 1,000 bootstrap samples from each sample, rejecting the null hypothesis if the value in the original sample was larger than the 950th order statistic of the values obtained in the bootstrap samples. The values given in the following tables represent the fraction of rejections for the 1,000 samples from the original distributions. The tabled values are also based on 1,000 samples from the original (Weibull) distributions.

To illustrate the effect of different observation distributions in the two samples, we generated the first sample of T_is again from the uniform density on $[0, 2]$, but the second sample from the decreasing density

$$g_2(t) = \tfrac{1}{4}(2 - t)^3, \quad t \in [0, 2];$$

see Tables 9.7 and 9.9. Note that in this case Theorem 9.7 does not apply, and we have to use Theorem 9.5 or 9.6. Nevertheless, the critical values using the bootstrap procedure were rather insensitive to the difference of the observation distributions G_1 and G_2.

The results of the experiments can be summarized in the following way. The test based on U_N has almost no power for different shape alternatives of the type shown as solid curves in Figure 9.12, even for sample sizes $m = n = 250$. The test based on W_N has somewhat more power here, but is clearly also not very good for this type of alternative, as already discussed in Andersen and Rønn, 1995 (they call this the "crossing alternatives," since the distribution functions indeed cross). The test, based on the MSLEs, has much more power here. The test, based on W_N, is surprisingly powerful for the alternatives that have the same shape but different baseline hazards, and the test, based on U_N, also has more power here. The other test, based on the MSLEs or MLEs, also has a reasonable power here. Finally, Tables 9.7 and 9.9 show that the observation distributions in the two samples can be different if we use the MSLE-type tests, in contrast with the other tests, considered here. In fact, it has a disastrous effect for the tests U_N and W_N; U_N even gives 100% rejection under the null hypothesis for several combinations of the parameters.

As a general rule one can say that the tests, based on U_N or W_N, can only have power if the corresponding moment functionals are different from zero. For U_N this functional is given by

$$\int_a^b \{F_1(t) - F_2(t)\}\, dG(t), \tag{9.66}$$

and for W_N it is given by

$$\int_a^b \{F_1(t)^2 - F_2(t)^2\}\, dG(t). \tag{9.67}$$

It is clear that F_1 and F_2 can be very different and still satisfy

$$\int_a^b \{F_1(t) - F_2(t)\}\, dG(t) = 0, \quad \text{or} \quad \int_a^b \{F_1(t)^2 - F_2(t)^2\}\, dG(t) = 0,$$

and in that case that test, based on U_N or W_N, respectively, will have no power; see Exercise 9.13. The MSLE test will not suffer from this drawback, since it involves a Kullback-Leibler type distance, and is locally (for example, if one would consider contiguous alternatives) equivalent to the squared L_2 distance

$$\int_a^b \frac{\{F_1(t) - F(t)\}^2}{F(t)\{1 - F(t)\}}\, dG_1(t) + \int_a^b \frac{\{F_2(t) - F(t)\}^2}{F(t)\{1 - F(t)\}}\, dG_2(t), \tag{9.68}$$

where F is the distribution function of the combined sample. Moreover, it allows the observation distributions to be different in the two samples, something the other test also does not allow.

The Weibull alternatives, considered in the simulation study, form a family for which the integrals, corresponding to the statistics U_N and W_N, are different under the alternatives considered there. So for these types of alternatives the tests U_N and W_N can be expected to have a power exceeding the level of the test. But if the first sample is generated from a Weibull distribution function F_1 with parameters $\alpha = 0.5$ and $\lambda = 0.7$ and the second sample is generated from a Weibull distribution function F_2 with parameters $\alpha = 1.8153$ and $\lambda = 0.7$, the distribution functions are very different (see Figure 9.14a), although we get:

$$\int_a^b \{F_1(t) - F_2(t)\}\, dt \approx -1.87 \cdot 10^{-6}, \quad a = 0.1, \quad b = 1.9. \tag{9.69}$$

Taking again the observations G_1 and G_2 to be uniform on $[0, 2]$, we get that the test based on the MSLE has power 0.993 for this alternative, whereas the tests based on U_N have power 0.048 (which is lower than the level 0.05), taking sample sizes $m = n = 250$.

If the first sample is generated from a Weibull distribution function F_1 with parameters $\alpha = 0.2$ and $\lambda = 0.8$ and the second sample is generated from a Weibull distribution function F_2 with parameters $\alpha = 0.767$ and $\lambda = 0.8$, the distribution functions are again rather different (see the right panel of Figure 9.14), although we get:

$$\int_a^b \left\{F_1(t)^2 - F_2(t)^2\right\} dt \approx 2.6 \cdot 10^{-6}, \quad a = 0.1, \quad b = 1.9.$$

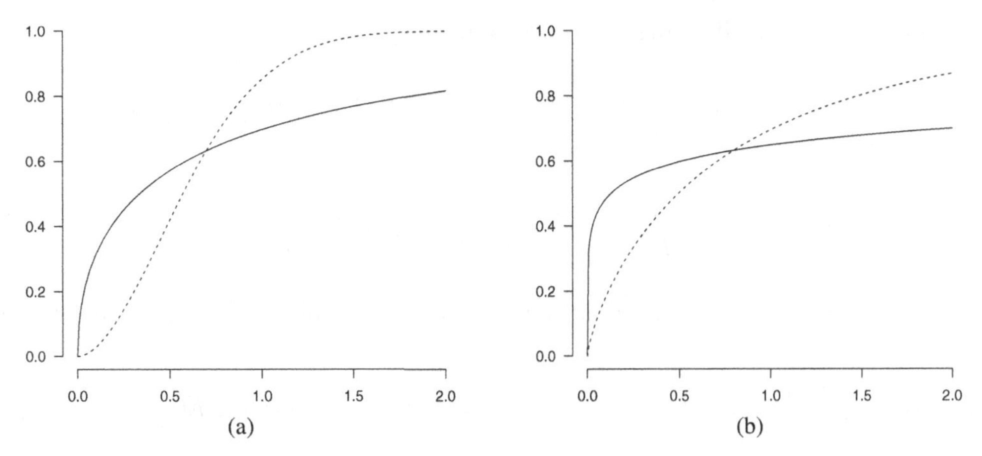

Figure 9.14 (a) The Weibull distribution function with parameters $\alpha = 0.5$ and $\lambda = 0.7$ (solid curve) and the Weibull distribution function with parameters $\alpha = 1.8153$ and $\lambda = 0.7$ (dashed). (b) Similarly, but with $\alpha = 0.2$ and $\lambda = 0.8$ (solid) and $\alpha = 0.767$ and $\lambda = 0.8$ (dashed).

Taking $g_1 = g_2 \equiv (1/2)1_{[0,2]}$ again, the SLR test has power 0.964 for this alternative, whereas the tests based on W_N have power 0.041 (which is again lower than the level 0.05), taking sample sizes $m = n = 250$ again.

9.4 Two-Sample Tests for Case 2 Interval Censored Data

We give a brief heuristic discussion of the extension of the test discussed in Section 9.3 to two-sample tests for interval censored data with two observation times per unobservable event time X_i. Research on this topic is still in progress.

Proceeding as in Section 9.3, we consider the statistic V_N, similar to (9.48), and defined by

$$
\begin{aligned}
V_N = \;& \frac{2m}{N} \int_{t\in[a,b]} \left\{ \tilde h_{N1,1}(t) \log \frac{\tilde F_{N1}(t)}{\tilde F_N(t)} + \tilde h_{N1,2}(t) \log \frac{1-\tilde F_{N1}(t)}{1-\tilde F_N(t)} \right\} dt \\
& + \frac{2m}{N} \int_{a<t<u<b} \tilde h_{N1}(t,u) \log \frac{\tilde F_{N1}(u) - \tilde F_{N1}(t)}{\tilde F_N(u) - \tilde F_N(t)}\, dt\, du \\
& + \frac{2n}{N} \int_{t\in[a,b]} \left\{ \tilde h_{N2,1}(t) \log \frac{\tilde F_{N2}(t)}{\tilde F_N(t)} + \tilde h_{N2,2}(t) \log \frac{1-\tilde F_{N2}(T_i)}{1-\tilde F_N(T_i)} \right\} dt \\
& + \frac{2n}{N} \int_{a<t<u<b} \tilde h_{N2}(t,u) \log \frac{\tilde F_{N1}(u) - \tilde F_{N1}(t)}{\tilde F_N(u) - \tilde F_N(t)}\, dt\, du, \qquad (9.70)
\end{aligned}
$$

where $\tilde F_{N1}$, $\tilde F_{N2}$ and $\tilde F_N$ are the maximum smoothed likelihood estimators (MSLEs) for the first, second and combined sample, respectively, and $\tilde h_{Ni,j}$ and $\tilde h_{Ni}$ are kernel estimates of the relevant observation densities, defined as follows:

$$\tilde h_{Ni,1}(t) = \int \delta_1 k_{b_N}(t-x)\,d\mathbb{P}_{Ni}(x,y,\delta_1\delta_2), \quad \tilde h_{Ni,2}(t) = \int \delta_3 k_{b_N}(t-x)\,d\mathbb{P}_{Ni}(x,y,\delta_1,\delta_2)$$

and

$$\tilde h_{Ni}(t,u) = \int \delta_2 k_{b_N}(t-x)k_{b_N}(u-y)\,d\mathbb{P}_{Ni}(x,y,\delta_1,\delta_2),$$

for $i=1,2$, where $i=1$ $(i=2)$ for the first (second) sample. In a similar way we define the estimates

$$\tilde g_{Ni,1}(t) = \int k_{b_N}(t-x)\,d\mathbb{P}_{Ni}(x,y,\delta_1,\delta_2), \quad \tilde g_{Ni,2}(t) = \int k_{b_N}(t-x)\,d\mathbb{P}_{Ni}(x,y,\delta_1,\delta_2)$$

and

$$\tilde g_{Ni}(t,u) = \int k_{b_N}(t-x)k_{b_N}(u-y)\,d\mathbb{P}_{Ni}(x,y,\delta_1,\delta_2),$$

of the observation densities g_{i1} of the T_k, g_{i2} of the U_k and g_i of the (T_k,U_k), respectively, in the ith sample. Note that $\tilde h_{Ni,1}$, $\tilde h_{Ni,2}$ and $\tilde h_{Ni}$ are estimates of the densities

$$h_{i,1}(t) = F_{0i}(t)g_{i1}(t), \qquad h_{i,2}(t) = \{1-F_{0i}(t)\}g_{i2}(t)$$

and

$$h_i(t,u) = \{F_{0i}(u) - F_{0i}(t)\}g_i(t,u),$$

where F_{0i} is the distribution function of the hidden variables in the ith sample.

It is discussed in Section 11.5 that, under the so-called separation hypothesis (which means that the observation intervals (T_i,U_i) do not become arbitrarily small), the MSLEs $\tilde F_{Ni}$, $i=1,2$, have the following asymptotic representation:

$$\begin{aligned}\tilde F_{Ni}(v) - F_{0i}(v) \sim\ & \frac{\{1-F_{0i}(v)\}\tilde h_{Ni,1}(v) - F_{0i}(v)\tilde h_{Ni,2}(v)}{\sigma_{1i}(v)\left(\{1-F_{0i}(v)\}g_1(v) + F_{0i}(v)g_2(v)\right)}\\ &+ \frac{d_{F_{0i}}(v)}{\sigma_{1i}(v)}\left\{\int_{t<v}\frac{\tilde h_{Ni}(t,v)}{F_{0i}(v)-F_{0i}(t)}\,dt - \int_{u>v}\frac{\tilde h_{Ni}(v,u)}{F_{0i}(u)-F_{0i}(v)}\,du\right\},\end{aligned} \tag{9.71}$$

where

$$d_{F_{0i}}(v) = \frac{F_{0i}(v)\{1-F_{0i}(v)\}}{\{1-F_{0i}(v)\}g_1(v) + F_{0i}(v)g_2(v)},$$

and σ_{1i} is given by:

$$\sigma_{1i}(v) = 1 + d_{F_{0i}}(v)\left\{\int_{u>v}\frac{g_i(v,u)}{F_{0i}(u)-F_{0i}(v)}\,du + \int_{t<v}\frac{g_i(t,v)}{F_{0i}(v)-F_{0i}(t)}\,dt\right\}. \tag{9.72}$$

We likewise have the representation

$$\tilde{F}_N(v) - F_{0,m,n}(v) \sim \frac{\{1 - F_{0,m,n}(v)\}\tilde{h}_{N,1}(v) - F_{0,m,n}(v)\tilde{h}_{N,2}(v)}{\sigma_{1,m,n}(v)(\{1 - F_{0,m,n}(v)\}g_1(v) + F_{0,m,n}(v)g_2(v))}$$
$$+ \frac{d_{F_{0,m,n}}(v)}{\sigma_{1,m,n}(v)} \left\{ \int_{t<v} \frac{\tilde{h}_N(t,v)}{F_{0,m,n}(v) - F_{0,m,n}(t)}\, dt \right.$$
$$\left. - \int_{u>v} \frac{\tilde{h}_N(v,u)}{F_{0,m,n}(u) - F_{0,m,n}(v)}\, du \right\}, \tag{9.73}$$

where

$$\sigma_{1,m,n}(v) = \frac{m}{N}\sigma_1 + \frac{n}{N}\sigma_2, \qquad F_{0,m,n} = \frac{m}{N}F_{01} + \frac{n}{N}F_{02}.$$

Under the null hypothesis of equality of the distribution functions of the hidden variables in the two samples we have:

$$F_{01} = F_{02} = F_{0,m,n} = F_0.$$

Using the representations (9.71) and (9.73), one can derive the asymptotic distribution of V_N under the null hypothesis of equality of the distributions of the hidden variables in the two samples in the separated case for this model.

In the nonseparated case the MLE itself is conjectured to have the following asymptotic behavior:

$$(n \log n)^{1/3} \left\{ \hat{F}_n(t_0) - F_0(t_0) \right\} / \left\{ \tfrac{3}{4} f_0(t_0)^2 / g(t_0, t_0) \right\}^{1/3} \xrightarrow{\mathcal{D}} 2Z,$$

where Z is the location of the maximum of two-sided Brownian motion minus t^2, see (4.31) in Section 4.7. Using (11.65), the approximation for $n\,\mathrm{var}(\tilde{F}_n(t_0))$, where $\tilde{F}_n$ is the SMLE, is given by:

$$\sigma_n^2 \stackrel{\text{def}}{=} E\,\theta_{t_0,h,F_0}(T_1, U_1, \Delta_{11}, \Delta_{12})^2$$
$$\sim h^{-1} d_{F_0}(t_0) \left\{ 1 + 2 d_{F_0}(t) g(t_0, t_0) / f_0(t_0) \right\}^{-1} \int K(u)^2\, du, \qquad h \downarrow 0,$$

and we expect the MSLE to have the same local asymptotic variance in the nonseparated case.

Since this matter is still under investigation at the present time, we only discuss here how the bootstrap procedure, used in Section 9.3 for the two-sample test for current status data, works in the present situation. We first have to face the problem of computing the MSLEs in (9.70). We use a simple EM algorithm for computing the MSLEs. This means that we use the iterations

$$f_i^{(k+1)}(v) = \left\{ \int_{t>v} \frac{\tilde{h}_{Ni,1}(t)}{F_i^{(k)}(t)}\, dt + \int_{t<v} \frac{\tilde{h}_{Ni,2}(t)}{1 - F_i^{(k)}(t)}\, dt + \int_{t<v<u} \frac{\tilde{h}_{N,i}(t,u)}{F_i^{(k)}(u) - F_i^{(k)}(t)}\, dt\, du \right\} f_i^{(k)}(v)$$

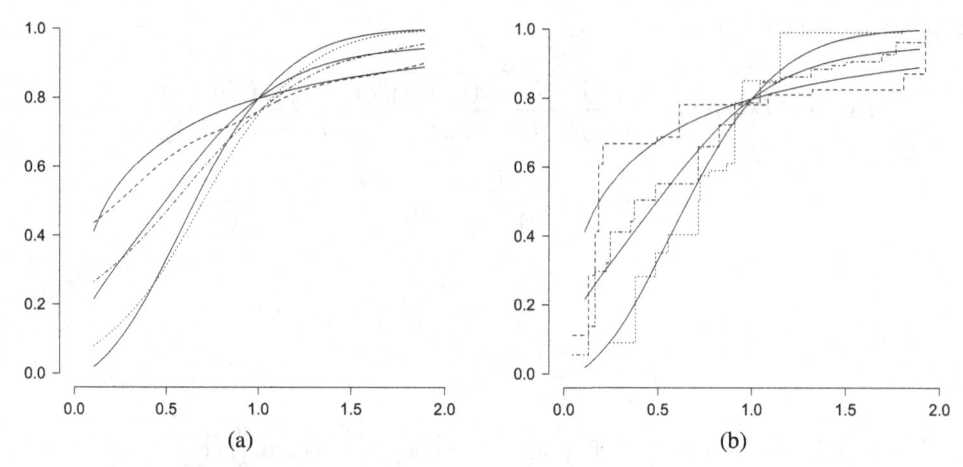

Figure 9.15 MSLEs and MLEs on $[a, b]$ for samples of size $m = n = 250$ from the Weibull densities (9.54) for interval censored data (see Figure 9.12 for the corresponding figure obtained with current status data). G_1 and G_2 are uniform on $[0, 2]$, and the interval $[a, b] = [0.1, 1.9]$. (a) The MSLE estimates; (b) the MLEs. The dashed curves give the estimates for the first sample ($\alpha_1 = 0.5$), the dotted curves the estimates for the second sample ($\alpha_2 = 2$), and the dashed-dotted curves the estimates for the combined samples. The solid curves give the corresponding actual distribution functions for these three situations. The bandwidth for the computation of the MSLEs was $b_N = 2N^{-1/5} \approx 0.57708$, where $N = m + n = 500$, and $\lambda = 1.6$.

on a grid of points v. The integrals are simply approximated by Riemann sums on this grid, and we put

$$F_i^{(k)}(v_j) = \sum_{\ell \le j} f_i^{(k)}(v_\ell)\,(v_\ell - v_{\ell-1}),$$

where the v_ℓ are points of the grid. A picture, comparable to Figure 9.12 for the current status model, is shown in Figure 9.15. Here we used as observation distribution the uniform distribution on the upper triangle of the unit square and an equidistant grid with steps of size 0.01 in both coordinates for the EM algorithm.

The critical values for a test of equality of the distributions F_{0i} in the two samples is estimated by resampling the indicators Δ_{i1} and Δ_{i2}, where the binomial distribution we used for the current status data is replaced by a trinomial distribution. Here we work again conditionally, keeping the observation times T_i and U_i fixed in the two samples, and for the trinomial distribution a smooth estimate of the distribution of the two samples under the null hypothesis is used, obtained by computing either the MSLE or the SMLE for the combined samples. It is to be expected that a procedure of this type will have higher power than the similar test for current status data.

9.5 Pointwise Confidence Intervals for the Current Status Model

In Banerjee and Wellner, 2005, confidence intervals for values of the distribution function and quantile function at a fixed point t are discussed for the current status model. There several methods are considered for constructing these intervals. These are distinguished in LR test based intervals, MLE based intervals and subsampling intervals. Also parametric fitting is discussed, but a preference for the LR based intervals is expressed.

The MLE based intervals for the distribution function are constructed in Banerjee and Wellner, 2005, in the following way, where we take the 95% intervals as an example. Starting with the limit distribution of the MLE $\hat{F}_n(t_0)$, based on a sample $(T_1, \Delta_1), \dots, (T_n, \Delta_n)$ (see Theorem 3.7), the 95% asymptotic confidence intervals are of the following form:

$$\left[\hat{F}_n(t_0) - 0.99818n^{-1/3}\hat{C}_n,\, \hat{F}_n(t_0) + 0.99818n^{-1/3}\hat{C}_n\right]. \tag{9.74}$$

Here 0.99818 is the 97.5th percentile of the Chernoff distribution (see Section 3.9, in particular Table 3.1 for the quantiles). Moreover, $\hat{C}_n$ is defined by

$$\hat{C}_n = \left(\frac{4\hat{f}_n(t_0)\hat{F}_n(t_0)\{1 - \hat{F}_n(t_0)\}}{\hat{g}_n(t_0)}\right)^{1/3},$$

where $\hat{f}_n$ and $\hat{g}_n$ are estimates of f_0 and the observation density g, respectively. Since $\hat{F}_n$ is a piecewise constant function, $\hat{f}_n$ cannot directly be deduced from $\hat{F}_n$, and other methods have to be used to construct this estimator. The estimate $\hat{g}_n$ could just be an ordinary kernel estimator and the natural estimate of f_0 is the SMLE introduced in Section 8.1

$$\int K_h(t - x)\, d\hat{F}_n(x),$$

where $\hat{F}_n$ is the MLE. These estimators are indeed used in Banerjee and Wellner, 2005, and they use likelihood based cross-validation to choose the bandwidth in the estimates for g and f_0.

The LR test based intervals are based on inverting the acceptance region of the LR test for the null hypothesis $F_0(t_0) = \theta$ and use the fact that

$$\begin{aligned} 2n\Bigg\{\int &\left\{\delta \log \hat{F}_n(t) + (1-\delta)\log\left(1 - \hat{F}_n(t)\right)\right\}\, d\mathbb{G}_n(t) \\ &- \int \left\{\delta \log \hat{F}_n^{(0)}(t) + (1-\delta)\log\left(1 - \hat{F}_n^{(0)}(t)\right)\right\}\, d\mathbb{G}_n(t)\Bigg\} \xrightarrow{\mathcal{D}} \mathbb{D}, \end{aligned}$$

where $\hat{F}_n$ and $\hat{F}_n^{(0)}$ are the nonrestricted MLE and the MLE restricted by the null hypothesis $F_0(t_0) = \theta$, respectively, and $\mathbb{D}$ has a distribution characterized in Banerjee and Wellner, 2001 (see Theorem 2.5 in that paper). We note here that, in contrast with the distribution of $\mathbb{Z}$, no analytic information is available for the distribution of $\mathbb{D}$. However, the 95th percentile of the distribution of $\mathbb{D}$ was determined by simulation to be approximately 2.26916.

In our current approach, it is more natural to base the confidence intervals on the SMLE or the MSLE more directly. Moreover, this approach will also allow us to give confidence intervals for the density and hazard, which is not possible without smoothing of the MLE. We first show here how to compute SMLE based confidence intervals for the distribution function F_0 and confidence intervals for the density f_0, based on the SMLE and the density estimate

$$\tilde{f}_{nh}(t) = \int K_h(t-x)\, d\hat{F}_n(x),$$

where $\hat{F}_n$ is the MLE.

The essential idea in our construction of the confidence intervals is the asymptotic representation of the SMLE, given in (11.51) in Section 11.3. This representation allows us to use the bootstrap in constructing confidence intervals for the distribution function F_0. Similar representations for the density estimate allow us to construct the confidence interval for the density f_0 and the hazard λ_0 in the same way.

There are some important differences between the approaches, based on the MLE and SMLE, respectively. How appropriate it is to use the MLE will largely depend on whether one expects that the distribution function will have jumps. If one expects the distribution function to be smooth, the jumps of the MLE will have no real meaning, nor will the corresponding jumps in the confidence intervals have meaning.

Second in the construction of the confidence intervals, based on the MLE, the bias of the MLE plays no role. But if one constructs confidence intervals, using the SMLE with an optimal bandwidth, the bias will play a role in the limiting distribution. There is an extensive literature on how to deal with the bias in nonparametric function estimation; some approaches use undersmoothing, other approaches oversmoothing. A recent paper discussing this literature and giving a solution for confidence bands is Hall and Horowitz, 2013. We will use undersmoothing, as suggested in Hall, 1992.

(A) Construction of SMLE-Based Confidence Intervals for the Distribution Function

Let F_0 be defined on an interval $[a, b]$ with $a < b$ satisfying $F_0(a) = 0$ and $F_0(b) = 1$. Then we can estimate F_0 by the SMLE, using the boundary correction:

$$\tilde{F}_{nh}(t) = \int \left\{ \mathbb{K}\left(\frac{t-x}{h}\right) + \mathbb{K}\left(\frac{t+x-2a}{h}\right) - \mathbb{K}\left(\frac{2b-t-x}{h}\right) \right\} d\hat{F}_n(x), \tag{9.75}$$

where $\hat{F}_n$ is the MLE, $\mathbb{K}(x) = \int_{-\infty}^{x} K(u)\, du$ and K is the usual symmetric kernel with compact support, such as the triweight kernel. See (11.37) in Section 11.3 for the boundary correction method. If $t \in [a+h, b-h]$ the SMLE is just given by

$$\tilde{F}_{nh}(t) = \int \mathbb{K}\left(\frac{t-x}{h}\right) d\hat{F}_n(x);$$

the other two terms in (9.75) are just there for correction at the left and right boundary.

Let $(T_1, \Delta_1), \dots, (T_n, \Delta_n)$ be a current status sample, where the T_i are generated from a distribution with distribution function G and density g, and where the Bernoulli random variable Δ_i equals 1 with probability $F_0(T_i)$. For the construction of the $1-\alpha$ confidence interval we take a number of bootstrap samples $(T_1^*, \Delta_1^*), \dots, (T_n^*, \Delta_n^*)$ with replacement from $(T_1, \Delta_1), \dots, (T_n, \Delta_n)$. For each such sample we compute the SMLE $\tilde{F}_{nh}^*$, using the same bandwidth h as used for the SMLE $\tilde{F}_{nh}$ in the original sample, and the same type of boundary correction. Next we compute at the points t:

$$Z^*_{n,h}(t)$$
$$= \frac{\tilde F^*_{nh}(t) - \tilde F_{nh}(t)}{\sqrt{n^{-2}\sum_{i=1}^n \left\{K_h(t-T_i^*) - K_h(t+T_i^*-2a) - K_h(2b-t-T_i^*)\right\}^2 \left(\Delta_i - \hat F_n^*(T_i^*)\right)^2}}, \tag{9.76}$$

where $\hat F_n^*$ is the ordinary MLE (not the SMLE!) of the bootstrap sample $(T_1^*, \Delta_1^*), \dots, (T_n^*, \Delta_n^*)$.

Let $U^*_\alpha(t)$ be the αth percentile of the B values $Z^*_{n,h}(t)$. Then, disregarding the bias for the moment, the following bootstrap $1-\alpha$ interval is suggested:

$$\left[\tilde F_{nh}(t) - U^*_{1-\alpha/2}(t) S_{nh}(t), \tilde F_{nh}(t) - U^*_{\alpha/2}(t) S_{nh}(t)\right], \tag{9.77}$$

where

$$S_{nh}(t) = n^{-2}\sum_{i=1}^n \left\{K_h(t-T_i) - K_h(t+T_i-2a) - K_h(2b-t-T_i)\right\}^2 \left(\Delta_i - \hat F_n(T_i)\right)^2.$$

The bootstrap confidence interval is inspired by the fact that the SMLE is asymptotically equivalent to the toy estimator

$$F_{nh}^{toy}(t) = \int \left\{\mathbb{K}_h(t-u) + \mathbb{K}_h(t+u-2a) - \mathbb{K}_h(2b-t-u)\right\} dF_0(u)$$
$$+ \frac{1}{n}\sum_{i=1}^n \frac{\left\{K_h(t-T_i) - K_h(t+T_i-2a) - K_h(2b-t-T_i)\right\}\left\{\Delta_i - F_0(T_i)\right\}}{g(T_i)},$$

see (11.53), which has a sample variance

$$S_n(t)^2 = \frac{1}{n^2}\sum_{i=1}^n \frac{\left\{K_h(t-T_i) - K_h(t+T_i-2a) - K_h(2b-t-T_i)\right\}^2\left\{\Delta_i - F_0(T_i)\right\}^2}{g(T_i)^2},$$

and also by Theorem 11.4, which tells us that, if $h \sim cn^{-1/5}$, under the conditions of that theorem, for each $t \in (a, b)$,

$$n^{2/5}\left\{\tilde F_{nh}(t) - F_0(t)\right\} \xrightarrow{\mathcal{D}} N\left(\mu, \sigma^2\right), \quad n \to \infty,$$

where

$$\mu = \tfrac{1}{2}c^2 f_0'(t)\int u^2 K(u)\, du$$

and

$$\sigma^2 = \frac{F_0(t)\{1-F_0(t)\}}{cg(t)}\int K(u)^2\, du;$$

see Section 11.3.

We now first study the behavior of intervals of type (9.77) for a situation where the bias plays no role (the uniform distribution) and compare the behavior of the intervals with the confidence intervals in Banerjee and Wellner, 2005, based on LR tests for the MLE.

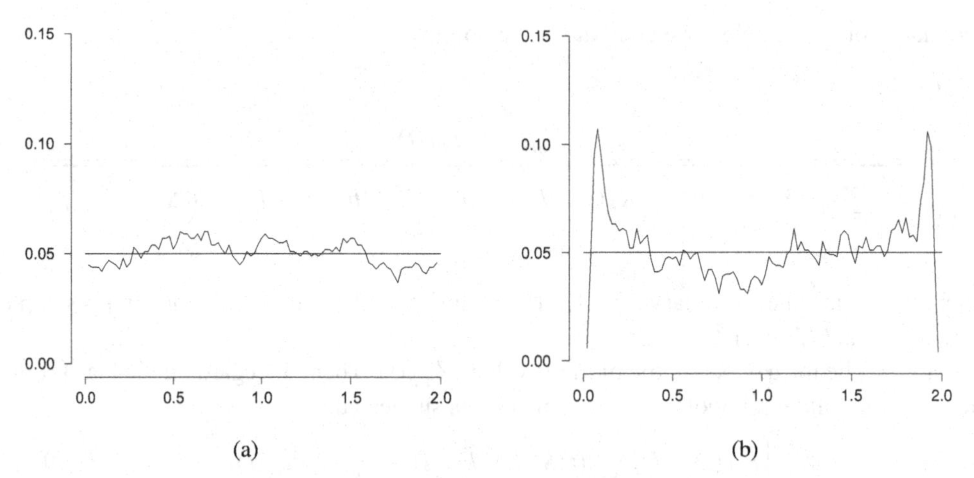

Figure 9.16 Proportion of times that $F_0(t_i)$, $t_i = 0.01, 0.02, \ldots$ is not in the 95% CIs in 1,000 samples $(T_1, \Delta_1) \ldots, (T_n, \Delta_n)$ using the SMLE and 1,000 bootstrap samples from the sample $(T_1, \Delta_1) \ldots, (T_n, \Delta_n)$. In (a), the SMLE is used with CIs given in (9.77). In (b) CIs are based on the LR test. The samples are uniformly distributed.

(B) Simulation for Uniform Distributions

We generated 1,000 samples $(T_1, \Delta_1), \ldots, (T_n, \Delta_n)$ by generating $T_1, \ldots, T_n$, $n = 1000$, from the uniform distribution on $[0, 2]$ and generated, independently, a sample $X_1, \ldots, X_n$, also from the uniform distribution on $[0, 2]$. If $X_i \le T_i$ we got a value $\Delta_i = 1$, otherwise $\Delta_i = 0$. For each such sample $(T_1, \Delta_1), \ldots, (T_n, \Delta_n)$ we generated 1,000 bootstrap samples, and computed the 25th and 975th percentile of the values (9.76) at the points $t_j = 0.02, 0.04, \ldots, 1.98$. On the basis of these percentiles we constructed the confidence intervals (9.77) for all of the (99) t_js and checked whether $F_0(t_j)$ belonged to it. The percentages of times that $F_0(t_j)$ did not belong to the interval are shown in Figure 9.16. We likewise computed the confidence interval, based on the LR test for the MLE proposed in Banerjee and Wellner, 2005, for each t_j, and also counted the percentages of times that $F_0(t_j)$ did not belong to the interval. The corresponding confidence intervals for one sample are shown in Figure 9.17.

(C) Simulation for Truncated Exponential Distributions

To investigate the role of the bias, we also generated 1,000 samples $(T_1, \Delta_1), \ldots, (T_n, \Delta_n)$ by generating $T_1, \ldots, T_n$, $n = 1000$, from the uniform distribution on $[0, 2]$ and, independently, samples $X_1, \ldots, X_n$, from the truncated exponential distribution on $[0, 2]$, with density

$$f_0(x) = \frac{e^{-x}}{1 - e^{-2}}, \; x \in [0, 2].$$

If $X_i \le T_i$ we get $\Delta_i = 1$, otherwise $\Delta_i = 0$. For each such sample $(T_1, \Delta_1), \ldots, (T_n, \Delta_n)$ we generated $B = 1000$ bootstrap samples, and computed the confidence intervals in the same way as for the uniform samples, discussed earlier, where the interval is of the form (9.77) and bias is neglected. This is compared in Figure 9.18 with the results for confidence

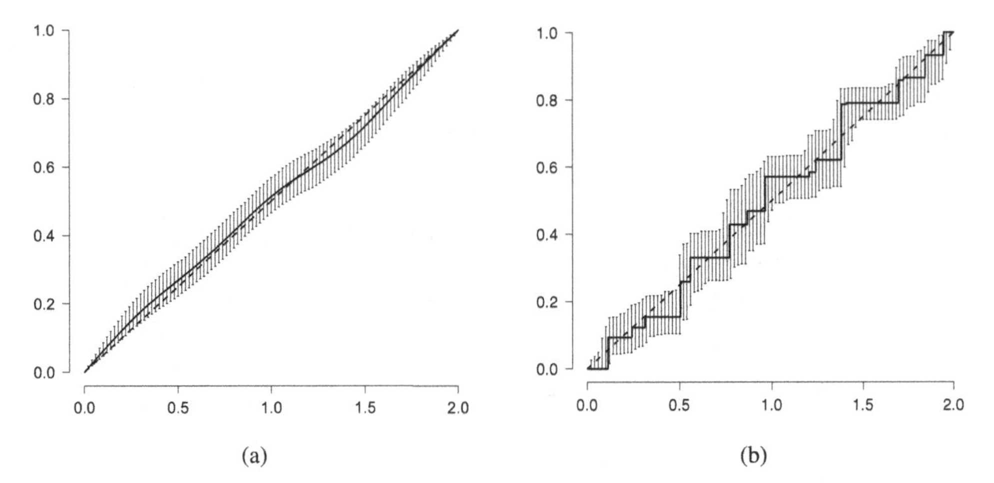

Figure 9.17 Uniform samples: 95% confidence intervals for $F_0(t_i)$, $t_i = 0.01, 0.02, \ldots$ for one sample $T_1, \ldots, T_n$. For (a) the SMLE and 1,000 bootstrap samples are used; F_0 is dashed and the SMLE solid. For (b) the LR test is used; F_0 is dashed and the MLE solid.

intervals of the form

$$\left[\tilde{F}_{nh}(t) - \beta(t) - U^*_{1-\alpha/2}(t)S_n(t), \tilde{F}_{nh}(t) - \beta(t) - U^*_{\alpha/2}(t)S_n(t)\right], \tag{9.78}$$

where $U^*_{\alpha/2}$, $U^*_{1-\alpha/2}$ and $S_n(t)$ are as in (9.77), and where $\beta(t)$ is the actual asymptotic bias, which is, for $t \in [h, 2-h]$, given by

$$\tfrac{1}{2} f_0'(t)h^2 \int u^2 K(u)\,du = -\frac{h^2 e^{-t} \int u^2 K(u)\,du}{2\{1-e^{-2}\}}.$$

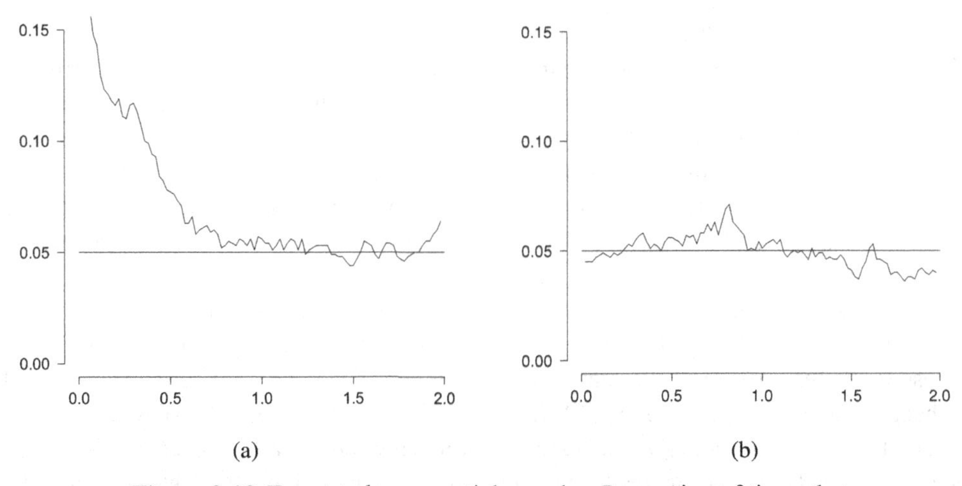

Figure 9.18 Truncated exponential samples. Proportion of times that $F_0(t_i)$, $t_i = 0.01, 0.02, \ldots$ is not in the 95% CIs in 1,000 samples $(T_1, \Delta_1) \ldots, (T_n, \Delta_n)$. In (a) the confidence intervals (9.77) are used, in (b) the bias corrected confidence intervals (9.78).

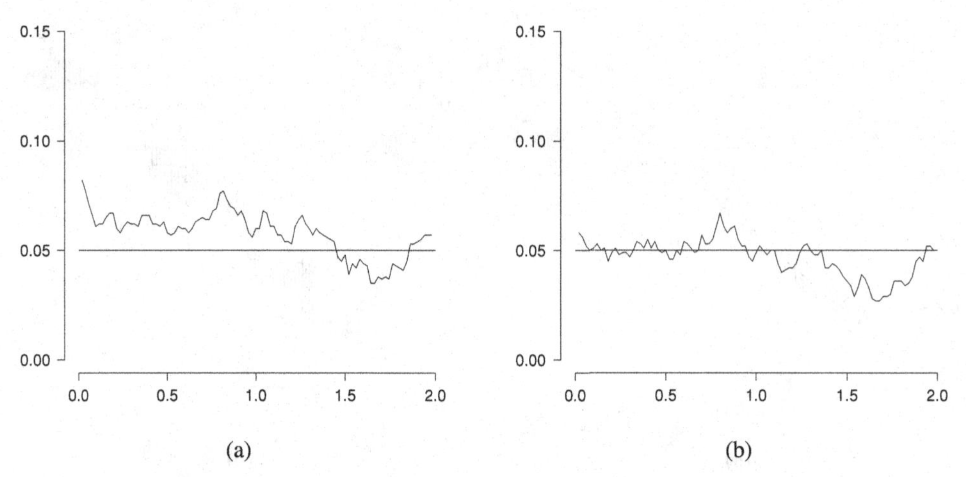

Figure 9.19 Truncated exponentials. Proportion of times that $F_0(t_i)$, $t_i = 0.01, 0.02, \dots$ is not in the CIs in 1,000 samples $(T_1, \Delta_1)\dots, (T_n, \Delta_n)$. In (a) the SMLE and (9.77) are used for $\alpha = 0.025$ with undersmoothing. In (b), (9.77) is used with $\alpha = 0.02$ instead of $\alpha = 0.025$ and the same undersmoothing as in (a).

For $t \notin [h, 2 - h]$ this expression is of the form

$$-\frac{h^2e^{-t}\left\{\int u^2K(u)\,du - 2\int_v^1 (u-v)^2K(u)\,du\right\}}{2\{1-e^{-2}\}}, \qquad v = \frac{t}{h},$$

if $t \in [0, h)$ and

$$-\frac{h^2e^{-t}\left\{\int u^2K(u)\,du - 2\int_v^1 (u-v)^2K(u)\,du\right\}}{2\{1-e^{-2}\}}, \qquad v = \frac{2-t}{h},$$

if $t \in (2 - h, 2]$, see (11.39).

It is seen in Figure 9.18 that if we use the bandwidth $2n^{-1/5}$ and do not use bias correction for the SMLE, the 95% coverage is off at the left end (where the bias is largest), but that the intervals are on target if we add the asymptotic bias to the intervals, as in (9.78). However, we cannot use the method of Figure 9.18b in practice, since the actual bias will usually not be available. We are faced here with a familiar problem in nonparametric confidence intervals and we can take several approaches. Two possible solutions are estimation of the bias or undersmoothing.

In the present case it turns out to be very difficult to estimate the bias term sufficiently accurately. Moreover, Hall, 1992, argues that undersmoothing has several advantages; one of these is that estimation of the bias term is no longer necessary. For the present model, we changed the bandwidth of the SMLE from $2n^{-1/5}$ to $2n^{-1/4}$ (if $n = 1000$) and computed the confidence intervals again by the bootstrap procedure, given earlier. This gave a remarkable improvement of the coverage at the left end, as is shown in Figure 9.19. Nevertheless, the undersmoothing has the tendency to make the confidence interval slightly anticonservative, as can be seen from Figure 9.19, so one might prefer to take, for example, the 20th and 980th percentile if one wants to have a coverage $\geq 95\%$. The effect of this method is shown in

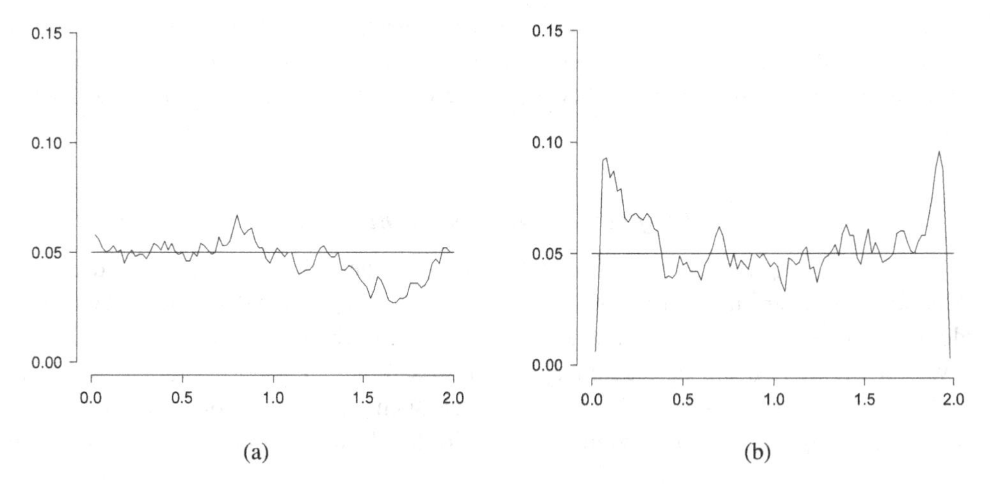

Figure 9.20 Truncated exponentials. Proportion of times that $F_0(t_i)$, $t_i = 0.01, 0.02, \ldots$ is not in the CIs in 1,000 samples $(T_1, \Delta_1) \ldots, (T_n, \Delta_n)$. Figure (a) uses the SMLE with the method of Figure 9.19b. In (b) the LR test for the MLE is used.

Figure 9.19b and the coverage of this method is compared to the coverage of the method, using the LR test, as in Banerjee and Wellner, 2005, in Figure 9.20. Undersmoothing, together with the method of Figure 9.19, will generally of course still produce narrower confidence intervals than the method, based on the LR test (which is based on cube root n asymptotics), under the appropriate smoothness conditions, as can be seen in Figure 9.21.

Another way of undersmoothing is to use a higher order kernel, for example a 4th order kernel, but still use a bandwidth of order $n^{-1/5}$. Since a 4th order kernel has necessarily negative parts, and since the estimate of F_0 will be close to zero or 1 at the boundary of

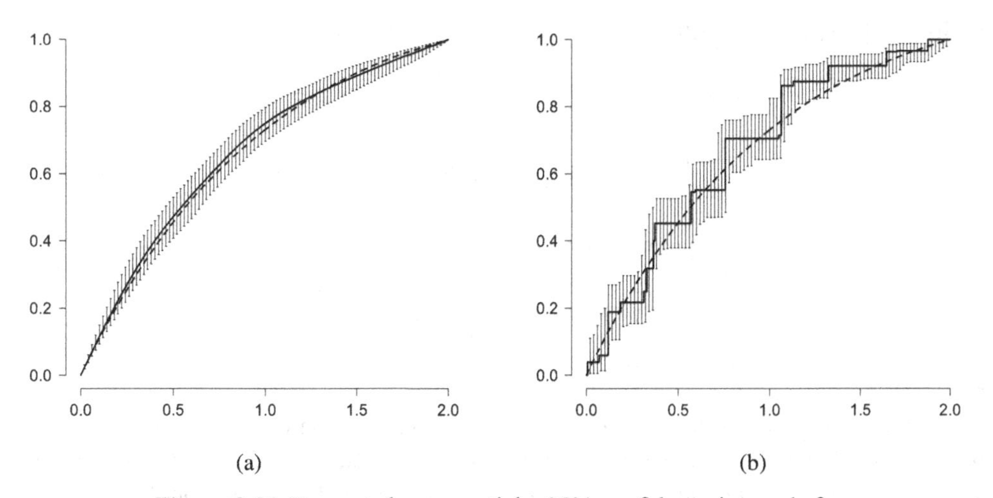

Figure 9.21 Truncated exponentials: 95% confidence intervals for $F_0(t_i)$, $t_i = 0.01, 0.02, \ldots$ for one sample $(T_1, \Delta_1) \ldots, (T_n, \Delta_n)$. In (a) the SMLE is used with undersmoothing and the method of Figure 9.19. Dashed: real F_0, solid: SMLE. In (b) the LR test for the MLE is used. Dashed: real F_0, solid: MLE.

the interval, this gives difficulties at the end of the interval. We also tried this method for constructing confidence intervals for the density, where it again gave rather unstable behavior. We therefore stuck to the earlier method, also in constructing confidence intervals for the density.

(D) Confidence Intervals for the Density

Let $(T_1, \Delta_1), \dots, (T_n, \Delta_n)$ be a current status sample, where the T_i are generated from a distribution with distribution function G and density g on $[a, b]$, and where the Bernoulli random variable $\Delta_i = 1$ with probability $F_0(T_i)$. For the construction of the $1 - \alpha$ confidence interval for the density we take again B bootstrap samples $(T_1^*, \Delta_1^*), \dots, (T_n^*, \Delta_n^*)$ with replacement from $(T_1, \Delta_1), \dots, (T_n, \Delta_n)$. For each such sample we compute the density estimate $\tilde{f}_{nh}^*$, using the same bandwidth h as used for the density estimate $\tilde{f}_{nh}$ in the original sample.

Since we want the bias to be (at least) of order $O(h^2)$ on the whole interval, if h is the chosen bandwidth, we cannot use the Schuster-type boundary correction we used for the SMLE, which is only $O(h)$ at the boundary (for the density estimate). We therefore used the boundary correction, given by (9.37), that is, for $t \in [a, a+h] \cup [b-h, b]$, we define

$$\tilde{f}_{nh}(t) = \int \frac{1}{h} K^t\left(\frac{t-x}{h}\right) d\hat{F}_n(x),$$

with, taking $u = (t-x)/h$,

$$K^t(u) = \begin{cases} \alpha\left(\dfrac{t-a}{h}\right) K(u) + \beta\left(\dfrac{t-a}{h}\right) uK(u), & t \in [a, a+h], \\[2ex] \alpha\left(\dfrac{b-t}{h}\right) K(u) - \beta\left(\dfrac{b-t}{h}\right) uK(u), & t \in [b-h, b], \end{cases}$$

and for $s \in [-1, 1]$, the coefficients $\alpha(s)$ and $\beta(s)$ are determined by

$$\alpha(s) \int_{-1}^{s} K(u)\, du + \beta(s) \int_{-1}^{s} uK(u)\, du = 1,$$

$$\alpha(s) \int_{-1}^{s} uK(u)\, du + \beta(s) \int_{-1}^{s} u^2 K(u)\, du = 0.$$

We now use that $\tilde{f}_{nh}$ is asymptotically equivalent to the toy estimator

$$f_{nh}^{toy}(t) = \int K_h^t(t-u)\, dF_0(u) + \frac{1}{n} \sum_{i=1}^{n} \frac{\left[K_h^t\right]'(t-T_i)\,\{\Delta_i - F_0(T_i)\}}{g(T_i)},$$

see (11.58).

In the simulation from the truncated exponential distribution on $[0, 2]$ we took $c = 2.0$ and $h = cn^{-1/6} \approx 0.632456$, using $1/6$ instead of $1/7$ in view of the undersmoothing. For example, the optimal bandwidth at $t = 1$, taking care of bias and variance would, according to Theorem 4.3, p. 366 of Groeneboom et al., 2010, actually be $c_1 n^{-1/7} \approx 1.07689$ if $n = 1000$, where $c_1 = 2.88897$, giving (for $n = 1000$) a bandwidth even larger than half the length of the interval.

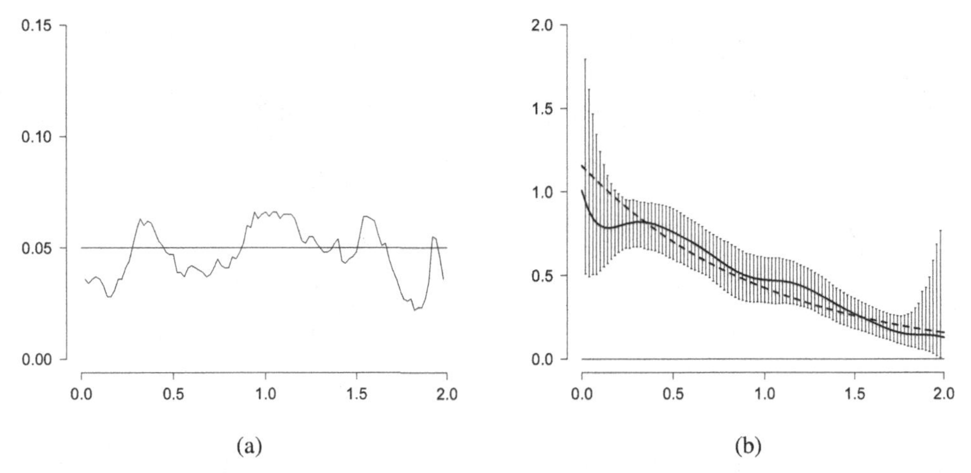

Figure 9.22 Truncated exponential samples. (a) Proportion of times that $f_0(t_i)$, $t_i = 0.01, 0.02, \ldots$ is not in the estimated 95% CIs in 1,000 samples, using the $\tilde{f}_{nh}$ with bandwidth $2n^{-1/6} \approx 0.632456$. (b) 95% confidence intervals, based on one sample and 1,000 bootstrap samples. Solid: density estimate. Dashed: real density.

Analogously to what we did for the SMLE, we computed in the bootstrap samples:

$$Z^*_{n,h}(t) = \frac{\tilde{f}^*_{nh}(t) - \tilde{f}_{nh}(t)}{\sqrt{n^{-2}\sum_{i=1}^n \left(\left[K_h^t\right]'(t - T_i^*)\right)^2 \left(\Delta_i - \hat{F}_n^*(T_i^*)\right)^2}},$$

where $\hat{F}_n^*$ is the ordinary MLE (and again not the SMLE!) of the bootstrap sample $(T_1^*, \Delta_1^*), \ldots, (T_n^*, \Delta_n^*)$, and where $\left[K_h^t\right]'(u)$ is defined by

$$\left[K_h^t\right]'(u) = h^{-1}\frac{d}{du}\left[K^t\right](u/h).$$

Let $U^*_\alpha(t)$ be the αth percentile of the B values $Z^*_{n,h}(t)$. Then the bootstrap $1-\alpha$ interval defined by:

$$\left[\tilde{F}_{nh}(t) - U^*_{1-\alpha/2}(t)S_{nh}(t), \tilde{F}_{nh}(t) - U^*_{\alpha/2}(t)S_{nh}(t)\right],$$

where

$$S_{nh}(t) = n^{-2}\sum_{i=1}^n \left(\left[K_h^t\right]'(t - T_i)\right)^2 \left(\Delta_i - \hat{F}_n(T_i)\right)^2.$$

In the simulations we took $B = 1000$ and the 15th and 985th order statistics of the values in the bootstrap samples instead of the 25th and 975th order statistics. We allowed negative values of the estimates in the bootstrap samples, since otherwise the coverage is not good at the right side of the interval. The results are shown in Figure 9.22.

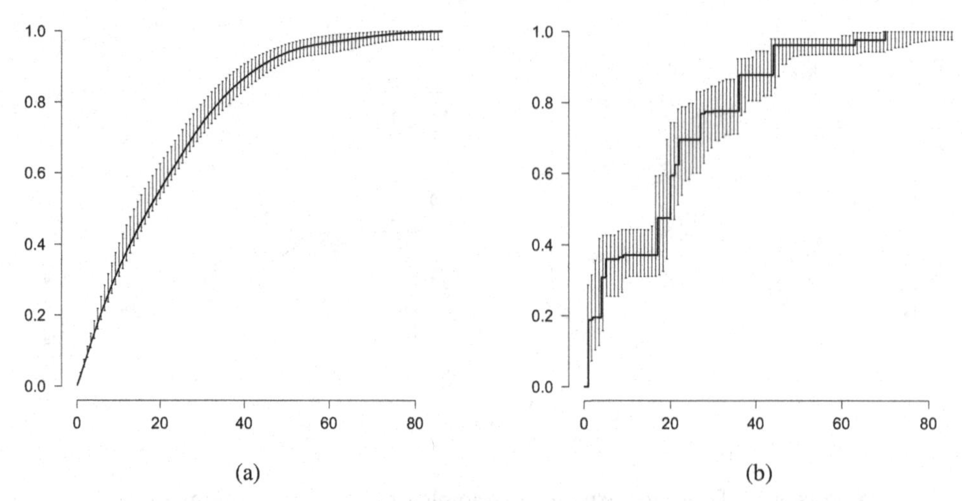

Figure 9.23 Hepatitis A data: 95% bootstrap confidence intervals for $F_0(t_i)$, $t_i = 0.87, 1.74, \ldots, 79.17$. (a) Based on the SMLE. The solid curve inside the confidence intervals is the SMLE (9.75), using bandwidth $h = Bn^{-1/4}$. (b) Based on the LR test for the MLE, as in Banerjee and Wellner, 2005. The solid curve inside the confidence intervals is the MLE.

(E) Confidence Intervals Based on the Hepatitis A Data

In Section 1.3 we discussed the hepatitis A data analyzed in Keiding, 1991. We now show the results of the methods presented here can be used in constructing confidence intervals for the distribution function and density in this model.

The SMLE $\tilde{F}_{nh}$ is given by (9.75), where $n = 850$, $[a, b] = [o, B]$, $B = 87$, $\hat{F}_n$ the MLE and $h = Bn^{-1/5} \approx 22.5754$. However, in the construction of the confidence intervals we use undersmoothing, and take instead $h = Bn^{-1/4} \approx 16.1126$. We compute the confidence intervals on an equidistant grid with distance 0.87 between successive points, with first point 0.87 and last point 85.26. The confidence intervals were computed in exactly the same way as the confidence intervals (9.77) in the simulation with the truncated exponential distribution.

The bootstrapping was done by first constructing a file with 850 values of (T_i, Δ_i), where within each group of equal T_is the number of Δ_is equal to 1 was given by the proportions of the data. As an example, for the first 16 (T_i, Δ_i)s all T_is are equal to 1, three Δ_is are equal to 1 and the other Δ_is are zero, see Table 1.3. Next we drew randomly (according to the discrete uniform distribution on the 850 point) with replacement from these values (T_i, Δ_i) a sample (T_i^*, Δ_i^*), $i = 1, \ldots, 850$, computed from these first the MLE and next the SMLE from the MLE. This was repeated 1,000 times, giving 1,000 values of (9.76). From these values the confidence intervals (9.77) were constructed. The lower and upper bounds were made monotone to the right by taking the maximum of neighboring values. See Figure 9.23 for the resulting confidence sets as well as those based on the LR test.

The confidence intervals for the density were computed with a bandwidth $h = Bn^{-1/6} \approx 27.5118$. For the remaining part the computation for the bootstrap confidence intervals followed exactly the same path as the computation of the confidence intervals for the density in the example with the truncated exponentials.

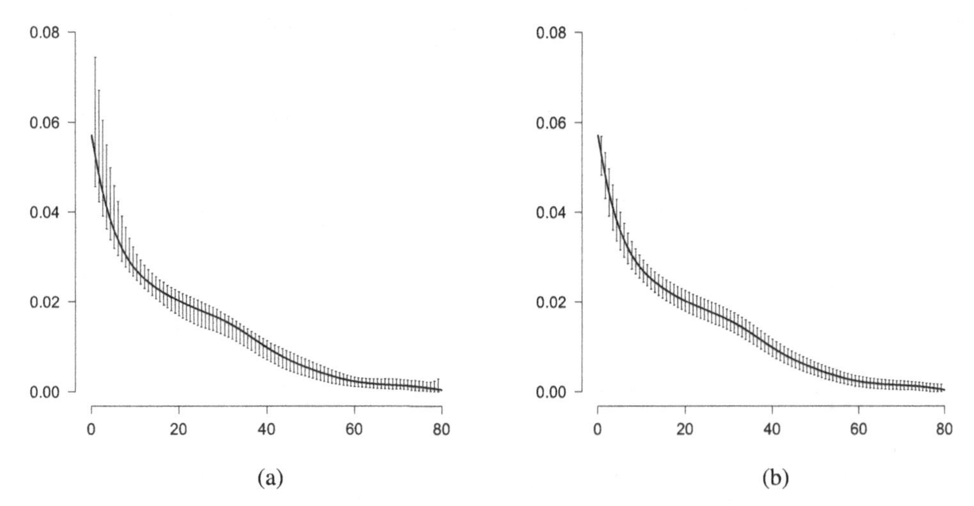

Figure 9.24 95% confidence intervals for $f_0(t_i)$, $t_i = 0.87, 1.74, \dots, 79.17$ for the hepatitis A data. The solid curve inside the confidence intervals is the density estimate, using bandwidth $h = Bn^{-1/6}$. In (a) the 95% bootstrap confidence intervals, obtained by bootstrapping, are shown; (b) shows the intervals obtained by using the asymptotic formula.

We also computed the confidence intervals, based on the asymptotic theory and obtained by plugging in estimates of the relevant parameters. The result is shown in Figure 9.24. The intervals are of the form

$$\left[\tilde{f}_{nh}(t) - 1.96n^{-2/7}\sqrt{\frac{\gamma(t)\tilde{F}_{n,h_1}(t)\{1-\tilde{F}_{n,h_1}(t)\}}{\hat{g}_{n,h_1}(t)}},\right.$$

$$\left.\tilde{f}_{nh}(t) + 1.96n^{-2/7}\sqrt{\frac{\gamma(t)\tilde{F}_{n,h_1}(t)\{1-\tilde{F}_{n,h_1}(t)\}}{\hat{g}_{n,h_1}(t)}}\right],$$

where $\gamma(t) = \int K'(u)^2\,du = 35/11$, if $t \in [h, B-h]$, and a rather complicated function of t if $t \in [0,h) \cup (B-h, B]$, which is specified in the following. The bandwidth $h_1 = Bn^{-1/5}$, and $\hat{g}_{n,h_1}(t)$ is an ordinary kernel estimate of the density g of the observation times T_i with the boundary correction (9.36).

The function γ is defined in the following way. If $t \in [0,h)$, $\gamma(t)$ is a function of t/h, and given by

$$\gamma(t) = \int_{u=-1}^{t/h}\left\{\frac{d}{du}\left\{K(u)\left(a\left(\frac{t}{h}\right) + b\left(\frac{t}{h}\right)u\right)\right\}\right\}^2 du.$$

If $t \in (B-h, B]$, $\gamma(t)$ is a function of $(B-t)/h$, and given by

$$\gamma(t) = \int_{u=-1}^{(B-t)/h}\left\{\frac{d}{du}\left\{K(u)\left(a\left(\frac{B-t}{h}\right) + b\left(\frac{B-t}{h}\right)u\right)\right\}\right\}^2 du.$$

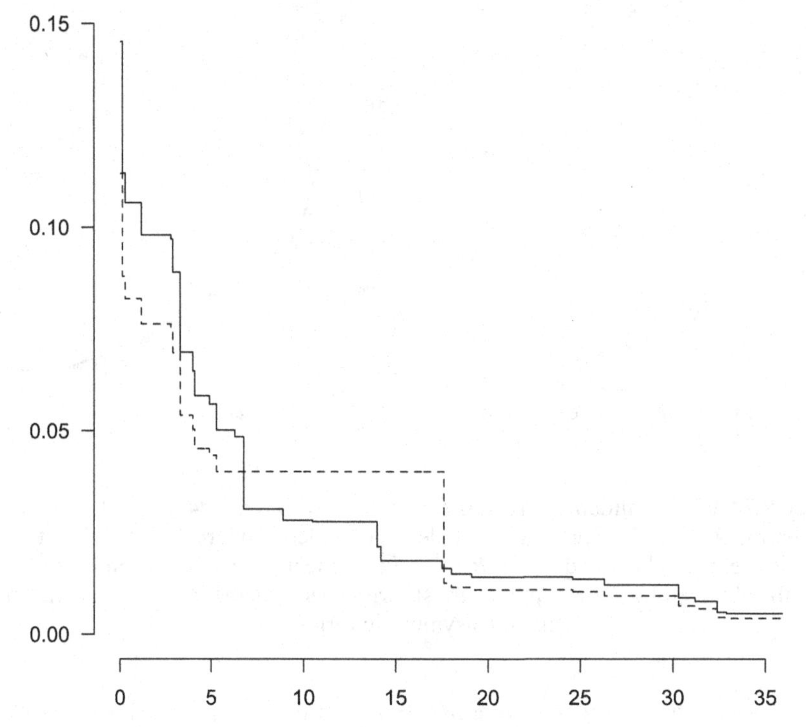

Figure 9.25 MLE (solid) and MLE, restricted to have value 0.04 at 17.5 (dashed), for the fertility data of Example 2.3.

The function γ is continuous on $[0, B]$, with a maximum approximately 120.597 at $t = 0$ and $t = B$. The most conspicuous differences between the bootstrap intervals and the asymptotic intervals are at the boundary, where the bootstrap intervals clearly exhibit skewness and are wider; the asymptotic intervals are symmetric with respect to the estimator (but are truncated at zero on the right-hand side). The asymptotic intervals do not lead to a very good coverage close to the boundary. For the asymptotic intervals we need to estimate the observation density g, which is not necessary in the bootstrap approach.

Banerjee and Wellner, 2001 in fact starts with the monotone density model as an example of a situation to which their methods could conceivably also be applied. However, they did not develop LR tests or confidence intervals for this model.

This raises the question whether the LR test for the monotone density case is actually similar to this test for the current status model. Will the limit distribution under the null hypothesis of the LR test be the same for the two models or will this distribution be different? And also, for the current status model we get a universal limit distribution, which can be used for the construction of confidence intervals; what can we expect for the monotone density case? The problem that makes computation more difficult is the restriction $\int f(x)dx = 1$ which the solution should satisfy.

The matter has been very recently resolved in Groeneboom and Jongbloed (2014), using "Lagrange-modified" cusum diagrams. The diagrams are used to compute the restricted MLEs. They are also used to prove that the limit distribution of the LR test under the null hypothesis is indeed the same as in the current status model, implying that we can use the same critical values for the confidence intervals.

Figure 9.25 shows a picture of the ordinary MLE and the restricted MLE for the fertility data of Example 2.3. It is seen that the condition $\int f(x)dx = 1$ induces a *global* change of the MLE, caused by the necessary redistribution of mass, in contrast with the only local change of the MLE for the current status model, if one computes the restricted MLE. See also Figure 2.6 in Section 2.2 for confidence intervals for the survival function in this model, based on LR tests for MLEs which also have a restriction at zero.

Exercises

9.1 Consider a Lebesgue measurable function H on the interval $[a, b]$ for $a < b$. The function H is called midpoint convex on $[a, b]$ if for all $u, v \in [a, b]$

$$H\left(\frac{u+v}{2}\right) \le \frac{H(u) + H(v)}{2}.$$

a) Show that convexity of H is equivalent with midpoint convexity of H. One implication is trivial, the other is a result due to Sierpinski.

b) Use (a) to show that H is convex on $[a, b]$ if and only if

$$H(x + y) - 2H(x) + H(x - y) \ge 0$$

for all $x \in (a, b)$ and $y \in \mathbb{R}$ such that $x \pm y \in [a, b]$.

9.2 Let $E_1, E_2, \ldots, E_n$ be independent and exponentially distributed with expectation 1. Define, for $1 \le i \le n$, the normalized spacings by

$$S_i = (n - i + 1)\left(E_{(i)} - E_{(i-1)}\right), \text{ where } E_{(0)} = 0.$$

a) Show that S_1 is standard exponentially distributed.

b) Show that $E_{(k)}$ given $E_{(1)}, \ldots, E_{(k-1)}$ has the same distribution as $E_{(k-1)} + \min_{1\le j\le n-k+1} E_i'$ where $E_1', E_2', \ldots$ are independent standard exponentially distributed random variables.

c) Show that S_k given $S_1, \ldots, S_{k-1}$ is standard exponentially distributed.

d) Conclude that

$$(S_1, S_2, \ldots, S_n) \stackrel{\mathcal{D}}{=} (E_1, \ldots, E_n).$$

9.3 Let E be a standard exponentially distributed random variable and H a continuous and strictly increasing cumulative hazard function on $[0, \infty)$. Denote its inverse by H^{-1}. Show that the random variable $X = H^{-1}(E)$ is distributed according to the distribution with cumulative hazard function H.

9.4 Use Exercises 9.2 and 9.3 to show that if g_0 has a convex cumulative hazard H_0,

$$S_1 \gtrsim S_2 \gtrsim \cdots \gtrsim S_n$$

where the S_is are defined as in (9.2) and $X \gtrsim Y$ means that X is stochastically bigger than Y so that for all $u \in \mathbb{R}$, $P(X > u) \ge P(Y > u)$.

9.5 Implement the Proschan-Pyke test of Example 9.1, use Monte Carlo simulation to approximate its 5% critical value and its power for Weibull alternatives of the type

$$G(x) = 1 - e^{-x^{\gamma}}$$

on $(0, \infty)$ for $\gamma = 1.1$ and $\gamma = 3$. Repeat this for various values of the sample size n.

9.6 Show that the function ϕ defined in (9.10) is convex and increasing on $[0, a]$.

9.7 Verify by inspecting the proof of Lemma 9.1 that the stochastic ordering result also holds if in the definition of T_n $(\mathbb{G}_n(x) - \hat{G}_n(x))^p$ would be used for some $1 < p \leq \infty$ rather than for $p = 1$.

9.8 Show that generating a realization of (9.20) based on a sample of independent standard exponential random variables corresponds to the procedure described in Lemma 9.1.

9.9 Show that the (log) likelihood ratio statistic for testing equality of two decreasing densities, (9.21), is scale invariant. This means that this statistic based on the original random variables $X_1, \ldots, X_{m+n}$ is the same as the statistic based on the scaled random variables $cX_1, \ldots, cX_{m+n}$ for any $c > 0$. What can be said in this respect about the (log) likelihood ratio test in the current status model, (9.47)?

9.10 In order to implement the likelihood ratio test for testing the null hypothesis of equality against the Lehmann alternative (9.82) in the current status setting, the ML estimator for F_1 under the alternative needs to be computed. Find a way to solve this computational problem.

9.11 Let I be a random variable with $P(I = 1) = \alpha$ and $P(I = 2) = 1 - \alpha$. Let g_1 and g_2 be densities on $[0, \infty)$ and F_1 and F_2 distribution functions on $[0, \infty)$. Then, construct the random variable Δ as follows:

$$T|I \sim g_I, \quad \Delta|T, I \sim \text{Bernoulli}(F_I(T)).$$

Determine $E\Delta$ and infer (9.61) within this setup, where the Δ_is are independent copies of Δ.

9.12 Show that the variance of (9.57) is indeed given by (9.58) if $F_1 = F_2$ and $G_1 = G_2$. Also derive the variance if $G_1 \neq G_2$. Recall that $\alpha_N = m/N$ and $\beta_N = 1 - \alpha_N = n/N$.

9.13 In view of (9.66), construct other combinations of F_1 and F_2 for which the test based on U_N can be expected to have power close to the level of the test.

9.14 Let K be the triweight kernel

$$K(u) = \frac{35}{32}\left(1 - u^2\right)^3 1_{[-1,1]}(u)$$

and let the density estimate $\tilde{f}_n$ at the boundary be given by

$$\tilde{f}_n(t) = \frac{1}{h_n}\int_{\mathbb{R}} K_{B,t}\left(\frac{t-x}{h_n}\right) \mathrm{d}F_{n0}(x),$$

where $K_{B,t}$ is defined by (9.37). Show that, using the notation of (9.37),
(i)

$$\alpha(x) = \frac{a(x)}{b(x)},$$

where, for $x \in [0, 1]$,

$$a(x) = 2048\left(16 - 64x + 160x^2 - 215x^3 + 140x^4 - 35x^5\right), \tag{9.79}$$

and

$$b(x) = (1+x)^6(5359 - 17750x + 24545x^2 - 18100x^3 + 7585x^4 - 1750x^5 + 175x^6). \quad (9.80)$$

(ii)

$$\beta(x) = \frac{c(x)}{b(x)},$$

where, for $x \in [0, 1]$,

$$c(x) = 80640(1-x)^4, \quad (9.81)$$

and $b(x)$ is given by (9.80).

Bibliographic Remarks

Birke and Dette, 2007, consider the problem of testing strict monotonicity against non-monotonicity of the regression function, using monotonic rearrangements as described in Section 8.4. Lemma 9.1 is related to the approximation in Durot, 2008, Section III. Tests of the same type as those introduced in Andersen and Rønn, 1995, and Sun, 2006, for testing equality of distributions in the current status setting are introduced in Zhang et al., 2001, and Zhang, 2006, for panel count data. There pseudo-maximum likelihood estimators are used. More on the k-sample problem with decreasing densities can be found in Durot et al., 2013. The (quasi-) likelihood ratio test for the two sample problem in the current status model is introduced in Groeneboom, 2012b. A likelihood ratio test is also considered in Chapter 3 of Kulikov, 2003, where the null hypothesis of equality of the distribution functions F_1 and F_2, generating the first and second sample, respectively, is tested against Lehmann alternatives of the form

$$F_2(t) = F_1(t)^{1+\theta},\ \theta \in (-1, \infty) \setminus \{0\}. \quad (9.82)$$

In Sun, 2006, the estimate of the variance of $N^{-1/2}$ times (9.57) severely overestimates the actual variance, and the proposed normalization will not give a standard normal distribution in the limit. For the proof of asymptotic normality of test statistic W_N defined in (9.63), Andersen and Rønn, 1995, rely in their proof on the master's thesis Hansen, 1991.

The bootstrap confidence intervals for the distribution function and density, considered in Section 9.5, are new (i.e., have not appeared in the literature at the time this book is written). The treatment of the bias is still a bit of an open problem, as it is in the bootstrap confidence intervals in ordinary density estimation; a completely satisfactory solution still does not seem to exist.

One can also try subsampling in the construction of confidence intervals, which is an automatic way of dealing with the bias, see Remark 3.7.2, p. 95, Politis et al., 1999. But the big problem with subsampling is that in the smaller bootstrap samples one has to take a larger bandwidth, which entails that the boundary effects enter at an earlier stage than in the original sample. So one has to compare a boundary corrected estimate in the subsample with a possibly not boundary corrected estimate in the original sample and in that sense one does not compare the same statistics. Nevertheless, pivoting by dividing by the square root of an estimate of the variance seems to annihilate this effect to a certain extent, but the matter is also still not completely solved at this time.

10

Asymptotic Theory of Smooth Functionals

In Chapter 3, asymptotic results were derived for the basic problems with monotonicity restrictions. The consistency results were global (sometimes uniform) whereas only pointwise asymptotic distributions were derived. These results were related to convex minorants and explicit representations of estimators. In this chapter, asymptotic results will be derived for quantities that depend more globally on the underlying distribution. These results can be obtained using so-called smooth functional theory.

In Section 10.1, we discuss the asymptotic distribution of the maximum likelihood estimator of the expected value of the underlying random variable in the deconvolution model. Computing this estimator requires some computational effort. We compare the asymptotics of this estimator to that of a natural and easy-to-compute competitor, the sample mean minus the expectation of the noise variable. The approach is based on smooth functional theory, where functionals of the underlying distribution of interest are approximated by smooth functionals of the observation distribution.

The remaining sections are devoted to smooth functionals for the interval censoring model. The theory is still moderately straightforward for the interval censoring case 1 model (or current status model) to be considered in Section 10.2, where we have an explicit expression for the score functions. This is rather different for the interval censoring case 2 model, where one can only say that these score functions are solutions of certain integral equations and the whole theory has to be developed from properties of these solutions. Important properties will be studied in Section 10.3. In Section 10.4 the properties are used to derive the asymptotic distribution of the MLE for smooth functionals in the interval censoring case 2 model. We only treat the so-called separated case, where the intervals between the two observation times cannot become arbitrarily small. This separated case is treated in full detail, since it seems to give the prototype for what to do in the case that the score functions are not explicitly given. It is believed that for the deconvolution model a similar analysis should be possible, but this theory still has to be developed.

The theory of the smooth functionals demonstrates that the usual rate $\sqrt{n}$ can be achieved for certain smooth functionals of the model, while at the same time the local rate is slower, for example $n^{1/3}$ for the MLE for the current status model. Also, the MLE will give efficient estimates of the smooth functionals for the interval censoring model. The same is conjectured to be true for the deconvolution model (and can be proved for special cases, such as the uniform and exponential deconvolution model).

10.1 Estimating the Expectation in Deconvolution Models

Consider the problem of estimating the expectation of a distribution function F in the deconvolution model as described in Sections 2.4 and 4.6. Recall that there is a sequence $X_1, X_2, \ldots$ of independent random variables with distribution function F, and independent of this a sequence of independent random variables $Y_1, Y_2, \ldots$ having (known) density function k. The observations then consist of $Z_1, Z_2, \ldots, Z_n$, where

$$Z_i = X_i + Y_i$$

are independent random variables from the density $g = g_F$, which is the convolution of the density k and the distribution function F:

$$g(z) = \int k(z - x)\, dF(x).$$

The first candidate for estimating $\mu_F = E_F X$ is a moment estimator given by

$$T_n = \frac{1}{n} \sum_{i=1}^{n} Z_1 - \int x k(x)\, dx.$$

Assume that $E_F X^2$ and $E_k Y^2$ are finite, implying that $\text{Var}(Z_1) = \text{Var}(X_1) + \text{Var}(Y_1)$ is finite as well. Then the central limit theorem immediately gives that

$$\sqrt{n}\,(T_n - \mu_F) \xrightarrow{\mathcal{D}} N(0, \text{Var}(Z_1)).$$

Estimator T_n will be the same for all densities k having the same first moment. It only uses that aspect of this kernel. In this section we will study an alternative estimator asymptotically, the estimator that is obtained by plugging in the maximum likelihood estimator

$$T_n' = \int x\, d\hat{F}_n(x). \tag{10.1}$$

This estimator uses more properties of the kernel function k (as the log likelihood, maximized by $\hat{F}_n$, depends on the density k not only via its first moment) and can be more efficient in general. A drawback is that it is computationally harder to determine than T_n, but as seen in Chapter 7, there are many ways to solve the computational problem. We start with a simple example, where the estimate of the first moment, using the MLE, coincides with a moment estimate.

Suppose now that the noise variables Y_i have a (known) normal $N(\mu, 1)$ distribution. The estimator T_n of the first moment of the distribution of the X_i is then given by

$$T_n = n^{-1} \sum_{i=1}^{n} Z_i - \mu. \tag{10.2}$$

The MLE of the unknown distribution function of the X_i is the distribution function $\hat{F}_n$, maximizing

$$\ell(F) = \int \log \int \phi(z - x - \mu)\, dF(x)\, d\mathbb{G}_n(z)$$

over F, where $\mathbb{G}_n$ is the empirical distribution function of the Z_i and ϕ is the standard normal density. Using this maximizer $\hat{F}_n$, the estimator T_n' defined in (10.1) can be computed.

Exercise 10.1 shows that, in fact, $T_n' = T_n$, so the two methods produce exactly the same (efficient) estimate here.

This relation does not hold for higher moments, however. We could, for example, estimate the variance of the X_i by a moment estimator,

$$U_n = n^{-1} \sum_{i=1}^{n} (Z_i - \bar{Z}_n)^2 - 1, \tag{10.3}$$

where $\bar{Z}_n$ is the mean of the Z_i, but also by

$$U_n' = \int x^2 \, d\hat{F}_n(x) - \left\{ \int x \, d\hat{F}_n(x) \right\}^2.$$

Here we do not get $U_n = U_n'$; for example, U_n can have negative values, in contrast with U_n' (see Exercise 10.2). On theoretical grounds, one would expect the MLE to produce an asymptotically efficient estimate of the variance, but looking at simulations, one also would expect this efficiency only to show up for huge sample sizes, because of the highly discrete character of the MLE, which only has very few points of mass for moderate sample sizes.

A usual method of producing estimates of F is to first estimate the characteristic function of the data in some way, and then use the fact that the characteristic function of the convolution is a product, meaning that one can divide by the characteristic function of the distribution of the known component of the deconvolution to obtain the characteristic function of the unknown component (see also Section 4.6). This does not necessarily produce efficient estimates of the smooth functionals, however, while for the MLE there is a general theory predicting the efficiency of the estimates of smooth functionals based on the MLE.

Dividing by the characteristic function of the known noise distribution becomes more difficult if this characteristic function has zeroes, as in the case of the uniform distribution (see Exercise 10.3). In this uniform deconvolution problem (also known as boxcar deconvolution) the moment estimator (10.2) also does not produce an efficient estimate of the first moment. We consider here the simplest case, where the distributions of the X_i and Y_i both have support $[0, 1]$ and the X_i have an absolutely continuous distribution function F_0. As discussed in Groeneboom and Wellner, 1992 (Exercise 2, Section 2.3, p. 61), the model is equivalent to the current status model in this case. This is seen in the following way.

Based on the observed $Z_1, \ldots, Z_n$, define for $i = 1, 2, \ldots, n$ the random variables $\Delta_i = 1_{\{Z_i \le 1\}}$ and

$$Z_i' = \begin{cases} Z_i, & \text{if } \Delta_i = 1, \\ Z_i - 1, & \text{if } \Delta_i = 0. \end{cases} \tag{10.4}$$

Then $Z_1', \ldots, Z_n'$ is distributed as a sample from a Uniform(0, 1) distribution; see Exercise 10.4. Moreover, the log likelihood for the unknown distribution function F can be written

$$\ell(F) = \sum_{i=1}^{n} \left\{ \Delta_i \log F(Z_i') + (1 - \Delta_i) \log\{1 - F(Z_i')\right\},$$

and we have, for $t, t+h \in (0,1)$

$$\mathbb{P}\{\Delta_i = 1,\ Z_i' \in [t, t+h]\} = \mathbb{P}\{Z_i \in [t, t+h]\} \sim h \int_0^t dF_0(u) = hF_0(t),\ h \downarrow 0,$$

and

$$\mathbb{P}\{\Delta_i = 0,\ Z_i' \in [t, t+h]\} = \mathbb{P}\{Z_i \in [1+t, 1+t+h]\} \sim h \int_t^1 dF_0(u)$$
$$= h\{1 - F_0(t)\},\ h \downarrow 0,$$

so we get factorization of the current status model for Δ_i and the corresponding observation Z_i'. This means, by Theorem 5.5 in Groeneboom and Wellner, 1992, that

$$\sqrt{n}\left\{\int x\, d\hat{F}_n(x) - \int x\, dF_0(x)\right\} \xrightarrow{\mathcal{D}} N(0, \sigma_{F_0}^2),$$

where

$$\sigma_{F_0}^2 = \int_0^1 F_0(t)\{1 - F_0(t)\}\, dt. \tag{10.5}$$

See also (10.7) and Theorem 10.1 in Section 10.2.

On the other hand, if we take the moment estimate T_n of (10.2) to estimate the first moment of the distribution of the X_i, we would get

$$\sqrt{n}\left\{T_n - \int x\, dF_0(x)\right\} \xrightarrow{\mathcal{D}} N(0, \sigma^2),$$

where

$$\sigma^2 = \frac{1}{12} + \mathrm{var}(X_1). \tag{10.6}$$

Since

$$\mathrm{var}(X_1) = 2\int_0^1 x\{1 - F_0(x)\}\, dx - \left\{\int_0^1 \{1 - F_0(x)\}\, dx\right\}^2,$$

a simple variational argument shows that $\sigma_{F_0}^2 < \sigma^2$, unless F_0 is the uniform distribution function, in which case $\sigma_{F_0}^2 = \sigma^2$ (see also Exercise 10.5). So in this case, the estimate of the first moment, based on the MLE, is more efficient than the moment estimate T_n, in contrast with the situation described in the normal deconvolution model.

10.2 Estimating Smooth Functionals in the Current Status Model

The current status (or interval censoring case 1) model is introduced in Section 2.3 and the pointwise cube root n asymptotics derived in Section 3.8. Smooth functional theory for the current status model can be considered to be one of the first successes of the use of the MLE, which is generally efficient for the estimation of these functionals. The asymptotic distribution of $\hat{F}_n(t_0)$, for fixed $t_0 \in \mathbb{R}_+$, is derived in Section 3.8, where $n^{1/3}$ is the obtained convergence rate. In part II, chapter 5, of Groeneboom and Wellner, 1992, it is shown that, under some extra conditions,

$$\sqrt{n}\left\{\mu(\hat{F}_n) - \mu(F)\right\} \xrightarrow{\mathcal{D}} N\left(0, \int \frac{F(x)[1-F(x)]}{g(x)}dx\right), \quad \text{as } n \to \infty \tag{10.7}$$

with g the density of the distribution of the observation times. Huang and Wellner, 1995, prove a similar result for a wider class of functionals. The proof in Groeneboom and Wellner, 1992, uses the convergence rate of the supremum distance between the MLE and the underlying distribution function, which is replaced by a simpler argument based on L_2-distance properties in Huang and Wellner's proof.

In studying information lower bounds for estimating smooth functionals $K(F)$ of F, we can define 'smooth" to mean "differentiable along a Hellinger differentiable path." The derivative is represented by the canonical gradient (or efficient influence function) κ_F at F, which is an element of the Hilbert space $L_2^0(F)$ of square integrable functions a with respect to the measure dF, satisfying $\int a\,dF = 0$. More formally, we consider functionals that, locally near a fixed distribution function F_0, allow the representation

$$K(F) = K(F_0) + \int \kappa_{F_0}(x)\,d(F - F_0)(x) + O\left(\|F - F_0\|^2\right) \tag{10.8}$$

where $\|\cdot\|$ denotes the L_2-norm of a function. Linear functionals like the mean value are smooth in this sense, since then $\kappa_{F_0}(x) = x - \int x\,dF_0(x)$, and the order term disappears.

Since there is only indirect information about the random variables X_i generated by F, the smooth functional $K(F)$ is only implicitly defined as $\theta(Q)$ in terms of the distribution Q of the observable random variables (T_i, Δ_i).

For interval censoring case 1, an explicit formula for the information lower bound for smooth functionals can be obtained. The fact that an explicit formula can be found is, however, not common. This property will be lost in the interval censoring case 2 model that will be considered in Sections 10.3 and 10.4.

Suppose the unobservable event times X_i have a distribution function F with support contained in $[0, M]$, and density f. The observation times T_i have an absolutely continuous distribution function G with density g; X_i and T_i are independent and F is dominated by G. Then it follows from Exercise 10.6 that the score operator L_F has the form

$$[L_F a](t, \delta) := E\{a(X)|T = t, \Delta = \delta\} = \frac{\delta \int_0^t a(x)\,dF(x)}{F(t)} + \frac{(1-\delta)\int_t^M a(x)\,dF(x)}{1 - F(t)} \tag{10.9}$$

with adjoint (see Exercise 10.7)

$$[L^* b](x) := E\{b(T, \Delta)|X = x\} = \int_{t=x}^{M} b(t, 1)\,g(t)\,dt + \int_{t=0}^{x} b(t, 0)\,g(t)\,dt. \tag{10.10}$$

Note that the adjoint operator L^* does not depend on F (given X, the distribution of the vector (T, Δ) does not depend on the distribution of X anymore), which is the reason for dropping the index F in the notation of the adjoint. We consider differentiability of functionals K at the point F. The key property that is needed for K of the type (10.8) to be a smooth functional is that κ_F is contained in the range of the operator L^*, so

$$\kappa_F \in \mathcal{R}(L^*).$$

If this holds, then the canonical gradient for the probability measures on the observation space is the unique element θ_F in $\overline{\mathcal{R}(L_F)}$ satisfying

$$L^*\theta_F = \kappa_F. \tag{10.11}$$

We consider the case $\theta_F \in \mathcal{R}(L_F)$, giving that $\theta_F = L_F a$ for some $a \in L_2^0(F)$. Then the score equation $L^* L_F a = \kappa_F$ has to be solved in $a \in L_2^0(F)$. It can be written as an equation in ϕ:

$$\int_{t=0}^{x} \frac{\phi(t)}{1 - F(t)} g(t)\, dt - \int_{t=x}^{M} \frac{\phi(t)}{F(t)} g(t)\, dt = -\kappa_F(x), \tag{10.12}$$

where ϕ is the integrated score function

$$\phi(x) = \int_0^x a(y)\, dF(y),\ x \in [0, M].$$

See Exercise 10.9. The advantage of solving for the integrated score function in this context will become clear in the following.

Suppose κ_F to be continuously differentiable. Then, taking derivatives,

$$\frac{\phi(x)}{1 - F(x)} g(x) + \frac{\phi(x)}{F(x)} g(x) = -\kappa_F'(x).$$

Hence, if $g > 0$,

$$\phi(x) = -\kappa_F'(x) \frac{F(x)[1 - F(x)]}{g(x)}, \tag{10.13}$$

yielding the canonical gradient

$$\theta_F(t, \delta) = \begin{cases} \kappa_F'(t) \dfrac{1 - F(t)}{g(t)}, & \text{if } \delta = 1 \\ \\ -\kappa_F'(t) \dfrac{F(t)}{g(t)}, & \text{if } \delta = 0 \end{cases} \tag{10.14}$$

and information lower bound:

$$\|\theta_F\|_{Q_{F,G}}^2 = \int \theta_F(t, \delta)^2\, dQ_{F,G}(t, \delta) = \int_0^M \left[\kappa_F'(x)\right]^2 \frac{F(x)[1 - F(x)]}{g(x)}\, dx.$$

We now want to relate this to the theory of the MLE $\hat{F}_n$. We have the following theorem, where we take the general interval $[a, b]$ instead of $[0, M]$.

Theorem 10.1 *Let the observation density g, with distribution function G, be continuous on the interval $[a, b]$, where $g(t) \geq c > 0$, for all $t \in [a, b]$. Furthermore, let F_0 have bounded support $[a, b]$, where F_0 is absolutely continuous, with a bounded derivative f_0 on $[a, b]$, which satisfies*

$$f_0(x) \geq c_1 > 0,\ x \in [a, b].$$

Moreover, let κ_{F_0} be a continuous function, so that $k(x) = \kappa_{F_0}'(x)$ is continuous on $[a, b]$ and

$$(k/g) \circ F_0^{-1} \text{ is Lipschitz on } [0, 1].$$

Finally, assume that the functional $K(F)$ of F satisfies:

$$K(F) - K(F_0) = \int \kappa_{F_0}(x)\,d(F - F_0)(x) + O\left(\|F - F_0\|^2\right), \tag{10.15}$$

where $\|F - F_0\|$ denotes L_2-distance with respect to Lebesgue measure. Then

$$\sqrt{n}\left\{K(\hat{F}_n) - K(F_0)\right\} \xrightarrow{\mathcal{D}} N\left(0, \|\theta_{F_0}\|_Q^2\right). \tag{10.16}$$

Here θ_{F_0} is defined as in (10.14) and $Q = Q_{F_0,G}$ is the (true underlying) probability measure of the observations (T_i, Δ_i).

Remark The proof follows a general method that is also used in the more complicated cases of interval censoring. See the rejoinder of the discussion in Groeneboom, 2013b, on the general aspects of this method.

Proof We may assume $\hat{F}_n(b) = 1$; see Exercise 10.8. The goal is to prove

$$\sqrt{n}\left\{K(\hat{F}_n) - K(F_0)\right\} = \sqrt{n}\int \theta_{F_0}\,d(Q_n - Q) + o_p(1), \tag{10.17}$$

since (10.16) clearly follows from this. We split the proof in several steps.

(i) *The nonlinear aspect of the functional is negligible.*
This means:

$$\sqrt{n}\left\{K(\hat{F}_n) - K(F_0)\right\} = \sqrt{n}\int \kappa_{F_0}\,d(\hat{F}_n - F_0) + o_p(1). \tag{10.18}$$

By (10.15) this is fulfilled if

$$\|\hat{F}_n - F_0\|_2 = o_p\left(n^{-1/4}\right).$$

It is proved in Van de Geer, 2000, that the Hellinger distance $h(q_{\hat{F}_n}, q_{F_0})$ satisfies

$$h\left(q_{\hat{F}_n}, q_{F_0}\right) = O_p\left(n^{-1/3}\right), \quad q_F(t,\delta) = \delta F(t) + (1-\delta)\{1 - F(t)\}, \tag{10.19}$$

from which $\|\hat{F}_n - F_0\|_2 = O_p(n^{-1/3})$ follows, which is stronger than what we need; see Exercise 10.11.

(ii) *Transformation to the observation space measure.*
For the dominant term at the right hand side of (10.18) we wish to show that

$$\sqrt{n}\int \kappa_{F_0}\,d(\hat{F}_n - F_0) = \sqrt{n}\int \theta_{F_0}\,d(Q_n - Q) + o_p(1).$$

A first step in this direction is to prove that

$$\int \kappa_{F_0}\,d(\hat{F}_n - F_0) = -\int \theta_{\hat{F}_n}\,dQ, \tag{10.20}$$

where

$$\theta_{\hat{F}_n}(t,\delta) = \begin{cases} \dfrac{\phi_{\hat{F}_n}(t)}{\hat{F}_n(t)}, & \delta = 1, \\ -\dfrac{\phi_{\hat{F}_n}(t)}{1 - \hat{F}_n(t)}, & \delta = 0, \end{cases} \tag{10.21}$$

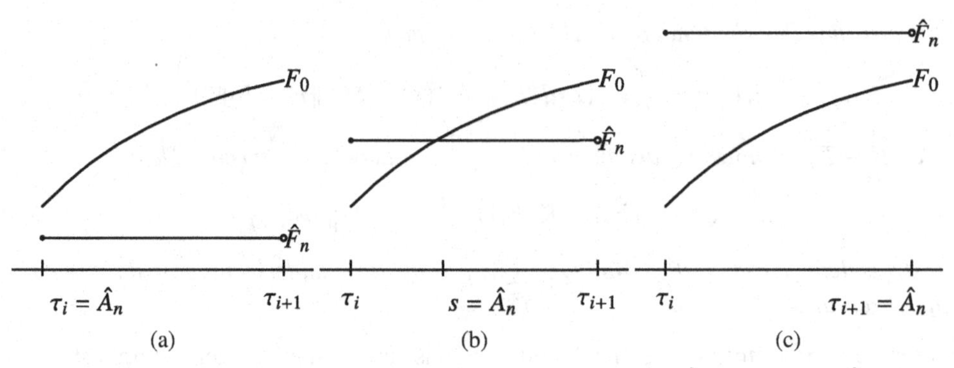

Figure 10.1 The three different possibilities for the function $\hat{A}_n$. (a) $F_0(t) > \hat{F}_n(\tau_i)$ for all t; (b) $F_0(s) = \hat{F}_n(\tau_i)$ for some s; (c) $F_0(t) < \hat{F}_n(\tau_i)$ for all t.

and where

$$\phi_F(t) = \frac{F(t)\{1 - F(t)\}}{g(t)}. \tag{10.22}$$

Note that $\theta_{\hat{F}_n}$ is no longer a canonical gradient, since $\theta_{\hat{F}_n}(t, \delta)$ has jumps, whereas the canonical gradient has to be absolutely continuous with respect to G. Relation (10.20) can be verified by straightforward calculation (see Exercise 10.12), but holds quite generally.

(iii) *Use that $\hat{F}_n$ is the MLE.*

We introduce a modification $\bar{\phi}_{\hat{F}_n}$ of $\phi_{\hat{F}_n}$ such that $\bar{\phi}_{\hat{F}_n}$ is absolutely continuous with respect to $\hat{F}_n$, and satisfies:

$$\int \{\bar{\theta}_{\hat{F}_n} - \theta_{\hat{F}_n}\}\, dQ = o_p\left(n^{-1/2}\right). \tag{10.23}$$

Here

$$\bar{\theta}_{\hat{F}_n}(t, \delta) = \begin{cases} \dfrac{\bar{\phi}_{\hat{F}_n}(t)}{\hat{F}_n(t)}, & \delta = 1, \\ -\dfrac{\bar{\phi}_{\hat{F}_n}(t)}{1 - \hat{F}_n(t)}, & \delta = 0. \end{cases} \tag{10.24}$$

To specify $\bar{\phi}_{\hat{F}_n}$, we define the map $\hat{A}_n$ on $[a, b]$ by

$$\hat{A}_n(t) = \begin{cases} \tau_i, & \text{, if } F_0(t) > \hat{F}_n(\tau_i), \quad t \in [\tau_i, \tau_{i+1}), \\ s, & \text{, if } F_0(s) = \hat{F}_n(s), \quad \text{for some } s \in [\tau_i, \tau_{i+1}), \\ \tau_{i+1}, & \text{, if } F_0(t) < \hat{F}_n(\tau_i), \quad t \in [\tau_i, \tau_{i+1}), \end{cases} \tag{10.25}$$

where the τ_i are successive points of jump of $\hat{F}_n$, and we define

$$\bar{\phi}_{\hat{F}_n}(t) = \phi_{\hat{F}_n}(\hat{A}_n(t)). \tag{10.26}$$

The definition of $\hat{A}_n$ is illustrated in Figure 10.1.

Then we have

$$\int \bar\theta_{\hat F_n}(t,\delta)\,dQ_n = 0, \tag{10.27}$$

and (10.23) now follows from Exercises 10.15 and 10.16. We can now write, using (10.20),

$$\begin{aligned}\int \kappa_{F_0}\,d(\hat F_n - F_0) &= -\int \theta_{\hat F_n}\,dQ = -\int \bar\theta_{\hat F_n}\,dQ + o_p\left(n^{-1/2}\right)\\ &= \int \bar\theta_{\hat F_n}\,d(Q_n - Q) + o_p\left(n^{-1/2}\right).\end{aligned}$$

(iv) *Asymptotic variance equals information lower bound.*
We finally have to show that

$$\int \bar\theta_{\hat F_n}\,d(Q_n - Q) = \int \theta_{F_0}\,d(Q_n - Q) + o_p\left(n^{-1/2}\right).$$

The fact that the class of functions that are integrated form a uniform Q-Donsker class, and convergence of $\hat F_n$ to F_0 in $L_2(G)$-norm or in supremum norm (Groeneboom and Wellner, 1992), can be used to show that

$$\int \{\bar\theta_{\hat F_n} - \theta_{F_0}\}\,d(Q_n - Q) = o_p\left(n^{-1/2}\right),$$

which is sufficient for our purposes.

In fact, the stronger result

$$\int \{\bar\theta_{\hat F_n} - \theta_{F_0}\}\,d(Q_n - Q) = O_p\left(n^{-2/3}\right)$$

can be proved. □

10.3 The Integral Equation for Interval Censoring Case 2

For the current status model, the asymptotic distribution of the ML estimator of smooth functionals of the event time distribution function F is derived in the previous section. A crucial step, that of solving the integral equation, can be taken explicitly in that context, leading to an expression for the (efficient) variance of the ML estimator. For the case 2 interval censoring model (see Section 4.7), it will be seen in this section that there is no explicit solution of the integral equation. As in Section 10.2, the smooth functional $K(F)$ is only implicitly defined as $\theta(Q)$ in terms of the distribution Q of the random variables $(T_i, U_i, \Delta_{i1}, \Delta_{i2})$ in the observation space. The key property that is needed is again

$$L^*\theta_F = \kappa_F \tag{10.28}$$

where the operators L_F and L^* (the adjoint of L_F) can be interpreted as conditional expectations and have the following form:

$$[L_F a](t,u,\delta_1,\delta_2) = E_F\{a(X)|(T_1,U_1,\Delta_{11},\Delta_{12}) = (t,u,\delta_1,\delta_2)\} = \theta_F(t,u,\delta_1,\delta_2) \tag{10.29}$$

and

$$[L^*b](x) = E\{b(T_1, U_1, \Delta_{11}, \Delta_{12})|X = x\} \tag{10.30}$$

$$= \int_{u=x}^{M}\int_{v=u}^{M} b(u, v, 1, 0)\, g(u, v)\, dv\, du + \int_{u=0}^{x}\int_{v=x}^{M} b(u, v, 0, 1)\, g(u, v)\, dv\, du$$

$$+ \int_{u=0}^{x}\int_{v=u}^{x} b(u, v, 0, 0)\, g(u, v)\, dv\, du. \tag{10.31}$$

We drop the index F for the adjoint operator L^* again, since this operator does not depend on F (see Exercise 10.10).

As in Section 10.2, the approach is not to consider the score functions themselves, but instead the integrated score functions $\phi_F(t) = \int_{[0,t]} a(x)\, dF(x)$ and to extend the definition of these integrated score functions to functions ϕ_F which are not absolutely continuous with respect to F, so not having the representation

$$\phi_F(t) = \int_{[0,t]} a(x)\, dF(x) \quad \text{or} \quad \phi_F(t) = \int_{(t,M]} a(x)\, dF(x).$$

This also means that the basic equations (10.28) and (10.30) are extended to functions θ_F, defined in terms of ϕ_F, where ϕ_F has both discrete and absolutely continuous parts. Taking

$$\phi_F(t) = \int_{(t,M]} a(x)\, dF(x),$$

the basic equation (10.28) becomes:

$$\kappa_F(x) = -\int_{u=0}^{x}\int_{v=u}^{x} \frac{\phi_F(v)}{1 - F(v)}\, g(u, v)\, dv\, du - \int_{u=0}^{x}\int_{v=x}^{M} \frac{\phi_F(v) - \phi_F(u)}{F(v) - F(u)}\, g(u, v)\, dv\, du$$

$$- \int_{u=x}^{M}\int_{v=u}^{M} \frac{\phi_F(u)}{F(u)}\, g(u, v)\, dv\, du. \tag{10.32}$$

Differentiating with respect to x yields equation

$$\phi_F(x) + d_F(x)\left[\int_{t=0}^{x} \frac{\phi_F(x) - \phi_F(t)}{F(x) - F(t)}\, g(t, x)\, dt - \int_{t=x}^{M} \frac{\phi_F(t) - \phi_F(x)}{F(t) - F(x)}\, g(x, t)\, dt\right]$$

$$= k(x) d_F(x), \tag{10.33}$$

where $k(x) = \frac{d}{dx}\kappa_F(x)$ and $d_F(x)$ is defined by

$$d_F(x) = \frac{F(x)\{1 - F(x)\}}{g_1(x)\{1 - F(x)\} + g_2(x)F(x)}. \tag{10.34}$$

Just as in the current status model, the solution ϕ_F contains a factor $F(1 - F)$. The structure of d_F already suggests this factor to be present. It can also be proved formally with hardly any extra effort. So we write

$$\phi = F(1 - F)\,\xi.$$

The function ξ is only defined on $\{x : 0 < F(x) < 1\} = (a, b)$. Plugging this formula into (10.33) and performing some reordering yields the following integral equation in ξ:

$$\xi(x) + c_F(x)\left[\int_{t=a}^{x} \frac{\xi(x) - \xi(t)}{F(x) - F(t)}\, g^*(t, x)\, dt - \int_{t=x}^{b} \frac{\xi(t) - \xi(x)}{F(t) - F(x)}\, g^*(x, t)\, dt\right] = k(x)c_F(x). \tag{10.35}$$

Here $c_F(x)$ is given by

$$c_F^{-1}(x) = \int_{t=0}^{x} \{1 - F(t)\}\, g(t, x)\, dt + \int_{t=x}^{M} F(t)\, g(x, t)\, dt \tag{10.36}$$

and

$$g^*(t, x) = F(t)[1 - F(t)]\, g(t, x).$$

We now first list the model conditions (M1) to (M3) (M stands for model). We suppose:

(M1) X_i is a nonnegative absolutely continuous random variable with distribution function F. Let $M > 0$. F is contained in the class

$$\mathcal{F}_M := \{F :\ \text{support}(F) \subset [0, M];\ F \ll \text{Lebesgue measure}\}.$$

F is the distribution on which we want to obtain information; however, we do not observe X_i directly.

(M2) Instead, we observe the pairs (U_i, V_i), with distribution function G. G is contained in $\mathcal{G}$, the collection of all two-dimensional distributions on $\{(u, v) : 0 \le u < v\}$, absolutely continuous with respect to two-dimensional Lebesgue measure. Let g denote the density of (U_i, V_i), with marginal densities and distribution functions g_1, G_1 and g_2, G_2 for U_i and V_i, respectively.

(M3) If both G_1 and G_2 put zero mass on some set A, then F has zero mass on A as well, so $F \ll G_1 + G_2$. This means that F does not have mass on sets in which no observations can occur.

Condition (M3) is needed to ensure consistency. Moreover, without this assumption the functionals we are interested in are not well defined. So discrete distribution functions F should be excluded from $\mathcal{F}_M$. Apart from the model conditions (M1) to (M3), some extra conditions will have to be introduced in order to make the proofs in this section possible (S for smoothness):

(S1) g_1 and g_2 are continuous, with $g_1(x) + g_2(x) > 0$ for all $x \in [0, M]$.

(S2) $(u, v) \mapsto g(u, v)$ is continuous on $[0, M]^2$.

(S3) $P_G(V - U < \epsilon_0) = 0$ for some ϵ_0 with $0 < \epsilon_0 \le 1/2\, M$, so g does not have mass close to the diagonal.

(S4) F is either a continuous distribution function with support $[0, M]$, or a piecewise constant distribution function with a finite number of jumps, all in $[0, M]$; F satisfies

$$F(u) - F(t) \ge c > 0, \quad \text{if } u - t \ge \epsilon_0.$$

(S5) $x \mapsto k(x) = \kappa_F'(x)$ is continuous on $[0, M]$.

Of course, (S2) implies continuity of g_1 and g_2, which is also stated in (S1). (S1) is the equivalent of continuity of g and $g > 0$ in the current status model. Note that (S1) implies that

d_F is bounded; see Exercise 10.17. Conditions (S3) and (S4) are needed to avoid singularity in the integral equation: if $F(x) - F(t)$ becomes very small, we have $g(t, x) = 0$. In fact, the domain of integration in (10.33) can be restricted to $[0, x - \epsilon_0]$ and $[x + \epsilon_0, M]$ respectively; ϵ_0 should be smaller than $\leq \frac{1}{2}M$ by (M3).

We will also need a Lipschitz property of ϕ_F and ξ_F, uniformly over $F \in \mathcal{F}_M$ for which we need two extra conditions (L for Lipschitz):

(L1) The partial derivatives $\partial_1 g(t, u)$ and $\partial_2 g(t, u)$ exist, except for at most a countable number of points, where left and right derivatives exist. The derivatives are bounded uniformly over t and u.

(L2) $k(x) = \kappa_F'(x)$ is differentiable, except for at most a countable number of points x, where left and right derivatives exist. The derivative is bounded uniformly over x.

Under these conditions we have uniqueness of the solutions ϕ_F and ξ_F, using the following fact from the theory of integral equations.

Lemma 10.1 *Let X be a normed linear space, and let $A : X \to X$ be a compact linear operator. Consider the homogeneous equation*

$$\phi - A\phi = 0. \tag{10.37}$$

If the homogeneous equation only has the trivial solution $\phi = 0$, then for each $f \in X$ the inhomogeneous equation

$$\phi - A\phi = f$$

has a unique solution $\phi \in X$ and this solution depends continuously on f.

Remark This is an immediate consequence of Theorem 3.4 in Kress, 1989. Note that hardly any restrictions are imposed on the space X.

Define, for $a < b$, $D[a, b]$ as the space of cadlag (right continuous with left limits) real valued functions on $[a, b]$. No topology (like the Skorohod topology) is given to $D[a, b]$.

Theorem 10.2 *The ϕ-equation (10.33) has a unique solution in $D([0, M])$. The ξ-equation (10.35) has a unique solution in $D[a, b]$, where*

$$a = \inf\{x : F(x) > 0\}, \qquad b = \sup\{x : F(x) < 1\}.$$

Proof The condition in Lemma 10.1 will be verified for the ϕ-equation; the proof for the ξ equation proceeds along similar lines. We define:

$$K(x, t) = \frac{d(x)\, D(x, t)}{1 + d(x) \int_0^M D(x, t)\, dt}, \tag{10.38}$$

with $D(x, t)$ defined as

$$D(x, t) = \begin{cases} \dfrac{g(t, x)}{F(x) - F(t)} & \text{if } t \leq x \\[2ex] \dfrac{g(x, t)}{F(t) - F(x)} & \text{if } t \geq x, \end{cases}$$

and consider the homogeneous equation

$$\phi(x) = \int_0^M K(x,t)\,\phi(t)\,dt \quad \text{for all } x \in [0, M].$$

This equation can only be satisfied for a function ϕ that at most has jumps at the same place as $K(x,\cdot)$. Moreover, a solution $\phi \le 0$ implies a solution $\phi \ge 0$. The homogeneous equation is equivalent to

$$\phi(x) + \left[d(x)\int_0^M D(x,t)\,dt\right]\phi(x) = d(x)\int_0^M D(x,t)\,\phi(t)\,dt \quad \text{for all } x \in [0, M]. \tag{10.39}$$

Suppose there exists a point x with $\phi(x) > 0$. If the supremum is attained, say at s, we get, since $D(s,t) \ge 0$,

$$d(s)\int_0^M D(s,t)\,\phi(t)\,dt \le \left[d(s)\int_0^M D(s,t)\,dt\right]\phi(s).$$

The right-hand side is strictly smaller than

$$\phi(s) + \left[d(s)\int_0^M D(s,t)\,dt\right]\phi(s),$$

which contradicts equation (10.39).

It may happen that ϕ jumps downward just before the supremum is attained, so

$$\sup_{x\in[0,M]} \phi(x) = \phi(s-) > \phi(s).$$

Then the same contradiction can be derived. Formally, the argument goes as follows. One can find a $\delta > 0$ such that $\phi(s-\delta) > 0$ and

$$d(s-\delta)\int_{\{t:\phi(t)>\phi(s-\delta)\}} D(s-\delta,t)\,\phi(t)\,dt \le \frac{1}{2}\phi(s-\delta).$$

Hence

$$d(s-\delta)\int_0^M D(s-\delta,t)\,\phi(t)\,dt \le \left[d(s-\delta)\int_0^M D(s-\delta,t)\,dt\right]\phi(s-\delta) + \frac{1}{2}\phi(s-\delta),$$

again contradicting (10.39). So the homogeneous equation is only solvable for $\phi = 0$. Hence the condition in Lemma 10.1 is satisfied. □

Since the solutions ϕ_F and ξ_F are contained in $D[0, M]$ and $D[a, b]$, respectively, we already know they are bounded. However, we will need boundedness of the solutions ϕ_F and ξ_F uniformly over a class of distribution functions F. The function k is held fixed all the time. The class we will use is

$$\mathcal{F} = \{F : F \text{ is a piecewise constant distribution function satisfying (S4)}\}. \tag{10.40}$$

We have the following property of the solution.

Lemma 10.2 *Let d_F and c_F be given by (10.34) and (10.36), respectively. Then*

$$\inf_{F\in\mathcal{F}}\ \inf_{x\in[0,M]} \left(g_1(x)[1-F(x)] + g_2(x)F(x)\right) > 0,$$

implying

$$\sup_{F \in \mathcal{F}} \|d_F\|_\infty < \infty.$$

Moreover,

$$\sup_{F \in \mathcal{F}} \|c_F\|_\infty < \infty.$$

Proof Let $x \in [0, M]$ be arbitrary. For x with $\epsilon_0 \le x \le 1 - \epsilon_0$ we have, using (S3) and (S4),

$$g_1(x)\{1 - F(x)\} + g_2(x)\,F(x) \ge c(g_1(x) + g_2(x)) > 0.$$

If $x < \epsilon_0$, we have $g_2(x) = 0$. Hence $g_1(x) > 0$ by (S1), implying $g_1(x)\,[1 - F(x)] > 0$. The argument for $x > 1 - \epsilon_0$ runs in a similar way.
For c_F the argument is similar. □

This gives us the uniform boundedness property of the solutions.

Lemma 10.3 *The classes $\{\phi_F : F \in \mathcal{F}\}$ and $\{\xi_F : F \in \mathcal{F}\}$ are uniformly bounded.*

Proof Let $F \in \mathcal{F}$. Define

$$I_F(x) := \int_{t=0}^{x} \frac{\phi_F(x) - \phi_F(t)}{F(x) - F(t)}\, h(t, x)\, dt$$

and

$$J_F(x) := -\int_{t=x}^{M} \frac{\phi_F(t) - \phi_F(x)}{F(t) - F(x)}\, h(x, t)\, dt.$$

So we have

$$\phi_F = d_F[k - I_F - J_F].$$

The argument is based on the observation that I_F and J_F have a reducing influence on the value of the extremum.

First suppose that the minimum and the maximum of ϕ_F are attained. Let

$$m = \mathrm{argmin}_{[0,M]}(\phi_F) \quad \text{and} \quad s = \mathrm{argmax}_{[0,M]}(\phi_F).$$

Since ϕ_F reaches its minimum at m, the integrands in the definitions of $I_F(m)$ and $J_F(m)$ do not change, implying that $I_F(m) \le 0$ and $J_F(m) \le 0$. Hence, for each $x \in [0, M]$,

$$\phi_F(x) \ge \phi_F(m) = d_F(m)k(m) - d_F(m)[I_F(m) + J_F(m)] \ge d_F(m)k(m).$$

Likewise, from $I_F(s) \ge 0$ and $J_F(s) \ge 0$ we derive

$$\phi_F(x) \le d_F(s)k(s)$$

for every x.

If the maximum is not attained, one can use a similar kind of argument as in Theorem 10.2, yielding

$$\phi_F(x) \le k(s-)\,d_F(s-) \quad \text{for all } x.$$

If the minimum is not attained, we have $\phi_F(x) \geq k(m-)\, d_F(m-)$ for all x. From boundedness of k and uniform boundedness of $\{d_F\colon F \in \mathcal{F}\}$, uniform boundedness of $\{\phi_F\colon F \in \mathcal{F}\}$ follows. □

Remark Note that if k is nonnegative, ϕ_F is nonnegative as well; likewise, $k \leq 0$ implies $\phi_F \leq 0$.

The following lemma shows that the solutions ϕ_F and ξ_F depend continuously on F.

Lemma 10.4 *For F_1 and $F_2 \in \mathcal{F}$, the respective solutions ϕ_{F_1} and ϕ_{F_2} to the integral equation (10.33) satisfy (for some constant $K > 0$)*

$$\left\|\phi_{F_1} - \phi_{F_2}\right\|_\infty \leq K \left\|F_1 - F_2\right\|_\infty .$$

Consequently,

$$\|F_n - F\|_\infty \to 0 \quad \textit{implies} \quad \left\|\phi_{F_n} - \phi_F\right\|_\infty \to 0.$$

The same holds for ξ_F.

For the proof of this, which relies on the compactness of the integral operator, we refer to Geskus and Groeneboom, 1996. We now arrive at the crucial lemma in this treatment of the smooth functionals for interval censoring case 2, which shows that the continuous part of the solution is absolutely continuous with respect to Lebesgue measure and that the discrete part of the solution is absolutely continuous with respect to the discrete part of $F \in \mathcal{F}$, uniformly for $F \in \mathcal{F}$.

Lemma 10.5 *Under the assumptions stated in this section, we have:*

(i) The derivative of ϕ_F at the points of continuity is bounded, uniformly over $F \in \mathcal{F}$ and the points of continuity, implying

$$|\phi_F(y) - \phi_F(x)| \leq K_1\, |y - x|$$

if y and x are in the same interval between jumps. K_1 is independent of F and x. The same holds when ϕ_F is replaced by ξ_F.

(ii) At the discontinuity points x of F,

$$|\phi_F(x) - \phi_F(x-)| \leq K_2\, |F(x) - F(x-)| \ \textit{and}\ |\xi_F(x) - \xi_F(x-)| \leq K_2\, |F(x) - F(x-)|$$

with K_2 independent of x and F.

Proof (i) At each continuity point x of F we have, using left or right derivatives if necessary:

$$\begin{aligned}
\phi_F'(x) &= d_F'(x)\phi_F(x)/d_F(x)\\
&\quad + d_F(x)\left\{k'(x) - \int_0^x \frac{\phi_F'(x)}{F(x)-F(t)}\, g(t,x)\,dt - \int_x^M \frac{\phi_F'(x)}{F(t)-F(x)}\, g(x,t)\,dt\right\}\\
&\quad - d_F(x)\left\{\int_0^x \frac{\phi_F(x)-\phi_F(t)}{F(x)-F(t)}\,\frac{\partial}{\partial x} g(t,x)\,dt - \int_x^M \frac{\phi_F(t)-\phi_F(x)}{F(t)-F(x)}\,\frac{\partial}{\partial x} g(x,t)\,dt\right\}.
\end{aligned} \tag{10.41}$$

Rewriting gives, using $\phi = F[1-F]\xi$,

$$\phi_F'(x)\left\{1 + d_F(x)\int_0^x \frac{1}{F(x)-F(t)}\,g(t,x)\,dt + d_F(x)\int_x^M \frac{1}{F(t)-F(x)}\,g(x,t)\,dt\right\}$$
$$= d_F'(x)\,\xi_F(x)\,[g_1(x)(1-F(x)) + g_2(x)F(x)] \tag{10.42}$$
$$+\, d_F(x)k'(x)$$
$$-\, d_F(x)\left\{\int_0^x \frac{\phi_F(x)-\phi_F(t)}{F(x)-F(t)}\frac{\partial}{\partial x}g(t,x)\,dt - \int_x^M \frac{\phi_F(t)-\phi_F(x)}{F(t)-F(x)}\frac{\partial}{\partial x}g(x,t)\,dt\right\}.$$

By Lemma 10.2, Lemma 10.3 and the conditions, the right hand side is bounded uniformly over x and F. Since the part between brackets on the left hand side is bounded away from zero, we get uniform boundedness of ϕ_F'.

(ii) At each point of jump x of F we get a similar expression. Defining $\Delta h(x)$ by $\Delta h(x) = h(x) - h(x-)$, we have

$$\frac{\Delta\phi_F(x)}{\Delta F(x)}\left\{1 + d_F(x-)\int_0^x \frac{1}{F(x)-F(t)}\,g(t,x)\,dt + d_F(x-)\int_x^M \frac{1}{F(t)-F(x)}\,g(x,t)\,dt\right\}$$
$$= \frac{\Delta d_F(x)}{\Delta F(x)}\xi_F(x)\,[g_1(x)(1-F(x)) + g_2(x)F(x)] \tag{10.43}$$
$$+\, d_F(x-)\int_0^x \frac{\phi_F(x-)-\phi_F(t)}{\{F(x)-F(t)\}\{F(x-)-F(t)\}}\,g(t,x)\,dt$$
$$+\, d_F(x-)\int_x^M \frac{\phi_F(t)-\phi_F(x-)}{\{F(t)-F(x)\}\{F(t)-F(x-)\}}\,g(x,t)\,dt,$$

with $\Delta d_F(x)/\Delta F(x)$ given by

$$\frac{\Delta d_F(x)}{\Delta F(x)} = \frac{[1-F(x)][1-F(x-)]g_1(x) \;-\; F(x)F(x-)g_2(x)}{\{g_1(x)[1-F(x)] + g_2(x)F(x)\}\{g_1(x)[1-F(x-)] + g_2(x)F(x-)\}}.$$

Again we have boundedness uniformly over the points of jump.

□

Finally, we turn back to the solvability of $\kappa_F = L^*L_F a$.

Theorem 10.3 *Let F be an absolutely continuous distribution function, with a density f bounded away from zero, say $f \ge C > 0$. Let the conditions (M1) to (M3), (S1) to (S5) and (L1) and (L2) be satisfied. Then the equation $\kappa_F = L^*L_F a$ is solvable.*

Proof The proof follows the same pattern as Lemma 10.5, part (i). The right hand side of equation (10.42) gets an extra term, since F is no longer piecewise constant:

$$d_F(x)\left\{\int_0^x \frac{\phi(x)-\phi(t)}{(F(x)-F(t))^2}\,f(x)g(t,x)\,dt + \int_x^M \frac{\phi(t)-\phi(x)}{(F(t)-F(x))^2}\,f(x)g(x,t)\,dt\right\},$$

which is bounded uniformly over x. So we get

$$|\phi(y)-\phi(x)| \le K\,|y-x| \le (K/C)\,|F(y)-F(x)|,$$

implying $d\phi/dF \in L_2^0(F)$.

□

10.4 Smooth Functional Estimation in the Interval Censoring Case 2 Model

Having derived important properties of the solutions of crucial integral equations in the previous section, these are now used to derive the asymptotic distribution of smooth functionals within the interval censoring case 2 problem. Write F_0 for the (true) underlying distribution function of the X_i, assumed to be continuous. Let $\hat{F}_n$ be the MLE of F_0, based on the sample of observations $(T_1, U_1, \Delta_{11}, \Delta_{12}), \ldots, (T_n, U_n, \Delta_{n1}, \Delta_{n2})$. It is obtained by maximizing the likelihood

$$\prod_{i=1}^{n} F(T_i)^{\Delta_{i1}}(F(U_i) - F(T_i))^{\Delta_{i2}}(1 - F(U_i))^{1-\Delta_{i1}-\Delta_{i2}}\, g(T_i, U_i) \qquad (10.44)$$

over the class of nondecreasing functions F with $F(0) = 0$ and $F(M) = 1$. The factor $\prod g(T_i, U_i)$ is of no importance in the maximization procedure, and can be neglected.

As also seen in Section 4.7, only the values of $\hat{F}_n$ at the observation times occur explicitly in the likelihood, and not even all of them do so. If $\Delta_{i1} = 1$, i.e., $X_i \leq T_i$, the corresponding U_i does not appear in (10.44). Likewise, if $X_j > U_j$, the corresponding T_j does not enter in the likelihood. The remaining observation points are called the relevant observation points. The order restriction on $\hat{F}_n$ makes it a function that is piecewise constant and uniquely defined on large parts of its domain. Generally the intervals of constancy contain several observation times. The only places where $\hat{F}_n$ is not uniquely defined are between two consecutive ordered relevant observation times for which $\hat{F}_n$ has a different value. How $\hat{F}_n$ is chosen there has no bearing on the properties that follow. Moreover, one can show that the total length of these intervals shrinks to zero as the sample size goes to infinity.

So, without loss of generality, we impose $\hat{F}_n$ to be piecewise constant everywhere, and only to have jumps at the observation points. Then it is uniquely determined everywhere between the first and the last point of jump. The EM algorithm described in Example 7.4, Section 7.2, or the ICM algorithm described in Example 7.6, Section 7.3, can be used to compute $\hat{F}_n$. Throughout we let $J_i = [\tau_i, \tau_{i+1})$, $i = 0, \ldots, m-1$ and $J_m = [\tau_m, \tau_{m+1}]$, with $\tau_0 = 0$, $\tau_{m+1} = M$ and τ_i is a point of jump of $\hat{F}_n$, $i = 1, \ldots, m$. So τ_1 and τ_m are the first and last point of jump of $\hat{F}_n$, respectively. Except for the case that all Δ_{i1}s are one, we always have $\hat{F}_n(0) = 0$ and $\hat{F}_n$ is uniquely determined from 0 to τ_1.

At the other end we may end up with a degenerate distribution, having $\hat{F}_n(t) < 1$ at all observation points. This occurs if the largest observation time corresponds to an event time beyond that observation time. For, if $U_{(n)}$ is this largest observation time, $[1 - F(U_{(n)})]$ in the likelihood formula should be larger than zero. Then the MLE is not uniquely determined beyond $U_{(n)}$. The asymmetry between the left hand side and the right hand side of $[0, M]$ is due to the right continuity of the MLE. However, for properties concerning the limit behavior, this matter does not play any role, since the probability of getting a defective distribution function tends to zero as $n \to \infty$, as is shown in the following lemma.

Lemma 10.6

$$\lim_{n\to\infty} P_{F_0}\{\hat{F}_n \text{ is defective}\} = 0.$$

Proof Let $U_{(n)}$ denote the largest observation time U_i, and let X_i be the corresponding (unobservable) event time. Then we have, using integration by parts (see also Exercise 10.18),

$$\begin{aligned} P_{F_0}\{\hat{F}_n \text{ is defective}\} &= \int_0^M P_{F_0}\{X_1 > v\}\, n g_2(v) G_2(v)^{n-1}\, dv \\ &= \int_0^M \{1 - F_0(v)\}\, n g_2(v) G_2(v)^{n-1}\, dv \\ &= \int_0^M G_2(v)^n \, dF_0(v). \end{aligned}$$

The result now follows from Lebesgue's dominated convergence theorem. □

As seen in Section 4.7, under uniqueness, necessary and sufficient conditions for a function to maximize (10.44) can be derived. Proposition 1.3 in Groeneboom and Wellner, 1992, gives an alternative formulation for these conditions. Given a sample

$$(T_1, U_1, \Delta_{11}, \Delta_{12}), \ldots, (T_n, U_n, \Delta_{n1}, \Delta_{n2}),$$

let $\mathcal{F}_n$ be the (random) class of distribution functions F satisfying

$$\begin{cases} F(T_i) > 0 & , \text{ if } X_i \le T_i, \\ F(U_i) - F(T_i) > 0 & , \text{ if } T_i < X_i \le U_i, \\ 1 - F(U_i) > 0 & , \text{ if } X_i > U_i, \end{cases}$$

and having mass concentrated on the set of observation points augmented with an extra point bigger than all observation points. It is easily seen that $\hat{F}_n$ belongs to this class. For distribution functions $F \in \mathcal{F}_n$, the following process $t \mapsto W_F(t)$ is properly defined:

$$\begin{aligned} W_F(t) = &\int_{u\in[0,t]} \delta_1\, F(u)^{-1} dQ_n(u, v, \delta_1, \delta_2) \\ &- \int_{u\in[0,t]} \delta_2\, \{F(v) - F(u)\}^{-1} dQ_n(u, v, \delta_1, \delta_2) \\ &+ \int_{v\in[0,t]} \delta_2\, \{F(v) - F(u)\}^{-1} dQ_n(u, v, \delta_1, \delta_2) \\ &- \int_{v\in[0,t]} (1 - \delta_1 - \delta_2)\, \{1 - F(v)\}^{-1} dQ_n(u, v, \delta_1, \delta_2), \\ &\qquad \text{for } t \ge 0, \end{aligned} \tag{10.45}$$

where Q_n is the empirical probability measure of the points $(T_i, U_i, \Delta_{i1}, \Delta_{i2})$, $i = 1, \ldots, n$.

We now state Proposition 1.3 in Part II of Groeneboom and Wellner, 1992.

Lemma 10.7 *The function $\hat{F}_n$ maximizes (10.44) over all $F \in \mathcal{F}_n$ if and only if*

$$\int_{[t,\tau_m]} dW_{\hat{F}_n}(t') \le 0, \quad \forall t \ge \tau_1, \tag{10.46}$$

and

$$\int_{[\tau_1,\tau_m]} \hat{F}_n(t)\, dW_{\hat{F}_n}(t) = 0. \tag{10.47}$$

Moreover, $\hat F_n$ is uniquely determined by (10.46) and (10.47).

It is actually a different formulation of Lemma 4.7. The following corollary is an immediate consequence.

Corollary 10.1 *Any function σ that is constant on the same intervals as $\hat F_n$ satisfies*

$$\begin{aligned}\int_{J_i}\sigma(u)\,dW_{\hat F_n}(u) &= \int_{u\in J_i}\sigma(u)\left\{\frac{\delta_1}{\hat F_n(u)}-\frac{\delta_2}{\hat F_n(v)-\hat F_n(u)}\right\}dQ_n(u,v,\delta_1,\delta_2)\\ &\quad+\int_{v\in J_i}\sigma(v)\left\{\frac{\delta_2}{\hat F_n(v)-\hat F_n(u)}-\frac{1-\delta_1-\delta_2}{1-\hat F_n(v)}\right\}dQ_n(u,v,\delta_1,\delta_2)\\ &=0,\end{aligned}$$

for $i=1,\ldots,m-1$.

Proof We use the following: if $0=a_0<\cdots<a_{m-1}$, $\sum_{j=i}^{m-1}x_j\le 0$, $j=1,\ldots,m-1$ and $\sum_{j=1}^{m-1}a_ix_i=0$, then $x_1=\ldots=x_{m-1}=0$. This easily follows by writing

$$\sum_{i=1}^{m-1}a_ix_i=\sum_{i=1}^{m-1}(a_i-a_{i-1})\sum_{j=i}^{k}x_j.$$

Taking $x_i=\int_{J_i}dW_{\hat F_n}$ and $a_i=\hat F_n(\tau_i)$, and using Lemma 10.7, we get from this:

$$\int_{J_i}dW_{\hat F_n}(t)=0,\ i=1,\ldots,m-1.$$

The result now follows, since σ is constant on the intervals J_i. □

Remark Lemma 10.7 characterizes maximization of the likelihood, in contrast with the so-called self-consistency equation which only yields a necessary but not a sufficient condition. If the points of jump of the MLE, and hence the intervals of constancy, were known, the problem would be reduced to a normal maximization problem without order restrictions. Lemma 10.7 and Corollary 10.1 have the partial derivatives of the loglikelihood appearing in the integrand. The fact that only the interval $[\tau_{ß1},\tau_m]$ is playing a role is caused by the extra restriction that the solution should have values between zero and one; see also Section 4.7.

One can also prove Corollary 10.1 by noting that, for any $F\ll\hat F_n$,

$$\int\{F(u)-\hat F_n(u)\}\,dW_{\hat F_n}(u)=0,$$

and by taking $F(u)-\hat F_n(u)$ proportional to $\sigma(u)$ for $u\in J_i$, and $F(u)-\hat F_n(u)=0$, elsewhere. One can interpret this result by saying that the derivative in directions absolutely continuous with respect to $\hat F_n$ is zero. This variational idea is also omnipresent in Chapter 2.

We also have uniform consistency of the MLE of F_0 (see Groeneboom and Wellner, 1992, part II, Section 4.3):

$$P\left\{\lim_{n\to\infty}\left\|\hat F_n-F_0\right\|_\infty=0\right\}=1. \tag{10.48}$$

In Section 10.3, we defined the set $\mathcal{F}$ as the set of piecewise constant distribution functions F on $[0, M]$, satisfying condition (S4):

$$F(u) - F(t) \ge c > 0, \quad \text{if } u - t \ge \epsilon_0. \tag{10.49}$$

For $F \in \mathcal{F}$, we now define the function q_F by

$$q_F(t, u, \delta_1, \delta_2) = \delta_1 F(t) + \delta_2\{F(u) - F(t)\} + (1 - \delta_1 - \delta_2)\{1 - F(u)\}, \tag{10.50}$$

and similarly

$$q_{F_0}(u, v, \delta_1, \delta_2) = \delta_1 F_0(u) + \delta_2\{F_0(v) - F_0(u)\} + (1 - \delta_1 - \delta_2)\{1 - F_0(v)\}. \tag{10.51}$$

We will also need the functions g_F, defined by

$$g_F = \frac{q_F - q_{F_0}}{q_F + q_{F_0}}, \tag{10.52}$$

for which Theorem 2.2 in Van de Geer, 1996, specialized to our situation, holds. The theorem is given as Lemma 10.8 for easy reference.

Lemma 10.8 *Let $\mathcal{G}$ denote the set $\{g_F : F \in \mathcal{F}\}$, and suppose that, for some $0 < \nu < 2$,*

$$\lim_{C\to\infty} \limsup_{n\to\infty} P\left\{\sup_{\epsilon>0} \epsilon^{\nu} H(\epsilon, \mathcal{G}, Q_n) > C\right\} = 0, \tag{10.53}$$

where $H(\epsilon, \mathcal{G}, Q_n)$ denotes the random ϵ-entropy of the set $\mathcal{G}$ with respect to the (random) L_2-distance, defined by

$$d_{Q_n}(g_1, g_2)^2 = \int (g_1 - g_2)^2 \, dQ_n.$$

Then for

$$\tau_n \ge n^{-1/(2+\nu)}, \; n = 1, 2, \ldots, \tag{10.54}$$

we have

$$\lim_{L\to\infty} \limsup_{n\to\infty} P\left\{h(q_{\hat{F}_n}, q_{F_0}) \ge L\tau_n\right\} = 0, \tag{10.55}$$

where $h(q_{\hat{F}_n}, q_{F_0})$ denotes the Hellinger distance between $q_{\hat{F}_n}$ and q_{F_0}.

Theorem 2.2 in Van de Geer, 1996 is in fact more general than Lemma 10.8, but we only need the present version of the result. The more general result is needed, for example, for the situation where the observation times can be arbitrarily close to each other (e.g., when the density of the observation times is strictly positive on the diagonal; see, for this, Geskus and Groeneboom, 1999).

We obtain the following corollary of Lemma 10.8.

Corollary 10.2 *Let, for distribution functions F_1 and F_2 in $\mathcal{F}$, $\|F_1 - F_2\|_{G_i}$ denote the L_2-distance, defined by*

$$\|F_1 - F_2\|_{G_i}^2 = \int_0^M \{F_1(x) - F_2(x)\}^2 \, dG_i(x), \; i = 1, 2,$$

where G_1 and G_2 denote the marginal distribution functions of, respectively, the first and the second observation time of the pair of observation times (T, U). Then

(i) The Hellinger distance $h(q_{\hat{F}_n}, q_{F_0})$ *satisfies*

$$h(q_{\hat{F}_n}, q_{F_0}) = O_p(n^{-1/3}).$$

(ii) The L_2*-distance* $\|\hat{F}_n - F_0\|_{G_i}$ *satisfies*

$$\|\hat{F}_n - F_0\|_{G_i} = O_p(n^{-1/3}),\ i = 1, 2.$$

Proof (i): For the functions g_F, defined by (10.52), we have

$$\begin{aligned} g_F(t,u,\delta_1,\delta_2) &= \frac{q_F(t,u,\delta_1,\delta_2) - q_{F_0}(t,u,\delta_1,\delta_2)}{q_F(t,u,\delta_1,\delta_2) + q_{F_0}(t,u,\delta_1,\delta_2)} \\ &= \frac{2q_F(t,u,\delta_1,\delta_2)}{q_F(t,u,\delta_1,\delta_2) + q_{F_0}(t,u,\delta_1,\delta_2)} - 1 \\ &= \frac{2\delta_1\, F(t)}{F(t)+F_0(t)} + \frac{2\delta_1\{F(u)-F(t)\}}{F(u)-F(t)+F_0(u)-F_0(t)} \\ &\quad + \frac{2(1-\delta_1-\delta_2)\{1-F(u)\}}{1-F(u)+1-F_0(u)} - 1. \end{aligned} \tag{10.56}$$

In dealing with the first and third term on the right hand side of (10.56) we use arguments similar to those in Section 3.1 of Van de Geer, 1996. The functions

$$u \mapsto \frac{F(t)F_0(t)}{F(t)+F_0(t)} \tag{10.57}$$

and

$$v \mapsto \frac{(1-F(u))(1-F_0(u))}{1-F(u)+1-F_0(u)} \tag{10.58}$$

are uniformly bounded monotone functions. We have, for $F_1, F_2 \in \mathcal{F}$,

$$\begin{aligned} &\int \left\{ \frac{F_1(t)}{F_1(t)+F_0(t)} - \frac{F_2(t)}{F_2(t)+F_0(t)} \right\}^2 \delta\, dQ_n \\ &\le 2 \int \left\{ \left\{ \frac{F_1(t)F_0(t)}{F_1(t)+F_0(t)} \right\}^{1/2} - \left\{ \frac{F_2(t)F_0(t)}{F_2(t)+F_0(t)} \right\}^{1/2} \right\}^2 d\,\tilde{Q}_n, \end{aligned}$$

where $\tilde{Q}_n$ is the random measure, defined by $d\tilde{Q}_n(t,u,\delta_1,\delta_2) = F_0(t)^{-1}\delta_1\, dQ_n(t,u,\delta_1,\delta_2)$. By the Markov inequality we have for $\tilde{Q}_n$:

$$P\left\{ \int d\tilde{Q}_n \ge K \right\} \le K^{-1} \int \frac{\delta_1}{F_0(t)}\, dQ_{F_0}(t,u,\delta_1,\delta_2) = K^{-1} \int dG(t,u) = K^{-1}$$

Hence, if we denote the class of functions $u \mapsto \delta\, F(u)/\{F(u)+F_0(u)\}$, $F \in \mathcal{F}$, by $\mathcal{G}_1$ and the class of functions (10.57) by $\tilde{\mathcal{G}}_1$ (also for $F \in \mathcal{F}$), we get

$$\sup_{\epsilon>0} \epsilon\, H(\epsilon, \mathcal{G}_1, Q_n) = \sup_{\epsilon>0} \epsilon\, H(\epsilon, \tilde{\mathcal{G}}_1, \tilde{Q}_n) = O_p(1),$$

using entropy results from Ball and Pajor, 1990, or Birman and Solomjak, 1967. In a similar way, denoting the class of functions

$$u \mapsto (1-\delta-\gamma)(1-F(u))/\{1-F(u)+1-F_0(u)\},\ F \in \mathcal{F},$$

by $\mathcal{G}_3$, we get

$$\sup_{\epsilon>0} \epsilon\, H(\epsilon, \mathcal{G}_3, Q_n) = O_p(1).$$

Finally, denoting the class of functions

$$(u,v) \mapsto \gamma(F(v)-F(u))/\{F(v)-F(u)+F_0(v)-F_0(u)\},\ F \in \mathcal{F} \tag{10.59}$$

by $\mathcal{G}_2$, we also get

$$\sup_{\epsilon>0} \epsilon\, H(\epsilon, \mathcal{G}_2, Q_n) = O_p(1).$$

This follows from the fact that, by (S3) and (S4), the denominator of the function (10.59) is bounded away from zero on the set where the distribution function G of the pair of observation times (T, U) puts its mass, together with the fact that the numerator of (10.59) contains the difference of uniformly bounded monotone functions.

Thus, denoting the class of functions $\{g_F : F \in \mathcal{F}\}$ by $\mathcal{G}$, we get

$$\sup_{\epsilon>0} \epsilon\, H(\epsilon, \mathcal{G}, Q_n) = O_p(1).$$

It follows that we can apply Lemma 10.8 with $\nu = 1$, and (i) now follows.
(ii): This follows from (i) and the inequalities (see Exercise 10.20)

$$(\hat F_n - F_0)^2 \le 4\left(\sqrt{\hat F_n} - \sqrt{F_0}\right)^2 \quad \text{and} \quad (\hat F_n - F_0)^2 \le 4\left(\sqrt{1-\hat F_n} - \sqrt{1-F_0}\right)^2.$$

□

Before formulating the main theorem, we still have two extra conditions.

(D1) $$K(F) - K(F_0) = \int \kappa_{F_0}(x)\, d(F-F_0)(x) + O(\|F-F_0\|_2^2),$$

for all distribution functions F with support contained in $[0, M]$, where $\|F - F_0\|_2$ is the L_2-distance between the distribution functions F and F_0 with respect to Lebesgue measure on $\mathbb{R}$.

(D2) The distribution function F_0 has a density bounded away from zero.

We are now ready to formulate our main theorem.

Theorem 10.4 *Let the following conditions on F_0, G and κ_{F_0} be satisfied: (M1) to (M3), (S1) to (S3), (S5), (L1) and (L2) of the preceding section and (D1) and (D2). Then we have*

$$\sqrt{n}\big(K(\hat F_n) - K(F_0)\big) \xrightarrow{\mathcal{D}} N\big(0, \|\theta_{F_0}\|_{Q_{F_0}}^2\big) \quad \text{as} \quad n \to \infty. \tag{10.60}$$

Proof Note that it is sufficient to show that

$$\sqrt{n}(K(\hat{F}_n) - K(F_0)) = \sqrt{n} \int \theta_{F_0}\, d(Q_n - Q_{F_0}) + o_p(1). \tag{10.61}$$

Then an application of the central limit theorem yields that the MLE of $K(F_0)$ has the desired asymptotically optimal behavior. The proof consists of the following steps, which we also took in Section 10.2 for the current status model.

(i) *The nonlinear aspect of the functional is negligible.*
By conditions (S1) and (D1), and Corollary 10.2, part (ii), we have

$$\sqrt{n}(K(\hat{F}_n) - K(F_0)) = \sqrt{n} \int \kappa_{F_0}\, d(\hat{F}_n - F_0) + o_p(1).$$

(ii) *Transformation to the observation space measure.*
For $F \in \mathcal{F}$, we have defined the functions ϕ_F and ξ_F as solutions to the integral equations (10.33) and (10.35), respectively. These solutions can be used to extend the definition of the canonical gradient to θ_F for $F \in \mathcal{F}$:

$$\theta_F(t, u, \delta_1, \delta_2) := \delta_1 \frac{\phi_F(u)}{F(u)} + \delta_2 \frac{\phi_F(u) - \phi_F(t)}{F(u) - F(t)} - (1 - \delta_1 - \delta_2) \frac{\phi_F(u)}{1 - F(u)}, \tag{10.62}$$

where $\phi_F(t)/F(t)$ and $\phi_F(u)/(1 - F(u))$ are defined to be zero if $F(t) = 0$ or if $F(u) = 1$, respectively. (At points where denominator the middle part of (10.62) becomes zero, we have $h(t, u) = 0$ as well, so there we need not define θ_F.) Note that these θ_F no longer have an interpretation as canonical gradient. In Lemma 10.9 the following will be shown:

$$\int \kappa_{F_0}\, d(F - F_0) = -\int \theta_F\, dQ_{F_0}.$$

Recall that the latter integral can be restricted to $\{0 \le t \le u - \epsilon_0 \le M\}$.

(iii) *Use that $\hat{F}_n$ is the MLE.*
We will use Corollary 10.1. Since $\phi_{\hat{F}_n}$ and $\xi_{\hat{F}_n}$ are not piecewise constant, we introduce the functions $\overline{\phi}_{\hat{F}_n}$ and $\overline{\xi}_{\hat{F}_n}$. These functions are constant on the same intervals $J_i = [\tau_i, \tau_{i+1})$ as $\hat{F}_n$. The values of $\overline{\phi}_{\hat{F}_n}(x)$ and $\overline{\xi}_{\hat{F}_n}$ on J_i is defined to be as $\overline{\phi}_{\hat{F}_n}(t) = \phi_{\hat{F}_n}(\hat{A}_n(t))$ and $\overline{\xi}_{\hat{F}_n}(t) = \xi_{\hat{F}_n}(\hat{A}_n(t))$, respectively, where $\hat{A}_n$ is defined as in (10.25). Let $\overline{\theta}_{\hat{F}_n}$ denote the function defined in (10.62), but with $\phi_{\hat{F}_n}$ replaced by $\overline{\phi}_{\hat{F}_n}$. Now Corollary 10.1 says

$$\int \overline{\theta}_{\hat{F}_n}\, dQ_n = 0,$$

yielding

$$-\sqrt{n} \int \theta_{\hat{F}_n}\, dQ_{F_0} = \sqrt{n} \int \overline{\theta}_{\hat{F}_n}\, d(Q_n - Q_{F_0}) + \sqrt{n} \int (\overline{\theta}_{\hat{F}_n} - \theta_{\hat{F}_n})\, dQ_{F_0}. \tag{10.63}$$

The second term will be shown to be $o_p(1)$ in Lemma 10.10. Note that, since points (T_i, U_i) with $U_i - T_i < \epsilon_0$ do not occur, the area of integration of Q_n can be taken to be $\{0 \le t \le u - \epsilon_0 \le M\}$ as well.

(iv) *Asymptotic variance equals information lower bound.*
The first term on the right hand side of (10.63) is further split into

$$\sqrt{n}\int \overline{\theta}_{\hat{F}_n} d(Q_n - Q_{F_0}) = \sqrt{n}\int \theta_{F_0} d(Q_n - Q_{F_0})$$
$$+ \sqrt{n}\int (\overline{\theta}_{\hat{F}_n} - \theta_{F_0}) d(Q_n - Q_{F_0}).$$

The last term will be shown to be $o_p(1)$ (Lemma 10.11).

This finishes the proof. □

The remaining part of this section is devoted to proving the lemmas quoted in (i), (iii) and (iv).

Lemma 10.9 *For any $F \in \mathcal{F}$ we have*

$$\int \kappa_{F_0}\, d(F - F_0) = -\int \theta_F\, dQ_{F_0}.$$

Proof Let, for $F \in \mathcal{F}$, $L_F : L_2(F) \to L_2(Q_F)$ be defined by (10.29) with adjoint L^*, defined by (10.30). As noted after definition (10.30), the structure of L^* does not depend on F. Furthermore, let $1 \in L_2(F)$ denote the constant function $1(x) \equiv 1$, $x \in \mathbb{R}$. Under L_F this transforms into the constant function $1(t, u, \delta, \gamma) \equiv 1$ on $L_2(Q_F)$. We have

$$\int \theta_F\, dQ_{F_0} = <\theta_F, 1>_{Q_{F_0}} = <\theta_F, L_{F_0}(1)>_{Q_{F_0}}$$
$$= <L^*(\theta_F), 1>_{F_0} = \int L^*(\theta_F)\, dF_0.$$

If we can prove

$$L^*(\theta_F) = \kappa_{F_0} - \int \kappa_{F_0}\, dF \quad \text{a.e.-}F_0,$$

we are done. This is shown as follows.

Recall that the integral equation was obtained by taking derivatives in the equation $\kappa_{F_0}(x) = [L^*\theta_F](x)$ for all $x \in [0, M]$, with $\theta_F = \theta_{F_0}$. Hence we get by integrating:

$$[L^*\theta_F](x) = [\kappa_{F_0}](x) + C, \qquad \text{for all } x \in [0, M].$$

For the constant C we have

$$C = \int C\, dF = \int L^*(\theta_F)\, dF - \int \kappa_{F_0}\, dF = <L^*(\theta_F), 1>_F - \int \kappa_{F_0}\, dF.$$

It is easily seen that θ_F is contained in $L_2^0(Q_F)$. We also have

$$<L^*(\theta_F), 1>_F = <\theta_F, L_F(1)>_{Q_F} = <\theta_F, 1>_{Q_F} = 0.$$

The result now follows. □

Note that this result can as well be proved by writing down the integrals. However, the proof presented here suggests that a similar result may hold more generally. Basically, what is needed is:

- $[L^*\theta_{\hat{F}_n}](x) = [\kappa_{F_0}](x) + C$ for all $x \in [0, M]$,
- $\theta_{\hat{F}_n} \in L_2(Q_{F_0})$,
- $\theta_{\hat{F}_n} \in L_2^0(Q_{\hat{F}_n})$.

The next lemma shows that we can replace $\theta_{\hat{F}_n}$ by the function $\overline{\theta}_{\hat{F}_n}$, belonging to the range of the score operator $L_{\hat{F}_n}$.

Lemma 10.10

$$\int \left(\overline{\theta}_{\hat{F}_n} - \theta_{\hat{F}_n}\right) dQ_{F_0} = O_p(n^{-2/3}).$$

Proof Let the function ψ_n be defined by

$$\begin{aligned}\psi_n(u, v) = &- [\overline{\theta}_{\hat{F}_n} - \theta_{\hat{F}_n}](u, v, 1, 0)\, F_0(u) - [\overline{\theta}_{\hat{F}_n} - \theta_{\hat{F}_n}](u, v, 0, 1)\, [F_0(v) - F_0(u)]\\ &+ [\overline{\theta}_{\hat{F}_n} - \theta_{\hat{F}_n}](u, v, 0, 0)\, [1 - F_0(v)].\end{aligned}$$

Using the decomposition $\phi_F = F(1-F)\xi_F$, and

$$F(v) - F(u) = -[(1 - F(v)) - (1 - F(u))],$$

we get

$$\begin{aligned}\psi_n(u, v) = &\frac{1 - \hat{F}_n(u)}{\hat{F}_n(v) - \hat{F}_n(u)}\,(\overline{\xi}_{\hat{F}_n}(u) - \xi_{\hat{F}_n}(u)) \cdot\\ &\cdot [F_0(v)\,(\hat{F}_n(u) - F_0(u)) + F_0(u)\,(F_0(v) - \hat{F}_n(v))]\\ &- \frac{\hat{F}_n(v)}{\hat{F}_n(v) - \hat{F}_n(u)}\,(\overline{\xi}_{\hat{F}_n}(v) - \xi_{\hat{F}_n}(v)) \cdot\\ &\cdot [(1 - F_0)(v)\,(\hat{F}_n(u) - F_0(u)) + (1 - F_0)(u)\,(F_0(v) - \hat{F}_n(v))].\end{aligned}$$

Applying the Cauchy-Schwarz inequality we get:

$$\begin{aligned}&\left|\sqrt{n}\int \left(\overline{\theta}_{\hat{F}_n} - \theta_{\hat{F}_n}\right) dQ_{F_0}\right|\\ &\le \sqrt{n}\,K\,\left\|\overline{\xi}_{\hat{F}_n} - \xi_{\hat{F}_n}\right\|_{G_1} \times \left[\left\|\hat{F}_n - F_0\right\|_{G_1} + \left\|\hat{F}_n - F_0\right\|_{G_2}\right]\\ &\quad + \sqrt{n}\,K\,\left\|\overline{\xi}_{\hat{F}_n} - \xi_{\hat{F}_n}\right\|_{G_2} \times \left[\left\|\hat{F}_n - F_0\right\|_{G_1} + \left\|\hat{F}_n - F_0\right\|_{G_2}\right]\end{aligned}$$

By Lemma 4 (i) of Geskus and Groeneboom, 1996, and (D2) we find

$$|\overline{\xi}_{\hat{F}_n}(u) - \xi_{\hat{F}_n}(u)| \le K\,|\hat{F}_n(u) - F_0(u)|, \tag{10.64}$$

and applying Corollary 10.2 finishes the proof. Property (10.64) is seen as follows. For example, if the interval $J_i \ni u$ has a point s where $\hat{F}_n$ and F_0 have equal value, we have

$$\begin{aligned}|\bar{\xi}_{\hat{F}_n}(u) - \xi_{\hat{F}_n}(u)| = |\xi_{\hat{F}_n}(s) - \xi_{\hat{F}_n}(u)| &\le K_1\,|s-u| \\ &\le (K_1/C)\,|F_0(s) - F_0(u)| \\ &= (K_1/C)\,|\hat{F}_n(s) - F_0(u)| \\ &= (K_1/C)\,|\hat{F}_n(u) - F_0(u)|.\end{aligned}$$

The same argument is used for the other two situations, with s replaced by τ_i or $\tau_{i+1}-$ and one $=$ sign replaced by a $<$ sign. □

Lemma 10.11

$$\sqrt{n}\int(\bar{\theta}_{\hat{F}_n} - \theta_{F_0})d(Q_n - Q_{F_0}) = o_p(1).$$

Proof Consider the class of functions

$$\mathcal{K} = \{\bar{\theta}_F - \theta_{F_0} : F \in \mathcal{F}\}.$$

By the uniform consistency result (10.48) we know that $\bar{\theta}_{\hat{F}_n} - \theta_{F_0} \in \mathcal{K}$ with probability tending to one as $n \to \infty$. Hence it suffices to consider distribution functions in $\mathcal{F}$.

Let $H(\epsilon, \mathcal{K}, Q_n)$ denote the random ϵ-entropy with respect to the L_2-distance d_n, defined by

$$d_n(k_1, k_2)^2 = \int (k_1 - k_2)^2\, dQ_n,\ k_1, k_2 \in \mathcal{K}.$$

Then

$$\sup_{\epsilon>0} \epsilon H(\epsilon, \mathcal{K}, Q_n) = O_p(1). \tag{10.65}$$

This is shown in the following way. The function $\bar{\theta}_F$ can be written

$$\begin{aligned}\bar{\theta}_F(u, v, \delta, \gamma) = -\delta\bar{\xi}_F(u)(1 - F(u) - \gamma\frac{\bar{\phi}_F(v) - \bar{\phi}_F(v)}{F(v) - F(u)} \\ +(1-\gamma-\delta)\bar{\xi}_F(v)F(v),\ u < v,\end{aligned} \tag{10.66}$$

where $\bar{\xi}_F(u) = \bar{\phi}_F(u)/\{F(u)(1 - F(u))\}$. By Lemma 4 of Geskus and Groeneboom, 1996, the functions ξ_F are of uniformly bounded variation, for $F \in \mathcal{F}$. This implies that the functions $\bar{\xi}_F$ are also of uniformly bounded variation, for $F \in \mathcal{F}$.

The middle part of (10.66) is composed of the functions $\bar{\phi}_F$ and F; both functions belong to a class of functions of uniformly bounded variation. This is clear for the distribution function F and for the functions $\bar{\phi}_F$ it again follows from Lemma 4 of Geskus and Groeneboom, 1996. In an approximating net for the functions $\{\bar{\phi}_F(v) - \bar{\phi}_F(v)\}/\{F(v) - F(u)\}$ we can take functions of the type

$$\frac{\bar{\phi}_m(v) - \bar{\phi}_m(v)}{F_l(v) - F_k(u)},$$

where $\bar{\phi}_m$ is an approximand to $\bar{\phi}_F$, and F_k and F_l are approximands to F such that $F_k \le F \le F_l$ ((F_k, F_l) is a bracket for F; see Section 3.4). Here we do not assume that $\bar{\phi}_m$ is related to F_k or F_l via $\bar{\phi}_m = \bar{\phi}_{F_k}$ or $\bar{\phi}_m = \bar{\phi}_{F_l}$; the only thing needed is that $\bar{\phi}_m$ belongs to a class of functions of uniformly bounded variation. We have, if $u - v \ge \epsilon_0$:

$$\left|\frac{\bar{\phi}_F(v) - \bar{\phi}_F(v)}{F(v) - F(u)} - \frac{\bar{\phi}_m(v) - \bar{\phi}_m(v)}{F_l(v) - F_k(u)}\right|$$
$$\le K\Big\{\left|\bar{\phi}_F(u) - \bar{\phi}_m(u)\right| + \left|\bar{\phi}_F(v) - \bar{\phi}_m(v)\right| + |F(u) - F_k(u)| + |F(v) - F_l(v)|\Big\},$$

for a constant $K > 0$, since, by the definition of $\mathcal{F}$,

$$F_l(v) - F_k(u) \ge F(v) - F(u) \ge c > 0, \text{ if } u - v \ge \epsilon_0.$$

Hence we get

$$\int\left\{\frac{\bar{\phi}_F(v) - \bar{\phi}_F(v)}{F(v) - F(u)} - \frac{\bar{\phi}_m(v) - \bar{\phi}_m(v)}{F_l(v) - F_k(u)}\right\}^2 \gamma\, dQ_n$$
$$\le 4K^2 \int\Big\{\left(\bar{\phi}_F(u) - \bar{\phi}_m(u)\right)^2 + \left(\bar{\phi}_F(v) - \bar{\phi}_m(v)\right)^2$$
$$+ \left(F(u) - F_k(u)\right)^2 + \left(F(v) - F_l(v)\right)^2\Big\}\,\gamma\, dQ_n. \tag{10.67}$$

Relation (10.65) now follows from (10.66), (10.67) and the uniformly bounded variation of the functions F, $\bar{\xi}_F$ and $\bar{\phi}_F$, for $F \in \mathcal{F}$, using the entropy results in Ball and Pajor, 1990 or Birman and Solomjak, 1967. Thus we get for the random entropy integral:

$$\int_0^{\epsilon} H(u, \mathcal{K}, Q_n)^{1/2}\, du = O_p(\epsilon^{1/2}),\ \epsilon > 0.$$

Using uniform consistency of $\hat{F}_n$, Lemma 3 of Geskus and Groeneboom, 1996, and the uniform boundedness of $\bar{\theta}_{\hat{F}_n}$, we get

$$\int \left\{\bar{\theta}_{\hat{F}_n} - \theta_{F_0}\right\}^2 dQ_n \to 0,$$

with probability one. The result now follows by a standard application of the chaining lemma; see, e.g., Pollard, 1984, p. 150 (15, Equicontinuity Lemma). □

Exercises

10.1 Show that in the normal deconvolution model of Section 10.1,

$$\int x\, d\hat{F}_n(x) = \frac{1}{n}\sum_{i=1}^{n} Z_i - \mu,$$

where $\hat{F}_n$ is the ML estimator of F_0 and $\mu = E_k Y_i$ is the expected value of the (normal) noise. Hint: in Lemma 4.6, necessary and sufficient conditions for the ML estimators are derived by taking derivatives of the log likelihood in the vertex directions $F_x = 1_{[x,\infty)}$. There are also other ways of deriving necessary conditions for optimality. Given the ML estimator $\hat{F}_n$, it is obvious that for $y \in \mathbb{R}$ the function

$$F_y(x) = \hat{F}_n(x - y)$$

is also a distribution function. Use this fact, a variational argument and the crucial relation between the normal density and its derivative,

$$k'(x) = (\mu - x)k(x)$$

to show the required equality.

10.2 Consider the normal deconvolution model with standard normal noise, so $\mu = 0$ and $\sigma^2 = 1$, and sample size $n = 2$.
   a) Show that the estimator of the variance of the underlying random variable X given in (10.3) reduces to

$$U_2 = \frac{1}{2}(Z_1 - Z_2)^2 - 1.$$

   b) Compute in this setting the probability that U_2 is negative, and conclude that this probability is nonzero.

10.3 Let Y be uniformly distributed on [0, 1]. Show that its characteristic function is given by

$$\phi(t) = Ee^{itY} = \frac{1}{t}(\sin t + i(\cos t - 1))$$

so that $\phi(t) = 0$ for $t = 2\pi k$ with $k = \pm 1, \pm 2, \ldots$.

10.4 Show that the random variables Z_i', $1 \le i \le n$, as defined in (10.4), are independent and standard uniformly distributed on [0, 1].

10.5 Show that in the uniform deconvolution problem, $\sigma^2 \ge \sigma_{F_0}^2$ with equality if and only if F_0 corresponds to the uniform distribution on [0, 1]. Here $\sigma_{F_0}^2$ and σ^2 are defined in (10.5) and (10.6). Hint: write $\psi(x) = F_0(x) - x$ and show that

$$\sigma^2 - \sigma_{F_0}^2 = \int_0^1 \psi(x)^2\,dx - \left(\int_0^1 \psi(x)\,dx\right)^2.$$

10.6 Let X have probability density f on $[0, \infty)$, T be independent of X with density g on $[0, \infty)$. Define $\Delta = 1_{[X \le T]}$. Show that the conditional densities of X, given that $T = t$ and $\Delta = 0$ (or $\Delta = 1$), are given by

$$f_{X|(T,\Delta)}(x|t, 0) = \frac{f(x)}{1 - F(x)} 1_{(t,\infty)}(x) \quad \text{and} \quad f_{X|(T,\Delta)}(x|t, 1) = \frac{f(x)}{F(x)} 1_{[0,t]}(x).$$

Use this to derive the second equality in (10.9).

10.7 The defining property of the adjoint is that

$$\langle\langle L_F(a), b\rangle_{L_2(Q)} = \langle\langle a, L^*(b)\rangle\rangle_{L_2(F)}.$$

Show that L_F and L^* defined in (10.9) and (10.10) satisfy this equality.

10.8 Let $\hat{F}_n$ be the MLE for the current status model, discussed in Theorem 10.1. Prove that

$$\mathbb{P}\left\{\hat{F}_n(b) = 1\right\} \xrightarrow{p} 1, \; n \to \infty$$

under the conditions of Theorem 10.1.

10.9 Deduce the score equation (10.12).

10.10 Verify the two conditional expectations given in (10.29) and (10.30).

10.11 Let $h\left(q_{\hat{F}_n}, q_{F_0}\right)$ be defined as in (10.19). Show that, under the conditions of Theorem 10.1,

$$\|\hat{F}_n - F_0\|_2^2 = O_p\left(n^{-2/3}\right).$$

Hint: use (10.19).

10.12 Verify (10.20).

10.13 Why does (10.27) hold?

10.14 Let $\bar{\phi}_{\hat{F}_n}$ be defined by (10.26). Show that

$$h\left(\bar{\phi}_{\hat{F}_n}, \phi_{\hat{F}_n}\right) = O_p\left(n^{-1/3}\right),$$

where h is the Hellinger distance.

10.15 With the notation introduced in Section 10.2, show that

$$\begin{aligned}\int \left\{\bar{\theta}_{\hat{F}_n} - \theta_{\hat{F}_n}\right\} dQ &= \int \left(\hat{F}_n - F_0\right) \frac{\bar{\phi}_{\hat{F}_n}(t) - \phi_{\hat{F}_n}(t)}{\hat{F}_n(t)\{1 - \hat{F}_n(t)\}}\, dG \\ &= \int \left\{\hat{F}_n(t) - F_0(t)\right\} \frac{g(t) - g(\hat{A}_n(t))}{g(\hat{A}_n(t))}\, dt.\end{aligned}$$

10.16 Deduce, using $\min_{t\in[a,b]} g(t) > 0$, from Exercise 10.15 that in the context of Section 10.2

$$\int \left\{\bar{\theta}_{\hat{F}_n} - \theta_{\hat{F}_n}\right\} dQ = O_p\left(n^{-2/3}\right). \tag{10.68}$$

This implies (10.23).

10.17 Show that condition (S1) in Section 10.3 implies the function d_F defined in (10.34) to be bounded.

10.18 In the context of the interval censoring case 2 model, show that the probability that the ML estimator based on a sample of size n is defective can be represented as

$$P_{F_0}\left\{\hat{F}_n \text{ is defective}\right\} = \int_0^M (1 - F_0(v))\, n g_2(v) G_2(v)^{n-1}\, dv.$$

Here F_0 is the distribution function of the event times X_i and G_2 is the marginal distribution function of U_i (with corresponding density function g_2). Hint: observe that the ML estimator is defective if and only if the observation indicators Δ_{i1} and Δ_{i2} corresponding to the index i, with the maximal value of U_i, equal zero.

10.19 Show that the formulations of the necessary and sufficient conditions for the (unique; so piecewise constant with jumps only possible at the recorded observation times) MLE in the case 2 interval censoring problem given in Lemma 10.7 and Lemma 4.7 are equivalent.

10.20 For $0 \le a, b \le 1$, show that

$$(a - b)^2 \le 4\left(\sqrt{a} - \sqrt{b}\right)^2.$$

Bibliographic Remarks

The pointwise asymptotic distribution of the MLE in the uniform deconvolution problem also follows from Example 11.2.3b, p. 226, in Van de Geer, 2000, where it is at the same time shown that (10.5) is the efficient asymptotic variance. The uniform (boxcar) deconvolution is also studied in Groeneboom and Jongbloed, 2003, Johnstone and Raimondo, 2004, and Feuerverger et al., 2008. Asymptotic result (10.7) was derived in part II, chapter 5 of Groeneboom and Wellner, 1992. The proof of Theorem 10.1 partly follows the approach in Geskus, 1997, p. 68 and 69. For the conditional expectation representation of the score operators, see Part I of Groeneboom and Wellner, 1992, and Bickel et al., 1998. The treatment of the integral equation for the separated case interval censoring case 2 problem, given in Section 10.3, where $\mathbb{P}\{U_i - T_i < \epsilon\} = 0$ for some $\epsilon > 0$, follows Geskus and Groeneboom, 1996. The approach to this problem seems to give the prototype for what to do in the case that the score functions are not explicitly given; this matter is further discussed in the discussion paper Groeneboom, 2014. The nonseparated case is considered in Groeneboom, 1996, and Geskus and Groeneboom, 1999. For general asymptotic theory for smooth functionals in statistical models, see Van der Vaart, 1991.

11

Pointwise Asymptotic Distribution Theory for Univariate Problems

In Chapter 3, pointwise asymptotic results are derived for estimators in some of the basic models involving monotonicity as described in Chapter 2. In this chapter, further asymptotic pointwise results will be derived, now for estimators introduced in Chapter 4 and Chapter 8. The first, in Section 11.1, gives the asymptotic distribution of the least squares estimator of a convex decreasing density, as introduced in Section 4.3. This needs to be derived solely from the characterization of the estimator, since an explicit representation of the estimator is lacking. The approach is based on the asymptotic behavior of the characterization. Section 11.2 is concerned with an interesting and useful tail bound for the maximum likelihood estimator in the current status model introduced in Section 2.3.

In Section 11.3, a local variant of smooth functional methods is applied to derive the asymptotic pointwise distribution of the smoothed maximum likelihood estimator (SMLE) in the current status model as introduced in Section 8.1. The $n^{1/3}$ rate of convergence for the plain MLE of the distribution function derived in Section 3.8 is replaced by the rate $n^{2/5}$ for the SMLE. For the interval censoring case 2 model of Section 4.7, the SMLE and the maximum smoothed likelihood estimator (MSLE) are considered in Section 11.4 and Section 11.5. Under the separation of inspection times hypothesis, the rates of convergence of these estimators are shown to be $n^{2/5}$, just as in the current status situation.

Finally, in Section 11.6, the problem of estimating a nondecreasing hazard rate under right censoring as introduced in Section 2.6 is considered. Also in this setting local smooth functional theory is applied to derive the asymptotic distribution on the SMLE.

11.1 The LS Estimator of a Convex Density

The least squares estimator of a convex decreasing density as introduced and studied in Section 4.3 cannot be expressed in terms of the empirical distribution as easily as, e.g., the maximum likelihood (or least squares) estimator of a decreasing density. The characterization in Lemma 4.4 is rather implicit. Consequently, methods based on explicit constructions for the estimators as applied in Chapter 3 cannot be applied to establish the asymptotics of this estimator. Instead, the asymptotic distribution of $\hat{g}_n$ is derived by taking its characterization to the limit.

An important result we will not prove here concerns the existence of a joint distribution that is related to Brownian motion. The asymptotic distributions of $\hat{g}_n(x)$ and $\hat{g}_n'(x)$ will be expressed in terms of this distribution. In fact, this invelope of integrated Brownian Motion $+t^4$ is a process that takes a prominent role the asymptotic behavior of estimators of convex functions and can in that sense be compared to the greatest convex minorant of Brownian motion $+t^2$ (see Section 3.9) in the context of estimating monotone functions.

Theorem 11.1 *(Theorem 2.1 and Corollary 2.1(ii) in Groeneboom et al., 2001b). Let* $X(t) = W(t) + 4t^3$ *where* $W(t)$ *is standard two-sided Brownian motion starting from* 0*, and let* Y *be the integral of* X*, satisfying* $Y(0) = 0$*. Thus* $Y(t) = \int_0^t W(s)ds + t^4$ *for* $t \geq 0$*. Then there exists an almost surely uniquely defined random continuous function* H *(invelope of integrated Brownian motion* $+t^4$*) satisfying the following conditions:*

(i) The function H *is everywhere above the function* Y*:*

$$H(t) \geq Y(t), \text{ for each } t \in \mathbb{R}. \tag{11.1}$$

(ii) H *has a convex second derivative, and, with probability one,* H *is three times differentiable at* $t = 0$.

(iii) The function H *satisfies*

$$\int_{\mathbb{R}} \{H(t) - Y(t)\}\, dH^{(3)}(t) = 0. \tag{11.2}$$

Note that whenever inequality (11.1) holds strictly on an interval, (11.2) implies that on this interval $H^{(3)}$ has to be constant, meaning that H will be a cubic polynomial on this interval.

In fact, the second derivative of H gives the random function that is, after rescaling, the asymptotic form of the convex least squares estimator and the third derivative gives the piecewise constant monotone estimate that is the derivative of this estimator. A picture of the realization of the process Y and the associated process H is given in Figure 11.1.

The limit result we will discuss in this section is Theorem 11.2. In contrast with the limit distribution of the Grenander estimator, we establish here the distributional convergence of a pair of estimators, the convex density itself and its derivative.

Theorem 11.2 *Suppose that* g_0 *is a convex decreasing density on* $[0, \infty)$ *and* x_0 *is such that* $g_0''(x_0) > 0$ *and that* g_0'' *is continuous in a neighborhood of* x_0*. Then the least squares estimator* $\hat{g}_n$*, studied in Section 4.3, satisfies*

$$\begin{pmatrix} n^{2/5} c_1(g_0)(\hat{g}_n(x_0) - g_0(x_0)) \\ n^{1/5} c_2(g_0)(\hat{g}_n'(x_0) - g_0'(x_0)) \end{pmatrix} \to_d \begin{pmatrix} H''(0) \\ H^{(3)}(0) \end{pmatrix}$$

where $(H''(0), H^{(3)}(0))$ *are the second and third derivatives at* 0 *of the invelope* H *of* Y *as described in Theorem 11.1 and*

$$c_1(g_0) = \left(\frac{24}{g_0^2(x_0) g_0''(x_0)} \right)^{1/5}, \quad c_2(g_0) = \left(\frac{24^3}{g_0(x_0) g_0''(x_0)^3} \right)^{1/5}. \tag{11.3}$$

The derivatives $\hat{g}_n'(x_0)$ *may be interpreted as left or right derivatives.*

Remark As noted in Theorem 11.2, the derivatives $\hat{g}_n'(x_0)$ may be interpreted as left or right derivatives. To avoid mentioning this matter again in the following text, we will, somewhat arbitrarily, settle on taking the right derivatives in the sequel.

Remark From Exercise 6.13, an asymptotic lower bound to the minimax risk for estimating g_0 at a fixed point can be derived. If the approach used in Example 6.2 is followed, this

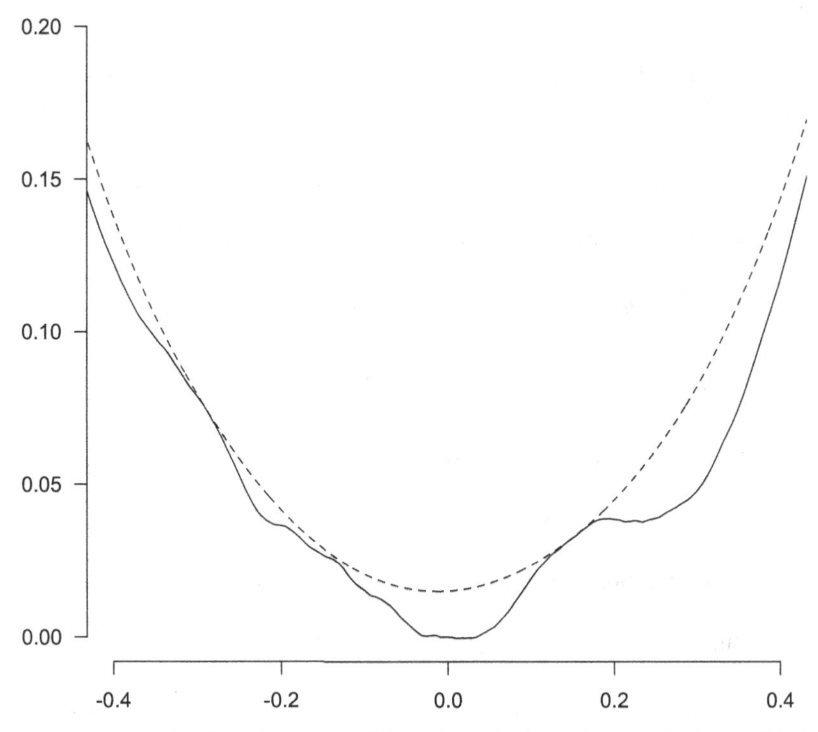

Figure 11.1 The functions Y (solid) and H (dashed) for standard two-sided Brownian motion on $[-0.4, 0.4]$.

bound equals (apart from a constant not depending on g_0) $1/c_1(g_0)$. See, e.g., Theorem 5.1 in Groeneboom et al., 2001a. This phenomenon can also be seen in the context of estimating a decreasing density using the Grenander estimator. Compare to this end the lower bound obtained in Example 6.2 and the asymptotic distribution of the Grenander estimator given in (3.12).

As usual, we first consider the rate of convergence that can be expected. Define a localized Y_n-process by

$$Y_n^{loc}(t) \equiv n^{4/5} \int_{x_0}^{x_0+n^{-1/5}t} \left\{ \mathbb{G}_n(v) - \mathbb{G}_n(x_0) - \int_{x_0}^{v} (g_0(x_0) + (u - x_0)g_0'(x_0))\, du \right\} dv \tag{11.4}$$

and a localized H_n-process by

$$H_n^{loc}(t) \equiv n^{4/5} \int_{x_0}^{x_0+n^{-1/5}t} \int_{x_0}^{v} \{\hat{g}_n(u) - g_0(x_0) - (u - x_0)g_0'(x_0)\}\, du\, dv + A_n t + B_n \tag{11.5}$$

where

$$A_n = n^{3/5} \left\{ \hat{G}_n(x_0) - \mathbb{G}_n(x_0) \right\} \quad \text{and} \quad B_n = n^{4/5} \{H_n(x_0) - Y_n(x_0)\}.$$

Noting that

$$A_n = n^{3/5}\left\{\hat{G}_n(x_0) - \hat{G}_n(x_n^-) - (\mathbb{G}_n(x_0) - \mathbb{G}_n(x_n^-))\right\},$$

where

$$x_n^- \equiv \max\{t \le x_0 \colon \hat{H}_n(t) = Y_n(t) \quad \text{and} \quad \hat{H}_n'(t) = Y_n(t)\},$$

we want to show that (A_n) is tight. We have

$$|A_n| = n^{3/5}\left|\int_{x_n^-}^{x_0} \hat{g}_n(u) - g_0(x_0) - (u - x_0)g_0'(x_0)\,du \right.$$
$$\left. - \int_{x_n^-}^{x_0} g_0(u) - g_0(x_0) - (u - x_0)g_0'(x_0)\,du - \int_{x_n^-}^{x_0} d(\mathbb{G}_n - G_0)(u)\right|,$$

which will be $O_p(1)$, provided we can show that

$$x_0 - \tau_n^- = O_p\left(n^{-1/5}\right), \tag{11.6}$$

where τ_n^- is the last jump point of $\hat{g}_n'$ before x_0,

$$\sup_{|t|\le M}\left|\hat{g}_n(x_0 + n^{-1/5}t) - g_0(x_0) - n^{-1/5}tg_0'(x_0)\right| = O_p(n^{-2/5}) \tag{11.7}$$

and

$$\sup_{|t|\le M}\left|\hat{g}_n'(x_0 + n^{-1/5}t) - g_0'(x_0)\right| = O_p(n^{-1/5}). \tag{11.8}$$

For B_n a similar calculation works.

However, before we start proving (11.6) to (11.8), we prove a technical lemma in the spirit of Kim and Pollard, 1990. As in Section 4.3, we denote by $\mathcal{T}_n$ the set of changes of slope of $\hat{g}_n$.

Lemma 11.1 *Let x_0 be an interior point of the support of g_0. Then: Let, for $0 < x \le y$, the random function $U_n(x, y)$ be defined by*

$$U_n(x, y) = \int_{[x,y]}\left\{z - \frac{1}{2}(x + y)\right\} d\left(\mathbb{G}_n - G_0\right)(z),\ y \ge x. \tag{11.9}$$

Then there exist constants $\delta > 0$ and $c_0 > 0$ such that, for each $\epsilon > 0$ and each x satisfying $|x - x_0| < \delta$:

$$|U_n(x, y)| \le \epsilon(y - x)^4 + O_p\left(n^{-4/5}\right), \quad 0 \le y - x \le c_0. \tag{11.10}$$

Proof Note that

$$\sup_{y:0\le y-x\le R}|U_n(x, y)| = \sup_{y:0\le y-x\le R}\left|\int f_{x,y}(z)\,d(\mathbb{G}_n - G_0)(z)\right|,$$

where

$$f_{x,y}(z) = (z - x)1_{[x,y]}(z) - \frac{1}{2}(y - x)1_{[x,y]}(z),\ y \ge x.$$

Since the collection of functions

$$\mathcal{F}_{x,R} = \{f_{x,y}(z) \colon x \le y \le x + R\} \text{ with envelope function}$$

$$F_{x,R}(z) = (z-x)1_{[x,x+R]}(z) + \frac{1}{2}R1_{[x,x+R]}(z)$$

is a VC-subgraph class,

$$\begin{aligned} EF_{x,R}^2(X_1) &= \frac{1}{3}R^3\{g_0(x_0) + O(1)\} + \frac{1}{4}R^2\{G_0(x+R) - G_0(x)\} \\ &= \frac{7}{12}R^3\{g_0(x_0) + O(1)\}, \end{aligned} \tag{11.11}$$

for x in some appropriate neighborhood $[x_0 - \delta, x_0 + \delta]$ of x_0. It now follows from Theorem 2.14.1 in Van der Vaart and Wellner, 1996, that

$$E\left\{\left(\sup_{f_{x,y}\in\mathcal{F}_{x,R}} \left|\int f_{x,y}(z)\, d(\mathbb{G}_n - G_0)(z)\right|\right)^2\right\} \le \frac{1}{n}KEF_{x,R}^2 = O\left(n^{-1}R^3\right)$$

for small values of R and a constant $K > 0$.

Hence there exists a $\delta > 0$ such that, for $\epsilon > 0$, $A > 0$ and $jn^{-1/5} \le \delta$:

$$\begin{aligned} &P\left\{\exists u \in \left[(j-1)n^{-1/5}, jn^{-1/5}\right) : n^{4/5}\left|U_n(x, x+u)\right| > A + \epsilon(j-1)^4\right\} \\ &\le cn^{8/5}E\left\{\|\mathbb{G}_n - G_0\|_{\mathcal{F}_{x,jn^{-1/5}}}\right\}^2 / \left\{A + \epsilon(j-1)^4\right\}^2 \\ &\le c'j^3 / \left\{A + \epsilon(j-1)^4\right\}^2 \end{aligned} \tag{11.12}$$

for constants $c, c' > 0$, independent of $x \in [x_0 - \delta, x_0 + \delta]$. The result now follows. See also Exercise 11.1. □

With the help of Lemma 11.1, we can now prove the first tightness result (11.6) needed.

Lemma 11.2 *Let x_0 be a point at which g_0 has a continuous and strictly positive second derivative. Let ξ_n be an arbitrary sequence of numbers converging to x_0 and define $\tau_n^- = \max\{t \in \mathcal{T}_n : t \le \xi_n\}$ and $\tau_n^+ = \min\{t \in \mathcal{T}_n : t > \xi_n\}$. Then,*

$$\tau_n^+ - \tau_n^- = O_p(n^{-1/5}).$$

Proof Let τ_n^- be the last point of change of slope of $H_n'' < \xi_n$ and τ_n^+ the first point of change of slope of $H_n'' \ge \xi_n$. Note that, since the number of changes of slope is bounded above by n by Lemma 4.1, we can only have strict changes of slope. Moreover, let τ_n be the midpoint of the interval $[\tau_n^-, \tau_n^+]$. Then, by the characterization of Lemma 4.2:

$$H_n(\tau_n) \ge Y_n(\tau_n).$$

Using (4.13), this can be written:

$$\frac{1}{2}\left\{Y_n(\tau_n^-) + Y_n(\tau_n^+)\right\} - \frac{1}{8}\left\{\mathbb{G}_n(\tau_n^+) - \mathbb{G}_n(\tau_n^-)\right\}\left(\tau_n^+ - \tau_n^-\right) \ge Y_n(\tau_n). \tag{11.13}$$

Replacing Y_n and $\mathbb{G}_n$ by their deterministic counterparts, and expanding the integrands at τ_n, we get for large n:

$$\int_{\tau_n}^{\tau_n^+}\{\tau_n^+ - x\}g_0(x)\,dx + \int_{\tau_n^-}^{\tau_n}\{x-\tau_n^-\}g_0(x)\,dx - \frac{1}{4}\left(\tau_n^+ - \tau_n^-\right)\int_{\tau_n^-}^{\tau_n^+} g_0(x)\,dx$$
$$= \int_{[\tau_n^-,\tau_n]}\left\{\frac{1}{2}\left(\tau_n^- + \tau_n\right) - x\right\} g_0(x)\,dx + \int_{[\tau_n,\tau_n^+]}\left\{x - \frac{1}{2}\left(\tau_n + \tau_n^+\right)\right\} f_0(x)\,dx$$
$$= -\frac{1}{384}g_0''(\tau_n)\left(\tau_n^+ - \tau_n^-\right)^4 + o_p\left(\tau_n^+ - \tau_n^-\right)^4,$$

using the consistency of $\hat{g}_n$ to ensure that τ_n belongs to a sufficiently small neighborhood of x_0 to allow this expansion. But, by Lemma 11.1 and the inequality (11.13), this implies:

$$-\frac{1}{384}g_0''(x_0)\left(\tau_n^+ - \tau_n^-\right)^4 + O_p\left(n^{-4/5}\right) + o_p\left(\tau_n^+ - \tau_n^-\right)^4 \geq 0.$$

Hence:

$$\tau_n^+ - \tau_n^- = O_p\left(n^{-1/5}\right).$$

□

Having established the order of the difference of successive points of changes of slope of H_n'', we can turn the consistency result into a rate result saying that there will, with high probability, be a point in an $O_p(n^{-1/5})$ neighborhood of x_0 where the difference between the estimator and the estimand will be of order $n^{-2/5}$. Lemma 11.3 has the exact statement.

Lemma 11.3 *Suppose $g_0'(x_0) < 0$, $g_0''(x_0) > 0$, and g_0'' is continuous in a neighborhood of x_0. Let ξ_n be a sequence converging to x_0. Then for any $\epsilon > 0$ there exists an $M > 1$ and a $c > 0$ such that the following holds with probability bigger than $1-\epsilon$. There are bend points $\tau_n^- < \xi_n < \tau_n^+$ of $\tilde{f}_n$ with $n^{-1/5} \leq \tau_n^+ - \tau_n^- \leq Mn^{-1/5}$ and for any of such points we have that*

$$\inf_{t\in[\tau_n^-,\tau_n^+]} |g_0(t) - \hat{g}_n(t)| < cn^{-2/5} \quad \textit{for all } n.$$

Proof Fix $\epsilon > 0$ and observe that Lemma 11.2, applied to the sequences $\xi_n \pm n^{-1/5}$, gives that there is an $M > 0$ such that with probability bigger than $1-\epsilon$, there exist jump points τ_n^- and τ_n^+ of $\hat{g}_n'$ satisfying $\xi_n - Mn^{-1/5} \leq \tau_n^- \leq \xi_n - n^{-1/5} \leq \xi_n + n^{-1/5} \leq \tau_n^+ \leq \xi_n + Mn^{-1/5}$ for all n.

Let $\tau_n^- < \tau_n^+$ be such points of jump. Fix $c > 0$ and consider the event

$$\inf_{t\in[\tau_n^-,\tau_n^+]} |g_0(t) - \hat{g}_n(t)| \geq cn^{-2/5}. \tag{11.14}$$

On this set we have:

$$\left|\int_{\tau_n^-}^{\tau_n^+} (g_0(t) - \hat{g}_n(t))(\tau_n^+ - t)\,dt\right| \geq \frac{1}{2}cn^{-2/5}(\tau_n^+ - \tau_n^-)^2.$$

On the other hand, the equality conditions in (4.9) imply:

$$0 = \int_{[\tau_n^-,\tau_n^+]} (\tau_n^+ - t)\, d(\hat G_n - \mathbb{G}_n)(t)$$

$$= \int_{\tau_n^-}^{\tau_n^+} \{\hat g_n(t) - g_0(t)\}\,(\tau_n^+ - t)\, dt - \int_{[\tau_n^-,\tau_n^+]} (\tau_n^+ - t)\, d(\mathbb{G}_n - G_0)(t).$$

Therefore, by (11.14),

$$\left|\int_{[\tau_n^-,\tau_n^+]} (\tau_n^+ - t)\, d(\mathbb{G}_n - G_0)(t)\right| \ge \frac{1}{2} c n^{-2/5} (\tau_n^+ - \tau_n^-)^2 \ge 2cn^{-4/5}. \tag{11.15}$$

But the collection of functions

$$\mathcal{F}_{x,R} = \{f_{x,y}(z) : x \le y \le x + R\} \quad \text{where} \quad f_{x,y}(z) = (y - z)1_{[x,y]}(z), \quad y \ge x$$

is a VC-subgraph class of functions with envelope function

$$F_{x,R}(z) = R1_{[x,x+R]}(z),$$

so that

$$EF_{x,R}^2(X_1) = R^2\{G_0(x + R) - G_0(x)\} = R^3\{g_0(x_0) + o(1)\} \tag{11.16}$$

for x in some appropriate neighborhood $[x_0 - \delta, x_0 + \delta]$ of x_0. Therefore, we get

$$\left|\int_{[\tau_n^-,\tau_n^+]} (\tau_n^+ - t)\, d(\mathbb{G}_n - G_0)(t)\right| = O_p(n^{-4/5}) + o_p\left((\tau_n^+ - \tau_n^-)^4\right) = O_p(n^{-4/5}).$$

So the probability of (11.14) can be made arbitrarily small by taking c sufficiently big. □

Using Lemma 11.3 monotonicity of the derivatives of the estimators and the limit density g_0, we obtain the local $n^{-2/5}$ consistency of the density estimators and $n^{-1/5}$ consistency of their derivatives, that is, the second tightness results (11.7) and (11.8) (see Exercise 11.2).

Lemma 11.4 *Suppose $g_0'(x_0) < 0$, $g_0''(x_0) > 0$, and g_0'' is continuous in a neighborhood of x_0. Then for each $M > 0$,*

$$\sup_{|t|\le M} |\hat g_n(x_0 + n^{-1/5}t) - g_0(x_0) - n^{-1/5} t g_0'(x_0)| = O_p(n^{-2/5}) \tag{11.17}$$

and, interpreting $\hat g_n'$ as left or right derivative,

$$\sup_{|t|\le M} |\hat g_n'(x_0 + n^{-1/5}t) - g_0'(x_0)| = O_p(n^{-1/5}). \tag{11.18}$$

We now have localized the process, giving the asymptotic distribution of $\hat g_n$, which is an important step in giving the proof of Theorem 11.2. However, unlike the situation with the Grenander estimator and the current status model, the limit situation is different. In case of the Grenander estimator and the MLE in the current status model, we get in the limit that the solution on $\mathbb{R}$ coincides with the solution on a finite interval around zero. For the solution, giving the asymptotic distribution of $\hat g_n$, it is not clear that a similar situation occurs. Here it seems that the points of touch of the invelope and the integrated Brownian motion $+\,t^4$ may change for the solution on each finite interval $[-c, c]$, if we vary c. But it is important to realize that the solutions on a finite interval $[-c, c]$ converge to a unique solution, as $c \to \infty$, as is proved in Groeneboom et al., 2001b. Using this fact, we get by the localization lemmas just proved the limit result Theorem 11.2.

Note that with the Grenander estimator the rate is $n^{-1/3}$, if the density is strictly decreasing, and that for the estimator of a convex function, the rate is $n^{-2/5}$ for the convex function itself and $n^{-1/5}$ for its (piecewise constant) derivative, so the rates encapsulate the rate for the Grenander estimator.

The proof of Theorem 11.1 is given in Theorem 2.1 and Corollary 2.1(ii) in Groeneboom et al., 2001b. The main issue is to prove the existence of a *unique* invelope, which is a cubic spline lying inside the integrated Brownian motion $+t^4$ and having a convex second derivative. It is conjectured that the points of touch of the invelope and the integrated Brownian motion $+t^4$ are isolated, but this still has not been proved. At present, there is still no analytical information on the form of the limit distribution available.

Remark Simulations indicate that the density of the distribution of the second derivative of the invelope at zero (and hence the limit distribution of the convex least squares estimate or convex density estimate) is not symmetric.

11.2 Tail Bounds for the MLE in the Current Status Model

Consider the current status model as introduced in Section 2.3, and denote the underlying distribution of the event times by F_0 and the MLE of F_0 based on a sample of size n by $\hat{F}_n$. It is well-known that $\|\hat{F}_n - F_0\|_p = O_p(n^{-1/3})$ for $p \in [1,2]$, see, e.g., Van de Geer, 2000, Example 7.4.3. One can prove this by first proving that the Hellinger distance is $O_p(n^{-1/3})$ and then deducing from this the result for the L_p-distance, for $p \in [1,2]$. The proof for the Hellinger distance uses, for example, an inequality of the type

$$h\left(\tfrac{1}{2}(\hat{p}_n + p_0), p_0\right)^2 \le \frac{1}{2}\int_{p_0>0} \log\frac{p_0+\hat{p}_n}{2p_0}\, d\left(\mathbb{P}_n - P_0\right), \tag{11.19}$$

where h is the Hellinger distance and

$$p_0(t,\delta) = \delta F_0 + (1-\delta)(1-F_0), \quad \hat{p}_n(t,\delta) = \delta\hat{F}_n + (1-\delta)(1-\hat{F}_n).$$

Inequality (11.19) is given in Lemma 4.1 of Van de Geer, 2000, and called a basic inequality, see (3.4) for a related inequality. The result for the Hellinger distance then follows from Theorem 7.4 in Van de Geer, 2000, by standard empirical process theory for classes of uniformly bounded monotone functions; see Example 7.4.3 in Van de Geer, 2000.

However, for later purposes in connection with the smoothed maximum likelihood estimator (the SMLE), one would like to have more precise information. In particular, the following result is needed, at an interior point t:

$$\int_{t-h_n}^{t+h_n} \left\{\hat{F}_n(x) - F_0(x)\right\}^2 dx = O_p\left(h_n n^{-2/3}\right), \tag{11.20}$$

if $h_n \downarrow 0$, as $n \to \infty$. This does not follow from the global bound on $\|\hat{F}_n - F_0\|_2$; see Exercise 11.3.

Now let f_0 be the density of F_0, with support $[0, M]$. Assuming that

$$0 < \inf_{t\in[0,M]} f_0(t) \le \sup_{t\in[0,M]} f_0(t) < \infty, \tag{11.21}$$

it follows from Durot, 2007, that for $p \in [1, 2)$,

$$\mathbb{E}\left|\hat{F}_n(t) - F_0(t)\right|^p \le Kn^{-p/3}, \ \forall t \in (0, M).$$

The bound (11.20) is unfortunately not implied by this result, since it does not cover the situation $p = 2$. We now set out to discuss a similar result for the L_p bound, for all $p \ge 1$. This can be done by an extension of the methods used in the proof of Lemma 3.5. We assume (11.21) to hold.

Note that, by Exercise 11.4 applied to $n^{1/3}(\hat{F}_n(t) - F_0(t))_+$,

$$\mathbb{E}\left\{n^{1/3}\left(\hat{F}_n(t) - F_0(t)\right)_+\right\}^p = \int_0^\infty \mathbb{P}\left\{n^{1/3}\{\hat{F}_n(t) - F_0(t)\} \ge x\right\} px^{p-1}\,dx,$$

and, that, by the switch relation (3.36),

$$\mathbb{P}\left\{n^{1/3}\{\hat{F}_n(t) - F_0(t)\} \ge x\right\} = \mathbb{P}\left\{U_n(a + n^{-1/3}x) \le t\right\}.$$

Therefore,

$$\mathbb{E}\left\{n^{1/3}\left(\hat{F}_n(t_0) - F_0(t_0)\right)_+\right\}^p = \int_0^\infty \mathbb{P}\left\{U_n(a_0 + n^{-1/3}x) \le t_0\right\} px^{p-1}\,dx. \tag{11.22}$$

Also, defining U as the inverse of F_0, that is

$$U(a) = \inf\{x \in \mathbb{R} : F_0(x) \ge a\}, \ \ 0 < a < 1,$$

it follows that

$$\begin{aligned}&\mathbb{P}\left\{U_n(a_0 + n^{-1/3}x) \le t_0\right\}\\ &= \mathbb{P}\left\{n^{1/3}\left(U_n(a_0 + n^{-1/3}x) - U(a_0 + n^{-1/3}x)\right) \le n^{1/3}\{t_0 - U(a_0 + n^{-1/3}x)\}\right\}.\end{aligned}$$

By (11.21), $U(a)$ is uniquely determined for $a \in (0, 1)$.

We have the following theorem, which gives a stronger version of Lemma 2 in Durot, 2007, in the present situation.

Theorem 11.3 *Suppose F_0 has a continuous density f_0 with support $[0, M]$ that satisfies (11.21). Also suppose that the observation distribution G has a continuous derivative g that stays away from zero and infinity on $[0, M]$. Then there exist constants $K_1, K_2 > 0$ such that, for every $a \in [0, 1]$ and $x > 0$,*

$$\mathbb{P}\left\{n^{1/3}\left|U_n(a) - U(a)\right| \ge x\right\} \le K_1 e^{-K_2 x^3}. \tag{11.23}$$

In the proof of this result we use the following lemma. Here $\mathbb{P}_n$ denotes the empirical measure of the observations (T_i, Δ_i), $1 \le i \le n$ and $\mathbb{G}_n$ the empirical distribution function of the observed T_i, $1 \le i \le n$.

Lemma 11.5 *Let V_n and $\bar{V}_n$ be defined by*

$$V_n(t) = \int_{u\in[0,x]} \delta\,d\mathbb{P}_n(u, \delta), \quad \bar{V}_n(t) = \int_{u\in[0,x]} F_0(u)\,d\mathbb{G}_n(u), \quad t \in [0, M],$$

and let $D_n = V_n - \bar{V}_n$. Then there exist constants $K_1, K_2 > 0$ such that, for each $j \geq 1$, $j \in \mathbb{N}$,

$$\begin{aligned}
&\mathbb{P}\left\{\exists y \in \left[(j-1)n^{-1/3}, jn^{-1/3}\right) : D_n(U(a)+y) - D_n(U(a))\right.\\
&\qquad \left.\leq -\int_{U(a)}^{U(a)+y} \{F_0(u) - F_0(U(a))\}\, d\mathbb{G}_n(u)\right\}\\
&\qquad \leq K_1 \exp\left\{-K_2(j-1)^3\right\}. \qquad (11.24)
\end{aligned}$$

Likewise, there exist constants $K_1, K_2 > 0$ such that, for each $j \geq 1$, $j \in \mathbb{N}$,

$$\begin{aligned}
&\mathbb{P}\left\{\exists y \in \left[-jn^{-1/3}, -(j-1)n^{-1/3}\right) : D_n(U(a)+y) - D_n(U(a))\right.\\
&\qquad \left.\leq -\int_{U(a)+y}^{U(a)} \{F_0(u) - F_0(U(a))\}\, d\mathbb{G}_n(u)\right\}\\
&\qquad \leq K_1 \exp\left\{-K_2(j-1)^3\right\}. \qquad (11.25)
\end{aligned}$$

Proof We only prove (11.24), since the proof of (11.25) is similar. We get from Doob's submartingale inequality and next Markov's inequality, conditionally on $T_1, \ldots, T_n$,

$$\begin{aligned}
&\mathbb{P}\left\{\exists y \in \left[(j-1)n^{-1/3}, jn^{-1/3}\right) : D_n(U(a)+y) - D_n(U(a))\right.\\
&\quad \left.\leq -\int_{U(a)}^{U(a)+y} \{F_0(u) - F_0(U(a))\}\, d\mathbb{G}_n(u) \;\middle|\; T_1, \ldots, T_n\right\}\\
&\quad \leq \mathbb{P}\left\{-D_n(U(a)+jn^{-1/3}) + D_n(U(a))\right.\\
&\quad \left.\geq \int_{U(a)}^{U(a)+(j-1)n^{-1/3}} \{F_0(u) - F_0(U(a))\}\, d\mathbb{G}_n(u) \;\middle|\; T_1, \ldots, T_n\right\}\\
&\quad \leq \mathbb{E}\exp\left\{-\theta \int_{U(a)}^{U(a)+(j-1)n^{-1/3}} \{F_0(u) - F_0(U(a))\}\, d\mathbb{G}_n(u)\right.\\
&\qquad\qquad \left.-\,\theta\{D_n(U(a)+jn^{-1/3}) - D_n(U(a))\} \;\middle|\; T_1, \ldots, T_n\right\}. \qquad (11.26)
\end{aligned}$$

This is often called exponential centering. Furthermore,

$$\begin{aligned}
&\mathbb{E}\left\{\exp\left\{-\theta\left\{D_n(U(a)+jn^{-1/3}) - D_n(U(a))\right\}\right\} \;\middle|\; T_1, \ldots, T_n\right\}\\
&= \mathbb{E}\left\{\exp\left\{-\frac{\theta}{n} \sum_{T_i \in [U(a), U(a)+jn^{-1/3}]} \{\Delta_i - F_0(T_i)\}\right\} \;\middle|\; T_1, \ldots, T_n\right\}\\
&= \prod_{T_i \in [U(a), U(a)+jn^{-1/3}]} \left\{\exp\left\{-\frac{\theta}{n}\{1 - F_0(T_i)\}\right\} F_0(T_i) + \exp\left\{\frac{\theta}{n} F_0(T_i)\right\} \{1 - F_0(T_i)\}\right\}.
\end{aligned}$$

Hence the conditional upper bound on the right hand side of (11.26) can be written:

$$\prod_{i=1}^{n}\Biggl\{\left\{\exp\left\{-\frac{\theta}{n}\{1-F_0(U(a))\}\right\}F_0(T_i)\right.$$
$$\left.+\exp\left\{\frac{\theta}{n}F_0(U(a))\right\}\{1-F_0(T_i)\}\right\}1_{[0,(j-1)n^{-1/3}]}(T_i-U(a))$$
$$+\left\{\exp\left\{-\frac{\theta}{n}\{1-F_0(T_i)\}\right\}F_0(T_i)\right.$$
$$\left.+\exp\left\{\frac{\theta}{n}F_0(T_i)\right\}\{1-F_0(T_i)\}\right\}1_{\left((j-1)n^{-1/3},\,jn^{-1/3}\right]}(T_i-U(a))$$
$$+1_{\mathbb{R}\setminus[0,jn^{-1/3}]}(T_i-U(a))\Biggr\}.$$

Taking the expectation over $T_1,\ldots,T_n$, the upper bound becomes:

$$\Biggl(\mathbb{E}\left\{\exp\left\{-\frac{\theta}{n}\{1-F_0(U(a))\}\right\}F_0(T_1)\right.$$
$$\left.+\exp\left\{\frac{\theta}{n}F_0(U(a))\right\}\{1-F_0(T_1)\}\right\}1_{[0,(j-1)n^{-1/3}]}(T_1-U(a))$$
$$+\mathbb{E}\left\{\exp\left\{-\frac{\theta}{n}\{1-F_0(T_1)\}\right\}F_0(T_1)\right.$$
$$\left.+\exp\left\{\frac{\theta}{n}F_0(T_1)\right\}\{1-F_0(T_1)\}\right\}1_{\left((j-1)n^{-1/3},\,jn^{-1/3}\right]}(T_1-U(a))$$
$$+\mathbb{E}1_{\mathbb{R}\setminus[0,jn^{-1/3}]}(T_1-U(a))\Biggr)^n. \tag{11.27}$$

Therefore,

$$\mathbb{E}\left\{\exp\left\{-\frac{\theta}{n}\{1-F_0(U(a))\}\right\}F_0(T_1)\right.$$
$$\left.+\exp\left\{\frac{\theta}{n}F_0(U(a))\right\}\{1-F_0(T_1)\}\right\}1_{[0,(j-1)n^{-1/3}]}(T_1-a)$$
$$=\int_{u=U(a)}^{U(a)+(j-1)n^{-1/3}}\left\{\exp\left\{-\frac{\theta}{n}\{1-F_0(U(a))\}\right\}F_0(u)\right.$$
$$\left.+\exp\left\{\frac{\theta}{n}F_0(U(a))\right\}\{1-F_0(u)\}\right\}dG(u)$$
$$=G(U(a)+(j-1)n^{-1/3})-G(U(a))-\frac{\theta}{n}\int_{u=U(a)}^{U(a)+(j-1)n^{-1/3}}\{F_0(u)-F_0(U(a))\}\,dG(u)$$
$$+\sum_{k=2}^{\infty}\frac{\theta^k}{n^k k!}\int_{u=U(a)}^{U(a)+(j-1)n^{-1/3}}\Bigl\{(-1)^k\{1-F_0(U(a))\}^kF_0(u)$$
$$+F_0(U(a))^k\{1-F_0(u)\}\Bigr\}\,dG(u).$$

and

$$\mathbb{E}\left\{\exp\left\{-\frac{\theta}{n}\{1-F_0(T_1)\}\right\}F_0(T_1)+\exp\left\{\frac{\theta}{n}F_0(T_1)\right\}\{1-F_0(T_1)\}\right\}1_{\left((j-1)n^{-1/3},\,jn^{-1/3}\right]}(T_1-a)$$

$$=\int_{u=U(a)+(j-1)n^{-1/3}}^{U(a)+jn^{-1/3}}\left\{\exp\left\{-\frac{\theta}{n}\{1-F_0(u)\}\right\}F_0(u)+\exp\left\{\frac{\theta}{n}F_0(u)\right\}\{1-F_0(u)\}\right\}dG(u)$$

$$=G(U(a)+jn^{-1/3})-G(U(a)+(j-1)n^{-1/3})$$

$$+\sum_{i=2}^{\infty}\frac{\theta^i}{n^i i!}\int_{u=U(a)+(j-1)n^{-1/3}}^{U(a)+jn^{-1/3}}\left\{(-1)^i\{1-F_0(u)\}^i F_0(u)+F_0(u)^i\{1-F_0(u)\}\right\}dG(u).$$

Hence it follows from the inequality $\log(1+x)\le x$, for all $x>-1$, that the unconditional upper bound (11.24) is bounded above by

$$\exp\left\{-\theta\int_{u=U(a)}^{U(a)+(j-1)n^{-1/3}}\{F_0(u)-F_0(U(a))\}\,dG(u)\right.$$

$$+\sum_{i=2}^{\infty}\frac{\theta^i}{n^{i-1}i!}\int_{u=U(a)}^{U(a)+(j-1)n^{-1/3}}\left\{(-1)^i\{1-F_0(U(a))\}^i F_0(u)\right.$$

$$\left.+F_0(U(a))^i\{1-F_0(u)\}\right\}dG(u)$$

$$+\sum_{i=2}^{\infty}\frac{\theta^i}{n^{i-1}i!}\int_{u=U(a)+(j-1)n^{-1/3}}^{U(a)+jn^{-1/3}}\left\{(-1)^i\{1-F_0(u)\}^i F_0(u)\right.$$

$$\left.\left.+F_0(u)^i\{1-F_0(u)\}\right\}dG(u)\right\}.$$

Let $c=\inf_{x\in[0,M]}f_0(x)\inf_{x\in[0,M]}g(x)$. Then, by the assumptions of Theorem 11.3, $c>0$ and the upper bound (11.25) is bounded above by

$$\exp\left\{-\theta\frac{c(j-1)^2}{2n^{2/3}}\right.$$

$$+\sum_{i=2}^{\infty}\frac{\theta^i}{n^{i-1}i!}\int_{u=U(a)}^{U(a)+(j-1)n^{-1/3}}\left\{(-1)^i\{1-F_0(U(a))\}^i F_0(u)+F_0(U(a))^i\{1-F_0(u)\}\right\}dG(u)$$

$$\left.+\sum_{i=2}^{\infty}\frac{\theta^i}{n^{i-1}i!}\int_{u=U(a)+(j-1)n^{-1/3}}^{U(a)+jn^{-1/3}}\left\{(-1)^i\{1-F_0(u)\}^i F_0(u)+F_0(u)^i\{1-F_0(u)\}\right\}dG(u)\right\},\tag{11.28}$$

using

$$\int_{u=U(a)}^{U(a)+(j-1)n^{-1/3}}\{F_0(u)-F_0(U(a))\}\,dG(u)\ge\frac{c(j-1)^2}{2n^{2/3}}.$$

The integrands

$$(-1)^i\{1-F_0(U(a))\}^i F_0(u)+F_0(U(a))^i\{1-F_0(u)\}g(u)$$

and

$$(-1)^i\{1-F_0(u)\}^i F_0(u)+F_0(u)^i\{1-F_0(u)\}g(u)$$

have a uniform upper bound c_1, as $U(a)\in[0,M]$ and $u\in[U(a),M]$. Hence, for $j\geq 2$

$$\begin{aligned}
&-\theta\frac{c(j-1)^2}{2n^{2/3}}\\
&\quad+\sum_{i=2}^{\infty}\frac{\theta^i}{n^{i-1}i!}\int_{u=U(a)}^{U(a)+(j-1)n^{-1/3}}\Big\{(-1)^i\{1-F_0(U(a))\}^i F_0(u)+F_0(U(a))^i\{1-F_0(u)\}\Big\}\,dG(u)\\
&\quad+\sum_{i=2}^{\infty}\frac{\theta^i}{n^{i-1}i!}\int_{u=U(a)+(j-1)n^{-1/3}}^{U(a)+jn^{-1/3}}\Big\{(-1)^i\{1-F_0(u)\}^i F_0(u)+F_0(u)^i\{1-F_0(u)\}\Big\}\,dG(u)\\
&\leq -\theta\frac{c(j-1)^2}{2n^{2/3}}+c_1jn^{-1/3}\sum_{i=2}^{\infty}\frac{\theta^i}{n^{i-1}i!}\\
&\leq -\theta\frac{c(j-1)^2}{2n^{2/3}}+2c_1(j-1)n^{-1/3}\sum_{i=2}^{\infty}\frac{\theta^i}{n^{i-1}i!}.
\end{aligned}$$

Let $t_j=(j-1)n^{-1/3}$. By Exercise 11.5, the function

$$\phi_{nj}:\theta\mapsto -\tfrac{1}{2}\theta ct_j^2+2c_1t_j\sum_{i=2}^{\infty}\frac{\theta^i}{n^{i-1}i!} \tag{11.29}$$

attains its maximum at the point

$$\theta_{nj}=n\log\left(1+\frac{ct_j}{4c_1}\right).$$

At this point we have:

$$\phi_{nj}(\theta_{nj})=\tfrac{1}{2}nt_j\left\{ct_j-(4c_1+ct_j)\log\left(1+\frac{ct_j}{4c_1}\right)\right\}\leq-\frac{nc^2t_j^3}{8c_1}=-\frac{c^2(j-1)^3}{8c_1},$$

where we use $\log(1+x)\leq x$ again in the inequality. So, going back to (11.26), the conclusion is:

$$\begin{aligned}
&\mathbb{P}\Big\{\exists y\in\big[(j-1)n^{-1/3},jn^{-1/3}\big):D_n(U(a)+y)-D_n(U(a))\\
&\qquad\leq-\int_{U(a)}^{U(a)+y}\{F_0(u)-F_0(U(a))\}\,d\mathbb{G}_n(u)\Big\}\\
&\qquad\leq\exp\left\{-\frac{c^2(j-1)^3}{8c_1}\right\}.
\end{aligned} \tag{11.30}$$

This, in turn, implies

$$
\begin{aligned}
&\mathbb{P}\left\{n^{1/3}\{U_n(a)-U(a)\}\ge j-1\right\}\\
&\le \mathbb{P}\left\{\inf_{y-U(a)\ge (j-1)n^{-1/3}x}\{V_n(y)-a\mathbb{G}_n(y)\}\le V_n(U(a))-a\mathbb{G}_n(U(a))\right\}\\
&\le \sum_{i=j}^{\infty}\exp\left\{-\frac{c^2(i-1)^3}{8c_1}\right\}=\exp\left\{-\frac{c^2(j-1)^3}{8c_1}\right\}\left\{\sum_{i=j}^{\infty}\exp\left\{-\frac{c^2\left\{(i-1)^3-(j-1)^3\right\}}{8c_1}\right\}\right\}\\
&\le \exp\left\{-\frac{c^2(j-1)^3}{8c_1}\right\}\sum_{i=j}^{\infty}\exp\left\{-\frac{c^2(i-j)}{8c_1}\right\}\\
&=\left(1-e^{-c^2/(8c_1)}\right)^{-1}\exp\left\{-\frac{c^2(j-1)^3}{8c_1}\right\}.
\end{aligned}
$$

We similarly find that there exist constants $K_1, K_2 > 0$ so that

$$
\mathbb{P}\left\{n^{1/3}\{U_n(a)-U(a)\}\le -(j-1)\right\}\le K_1\exp\left\{-K_2(j-1)^3\right\}.
$$

The statement of the lemma now follows. □

We now proceed with the proof of Theorem 11.3. Let $a \in [0, 1)$. If $n^{1/3}\left|U_n(a) - U(a)\right| \ge x$ there exists a y such that $n^{1/3}|y - U(a)| \ge x$ and $V_n(y) - a\mathbb{G}_n(y) \le V_n(U(a)) - a\mathbb{G}_n(U(a))$. Hence

$$
\mathbb{P}\left\{n^{1/3}\left|U_n(a)-U(a)\right|\ge x\right\}\le \mathbb{P}\left\{\inf_{|y-U(a)|\ge n^{-1/3}x}\{V_n(y)-a\mathbb{G}_n(y)\}\le V_n(U(a))-a\mathbb{G}_n(U(a))\right\}.
$$

We now get

$$
\begin{aligned}
&P\left\{\inf_{y-U(a)\ge n^{-1/3}x}\{V_n(y)-a\mathbb{G}_n(y)\}\le V_n(U(a))-a\mathbb{G}_n(U(a))\right\}\\
&\le P\left\{\exists y\ge U(a)+n^{-1/3}x : D_n(y)-D_n(U(a))\le -\int_{U(a)}^{y}\{F_0(u)-F_0(U(a))\}\,d\mathbb{G}_n(u)\right\}\\
&\le \sum_{j=i}^{\infty}\mathbb{P}\left\{\exists y\in\left[(j-1)n^{-1/3}, jn^{-1/3}\right) : D_n(U(a)+y)-D_n(U(a))\right.\\
&\left.\le -\int_{U(a)}^{U(a)+y}\{F_0(u)-F_0(U(a))\}\,d\mathbb{G}_n(u)\right\},
\end{aligned}
$$

where $x \in [(i-1)n^{-1/3}, in^{-1/3}]$. By (11.24), this is bounded above by

$$
K_1\sum_{j=i}^{\infty}e^{-K_2(j-1)^3}\le K_1e^{-K_2(i-1)^3}\sum_{j=1}^{\infty}e^{-K_2\{(j-1)^3-(i-1)^3\}}=K_1'e^{-K_2(i-1)^3}.
$$

This is obviously equivalent to an upper bound of the form $K_1'e^{-K_2'x^3}$ for the probability that $n^{1/3}\{U_n(a) - U(a)\} \ge x$, for $x > 0$. Since a similar bound holds for the probability that $n^{1/3}\{U_n(a) - U(a)\} \le -x$, Theorem 11.3 now follows.

Remark Theorem 2.1 of Groeneboom et al., 1999, says that

$$\mathbb{P}\left\{n^{1/3}\{U_n(a) - U(a)\} > x\right\} \le 2e^{-Cx^3} \tag{11.31}$$

for processes similar to the processes U_n and U used here, but connected to the Grenander estimator of a decreasing density f. Here the positive constant C only depends on the density f, but not on a (see also Section 13.1, Theorem 13.2).

Using Theorem 11.3 we get from (11.22) that

$$\mathbb{E}\left\{n^{1/3}\left(\hat{F}_n(t) - F_0(t)\right)_+\right\}^p \le K, \tag{11.32}$$

for a constant $K > 0$, uniformly in the chosen point $t \in [0, M]$. By similar methods, it can be shown that

$$\mathbb{E}\left\{n^{1/3}\left(F_0(t) - \hat{F}_n(t)\right)_+\right\}^p \le K', \tag{11.33}$$

for a constant $K' > 0$, uniformly in the chosen point $t \in [0, M]$. Combining these inequalities leads to

$$\mathbb{E}\left\|n^{1/3}\{\hat{F}_n - F_0\}\right\|_p \le K_p < \infty, \tag{11.34}$$

for a constant K_p, for any $p \ge 1$. Moreover, by Exercise 11.6, (11.20) also follows from (11.32) and (11.33).

11.3 The SMLE in the Current Status Model

In Section 8.1 and 8.5, smooth estimators are introduced for the distribution function of the event times in the current status model. One of these, the SMLE of Section 8.1, is defined by first computing the ordinary ML estimator $\hat{F}_n$ (see Section 2.3) and then smoothing this using a smoothing kernel. More specifically, we define the SMLE $\hat{F}_{nh}^{(SML)}(t)$ at points t away from the boundary by

$$\hat{F}_{nh}^{(SML)}(t) = \int \mathbb{K}_h(t - x)\,d\hat{F}_n(x), \quad \mathbb{K}_h(x) = \mathbb{K}(x/h), \tag{11.35}$$

where $\mathbb{K}$ is an integrated kernel, defined by

$$\mathbb{K}(x) = \int_{-\infty}^{x} K(y)\,dy, \tag{11.36}$$

and K is a symmetric kernel of the usual kind, used in density estimation. We will assume that K has support $[-1, 1]$.

A tricky part of the theory is the handling of the boundary. Denoting the interval of support of the distribution corresponding to F by $[a, b]$, we propose to do this via an asymmetric

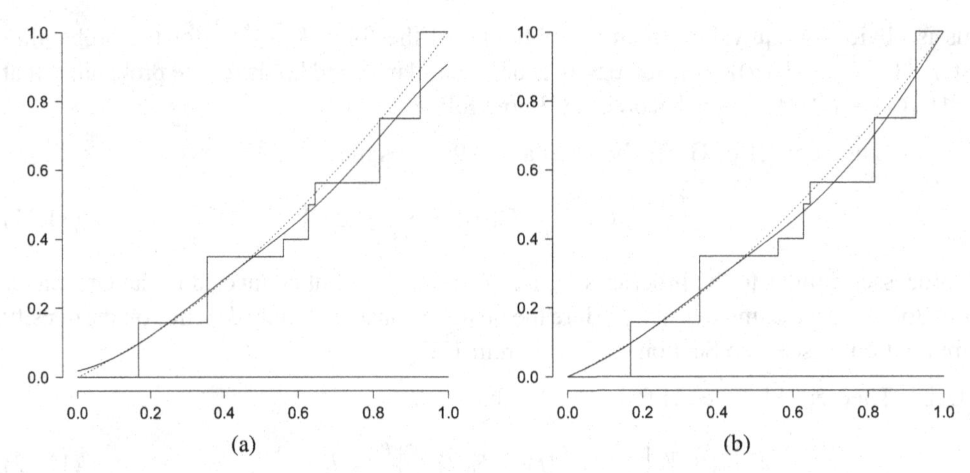

Figure 11.2 MLE (solid) and SMLE (dashed) for a sample of size $n = 1000$ for current status data from the distribution function $F_0(x) = (x + x^2)/2$ (dotted) on $[0, 1]$, taking the bandwidth $h = n^{-1/5}$ and without (a) and with (b) the boundary correction. The observation distribution is uniform.

version of Schuster's method (see Schuster, 1985), by defining

$$\hat{F}_{nh}^{(SML)}(t) = \int \left\{ \mathbb{K}\left(\frac{t-x}{h}\right) + \mathbb{K}\left(\frac{t+x-2a}{h}\right) - \mathbb{K}\left(\frac{2b-t-x}{h}\right) \right\} d\hat{F}_n(x) \quad (11.37)$$

for the estimation of the distribution function. The corresponding density estimate is obtained by differentiating with respect to t:

$$\hat{f}_{nh}^{(SML)}(t) = \int \left\{ K\left(\frac{t-x}{h}\right) + K\left(\frac{t+x-2a}{h}\right) + K\left(\frac{2b-t-x}{h}\right) \right\} d\hat{F}_n(x) \quad (11.38)$$

and equals the boundary correction in Schuster, 1985. Note, however, that the SMLE, boundary corrected via the asymmetric Schuster method, has a bias of order $O(h^2)$ (see the following), while density estimate, corrected by Schuster's method, has a bias of order $O(h)$, unless the derivative of the density is zero at the boundary (known as the shoulder condition).

Note that (11.37) reduces to (11.35) if $a + h \le t \le b - h$, and that

$$\begin{aligned}\hat{F}_{nh}^{(SML)}(a) &= \int \left\{ \mathbb{K}\left(\frac{a-x}{h}\right) + \mathbb{K}\left(\frac{x-a}{h}\right) - \mathbb{K}\left(\frac{2b-a-x}{h}\right) \right\} d\hat{F}_n(x)\\ &= \int \left\{ 1 - \mathbb{K}\left(\frac{2b-a-x}{h}\right) \right\} d\hat{F}_n(x) = 0,\end{aligned}$$

using $\mathbb{K}((2b-a-x)/h) = 1$, if $x \in [a, b]$ and $h < (b-a)/2$. A picture of the MLE, together with the SMLE, both uncorrected and corrected for the boundary effects, using the asymmetric Schuster-type method, is shown in Figure 11.2.

The bias for $t \in [a, a+h]$ is given by:

$$\int \left\{ \mathbb{K}\left(\frac{t-x}{h}\right) + \mathbb{K}\left(\frac{t+x-2a}{h}\right) - \mathbb{K}\left(\frac{2b-t-x}{h}\right) \right\} f_0(x)\,dx - F_0(t).$$

Integrating by parts yields, again for $t \in [a, a+h]$:

$$\begin{aligned}
&\int \left\{ \mathbb{K}\left(\frac{t-x}{h}\right) + \mathbb{K}\left(\frac{t+x-2a}{h}\right) - \mathbb{K}\left(\frac{2b-t-x}{h}\right) \right\} f_0(x)\,dx \\
&= \int \left\{ K_h\left(\frac{t-x}{h}\right) - K_h\left(\frac{t+x-2a}{h}\right) \right\} F_0(x)\,dx \\
&= \int_{u=-1}^{(t-a)/h} K(u) F_0(t-hu)\,du - \int_{u=(t-a)/h}^{1} K(u) F_0(2a-t+hu)\,du.
\end{aligned}$$

Let $v = (t-a)/h$. Then the bias equals:

$$\begin{aligned}
&\tfrac{1}{2}h^2 f_0'(a) \left\{ \int_{u=-1}^{v} (u-v)^2 K(u)du - \int_{u=v}^{1} (u-v)^2 K(u)du \right\} - \tfrac{1}{2} f_0'(a) v^2 + o\left(h^2\right) \\
&= \tfrac{1}{2}h^2 f_0'(a) \left\{ \int_{u=-1}^{v} (u-v)^2 K(u)du - \int_{u=v}^{1} (u-v)^2 K(u)du - \int_{u=-1}^{1} v^2 K(u)du \right\} + o\left(h^2\right) \\
&= \tfrac{1}{2}h^2 f_0'(a) \left\{ \int u^2 K(u)du - 2\int_{u=v}^{1} (u-v)^2 K(u)du \right\} + o\left(h^2\right),
\end{aligned} \tag{11.39}$$

if f_0 is twice differentiable on $[0, h)$, uniformly for $t \in [a, a+h]$. Note that the bias is of order $\frac{1}{2}h^2 f_0'(t) \int u^2 K(u)\,du + o(h^2)$ at $t = h$.

It can be proved in a similar way that the bias of the estimate defined by (11.37) is also of order $O(h^2)$ uniformly for $t \in [b-h, b]$, so we get that the bias is of order $O(h^2)$ uniformly for $t \in [a, b]$. This means that, in fact, the modified SMLE $\hat{F}_{nh}^{(SML)}(t)$, as defined by (11.37), converges pointwise at rate $O(n^{-2/5})$ to $F_0(t)$ for each $t \in [a, b]$ (but not uniformly in t, because we have to deal with Gumbel-type extreme value behavior if we look at the maximum distance to $F_0(t)$, for $t \in [a, b]$).

The way the pointwise distribution theory for this estimator can be treated is the prototype of local smooth functional theory: for sufficiently large bandwidth the estimator will be asymptotically normal and the local behavior of the MLE, with the nonnormal limit distribution, will be washed away by the smoothing operation. To be more specific, from Section 3.8 it is known that the behavior of the ML estimator is determined locally in neighborhoods with lengths of order $n^{-1/3}$. In those intervals, estimates based on local averages will not be monotone. In this section, neighborhoods with length of the order $n^{-1/5}$ are used. Local averages over this type of neighborhoods tend to be monotone. See also the heuristic reasoning for the (related) Grenander estimator given in Section 3.2.

We now assume for simplicity that $a + h \le t \le b - h$, where t is the point at which we want to study the asymptotic behavior of the SMLE, and first define the (canonical) score function or efficient influence function in the hidden space:

$$\mathbb{K}_h(t-x)+\mathbb{K}_h(t+x-2a)-\mathbb{K}_h(2b-t-x),\quad \mathbb{K}_h(x)=\int_{-\infty}^{x/h}K(u)\,du$$

for a symmetric positive kernel K with support $[-1,1]$. The images of the score operator in the observation space are given by:

$$\theta_{t,h,F}(u,\delta)=E_F\left\{\alpha(X)|(T,\Delta)=(u,\delta)\right\}, \tag{11.40}$$

where α is a score in the hidden space. Note that

$$\theta_{t,h,F}(u,1)=E\left\{\alpha(X)|(T,\Delta)=(u,1)\right\}=\frac{\int_{x\le u}\alpha(x)\,dF(x)}{F(u)},$$

and

$$\theta_{t,h,F}(u,0)=E\left\{\alpha(X)|(T,\Delta)=(u,0)\right\}=\frac{\int_{x>u}\alpha(x)\,dF(x)}{1-F(u)}.$$

With this notation, we want to solve the equation

$$\begin{aligned}E\left\{\theta_{t,h,F}(T,\Delta)\;\middle|\;X=x\right\}&=\int_{u\ge x}\theta_{h,F}(u,1)g(u)\,du+\int_{u<x}\theta_{t,h,F}(u,0)g(u)\,du\\&=\mathbb{K}_h(t-x)+\mathbb{K}_h(t+x-2a)-\mathbb{K}_h(2b-t-x)\\&\quad-\int\left\{\mathbb{K}_h(t-u)+\mathbb{K}_h(t+u-2a)-\mathbb{K}_h(2b-t-u)\right\}\,dF(u)\end{aligned} \tag{11.41}$$

where g is the density of the observation time distribution. If $\theta_{t,h,F}$ is a solution of this equation, we get:

$$\begin{aligned}&\int\left\{\mathbb{K}_h(t-u)+\mathbb{K}_h(t+u-2a)-\mathbb{K}_h(2b-t-u)\right\}\,d\big(F-F_0\big)(x)=-\int\theta_{h,F}(u,\delta)\,dP_0(u,\delta)\\&=-\int\theta_{h,F}(u,1)\,F_0(u)\,dG(u)-\int\theta_{h,F}(u,0)\left\{1-F_0(u)\right\}dG(u),\end{aligned} \tag{11.42}$$

where P_0 is the underlying probability measure for the pairs (T_i,Δ_i). This is the first step in reducing the integral $\int\left\{\mathbb{K}_h(t-u)+\mathbb{K}_h(t+u-2a)-\mathbb{K}_h(2b-t-u)\right\}\,d\big(\hat{F}_n-F_0\big)(x)$ to an integral in the observation space for which asymptotic normality can be proved.

Lemma 11.6 *The equation (11.41) is solved by*

$$\theta_{t,h,F}(u,\delta)=\frac{\left\{\delta-F(u)\right\}\left\{K_h(t-u)-K_h(t+u-2a)-K_h(2b-t-u)\right\}}{g(u)}.$$

Proof We have:

$$\int_{u\ge x}\theta_{t,h,F}(u,1)g(u)\,du+\int_{u<x}\theta_{t,h,F}(u,0)g(u)\,du$$
$$=\int_{u\ge x}\{1-F(u)\}\frac{\{K_h(t-u)-K_h(t+u-2a)-K_h(2b-t-u)\}}{g(u)}g(u)\,du$$
$$-\int_{u<x}F(u)\frac{\{K_h(t-u)-K_h(t+u-2a)-K_h(2b-t-u)\}}{g(u)}g(u)\,du$$
$$=\int_{u\ge x}\{K_h(t-u)-K_h(t+u-2a)-K_h(2b-t-u)\}\,du$$
$$-\int F(u)\{K_h(t-u)-K_h(t+u-2a)-K_h(2b-t-u)\}\,du$$
$$=\mathbb{K}_h(t-x)+\mathbb{K}_h(t+x-2a)-\mathbb{K}_h(2b-t-x)$$
$$-\int\{\mathbb{K}_h(t-x)+\mathbb{K}_h(t+x-2a)-\mathbb{K}_h(2b-t-x)\}\,dF(u),$$

using integration by parts in the last step. □

As in the treatment of the global smooth functionals in Section 10.2, we now define

$$\theta_{t,h,\hat F_n}(u,\delta)=\frac{\{\delta-F(u)\}\{K_h(t-u)-K_h(t+u-2a)-K_h(2b-t-u)\}}{g(u)}. \tag{11.43}$$

Note that $\theta_{t,h,\hat F_n}$ no longer has the interpretation (11.40). But we retain relation (11.42), that is, we have:

$$\int\{\mathbb{K}_h(t-u)+\mathbb{K}_h(t+u-2a)-\mathbb{K}_h(2b-t-u)\}\,d\big(\hat F_n-F_0)(x)$$
$$=-\int\theta_{t,h,\hat F_n}(u,\delta)\,dP_0(u,\delta), \tag{11.44}$$

where P_0 is the underlying probability measure for the pairs (T_i,Δ_i). We now turn this into a representation as an integral with respect to the empirical measure $\mathbb{P}_n-P_0$ in the observation space.

We have, by (11.43),

$$\int\theta_{t,h,\hat F_n}(u,\delta)\,d\mathbb{P}_n(u,\delta)$$
$$=\int\frac{\{\delta-F(u)\}\{K_h(t-u)-K_h(t+u-2a)-K_h(2b-t-u)\}}{g(u)}\,d\mathbb{P}_n(u,\delta).$$

We next define, analogous to the treatment in Section 10.2,

$$\psi_{t,h}(u)=\frac{K_h(t-u)-K_h(t+u-2a)-K_h(2b-t-u)}{g(u)}, \tag{11.45}$$

and

$$\bar{\psi}_{t,h}(u) = \begin{cases} \psi_{t,h}(\tau_i), & \text{, if } F_0(u) > \hat{F}_n(\tau_i),\ u \in [\tau_i, \tau_{i+1}), \\ \psi_{t,h}(s), & \text{, if } F_0(s) = \hat{F}_n(s),\ \text{ for some } s \in [\tau_i, \tau_{i+1}), \\ \psi_{t,h}(\tau_{i+1}), & \text{, if } F_0(u) < \hat{F}_n(\tau_i),\ u \in [\tau_i, \tau_{i+1}), \end{cases}$$

where the τ_i are successive points of jump of $\hat{F}_n$. Note that we follow exactly the same method of proof as used for the global smooth functional estimation; see also Figure 10.1.

We then get, by the equality condition in the characterization (8.15) of the MLE:

$$\int \{\delta - \hat{F}_n(u)\}\bar{\psi}_{t,h}(u)\, d\mathbb{P}_n(u,\delta) = 0.$$

Furthermore,

$$\begin{aligned}
&\int \{\delta - \hat{F}_n(u)\}\bar{\psi}_{t,h}(u)\, d\left(\mathbb{P}_n - P_0\right)(u,\delta) = -\int \{\delta - \hat{F}_n(u)\}\bar{\psi}_{t,h}(u)\, dP_0(u,\delta) \\
&\quad = -\int \{F_0(u) - \hat{F}_n(u)\}\bar{\psi}_{t,h}(u)\, dG(u) \\
&\quad = -\int \{F_0(u) - \hat{F}_n(u)\}\psi_{t,h}(u)\, dG(u) - \int \{F_0(u) - \hat{F}_n(u)\}\left\{\bar{\psi}_{t,h}(u) - \psi_{t,h}(u)\right\} dG(u) \\
&\quad = \int \{\mathbb{K}_h(t-u) + \mathbb{K}_h(t+u-2a) - \mathbb{K}_h(2b-t-u)\}\, d\left(\hat{F}_n - F_0\right)(u) \\
&\qquad - \int \{F_0(u) - \hat{F}_n(u)\}\left\{\bar{\psi}_{t,h}(u) - \psi_{t,h}(u)\right\} dG(u).
\end{aligned} \tag{11.46}$$

So we find:

$$\begin{aligned}
\hat{F}_{n,h}^{(SML)}(t) &= \int \{\mathbb{K}_h(t-u) + \mathbb{K}_h(t+u-2a) - \mathbb{K}_h(2b-t-u)\}\, d\hat{F}_n(u) \\
&= \int \{\mathbb{K}_h(t-u) + \mathbb{K}_h(t+u-2a) - \mathbb{K}_h(2b-t-u)\}\, dF_0(u) \\
&\quad + \int \{\mathbb{K}_h(t-u) + \mathbb{K}_h(t+u-2a) - \mathbb{K}_h(2b-t-u)\}\, d\left(\hat{F}_n - F_0\right)(u) \\
&= \int \{\mathbb{K}_h(t-u) + \mathbb{K}_h(t+u-2a) - \mathbb{K}_h(2b-t-u)\}\, dF_0(u) \\
&\quad + \int \{\delta - \hat{F}_n(u)\}\psi_{t,h}(u)\, d\left(\mathbb{P}_n - P_0\right)(u,\delta) \\
&\quad + \int \{F_0(u) - \hat{F}_n(u)\}\left\{\bar{\psi}_{t,h}(u) - \psi_{t,h}(u)\right\} dG(u),
\end{aligned} \tag{11.47}$$

and the asymptotic behavior of $\hat{F}_{n,h}^{(SML)}$ will (almost) have been pinned down if we can deal with the remainder term

$$\int \{F_0(u) - \hat{F}_n(u)\}\left\{\bar{\psi}_{t,h}(u) - \psi_{t,h}(u)\right\} dG(u). \tag{11.48}$$

By the Cauchy-Schwarz inequality,

$$\left|\int \{F_0(u) - \hat{F}_n(u)\}\{\bar{\psi}_{t,h}(u) - \psi_{t,h}(u)\}\,dG(u)\right|$$
$$\lesssim \left\|(\hat{F}_n - F_0)1_{[t-h,t+h]}\right\|_{2,G}\left\|(\bar{\psi}_{t,h} - \psi_{t,h})1_{[t-h,t+h]}\right\|_{2,G}, \tag{11.49}$$

where $\lesssim$ denotes smaller than or equal to, up to a fixed positive multiplicative constant, and

$$\|\phi\|_{2,G}^2 = \int \phi(x)^2\,dG(x).$$

By the choice of the function $\bar{\psi}_{t_0,h}$,

$$\left|\bar{\psi}_{t_0,h}(x) - \psi_{t_0,h}(x)\right| \le \frac{K}{h^2}\left|\hat{F}_n(x) - F_0(x)\right|$$

(compare this with the treatment in Section 10.2), and hence, by the Cauchy-Schwarz inequality and (11.34),

$$\left\|\bar{\psi}_{t_0,h} - \psi_{t_0,h}\right\|_{2,G} = O_p\left(h^{-3/2}\|\hat{F}_n - F_0\|_{2,G}\right).$$

Since, for the same reason,

$$\left\|(\hat{F}_n - F_0)1_{[t_0-h,t_0+h]}\right\|_{2,G} = O_p\left(h^{1/2}\|\hat{F}_n - F_0\|_{2,G}\right),$$

we get from (11.49)

$$\left|\int \{F_0(u) - \hat{F}_n(u)\}\{\bar{\psi}_{t_0,h}(u) - \psi_{t_0,h}(u)\}\,dG(u)\right| = O_p\left(h^{-1}n^{-2/3}\right).$$

So we get, if $h = h_n \asymp n^{-1/5}$,

$$\left|\int \{F_0(u) - \hat{F}_n(u)\}\{\bar{\psi}_{t_0,h}(u) - \psi_{t_0,h}(u)\}\,dG(u)\right| = O_p\left(n^{1/5-2/3}\right)$$
$$= O_p\left(n^{-7/15}\right) = o_p\left(n^{-2/5}\right). \tag{11.50}$$

Hence we have obtained, using (11.46),

$$\hat{F}_{n,h}^{(SML)}(t) - \int \{\mathbb{K}_h(t-u) + \mathbb{K}_h(t+u-2a) - \mathbb{K}_h(2b-t-u)\}\,dF_0(u)$$
$$= \int \{\delta - \hat{F}_n(u)\}\bar{\psi}_{t_0,h}(u)\,d\left(\mathbb{P}_n - P_0\right)(u,\delta) + o_p\left(n^{-2/5}\right),$$

that is, we have an asymptotic representation of the relevant functional in the hidden space by an empirical integral in the observation space. It is proved in Lemma A.7 on p. 382 of Groeneboom et al., 2010, using entropy methods from empirical process theory, that

$$n^{2/5}\int \bar{\psi}_{t,h}(u)\{\hat{F}_n(u) - F_0(u)\}\,d\left(\mathbb{P}_n - P_0\right)(u,\delta) = o_p(1),$$

and

$$n^{2/5}\int \{\bar{\psi}_{t,h}(u) - \psi_{t,h}(u)\}\{\delta - F_0(u)\}\,d\left(\mathbb{P}_n - P_0\right)(u,\delta) = o_p(1)$$

(this is indeed what is *proved* in Lemma A.7 on p. 382 of Groeneboom et al., 2010, although in the statement of the lemma the $o_p(1)$ is printed as $O_p(1)$). Combining these results yields

$$\hat{F}_{n,h}^{(SML)}(t) - \int \{\mathbb{K}_h(t-u) + \mathbb{K}_h(t+u-2a) - \mathbb{K}_h(2b-t-u)\}\, dF_0(u)$$
$$= \int \{\delta - F_0(u)\}\psi_{t,h}(u)\, d\left(\mathbb{P}_n - P_0\right)(u,\delta) + o_p\left(n^{-2/5}\right).$$

where $\psi_{t,h}$ is defined by (11.45).

So we obtain:

$$\hat{F}_{n,h}^{(SML)}(t) = \int \{\mathbb{K}_h(t-u) + \mathbb{K}_h(t+u-2a) - \mathbb{K}_h(2b-t-u)\}\, dF_0(u)$$
$$+ \int \{\delta - F_0(u)\}\psi_{t,h}(u)\, d\mathbb{P}_n(u,\delta) + o_p\left(n^{-2/5}\right), \tag{11.51}$$

and, if $h \sim cn^{-2/5}$, then $n^{2/5}\int\{\delta - F_0(u)\}\psi_{t,h}(u)\, d\left(\mathbb{P}_n - P_0\right)(u,\delta)$ is asymptotically normal with mean zero and variance

$$h\,\mathbb{E}\,\frac{\{K_h(t-T_1) - K_h(t+T_1-2a) - K_h(2b-t-T_1)\}^2 F_0(T_1)\{1-F_0(T_1)\}}{g(T_1)^2}$$
$$\sim h\int \frac{\{K_h(t-u) - K_h(t+u-2a) - K_h(2b-t-u)\}^2 F_0(u)\{1-F_0(u)\}}{g(u)^2}\, dG(u)$$
$$\sim \frac{F_0(t)\{1-F_0(t)\}}{cg(t)}\int\left\{K(u) - K\left(\frac{2(t-a)}{h} - u\right) - K\left(\frac{2(b-t)}{h} + u\right)\right\}^2 du$$
$$\longrightarrow \frac{F_0(t)\{1-F_0(t)\}}{cg(t)}\int K(u)^2\, du, \quad h \downarrow 0, \tag{11.52}$$

for each $t \in (a,b)$.

Remark We note that the argument for treating the remainder term (11.48) in Groeneboom et al., 2010, is different. Apart from the fact that the boundary correction is not incorporated, it is used that

$$\|F_0 - \hat{F}_n\|_\infty = O_p\left(n^{-1/3}\log n\right),$$

and that

$$\max_i |\tau_{i+1} - \tau_i| = O_p\left(n^{-1/3}\log n\right),$$

where τ_i and τ_{i+1} are successive point of jump of $\hat{F}_n$ (see (2.3) and (2.4) on p. 355 of Groeneboom et al., 2010). The argument is replaced here by an L_2-bound and the Cauchy-Schwarz inequality, using Theorem 11.3. In this way the treatment of the global and local smooth functional theory becomes completely similar.

The preceding leads to the following theorem (Theorem 4.2 on p. 365 of Groeneboom et al., 2010).

Theorem 11.4 *Let the distribution corresponding to* F_0 *have support* $[0, M]$ *and let* F_0 *have a density* f_0 *staying away from zero on* $(0, M)$*. Furthermore, let* G *have a density* g *with* *a support that contains* $[0, M]$ *and let* g *stay away from zero on* $[0, M]$, *with a bounded derivative* g'. *Finally, let* t *be an interior point of* $[0, M]$ *such that* f_0 *has a continuous derivative* f_0' *at* t. *Then, if* $h \sim cn^{-1/5}$ *and the SMLE* $\hat{F}_{n,h}^{(SML)}$ *is defined by (11.35),*

$$n^{2/5}\left\{\hat{F}_{n,h}^{(SML)}(t) - F_0(t)\right\} \xrightarrow{\mathcal{D}} N\left(\mu, \sigma^2\right),$$

where

$$\mu = \tfrac{1}{2}c^2 f_0'(t)\int u^2 K(u)\,du$$

and

$$\sigma^2 = \frac{F_0(t)\{1 - F_0(t)\}}{cg(t)}\int K(u)^2\,du.$$

Remark In Theorem 4.2 on p. 365 of Groeneboom et al., 2010, there is the extra condition that $f_0'(t) \neq 0$. This condition is not needed for the validity of Theorem 11.4, but only to ensure that a bandwidth of order $n^{-1/5}$ is the optimal choice. If $f_0'(t) = 0$, the squared bias vanishes with respect to the variance and in that situation one can choose a larger bandwidth to obtain a faster convergence than order $n^{-2/5}$. For more details, see Groeneboom et al., 2010.

Theorem 11.4 was used in Section 9.5, in constructing pointwise confidence intervals using the classical bootstrap by resampling with replacement from the sample $(T_1, \Delta_1), \dots, (T_n, \Delta_n)$. It is not immediately obvious that the classical bootstrap will work. There is, for example, the result in Kosorok, 2008a, which states that the classical bootstrap is inconsistent for the MLE itself. But (11.51) tells us (see Exercises 11.12 and 11.13) that the SMLE is asymptotically equivalent to

$$\begin{aligned} F_{nh}^{toy}(t) &= \int \{\mathbb{K}_h(t-u) + \mathbb{K}_h(t+u-2a) - \mathbb{K}_h(2b-t-u)\}\,dF_0(u) \\ &\quad + \frac{1}{n}\sum_{i=1}^{n}\frac{\{K_h(t-T_i) - K_h(t+T_i-2a) - K_h(2b-t-T_i)\}\{\Delta_i - F_0(T_i)\}}{g(T_i)}, \end{aligned} \tag{11.53}$$

and this toy estimator satisfies the conditions that allow us to use the classical bootstrap.

The sample variance of the toy estimator (11.53) (it is a toy estimator because it contains the density g and the distribution function F_0, which have to be estimated, something we do not have to do if the SMLE is used) is given by

$$S_n(t)^2 = \frac{1}{n^2}\sum_{i=1}^{n}\frac{\{K_h(t-T_i) - K_h(t+T_i-2a) - K_h(2b-t-T_i)\}^2\{\Delta_i - F_0(T_i)\}^2}{g(T_i)^2}.$$

Note that, for $t \in [a+h, b-h]$,

$$\mathbb{E}S_n(t)^2 = \frac{1}{n}\int \frac{K_h(t-x)^2 F_0(x)\{1-F_0(x)\}}{g(x)}\,dx \sim \frac{F_0(t)\{1-F_0(t)\}}{nh\,g(t)}\int K(u)^2\,du.$$

By similar methods one can prove the following result for the density estimate:

$$\hat{f}_{nh}^{SML}(t) = \int K_h(t-x)\,d\hat{F}_n(x).$$

Theorem 11.5 (Groeneboom et al., 2010, Theorem 4.3, p. 366.) *Fix $t > 0$ such that f_0'' is continuous at t_0. Suppose f_0 has compact support $[0, M]$ and stays strictly away from zero on $[0, M]$ and suppose that the observation density g also stays away from zero on $[0, M]$ and has a bounded derivative on $[0, M]$. Let $h = cn^{-1/7}$ ($c > 0$) be the bandwidth used in the definition of $\hat{f}_{nh}^{SML}$. Then*

$$n^{2/7}\left(\hat{f}_{nh}^{SML}(t) - f_0(t)\right) \xrightarrow{\mathcal{D}} N(\nu, \tau^2),$$

where

$$\nu = \frac{1}{2}c^2 f_0''(t)\int u^2 K(u)\,du, \quad \tau^2 = \frac{F_0(t)(1 - F_0(t))}{c^3 g(t)}\int K'(u)^2 du. \tag{11.54}$$

For $t \notin [a + h, b - h]$, we use a boundary correction of another type than used for the SMLE. If we use the Schuster boundary correction

$$\int \{K_h(t - x) + K_h(t + x - 2a) + K_h(2b - t - x)\}\, d\hat{F}_n(x),$$

the bias is of order $O(h)$ instead of $O(h^2)$ under the usual smoothness condition, unless $f_0'(a) = f_0'(b) = 0$. We therefore use the boundary correction similar to the boundary correction given by (9.35) and (9.36). This means that we replace the kernel K by the kernel

$$K^t(u) = \begin{cases} \{\alpha((t - a)/h)K(u) + \beta((t - a)/h)uK(u)\}1_{[-1,(t-a)/h)}(u), & t \in [a, a + h), \\ K(u), & t \in [a + h, b - h], \\ \{\alpha((b - t)/a)K(u) - \beta((b - t)/a)uK(u)\}1_{[(t-b)/h,1]}(u), & t \in (b - h, b], \end{cases} \tag{11.55}$$

where, for $s \in [-1, 1]$, the coefficients $\alpha(s)$ and $\beta(s)$ are determined by

$$\alpha(s)\int_{-1}^{s} K(u)\,\mathrm{d}u + \beta(s)\int_{-1}^{s} uK(u)\,\mathrm{d}u = 1,$$

$$\alpha(s)\int_{-1}^{s} uK(u)\,\mathrm{d}u + \beta(s)\int_{-1}^{s} u^2K(u)\,\mathrm{d}u = 0.$$

With these definitions, we further define

$$\hat{f}_{nh}^{SML}(t) = \int K_h^t(t - x)\,d\hat{F}_n(x), \quad K_h^t(u) = h^{-1}K_t(u/h). \tag{11.56}$$

Instead of the equation (11.41) we now get the equation

$$\begin{aligned} E\left\{\theta_{t,h,F}(T, \Delta) \,\middle|\, X = x\right\} &= \int_{u \ge x} \theta_{h,F}(u, 1)g(u)\,du + \int_{u<x} \theta_{t,h,F}(u, 0)g(u)\,du \\ &= K_h^t(t - x) - \int K_h^t(t - u)\,dF(u). \end{aligned} \tag{11.57}$$

We now have the following lemma, analogous to Lemma 11.6.

Lemma 11.7 *Equation (11.57) is solved by*

$$\theta_{t,h,F}(u, \delta) = \frac{\{\delta - F(u)\}\left[K_h^t\right]'((t - u)/h)}{g(u)},$$

where

$$\left[K_h^t\right]'(x) = h^{-1}\frac{d}{dx}K^t(x/h).$$

Proof The proof proceeds along the same lines as the proof of Lemma 11.6. We have:

$$\begin{aligned}
&\int_{u\ge x}\theta_{t,h,F}(u,1)g(u)\,du + \int_{u<x}\theta_{t,h,F}(u,0)g(u)\,du\\
&= \int_{u\ge x}\frac{\{1-F(u)\}\left[K_h^t\right]'(t-u)}{g(u)}g(u)\,du - \int_{u<x}\frac{F(u)\left[K_h^t\right]'(t-u)}{g(u)}g(u)\,du\\
&= \int_{u\ge x}\left[K_h^t\right]'(t-u)\,du - \int F(u)\left[K_h^t\right]'(t-u)\,du\\
&= K_h^t(t-x) - \int K_h^t(t-u)\,dF(u),
\end{aligned}$$

using integration by parts in the last step. □

From this lemma we get, by an analysis similar to the one used for the SMLE, that $\hat f_{nh}^{(SMLE)}$ is asymptotically equivalent to the toy estimator

$$f_{nh}^{toy}(t) = \int K_h^t(t-u)\,dF_0(u) + \frac{1}{n}\sum_{i=1}^n \frac{\left[K_h^t\right]'(t-T_i)\,\{\Delta_i - F_0(T_i)\}}{g(T_i)}. \tag{11.58}$$

From this representation we see that the asymptotic variance of $f_{nh}^{SML}(t)$ is given by:

$$\frac{F_0(t)\{1-F_0(t)\}}{nh_n^3}\int \left[K^t\right]'(u)^2\,du \sim \frac{F_0(t)\{1-F_0(t)\}}{nh_n^3}\int K'(u)^2\,du, \quad n\to\infty,$$

for each $t\in(a,b)$, if $h_n \sim cn^{-1/7}$. The asymptotic equivalence with the toy estimator is used in the justification of the bootstrap confidence intervals for the density in Section 9.5.

We likewise get the following result for the estimate of the hazard, defined by:

$$\hat\lambda_n^{SML}(t) = \frac{\hat f_n^{SML}(t)}{1-\hat F_n^{SML}(t)}.$$

Theorem 11.6 (Groeneboom et al., 2010, Corollary 4.4, p. 366.) *Let the conditions of Theorem 11.5 be satisfied. Then we have, for the estimate $\hat\lambda_n^{SML}$ of the hazard λ_0,*

$$n^{2/7}\left(\hat\lambda_n^{SML}(t) - \lambda_0(t)\right) \xrightarrow{\mathcal{D}} N(\mu_\lambda, \sigma_\lambda^2),$$

where

$$\begin{aligned}
\mu_\lambda &= \frac{c^2}{2(1-F_0(t))}\left\{f_0''(t) + \frac{f_0(t)f_0'(t)}{1-F_0(t)}\right\}\int u^2K(u)\,du,\\
\sigma_\lambda^2 &= \frac{F_0(t)}{c^3 g(t)(1-F_0(t))}\int K'(u)^2du.
\end{aligned} \tag{11.59}$$

11.4 The SMLE for Interval Censoring Case 2

The case 2 interval censoring model is introduced in Section 4.7 and the SMLE for the event time distribution is considered in Section 8.6. Compared with the current status model, the asymptotic theory for the SMLE in the interval censoring case 2 model is still incomplete. Recall that the smoothed MLE (SMLE) for the interval censoring model case 2 is defined by

$$\tilde{F}_n^{SML}(t) = \int \mathbb{K}((t-u)/h_n)\, d\hat{F}_n(u), \tag{11.60}$$

where the integrated kernel $\mathbb{K}$ is defined by (11.36) again. The corresponding estimator of the density f_0 of the underlying distribution function F_0 is given by

$$\tilde{f}_n^{SML}(t) := \frac{d}{dt}\tilde{F}_n^{SML}(t) = \frac{1}{h_n}\int K\left((t-u)/h_n\right) d\hat{F}_n(u),$$

where K is a symmetric kernel function.

Let g be the density of (T_i, U_i), with first marginal density g_1 and second marginal density g_2, and let $\phi_{t,h,F}$ be a solution of the integral equation (in ϕ):

$$\phi(u) = d_F(u)\left\{k_{t,h}(u) + \int_{v>u} \frac{\phi(v)-\phi(u)}{F(v)-F(u)}\, g(u,v)\, dv - \int_{v<u} \frac{\phi(u)-\phi(v)}{F(u)-F(v)}\, g(v,u)\, dv\right\},$$

where

$$d_F(u) = \frac{F(u)\{1-F(u)\}}{g_1(u)\{1-F(u)\} + g_2(u)F(u)},$$

and the function $k_{t,h}$ is defined by

$$k_{t,h}(u) = h^{-1}K\left((t-u)/h\right). \tag{11.61}$$

Moreover, let the function $\theta_{t,h,F}$ be defined by

$$\theta_{t,h,F}(u,v,\delta_1,\delta_2) = -\frac{\delta_1\phi_{t,h,F}(u)}{F(u)} - \frac{\delta_2\{\phi_{t,h,F}(v)-\phi_{t,h,F}(u)\}}{F(v)-F(u)} + \frac{\delta_3\phi_{t,h,F}(v)}{1-F(v)}, \tag{11.62}$$

where $u < v$. Then, as in Geskus and Groeneboom, 1997 (the separated case, where the two observation times for each subject cannot be arbitrarily close), we have the representation

$$\begin{aligned}
\int \mathbb{K}((t-u)/h)\, d(\hat{F}_n - F_0)(u) &= \int \theta_{t,h,\hat{F}_n}(u,v,\delta_1,\delta_2)\, dP_0(u,v,\delta_1,\delta_2)\\
&= \int \frac{\phi_{t,h,\hat{F}_n}(u)}{\hat{F}_n(u)} F_0(u)g_1(u)\, du\\
&\quad + \int \frac{\phi_{t,h,\hat{F}_n}(v)-\phi_{t,h,\hat{F}_n}(u)}{\hat{F}_n(v)-\hat{F}_n(u)}\{F_0(v)-F_0(u)\}g(u,v)\, du\, dv\\
&\quad - \int \frac{\phi_{t,h,\hat{F}_n}(v)}{1-\hat{F}_n(v)}\{1-F_0(v)\}g_2(v)\, dv.
\end{aligned}$$

In Geskus and Groeneboom, 1999, a similar representation is used for the nonseparated case (where the two observation times can be arbitrarily close), using a pair of functions. For

$F = F_0$ the integral equation becomes

$$\phi_{t,h,F_0}(u) = d_{F_0}(u)\left\{k_{t,h}(u) + \int_{v>u} \frac{\phi_{t,h,F_0}(v) - \phi_{t,h,F_0}(u)}{F_0(v) - F_0(u)}\, g(u,v)\,dv \right.$$
$$\left. - \int_{v<u} \frac{\phi_{t,h,F_0}(u) - \phi_{t,h,F_0}(v)}{F_0(u) - F_0(v)}\, g(v,u)\,dv \right\}. \tag{11.63}$$

We now first treat (heuristically) the properties of the solution in the nonseparated case. The dominating part of the solution will be defined on a shrinking neighborhood of t and assuming

$$\phi_{t,h,F_0}(v) - \phi_{t,h,F_0}(u) \sim (v-u)\phi'_{t,h,F_0}(v),$$

we get

$$\int_{v>u} \frac{\phi_{t,h,F_0}(v) - \phi_{t,h,F_0}(u)}{F_0(v) - F_0(u)}\, g(u,v)\,dv \sim \frac{g(t,t)}{f_0(t)} \int_{v>u} \phi'_{t,h,F_0}(v)\,dv = -\frac{g(t,t)\phi_{t,h,F_0}(u)}{f_0(t)}.$$

Likewise, assuming

$$\phi_{t,h,F_0}(u) - \phi_{t,h,F_0}(v) \sim (u-v)\phi'_{t,h,F_0}(v),$$

we get

$$-\int_{v<u} \frac{\phi_{t,h,F_0}(u) - \phi_{t,h,F_0}(v)}{F_0(u) - F_0(v)}\, g(v,u)\,dv \sim -\frac{g(t,t)}{f_0(t)} \int_{v<u} \phi'_{t,h,F_0}(v)\,dv = -\frac{g(t,t)\phi_{t,h,F_0}(u)}{f_0(t)}.$$

So we end up with the approximate equation

$$\phi_{t,h,F_0}(u)\left\{1 + \frac{2d_{F_0}(t)g(t,t)}{f_0(t)}\right\} \sim d_{F_0}(u)k_{t,h}(u),$$

implying

$$\phi_{t,h,F_0}(u) \sim \frac{d_{F_0}(t)k_{t,h}(u)}{1 + 2d_{F_0}(t)g(t,t)/f_0(t)}. \tag{11.64}$$

Using the theory in Geskus and Groeneboom, 1999, we get that the solution ϕ_{t,h,F_0} gives as an approximation for $n\,\mathrm{Var}(\tilde F_n(t))$

$$\begin{aligned}\sigma_n^2 &:= E\,\theta_{t,h,F_0}(T_1, U_1, \Delta_{11}, \Delta_{12})^2\\ &= \int \frac{\phi_{t,h,F_0}(u)^2}{F_0(u)} g_1(u)\,du + \int \frac{\left\{\phi_{t,h,F_0}(v) - \phi_{t,h,F_0}(u)\right\}^2}{F_0(v) - F_0(u)} g(u,v)\,du\,dv\\ &\quad + \int \frac{\phi_{t,h,F_0}(v)^2}{1 - F_0(v)} g_2(v)\,dv \sim h^{-1} d_{F_0}(t)\left\{1 + 2d_{F_0}(t)g(t,t)/f_0(t)\right\}^{-1} \int K(u)^2\,du, \quad h \downarrow 0.\end{aligned} \tag{11.65}$$

The approximation seems to work rather well, as can be seen in Table 11.1, where the actual variance for samples of size $n = 1000$ is estimated by generating 10,000 samples of size $1,000$ from a Uniform $(0, 1)$ distribution F_0 and a uniform observation distribution H on

Table 11.1 *Estimates of the Actual Variances var($\tilde{F}_n(t)$) (times n) and σ_n^2 of (11.65), Where $h_n = n^{-1/5}$, for Sample Size $n = 1000$*

t	$n\text{var}(\tilde{F}_n(t))$	σ_n^2 of (11.65)	ratio
0.1	0.146489	0.146158	1.002263
0.2	0.262056	0.259837	1.008541
0.3	0.334990	0.341036	0.982272
0.4	0.380357	0.389755	0.975887
0.5	0.399258	0.405995	0.983406
0.6	0.386292	0.389755	0.991114
0.7	0.342651	0.341036	1.004736
0.8	0.261457	0.259837	1.006235
0.9	0.145304	0.146158	0.994155

Note: The estimates of the actual variances were based on 10,000 samples of size 1,000 from a Uniform (0, 1) distribution F_0 and a uniform observation distribution H on the upper triangle of the unit square.

the upper triangle of the unit square. Note that we have

$$\sigma_n^2 \sim \frac{t(1-t)}{2h_n} \int K(u)^2 \, du, \; h_n \downarrow 0,$$

in this case.

A picture of ϕ_{t,h_n,F_0} for the Uniform(0, 1) distribution F_0 and $h_n = n^{-1/5}$ is shown in Figure 11.3; the function was computed by solving the corresponding matrix equation on a 1000×1000 grid. Note that we apply the smooth functional theory (which is also discussed in Groeneboom, 1996) not for a fixed functional, but for changing functionals on shrinking intervals (in the hidden space). The reason that this can be done is that the bandwidth h is chosen to be of a larger order than the critical rate $n^{-1/3}$, and that then a different type of asymptotics sets in; with asymptotic normality, and so on, instead of the nonstandard asymptotics of the MLE itself. This method is also used in Section 11.3, for the current status model.

In analogy with Theorem 11.4, we expect the following result to hold, using the conditions on the underlying distributions, discussed in Geskus and Groeneboom, 1997, and Geskus and Groeneboom, 1999. To avoid messy notation, we will denote the smoothed MLE by $\tilde{F}_n$ instead of $\tilde{F}_n^{SML}$ in the remainder of this section.

Conjecture 11.1 *Let the conditions of Theorem 1, p. 212, in Geskus and Groeneboom, 1997 (separated case), or Theorem 3.2, p. 647, in Geskus and Groeneboom, 1999 (nonseparated case), be satisfied. Moreover, let the joint density g of the joint density of (T_i, U_i) have a continuous bounded second total derivative in the interior of its domain and let f_0 have a continuous derivative at the interior point t of the support of f_0, and let $\tilde{F}_n$ be the smoothed MLE, defined by (11.60). Then, if $h_n \asymp n^{-1/5}$, we have*

$$\sqrt{n}\left\{\tilde{F}_n(t) - F_0(t) - \tfrac{1}{2}h_n^2 f_0'(t) \int u^2 K(u)\, du\right\} \Big/ \sigma_n \xrightarrow{\mathcal{D}} N(0,1), \; n \to \infty,$$

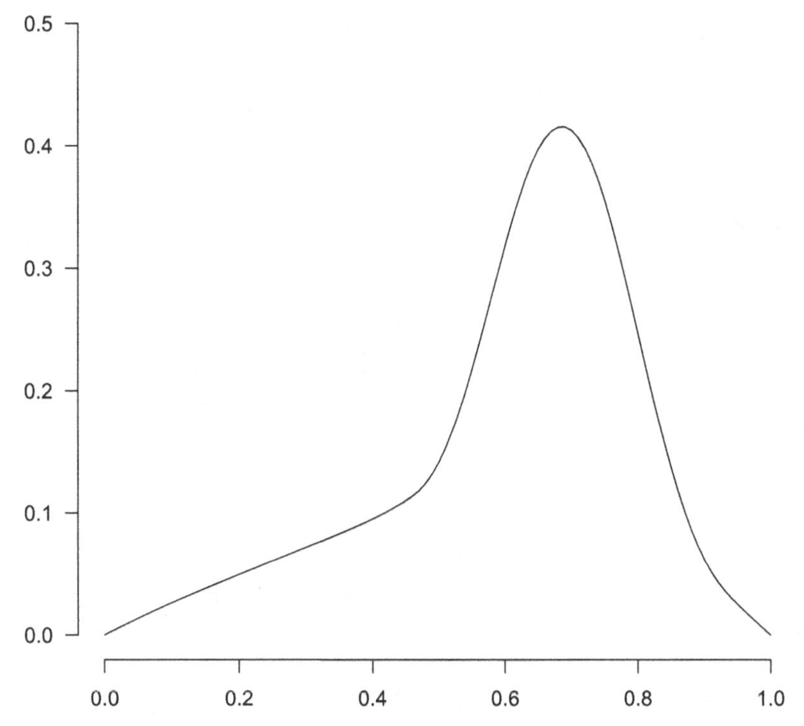

Figure 11.3 The function $u \mapsto \phi_{t,h_n,F_0}(u),\ u \in [0,1]$, for $t = 0.7$, $h_n = n^{-1/5}$, $n = 1000$, the Uniform(0, 1) distribution F_0 and a uniform observation distribution H on the upper triangle of the unit square.

where $N(0,1)$ is the standard normal distribution and σ_n^2 is defined by

$$\sigma_n^2 = E\,\theta_{t,h_n,F_0}(T_1, U_1, \Delta_{11}, \Delta_{12})^2, \tag{11.66}$$

with θ_{t,h_n,F_0} given by (11.62).

Note that Conjecture 11.1 covers both the separated and the nonseparated case. The functions ϕ_{t,h_n,F_0}, defining the function θ_{t,h_n,F_0} and hence also the variance σ_n^2, are of a rather different nature for the separated case and the nonseparated case. For an example of this, see Figure 11.4.

The variance σ_n^2 can be estimated by

$$\hat\sigma_n^2 = \int \tilde\theta_{t,h_n,\tilde F_n}(t, u, \delta_1, \delta_2)\, d\mathbb{P}_n(u, v, \delta_1, \delta_2),$$

where

$$\tilde\theta_{t,h_n,\tilde F_n}(u, v, \delta_1, \delta_2) = \frac{\delta_1 \tilde\phi_{t,h_n,\tilde F_n}(u)}{\tilde F_n(u)} + \frac{\delta_2\{\tilde\phi_{t,h_n,\tilde F_n}(v) - \tilde\phi_{t,h_n,\tilde F_n}(u)\}}{\tilde F_n(v) - \tilde F_n(u)} - \frac{\delta_3 \tilde\phi_{t,h_n,\tilde F_n}(v)}{1 - \tilde F_n(v)},$$

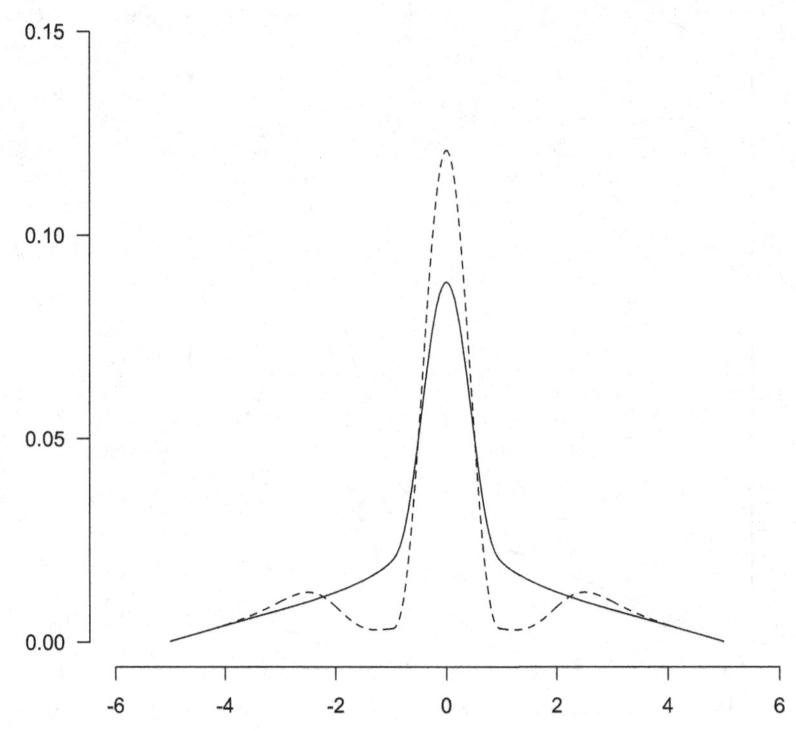

Figure 11.4 The function $u \mapsto \phi_{t,h,F_0}(t - hu)$, $u \in [-5, 5]$, for $t = 0.5$, $h = 0.1$, the Uniform(0, 1) distribution F_0 and (nonseparated case) a uniform observation distribution H on the upper triangle of the unit square (solid curve) and the function $u \mapsto \phi_{t,h,F_0}(t - hu)$ for the (separated) case where the observation distribution H is uniform on the triangle with vertices $(0, \epsilon)$, $(0, 1)$ and $(1 - \epsilon, 1)$, where $\epsilon = 0.2$ (dashed).

for $u < v$, where $\tilde{\phi}_{t,h_n,\tilde{F}_n}$ solves the integral equation

$$\phi(u) = d_{\tilde{F}_n(u)}(u)\left\{K_{t,h_n}(u) + \int_{v>u} \frac{\phi(v) - \phi(u)}{\tilde{F}_n(v) - \tilde{F}_n(u)}\, g_n(u, v)\, dv \right.$$
$$\left. - \int_{v<u} \frac{\phi(u) - \phi(v)}{\tilde{F}_n(u) - \tilde{F}_n(v)}\, g_n(v, u)\, dv \right\}, \tag{11.67}$$

where g_n is a kernel estimate of the density g and where

$$d_{\tilde{F}_n(u)}(u) = \frac{\tilde{F}_n(u)\{1 - \tilde{F}_n(u)\}}{g_{n1}(u)\{1 - \tilde{F}_n(u)\} + g_{n2}(u)\tilde{F}_n(u)},$$

$$g_{n1}(u) = \int g_n(u, v)\, dv, \quad g_{n2}(u) = \int g_n(v, u)\, dv.$$

For g_n chosen as in the theorem, the distribution function $\tilde{F}_n$ will be strictly increasing with probability tending to one. Since $\tilde{F}_n$ is also continuously differentiable, the equation (11.67) will have an absolutely continuous solution $\tilde{\phi}_{t,h_n,\tilde{F}_n}$, and we do not have to take recourse to a solution pair, as in Geskus and Groeneboom, 1999, which deals separately with a discrete and absolutely continuous part in treating the MLE itself.

In the corresponding result for the current status model we have explicit expressions, and we briefly discuss the analogy here, using a notation of the same type. Let $\tilde F_n^{(CS)}$ be the smoothed MLE for the current status model, defined by (11.35), but now using the MLE $\hat F_n$ in the current status model. In this case the function $\theta_{t,h,F}$, representing the functional in the observation space, is given by

$$\theta_{t,h,F}^{(CS)}(u,\delta) = -\frac{\delta\phi_{t,b,F}^{(CS)}(u)}{F(u)} + \frac{(1-\delta)\phi_{t,b,F}^{(CS)}(u)}{1-F(u)},\ u \in (0,1). \tag{11.68}$$

where ϕ is given by:

$$\phi_{t,b,F}^{(CS)}(u) = \frac{F(u)\{1-F(u)\}}{g(u)} k_{t,h}(u),$$

and $k_{t,h}$ is defined by (11.61). Moreover, g is the density of the (one-dimensional) observation distribution. The solution $\phi_{t,h_n,F_0}^{(CS)}$ gives as an approximation for $n\,\mathrm{var}(\tilde F_n(t))$:

$$E\,\theta_{t,h_n,F_0}^{(CS)}(T_1,\Delta_1)^2 = \int \frac{\phi_{t,h_n,F_0}^{(CS)}(u)^2}{F_0(u)} g(u)\,du + \int \frac{\phi_{t,h_n,F_0}^{(CS)}(u)^2}{1-F_0(u)} g(u)\,du$$

$$= \int \frac{F_0(u)\{1-F_0(u)\}k_{t,h_n}(u)^2}{g(u)}\,du \sim \frac{F_0(t)\{1-F_0(t)\}}{h_n g(t)} \int K(u)^2\,du,\ h_n \to 0.$$

Moreover,

$$\lim_{h\downarrow 0} h E\,\theta_{t,h,F_0}^{(CS)}(T_1,\Delta_1)^2 = \frac{F_0(t)\{1-F_0(t)\}}{g(t)} \int K(u)^2\,du,$$

so in this case we obtain the central limit theorem

$$\sqrt{n}\left\{\tilde F_n(t) - F_0(t) - \tfrac{1}{2}h_n^2 f_0'(t)\int u^2 K(u)\,du\right\} \Big/ \sigma_n \xrightarrow{\mathcal{D}} N(0,1),\ n \to \infty,$$

where

$$\sigma_n^2 = E\,\theta_{t,h_n,F_0}^{(CS)}(T_1,\Delta_1)^2 \sim \frac{F_0(t)\{1-F_0(t)\}}{h_n g(t)} \int K(u)^2\,du;$$

see Theorem 11.4.

It is seen here that the asymptotic variance of the SMLE is equal to the asymptotic variance of the MSLE in the current status model, and we also expect this to be true in the interval censoring case 2 situation. Actually, the dominating part of the solution of the equation (11.63) in the separated case seems to be given by

$$\phi_1(u) = d_F(u)\left\{1 + d_F(u)\int_{v>u} \frac{g(u,v)}{F(v)-F(u)}\,dv + d_F(u)\int_{v<u} \frac{g(v,u)}{F(u)-F(v)}\,dv\right\}^{-1} h^{-1}K((t-u)/h),$$

and plugging this into the expression for the variance (11.65), we get:

$$\lim_{h\downarrow 0} hE\,\theta_{t,h,F_0}(T_1,U_1,\Delta_{11},\Delta_{12})^2$$

$$= \lim_{h\downarrow 0} h\left\{\int \frac{\phi_0(u)^2}{F_0(u)} g_1(u)\,du + \int \frac{\{\phi_0(v)-\phi_0(u)\}^2}{F_0(v)-F_0(u)} g(u,v)\,du\,dv + \int \frac{\phi_0(v)^2}{1-F_0(v)} g_2(v)\,dv\right\}$$

$$\sim d_F(t)^2\left\{1 + d_F(t)\int_{v>t}\frac{g(t,v)}{F(v)-F(t)}\,dv + d_F(t)\int_{v<t}\frac{g(v,t)}{F(t)-F(v)}\,dv\right\}^{-2}$$

$$\cdot\left\{\frac{g_1(t)\{1-F_0(t)\}+f_2(t)F_0(t)}{F_0(t)\{1-F_0(t)\}} + \int_{u<t}\frac{g(u,t)}{F_0(t)-F_0(u)}\,du + \int_{t<u}\frac{g(t,u)}{F_0(u)-F_0(t)}\,du\right\}$$

$$\cdot\int K(u)^2\,du$$

$$= d_F(t)\left\{1 + d_F(t)\int_{v>t}\frac{g(t,v)}{F(v)-F(t)}\,dv + d_F(t)\int_{v<t}\frac{g(v,t)}{F(t)-F(v)}\,dv\right\}^{-1}\int K(u)^2\,du,$$

which is in fact the asymptotic variance of the MSLE for the separated case, see Theorem 11.7 in Section 11.5. Although we expect this also to hold for the nonseparated case, we do not have a similar heuristic argument for that situation.

This leads to the last conjecture of this section.

Conjecture 11.2 *Let $\tilde F_n$ be the SMLE for the interval censoring case 2 model, and let the conditions of Conjecture 11.1 for the separated case be satisfied. Then, if $h_n \sim cn^{-1/5}$, we have*

$$n^{2/5}\left\{\tilde F_n(t) - F_0(t)\right\} \xrightarrow{\mathcal{D}} N\left(\mu,\sigma^2\right),\ n\to\infty,$$

where $N(\mu,\sigma^2)$ is a normal distribution, with

$$\mu = \tfrac{1}{2}c^2 f_0'(t)\int u^2K(u)\,du, \tag{11.69}$$

and

$$\sigma^2 = \frac{d_{F_0}(t)}{c}\left\{1 + d_{F_0}(t)\int_{v>t}\frac{g(t,v)}{F_0(v)-F_0(t)}\,dv + d_{F_0}(t)\int_{v<t}\frac{g(v,t)}{F_0(t)-F_0(v)}\,dv\right\}^{-1}\int K(u)^2\,du. \tag{11.70}$$

Remark According to this conjecture, the asymptotic bias is the same as for the current status model (Theorem 11.4 in Section 11.3) and the asymptotic variance coincides with the asymptotic variance of the MSLE (Theorem 11.7 in Section 11.5).

11.5 The MSLE for Interval Censoring Case 2

Just like the SMLE, the MSLE for the interval censoring model case 2 is introduced in Section 8.6. Let $\mathbb{P}_n$ be the empirical measure of the quadruples $(T_i, U_i, \Delta_{i1}, \Delta_{i2})$, for $i = 1,\ldots,n$. Let $\tilde g_{nj}$ and $\tilde g_n$ be estimates of the densities g_{0j}, $j = 1, 2$, and the two-dimensional density g_0, where

$$\tilde{g}_{n1}(t) = \int K_{h_n}(t-v)\,\delta_1\,d\mathbb{P}_n(v,w,\delta_1,\delta_2),$$

$$\tilde{g}_{n2}(u) = \int K_{h_n}(u-w)\,(1-\delta_1-\delta_2)\,d\mathbb{P}_n(v,w,\delta_1,\delta_2),$$

$$\tilde{g}_{n}(t,u) = \int K_{h_n}(t-v)\,K_{h_n}(u-w)\,\delta_2\,d\mathbb{P}_n(v,w,\delta_1,\delta_2),$$

and

$$g_{01}(t) = F_0(t)g_1(t), \quad g_{02}(u) = \{1-F_0(u)\}g_2(u), \quad g_0(t,u) = \{F_0(u)-F_0(t)\}g(t,u).$$

The MSLE is then defined as maximizer of the smoothed log likelihood

$$\int \tilde{g}_{n1}(t)\log F(t)\,dt + \int \tilde{g}_n(t,u)\log\{F(u)-F(t)\}\,dt\,du + \int \tilde{g}_{n2}(t)\log\{1-F(t)\}\,dt \tag{11.71}$$

over distribution functions F.

Parameterizing by the density f, we have to maximize

$$\int \tilde{g}_{n1}(t)\log\int_0^t f(v)\,dv\,dt + \int \tilde{g}_n(t,u)\log\int_t^u f(v)\,dv\,dt\,du + \int \tilde{g}_{n2}(t)\log\int_t^\infty f(v)\,dv\,dt,$$

under the side condition $f \ge 0$ and

$$\int_0^\infty f(v)\,dv = 1.$$

Pointwise maximization of the integrand, ignoring the positivity requirement, yields the equations

$$\int_{t>v}\frac{\tilde{g}_{n1}(t)}{F(t)}\,dt + \int_{t<v}\frac{\tilde{g}_{n2}(t)}{1-F(t)}\,dt + \int_{t<v<u}\frac{\tilde{g}_n(t,u)}{F(u)-F(t)}\,dt\,du = 1,\ v>0, \tag{11.72}$$

which gives the self-consistency equation,

$$f(v) = \left\{\int_{t>v}\frac{\tilde{g}_{n1}(t)}{F(t)}\,dt + \int_{t<v}\frac{\tilde{g}_{n2}(t)}{1-F(t)}\,dt + \int_{t<v<u}\frac{\tilde{g}_n(t,u)}{F(u)-F(t)}\,dt\,du\right\} f(v),\ v>0,$$

which can be used for an EM-type algorithm maximizer of the smoothed log likelihood over a rich class of functions f.

The MSLE minimizes the Kullback-Leibler distance

$$\mathcal{K}(\tilde{Q}_n, \tilde{P}_{n,F}) \tag{11.73}$$

over distribution functions F, where $\tilde{Q}_n$ is a smoothed version of $\mathbb{P}_n$, defined by

$$\begin{aligned}&\int \psi(t,u,\delta_1,\delta_2)\,d\tilde{Q}_n(t,u,\delta_1,\delta_2)\\ &\quad= \int \psi(t,u,1,0)\,d\tilde{Q}_n(t,u,1,0) + \int \psi(t,u,0,1)\,d\tilde{Q}_n(t,u,1,0)\\ &\qquad+ \int \psi(t,u,0,0)\,d\tilde{Q}_n(t,u,0,0),\end{aligned} \tag{11.74}$$

where ψ is a bounded measurable function, and the three measures on the right hand side are smoothed versions of the measures $\mathbb{P}_n(t,u,1,0)$, $\mathbb{P}_n(t,u,0,1)$ and $\mathbb{P}_n(t,u,0,0)$, respectively. Furthermore, $\tilde{P}_{n,F}$ is defined by

$$\int \psi(t,u,\delta_1,\delta_2)\,d\tilde{P}_{n,F}(t,u,\delta_1,\delta_2)$$
$$= \int \Big\{\psi(t,u,1,0)F(t) + \psi(t,u,0,1)\{F(u)-F(t)\}$$
$$+ \psi(t,u,0,0)\{1-F(u)\}\Big\}\,d\tilde{G}_n(t,u), \tag{11.75}$$

where $d\tilde{G}_n$ is given by

$$d\tilde{G}_n(t,u) = d\tilde{Q}_n(t,u,1,0) + d\tilde{Q}_n(t,u,0,1) + d\tilde{Q}_n(t,u,0,0).$$

Minimizing (11.73) is equivalent to maximizing the smoothed log likelihood (11.71) over F.

In order to formulate a theorem establishing the asymptotic distribution of the MSLE, some preliminary definitions are needed. The first definition is, for v in the interior of the support of F_0,

$$d_{F_0}(v) = \frac{F_0(v)\{1-F_0(v)\}}{g_1(v)\{1-F_0(v)\} + F_0(v)g_2(v)}.$$

This leads to the definition

$$\sigma_1(v) = 1 + d_{F_0}(v)\left\{\int_{t<v} \frac{g(t,v)}{F_0(v)-F_0(t)}\,dt + \int_{w>v} \frac{g(v,w)}{F_0(w)-F_0(v)}\,dw\right\}, \tag{11.76}$$

and finally to

$$\sigma(v)^2 = \frac{d_{F_0}(v)}{\sigma_1(v)} \int K(u)^2\,du, \tag{11.77}$$

a quantity that will appear in the asymptotic variance. Furthermore,

$$\beta_1(v) = \frac{1}{2\sigma_1(v)}\left\{\frac{\{1-F_0(v)\}h_1''(v) - F_0(v)h_2''(v)}{g_1(v)\{1-F_0(v)\} + F_0(v)g_2(v)}\int u^2K(u)\,du\right.$$
$$\left. + d_{F_0}(v)\left\{\int_{t=0}^{v} \frac{\frac{\partial^2}{\partial v^2}h_0(t,v)}{F_0(v)-F_0(t)}\,dt - \int_{u=v}^{M} \frac{\frac{\partial^2}{\partial v^2}h_0(v,u)}{F_0(u)-F_0(v)}\,du\right\}\int u^2K(u)\,du\right\} \tag{11.78}$$

is needed to define the asymptotic bias $\beta(v)$:

$$\beta(v) = \beta_1(v) + \frac{d_{F_0}(v)}{\sigma_1(v)}\left\{\int_{u=0}^{v} \frac{g(u,v)\beta_1(u)}{F_0(v)-F_0(u)}\,du + \int_{u=v}^{M} \frac{g(v,u)\beta_1(u)}{F_0(u)-F_0(v)}\,du\right\}. \tag{11.79}$$

The following result is proved in Groeneboom, 2014.

Theorem 11.7 *Let conditions (S1) to (S4) and (L1) and (L2) of Section 10.3 be satisfied. Moreover, let F_0 be twice differentiable, with a bounded continuous derivative f_0 on the interior of $[0, M]$, which is bounded away from zero on $[0, M]$, with a finite positive right limit at 0 and a positive left limit at M. Also, let f_0 have a bounded continuous derivative on $(0, M)$ and let g_1 and g_2 be twice differentiable on the interior of their supports S_1 and*

S_2, respectively. Furthermore, let the joint density g of the pair of observation times (T_i, U_i) have a bounded (total) second derivative on $\{(x, y) : 0 < x < y < M\}$. Suppose that X_i is independent of (T_i, U_i)

Then, choosing the bandwidth $h_n \asymp n^{-1/5}$, we have, for each $v \in (0, M)$,

$$\sqrt{nh_n}\left\{\hat{F}_n(v) - F_0(v) - \beta(v)h_n^2\right\} \xrightarrow{\mathcal{D}} N\left(0, \sigma(v)^2\right),$$

where $N\left(0, \sigma(v)^2\right)$ is a normal distribution with first moment zero and variance $\sigma(v)^2$ given in (11.77) and the bias $\beta(v)$ is given by (11.79).

The proof is based on a version of the implicit function theorem in Banach spaces, which leads to a non-linear integral equation, characterizing the MSLE asymptotically. Subsequently, it is shown that the solution of the non-linear integral equation is asymptotically equivalent to the solution of a linear integral equation. Finally, it is shown that the 'off-diagonal' elements of the linear integral equation give a contribution of lower order, implying that the MSLE is asymptotically equivalent to a toy estimator for which the bias and variance can be computed explicitly. A further discussion of this can be found in Groeneboom, 2013b. Note that the bias is considerably more complicated than the bias of the SMLE of Section 11.4, but that the asymptotic variance is expected to be the same as that of the SMLE.

A picture of an observation density, satisfying the conditions of Theorem 11.7, is shown in Figure 11.5; g is defined by:

$$g(x, y) = 12(y - x - \epsilon)^2/(2 - \epsilon)^4, \; x + \epsilon < y, \tag{11.80}$$

on the triangle with vertices $(0, \epsilon)$, $(0, 2)$ and $(2 - \epsilon, 2)$, where $\epsilon = 0.1$.

For the non-separated case, where the intervals (T_i, U_i) can be arbitrarily small, there presently does not exist a result, corresponding to Theorem 11.7.

11.6 Estimation of a Nondecreasing Hazard in the Right Censoring Model: SMLE

In Section 2.6 the MLE for a nondecreasing hazard in the right censoring model is introduced. We now introduce an SMLE for a decreasing hazard rate in the presence of right censoring and analyze this estimator asymptotically, using similar methods as in the preceding sections to analyze the SMLE. The data observed are independent and identically distributed copies (T_i, Δ_i) of random variables

$$T_i = X_i \wedge C_i, \quad \Delta_i = 1_{\{X_i \le T_i\}}.$$

The ML estimator $\hat{\lambda}_n$ of the hazard rate λ_0 of X is defined as in Section 2.6. Then, taking bandwidth $h > 0$ and a kernel function K satisfying the usual conditions, the SMLE of λ_0 can be defined as

$$\tilde{\lambda}_{nh}^{SML}(t) = \frac{1}{h}\int K\left(\frac{t-y}{h}\right)\hat{\lambda}_n(y)\,dy.$$

We assume that the underlying distribution function F_0 of X is concentrated on $[0, \infty)$ and that both λ_0 and λ_0' are finite, continuous and strictly positive on $[0, b]$, where $\lambda_0'(0)$ is the right derivative at zero and $\lambda_0'(b)$ is the left derivative at b. Since $\lambda_0(b) < \infty$, we have $F_0(b) < 1$.

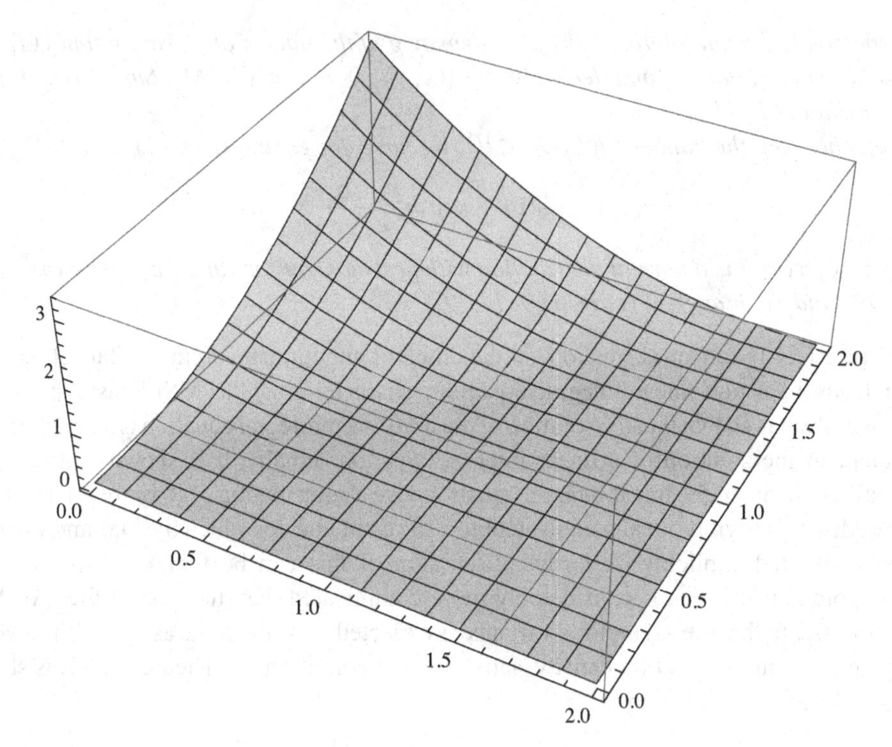

Figure 11.5 The bivariate observation density g on $[0, 2]^2$, where $\epsilon = 0.1$.

Differentiation of log likelihood (2.34) with respect to $\lambda(T_{(i)})$ gives:

$$\frac{\Delta_{(i)}}{\lambda(T_{(i)})} - (n-i)\{T_{(i+1)} - T_{(i)}\}.$$

This implies relations of the form

$$\sum_{i=1}^{n-1} a(T_{(i)})\left\{\Delta_{(i)} - \hat\lambda_n(T_{(i)})\{T_{(i+1)} - T_{(i)}\}(n-i)\right\} = 0,$$

for functions a which are constant on the same intervals as $\hat\lambda_n$. The intervals correspond to regions where the greatest convex minorant of the cusum diagram has a constant derivative. This implies for such functions that

$$\begin{aligned}
&\sum_{i=1}^{n-1} a(T_{(i)})\left\{\Delta_{(i)} - \{\Lambda_n(T_{(i+1)}) - \Lambda_n(T_{(i)})\}(n-i)\right\} \\
&= \sum_{i=1}^{n-1} a(T_{(i)})\left\{\Delta_{(i)} - n\{\Lambda_n(T_{(i+1)}) - \Lambda_n(T_{(i)})\}\int_{t>T_{(i)}} d\mathbb{P}_n(t,\delta)\right\} \\
&= \sum_{i=1}^{n-1} a(T_{(i)})\left\{\Delta_{(i)} - n\, d\Lambda_n(T_{(i)})\int_{t>T_{(i)}} d\mathbb{P}_n(t,\delta)\right\} = 0, \qquad (11.81)
\end{aligned}$$

where $d\Lambda_n(T_{(i)}) = \lambda_n(T_{(i)})(T_{(i+1)} - T_{(i)})$.

The continuous variant of the sum in (11.81) is:

$$\int a(x) f_0(x)\{1-G(x)\}\,dx$$
$$-\int a(x)\left\{\int_{y>x}\{f_0(y)\{1-G(y)\}+g(y)\{1-F_0(y)\}\}\,dy\right\}\,d\Lambda(x)$$
$$=\int\left\{a(x)\{1-G(x)\}-\int_{u=0}^{x} a(u)\{1-G(u)\}\,d\Lambda(u)\right\}\,dF_0(x). \tag{11.82}$$

In accordance with the methods used in the preceding section, this means that we look for a function a (approximately) satisfying

$$a(x)\{1-G(x)\}-\int_0^x a(u)\{1-G(u)\}\,d\hat{\Lambda}_n(u)=\frac{K_h(t-x)}{1-F_0(x)}-\int K_h(t-y)\,d\hat{\Lambda}_n(y) \tag{11.83}$$

and a piecewise constant modification $\bar{a}$ of a, satisfying (11.81). Note that if a satisfies (11.83), it also satisfies

$$a'(x)\{1-G(x)\}-g(x)a(x)-a(x)\{1-G(x)\}\hat{\lambda}_n(x)=\frac{d}{dx}\left(\frac{K_h(t-x)}{1-F_0(x)}\right), \tag{11.84}$$

except at points x where $\hat{\lambda}_n(x)$ has a jump.

We have the following lemma.

Lemma 11.8 *Suppose* $0<t-h<t+h<a$. *Then the unique solution* a_n, *satisfying (11.84) and such that* $a_n(x)=0$ *for* $x\ge t+h$, *is given by:*

$$a_n(x)=\frac{K_h(t-x)}{\{1-G(x)\}\{1-F_0(x)\}}-\frac{1}{\{1-G(x)\}\{1-\hat{F}_n(x)\}}\int_x^M\frac{K_h(t-u)}{1-F_0(u)}\,d\hat{F}_n(u). \tag{11.85}$$

Proof If $|x-t|>h$, the right hand side of (11.84) is zero, and we get the equation

$$a'(x)-\left\{\frac{g(x)}{1-G(x)}+\hat{\lambda}_n(x)\right\}a(x)=0, \tag{11.86}$$

which has as general solution

$$c\exp\left\{-\log\{1-G(x)\}+\hat{\Lambda}_n(x)\right\}=\frac{c}{\{1-G(x)\}\{1-\hat{F}_n(x)\}}, \tag{11.87}$$

where

$$\hat{F}_n(x)=1-\exp\{-\hat{\Lambda}_n(x)\};$$

see Exercise 11.16.

Hence the general solution of (11.84) is given by:

$$a_n(x)=\frac{c}{\{1-G(x)\}\{1-\hat{F}_n(x)\}}$$
$$-\frac{1}{\{1-G(x)\}\{1-\hat{F}_n(x)\}}\int_x^M\{1-\hat{F}_n(u)\}\frac{d}{dx}\left(\frac{K_h(t-x)}{1-F_0(x)}\right)\,du.$$

Note that

$$\frac{1}{\{1-G(x)\}\{1-\hat{F}_n(x)\}}\int_0^x\{1-\hat{F}_n(u)\}\frac{d}{dx}\left(\frac{K_h(t-x)}{1-F_0(x)}\right)du$$
$$=\frac{K_h(t-x)}{\{1-G(x)\{1-F_0(x)\}}+\frac{1}{\{1-G(x)\}\{1-\hat{F}_n(x)\}}\int_0^x\frac{K_h(t-u)}{1-F_0(u)}\,d\hat{F}_n(u),$$

which implies, choosing c in such a way that $a_n(x)=0$ if $x\ge t+h$,

$$a_n(x)=\frac{K_h(t-x)}{\{1-G(x)\}\{1-F_0(x)\}}-\frac{1}{\{1-G(x)\}\{1-\hat{F}_n(x)\}}\int_x^M\frac{K_h(t-u)}{1-F_0(u)}\,d\hat{F}_n(u).$$

□

Now, using $\sup_{x\in[0,a]}|\hat{F}_n(x)-F_0(x)|=O_p(n^{-1/2})$ (Exercise 11.15),

$$a_n(x)\{1-G(x)\}-\int_0^x a_n(u)\{1-G(u)\}\,d\hat{\Lambda}_n(u)$$
$$=\frac{K_h(t-x)}{1-F_0(x)}-\frac{1}{1-\hat{F}_n(x)}\int_x^M\frac{K_h(t-u)}{1-F_0(u)}\,d\hat{F}_n(u)$$
$$-\int_0^x\frac{K_h(t-u)}{1-F_0(u)}\,d\hat{\Lambda}_n(u)+\int_0^x\frac{1}{1-\hat{F}_n(u)}\int_u^M\frac{K_h(t-v)}{1-F_0(v)}\,d\hat{F}_n(v)\,d\hat{\Lambda}_n(u)$$
$$=\frac{K_h(t-x)}{1-F_0(x)}-\int K_h(t-u)\,d\hat{\Lambda}_n(u)-\int_0^x\frac{\hat{F}_n(u)K_h(t-u)}{1-\hat{F}_n(u)}\,d\hat{\Lambda}_n(u)$$
$$-\frac{\hat{F}_n(x)}{1-\hat{F}_n(x)}\int_x^M K_h(t-u)\,d\hat{\Lambda}_n(u)+\int_{v=0}^M\frac{K_h(t-v)\hat{F}_n(x\wedge v)}{1-\hat{F}_n(x\wedge v)}\,d\hat{\Lambda}_n(v)+O_p\left(n^{-1/2}\right)$$
$$=\frac{K_h(t-x)}{1-F_0(x)}-\int K_h(t-u)\,d\hat{\Lambda}_n(u)+O_p\left(n^{-1/2}\right).$$

Let a_0 be the solution of the integral equation

$$a(x)\{1-G(x)\}-\int_0^x a(u)\{1-G(u)\}\,d\Lambda_0(u)=\frac{K_h(t-x)}{1-F_0(x)}-\int K_h(t-y)\,d\Lambda_0(y). \tag{11.88}$$

Then it can be shown in a similar way that

$$a_0(x)=\frac{K_h(t-x)}{\{1-G(x)\}\{1-F_0(x)\}}-\frac{\int_x^M K_h(t-y)\,d\Lambda_0(y)}{\{1-G(x)\}\{1-F_0(x)\}}. \tag{11.89}$$

The solutions (11.85) and (11.89), where $n=1000$ for F_0 and G both uniform and $t=0.5$, $h=0.2$, are shown in Figure 11.6. Note that $a_n(x)$ and $a_0(x)$ are zero if $x\ge t+h$.

Lemma 11.9 *Let the function a_n be defined by (11.85) and let*

$$\theta_{t,h,\hat{F}_n}(x,\delta)=\delta a_n(x)-\int_{y=0}^x a_n(y)\,d\hat{\Lambda}_n(y). \tag{11.90}$$

Then

$$\int\theta_{t,h,\hat{F}_n}(x,\delta)\,dP_0(x,\delta)=-\int K_h(t-x)\,d(\hat{\Lambda}_n-\Lambda_0)(x)+O_p\left(n^{-1/2}\right). \tag{11.91}$$

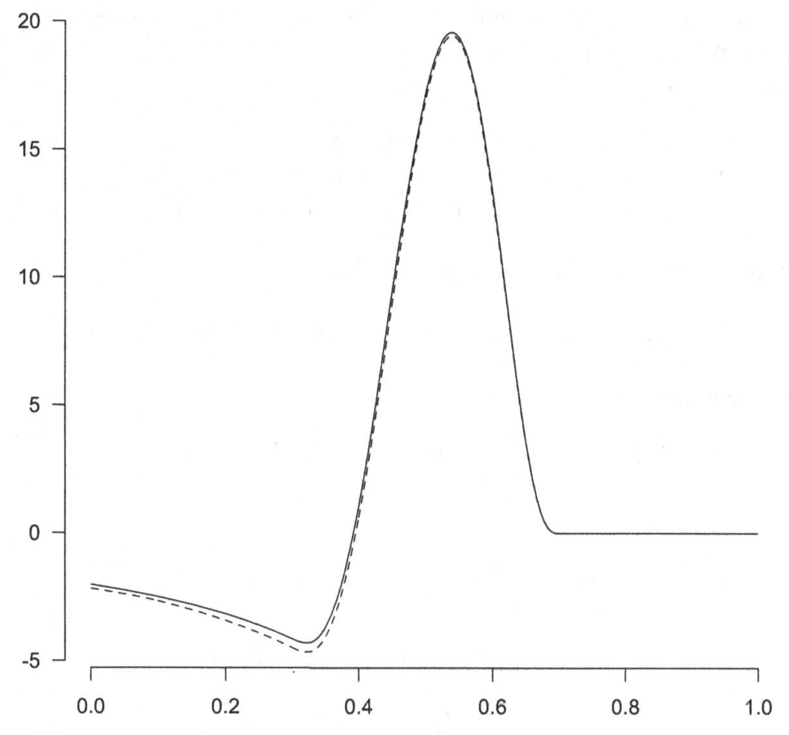

Figure 11.6 The functions a_n (dashed) and a_0 for $h = 0.2$, when $t = 0.5$ for a sample of size $n = 1000$; F and G are the standard uniform distribution functions.

Proof Observe that

$$
\begin{aligned}
&\int \theta_{t,h,\hat{F}_n}(x,\delta)\,dP_0(x,\delta)\\
&= \int a_n(x) f_0(x)\{1 - G(x)\}\,dx - \int \int_{y=0}^{x} a_n(y)\,d\hat{\Lambda}_n(y)\,dP_0(x,1)\\
&\quad - \int \int_{y=0}^{x} a_n(y)\,d\hat{\Lambda}_n(y)\,dP_0(x,0)\\
&= \int a_n(x) f_0(x)\{1 - G(x)\}\,dx - \int \int_{y=0}^{x} a_n(y)\,d\hat{\Lambda}_n(y) f_0(x)\{1 - G(x)\}\,dx\\
&\quad - \int \int_{y=0}^{x} a_n(y)\,d\hat{\Lambda}_n(y) g(x)\{1 - F_0(x)\}\,dx\\
&= \int \left\{a_n(x)\{1 - G(x)\} - \int_{y=0}^{x} \{1 - G(y)\} a_n(y)\,d\hat{\Lambda}_n(y)\right\} dF_0(x)\\
&= \int \frac{K_h(t-x)}{1 - F_0(x)}\,dF_0(x) - \int K_h(t-y)\,d\hat{\Lambda}_n(y) + O_p\left(n^{-1/2}\right)\\
&= \int K_h(t-x) d\left(\Lambda_0 - \hat{\Lambda}_n\right)(x) + O_p\left(n^{-1/2}\right).
\end{aligned}
$$

□

Now consider a piecewise constant modification $\bar{a}_n$ of a_n, which is constant on the same intervals as $\hat{\lambda}_n$. We choose the value of $\bar{a}_n$ in the same way as was done in Section 11.3 for $\bar{\psi}_{t,h}$ with respect to $\psi_{t,h}$:

$$\bar{a}_n(u) = \begin{cases} \bar{a}_n(\tau_i), & \text{, if } \lambda_0(t) > \hat{\lambda}_n(\tau_i),\ t \in [\tau_i, \tau_{i+1}), \\ a_n(s), & \text{, if } \lambda_0(s) = \hat{\lambda}_n(s), \text{ for some } s \in [\tau_i, \tau_{i+1}), \\ a_n(\tau_{i+1}), & \text{, if } \lambda_0(t) < \hat{\lambda}_n(\tau_i),\ t \in [\tau_i, \tau_{i+1}), \end{cases}$$

where the τ_i are successive points of jump of $\hat{\lambda}_n$. Then

$$\int \delta \bar{a}_n(x)\, d\mathbb{P}_n(x,\delta) - \int \bar{a}_n(x) \int_{u>x} d\mathbb{P}_n(u,\delta)\, d\hat{\Lambda}_n(x) = 0.$$

This can also be written as:

$$\int \left\{ \delta \bar{a}_n(x) - \int_{y<x} \bar{a}_n(y)\, d\hat{\Lambda}_n(y) \right\} d\mathbb{P}_n(x,\delta) = 0.$$

So we find:

$$\begin{aligned} \int K_h(t-x) d\big(\hat{\Lambda}_n - \Lambda_0\big)(x) &= \int \theta_{t,h,\hat{\Lambda}_n}(x,\delta)\, dP_0(x,\delta) \\ &= \int \bar{\theta}_{t,h,\hat{\Lambda}_n}(t,\delta)\, d\big(\mathbb{P}_n - P_0\big)(x,\delta) \\ &\quad + \int \left\{ \bar{\theta}_{t,h,\hat{\Lambda}_n}(t,\delta) - \theta_{t,h,\hat{\Lambda}_n}(t,\delta) \right\} dP_0(x,\delta), \end{aligned} \tag{11.92}$$

where

$$\bar{\theta}_{t,h,\hat{\Lambda}_n}(t,\delta) = \delta \bar{a}_n(x) - \int_{y<x} \bar{a}_n(y)\, d\hat{\Lambda}_n(y).$$

We will need the following lemma.

Lemma 11.10 *For all $u \in [0,b]$ and all $p \geq 1$, we have:*

$$\mathbb{E}\left\{ n^{1/3} \big(\hat{\lambda}_n(u) - \lambda_0(u)\big)_+ \right\}^p \leq K, \tag{11.93}$$

for a constant $K > 0$, uniformly in the chosen point $u \in [0,b]$. Similarly,

$$\mathbb{E}\left\{ n^{1/3} \big(\lambda_0(u) - \hat{\lambda}_n(u)\big)_+ \right\}^p \leq K', \tag{11.94}$$

for a constant $K' > 0$, uniformly in the chosen point $u \in [0,b]$.

Proof In proving Lemma 11.10 we use the switch relation and the process inverse to $x \mapsto \hat{\lambda}_n(x)$. To this end, a cumulative sum diagram $\{(Z_n(t), V_n(t)) : t \in [0, T_{(n-1)})\}$ is defined. Here

$$Z_n(t) = \frac{1}{n} \sum_{j : T_{(j)} \leq t,\, j<n} w\big(T_{(j)}\big),$$

with

$$w\big(T_{(j)}\big) = n\big\{T_{(j+1)} - T_{(j)}\big\} \int_{s \geq T_{(j+1)}} d\mathbb{P}_n(s,\delta),\ j = 1, \ldots, n-1,$$

and

$$V_n(t) = \frac{1}{n} \sum_{j:T_{(j)} \le t, j<n} \Delta_j.$$

The coordinate t runs through the interval $[0, T_{(n-1)}]$ and the first point of the cumulative sum diagram is $P_0 = (0, 0)$. The process Z_n is close to the process

$$Z(t) = \int_0^t (1-u)\, dH^{-1}(u),\ t \ge 0,$$

where H is the distribution of the observation times $T_i = X_i \wedge C_i$ and H^{-1} its inverse. We now define the process U_n by

$$U_n(a) = \mathrm{argmin}\{t \ge 0\colon V_n(t) - aZ_n(t)\}. \tag{11.95}$$

Then we have the switch relation

$$\hat\lambda_n(t) \ge a \iff U_n(a) \le t. \tag{11.96}$$

This implies that, if $a = \lambda_0(t)$,

$$\mathbb{P}\left\{n^{1/3}\{\hat\lambda_n(t) - \lambda_0(t)\} \ge x\right\} = \mathbb{P}\left\{\hat\lambda_n(t) \ge a + n^{-1/3}x\right\} = \mathbb{P}\left\{U_n(a + n^{-1/3}x) \le t\right\}. \tag{11.97}$$

As in Section 3.8,

$$\mathbb{P}\left\{n^{1/3}\{U_n(a) - U(a)\} > x\right\} \le 2e^{-Cx^3} \tag{11.98}$$

where C only depends on the underlying distributions, and not on $a \in [0, a_0]$, where $a_0 = \lambda(b)$. The result now follows from

$$\mathbb{E}\left\{n^{1/3}\left(\hat\lambda_n(u) - \lambda_0(u)\right)_+\right\}^p = \int_{x=0}^{\infty} \mathbb{P}\left\{n^{1/3}\left(\hat\lambda_n(u) - \lambda_0(u)\right) \ge x\right\} px^{p-1}\, dx,$$

and

$$\mathbb{E}\left\{n^{1/3}\left(\lambda_0(u) - \hat\lambda_n(u)\right)_+\right\}^p = \int_{x=0}^{\infty} \mathbb{P}\left\{n^{1/3}\left(\lambda_0(u) - \hat\lambda_n(u)\right) \ge x\right\} px^{p-1}\, dx;$$

see also Exercise 11.4. □

Lemma 11.11

$$\int_{x=t-h}^{t+h} \{\hat\lambda_n(x) - \lambda_0(x)\}^2\, dx = O_p\left(hn^{-2/3}\right).$$

Proof This follows immediately from Lemma 11.10 and Markov's inequality. See also Exercise 11.6 for a related argument in the current status model. □

For the last term in (11.92) we now have the following result.

Lemma 11.12 *Let $h = h_n \asymp n^{-1/5}$. Then:*

$$\int \left\{\bar\theta_{t,h,\hat\Lambda_n}(t,\delta) - \theta_{t,h,\hat\Lambda_n}(t,\delta)\right\} dP_0(x,\delta) = O_p\left(n^{-7/15}\right).$$

Proof Note that

$$\int \left\{\bar\theta_{t,h,\hat\Lambda_n}(t,\delta) - \theta_{t,h,\hat\Lambda_n}(t,\delta)\right\} dP_0(x,\delta)$$

$$= \int \left\{(\bar a_n(x) - a_n(x))\{1-G(x)\} - \int_{y=0}^{x} (\bar a_n(y) - a_n(y))\{1-G(y)\}\, d\hat\Lambda_n(y)\right\} dF_0(x)$$

$$+ O_p\left(n^{-1/2}\right)$$

$$= \int (\bar a_n(x) - a_n(x))\{1-G(x)\}\{1-F_0(y)\}\, d\Lambda_0(x)$$

$$- \int (\bar a_n(y) - a_n(y))\{1-G(y)\}\{1-F_0(y)\}\, d\hat\Lambda_n(y) + O_p\left(n^{-1/2}\right)$$

$$= \int_{y=0}^{t+h} (\bar a_n(y) - a_n(y))\left(\lambda_0(y) - \hat\lambda_n(y)\right)\{1-G(y)\}\{1-F_0(y)\}\, dy + O_p\left(n^{-1/2}\right),$$

and, by the Cauchy-Schwarz inequality,

$$\left|\int_{y=0}^{t+h} (\bar a_n(y) - a_n(y))\left(\lambda_0(y) - \hat\lambda_n(y)\right)\{1-G(y)\}\{1-F_0(y)\}\, dy\right|$$

$$\le K\|(\bar a_n - a_n)1_{[0,t+h]}\|_2\|(\hat\lambda_n - \lambda_0)1_{[0,t+h]}\|_2$$

$$\sim K\|(\bar a_n - a_n)1_{[t-h,t+h]}\|_2\|(\hat\lambda_n - \lambda_0)1_{[t-h,t+h]}\|_2.$$

Therefore

$$\|(\bar a_n - a_n)1_{[t-h,t+h]}\|_2^2 \sim \int_{t-h}^{t+h} \left\{\frac{K_h(t-x)}{\{1-G(x)\}\{1-F_0(x)\}} - \frac{\bar K_h(t-x)}{\{1-\bar G(x)\}\{1-\bar F_0(x)\}}\right\}^2 dx$$

$$\le Kh^{-4}\int_{x=t-h}^{t+h} \{\hat\lambda_n(x) - \lambda_0(x)\}^2\, dx = O_p\left(n^{-2/3}h^{-3}\right),$$

for a constant $K > 0$, where we write $\bar K_h(t-x)$, $\bar G(x)$ and $\bar F_0(x)$ for the values of these functions at the chosen point for the definition of $\bar a_n$ in the interval containing x; we also use

$$\|(\hat\lambda_n - \lambda_0)1_{[t-h,t+h]}\|_2 = O_p\left(h^{1/2}n^{-1/3}\right).$$

So the conclusion is:

$$\int \left\{\bar\theta_{t,h,\hat\Lambda_n}(t,\delta) - \theta_{t,h,\hat\Lambda_n}(t,\delta)\right\} dP_0(x,\delta) = O_p\left(n^{-2/3}h^{-1}\right)$$

$$= O_p\left(n^{-2/3+1/5}\right) = O_p\left(n^{-7/15}\right).$$

□

We now have the following asymptotic representation of $\int K_h(t-x)\, d(\hat\Lambda_n - \Lambda_0)(x)$ in the observation space.

Lemma 11.13 *Let Λ_0 be the cumulative hazard function, corresponding to F_0, i.e.,*

$$\Lambda_0(x) = -\log\{1 - F_0(x)\}.$$

Then, if $h \asymp n^{-1/5}$,

$$\int K_h(t-x)d\big(\hat{\Lambda}_n - \Lambda_0\big)(x) = \int \theta_{t,h,\Lambda_0}(x,\delta)\,d\big(\mathbb{P}_n - P_0\big)(x,\delta) + o_p\left(n^{-2/5}\right), \quad (11.99)$$

where

$$\theta_{t,h,\Lambda_0}(x,\delta) = \delta a_{0,h}(x) - \int_0^x a_{0,h}(y)\,d\Lambda_0(y), \quad (11.100)$$

and the function $a_{0,h}$ *solves the following integral equation in* a:

$$a(x)\{1-G(x)\} - \int_0^x a(u)\{1-G(u)\}\,d\Lambda_0(u) = \frac{K_h(t-x)}{1-F_0(x)} - \int K_h(t-y)\,d\Lambda_0(y). \quad (11.101)$$

The asymptotic variance is given by

$$\lim_{h\downarrow 0} h\left\{\int\left\{a_{0,h}(x) - \int_0^x a_{0,h}(u)\,d\Lambda_0(u)\right\}^2\{1-G(x)\}\,dF_0(x) + \int\left\{\int_0^x a_{0,h}(u)\,d\Lambda_0(u)\right\}^2\{1-F_0(x)\}\,dG(x)\right\}$$
$$= \frac{\lambda(t)}{\{1-F_0(t)\}\{1-G(t)\}}\int K(u)^2\,du.$$

The asymptotic representation in Lemma 11.13 leads to the following result.

Theorem 11.8 *Let the hazard* λ *be twice continuously differentiable at* t_0, *with* λ *and* λ' *strictly positive, and let* $\hat{\lambda}_n$ *be its MLE under the restriction that* λ *is increasing. Moreover, let the distribution function* F_0 *of the variables of interest and the distribution function* G *of the censoring variable be absolutely continuous, and let* $\hat{\lambda}_{nh}^{SML}$ *be the smoothed maximum likelihood estimator of* λ, *defined by*

$$\hat{\lambda}_{nh}^{SML}(x) = \int K_h(x-y)\,\hat{\lambda}_n(y)\,dy,$$

where K *is a symmetric positive kernel with support* $[-1,1]$, *like the triweight kernel. Then, if* $h_n \asymp n^{-1/5}$,

$$\sqrt{nb_n}\left\{\hat{\lambda}_{nh}^{SML}(t_0) - \lambda(t_0)\right\} \xrightarrow{\mathcal{D}} N\left(\mu,\sigma^2\right),$$

where

$$\mu = \tfrac{1}{2}\lambda''(t_0)\int u^2 K(u)\,du,$$

and

$$\sigma^2 = \frac{\lambda(t_0)}{\{1-F_0(t_0)\}\{1-G(t_0)\}}\int K(u)^2\,du.$$

Remark Note that this shows that the SMLE and the estimator $\int K_h(t-x)\,d\hat{\Lambda}_n^{NA}(x)$, where $\hat{\Lambda}_n^{NA}$ is the Nelson-Aalen estimator of the cumulative hazard, have the same asymptotic distribution.

In the example where both F_0 and G are standard uniform distribution functions, we get:

$$\mu = \frac{1}{(1-t_0)^3}\int u^2 K(u)\,du \quad \text{and} \quad \sigma^2 = \frac{\int K(u)^2\,du}{(1-t_0)^3}.$$

Remark Another method of proving Theorem 11.8 is to first show that $\hat{\Lambda}_n$ (the integrated estimator $\hat{\lambda}_n$) has a supremum distance of order $n^{-2/3}\log n$ to the Nelson-Aalen estimator and next use integration by parts. A method of this type was used in Theorem 3.1 on p. 183 of Groeneboom and Jongbloed, 2013a, for a kernel estimate, based on an isotonic estimate of the hazard if there is no censoring. On the basis of this result Theorem 11.8 was conjectured by Nane, 2013.

The present method, using the integral equation approach, seems to have more potential for generalizing to other models, such as the Cox regression model, since the bounds for the supremum distance for the integrated functions may not be available in these situations.

Exercises

11.1 Derive (11.10) from (11.12).

11.2 Prove Lemma 11.4.

11.3 Construct a sequence of distribution functions (F_n), $n = 1, 2, \ldots$, and a distribution function F_0 on $[0, 1]$ such that

$$\int_0^1 (F_n(x) - F_0(x))^2\,dx = n^{-2/3} \quad \text{and} \quad \int_{1/2-h_n}^{1/2+h_n} (F_n(x) - F_0(x))^2\,dx = n^{-2/3}$$

for a sequence $0 \le h_n \to 0$.

11.4 Let X be a bounded random variable with distribution function F and $p > 1$. Define $X_+ = \max\{X, 0\}$. Show that

$$EX_+^p = \int_0^\infty (1 - F(x))px^{p-1}\,dx.$$

11.5 Show that the function ϕ_{nj} defined in (11.29) attains its maximum at $\theta_{nj} = n\log(1 + ct_j/(4c_1))$.

11.6 Let (h_n) be a sequence of vanishing positive numbers. Show that (11.32) and (11.33) imply that for each $\epsilon > 0$, there exists a $C > 0$ such that

$$P\left(\int_{t-h_n}^{t+h_n} \left(\hat{F}_n(x) - F_0(x)\right)^2 dx > Ch_n n^{-2/3}\right) < \epsilon$$

for all n (i.e., (11.20)). Hint: use Markov's inequality and Fubini.

11.7 Verify that the boundary corrected estimator of the distribution function, (11.37), reduces to the traditional estimator (11.35) if $a + h \le t \le b - h$.

11.8 Prove (11.50).

11.9 The SMLE for the distribution function in the current status model satisfies an equality of the type given in (10.20). Show that

$$\hat{F}_{n,h}^{SML}(t) - \int \mathbb{K}_h(t-u)\,dF_0(u) = \int \mathbb{K}_h(t-u)\,d\left(\hat{F}_n - F_0\right)(u) = -\int \theta_{\hat{F}_n}(u, h, \delta)\,dP_0(u, \delta).$$

11.10 Let the conditions of Theorem 11.4 be satisfied and suppose that $F_0(a) = 0$, $F_0(b) = 1$ and that f_0' is continuous in a neighborhood of b with a left limit $f_0'(b)$ at b. Show that the bias of the SMLE for t in $(b-h, b]$, using the asymmetric Schuster correction, is given by:

$$\tfrac{1}{2}h^2 f_0'(t)\left\{\int u^2 K(u)\,du - 2\int_{(b-t)/h}^{1}\left(\frac{b-t}{h} - u\right)^2 K(u)\,du\right\} + o\left(h^2\right), \qquad h \downarrow 0.$$

Note that this means that the leading term of the bias converges to zero, as $t \uparrow b$.

11.11 Let the conditions of the preceding exercise be satisfied. Show that the SMLE satisfies:
(a) For $t \in [a, a+h)$ we have:

$$\hat F_{nh}^{(SML)}(t) = \int_{x=a}^{b} \{K_h(t-x) - K_h(t+x)\}\,\hat F_n(x)\,dx.$$

(b) For $t \in [a+h, b-h]$ we have:

$$\hat F_{nh}^{(SML)}(t) = \int_{x=a}^{b} K_h(t-x)\hat F_n(x)\,dx.$$

(c) For $t \in (b-h, b]$ we have:

$$\hat F_{nh}^{(SML)}(t) = 1 - \int_{x=a}^{b} \{K_h(t-x) - K_h(2b-t-x)\}\left\{1 - \hat F_n(x)\right\}dx.$$

11.12 Deduce from Exercise 11.11 that, under the conditions of that exercise, the SMLE is asymptotically equivalent to the following toy estimator:
(a) For $t \in [a, a+h)$ we have:

$$F_{nh}^{toy}(t) = \int \{K_h(t-x) - K_h(t+x)\}\,F_0(x)\,dx + \int \{K_h(t-x) - K_h(t+x)\}\,\frac{\delta - F_0(x)}{g(x)}\,d\mathbb{P}_n(x,\delta).$$

(b) For $t \in [a+h, b-h]$ we have:

$$F_{nh}^{toy}(t) = \int K_h(t-x)F_0(x)\,dx + \int K_h(t-x)\frac{\delta - F_0(x)}{g(x)}\,d\mathbb{P}_n(x,\delta).$$

(c) For $t \in (b-h, b]$ we have:

$$F_{nh}^{toy}(t) = 1 - \int \{K_h(2b-t-x) - K_h(t-x)\}\{1 - F_0(x)\}\,dx + \int \{K_h(t-x) - K_h(2b-t-x)\}\,\frac{\delta - F_0(x)}{g(x)}\,d\mathbb{P}_n(x,\delta).$$

11.13 Prove that Theorem 11.4 also holds for the toy estimator of the preceding exercise.

11.14 Show that, for $t \in [a+h, b-h]$, the estimator of the density in the current status model, based on the SMLE, satisfies:

$$\hat f_n^{SML}(t) - \int K_h(t-u)\,dF_0(u) = \int K_h(t-u)\,d\left(\hat F_n - F_0\right)(u) = \int \frac{K_h'(t-u)}{g(u)}\{\delta - \hat F_n(u)\}\,dP_0(u,\delta).$$

11.15 Show that under the conditions, given at the start of Section 11.6:

$$\sup_{x\in[0,a]} \left|\hat{F}_n(x) - F_0(x)\right| = O_p\left(n^{-1/2}\right),$$

where $\hat{F}_n$ is defined by

$$\hat{F}_n(x) = 1 - \exp\left\{-\Lambda_n(x)\right\}.$$

11.16 Show that the first order differential equation (11.86) has solution (11.87).

11.17 Prove that (11.89) is the solution of the integral equation (11.88).

11.18 The density estimate $\hat{f}_{nh}$ of the density f_0 in the current status model can, for $t \in [0, h]$, be defined by

$$\hat{f}_{nh}(t) = \alpha(t/h)\int K_h(t-x)\,d\hat{F}_n(x) + \beta(t/h)\int \frac{t-x}{h}K_h(t-x)\,d\hat{F}_n(x),$$

where we assume that 0 is the left boundary of the interval on which we do the estimation, and where α and β are the weights of the boundary kernel, defined by (9.38). Deduce formulas for the bias and variance, analogous to (11.54) for the asymptotic bias and variance. Same question for $t \in [B-h, B]$, where B is the right endpoint of the estimation interval.

11.19 Under the same conditions as Exercise 11.18 an estimate of $f_0'(t)$, for $t \in [0, h]$, is given by:

$$\hat{f}_{nh}'(t) = h^{-1}\left\{\alpha'(t/h)\int K_h(t-x)\,d\hat{F}_n(x) + \beta'(t/h)\int \frac{t-x}{h}K_h(t-x)\,d\hat{F}_n(x)\right\}$$
$$+ h^{-1}\left\{\alpha(t/h)\int K_h'(t-x)\,d\hat{F}_n(x) + \beta(t/h)\int \frac{t-x}{h}K_h'(t-x)\,d\hat{F}_n(x)\right\}$$
$$+ h^{-1}\beta(t/h)\int K_h(t-x)\,d\hat{F}_n(x).$$

Show that, under the conditions of Theorem 11.5, this is a consistent estimate of $f_0'(t)$. Deduce the same type of result for $f_0'(t)$, if $t \in [B-h, B]$, is B is the upper end of the observation interval and the boundary kernel is used on $[B-h, B]$.

Remark Note that for the triweight kernel $K(u) = \frac{35}{32}\left(1-u^2\right)^3 1_{[-1,1]}(u)$ the derivatives α' and β' can be deduced from Exercise 9.14.

Bibliographic Remarks

The asymptotic distribution theory for the maximum likelihood and least squares estimator of a convex density as well as the least squares estimator of a convex regression function is derived in Groeneboom et al., 2001a. The asymptotic distribution is studied in Groeneboom et al., 2001b. This distribution also emerges in Jongbloed and van der Meulen, 2009, as asymptotic distribution in a deconvolution model where the underlying density to be estimated is decreasing. Also, in Balabdaoui et al., 2009, it appears as limiting distribution of the maximum likelihood estimator of a log concave density, the problem discussed in Section 4.4. The tail bound for the MLE in the current status model derived in Section 11.2 strengthens a result in Durot, 2007. The SMLE and MLSE in the current status model are both introduced and studied asymptotically in Groeneboom et al., 2010. The Schuster-type method for boundary correction is quite suitable to solve this problem. Within the plain density estimation context, many methods are known to solve boundary problems of kernel estimators; see, e.g., Zhang et al., 1999. The right censoring problem is well studied overall. The smooth monotone estimator of the hazard rate (without right censoring) was introduced in Groeneboom and Jongbloed, 2013b. In Nane, 2013, the problem of estimating a smooth monotone hazard within the Cox proportional hazard model is considered.

12

Pointwise Asymptotic Distribution Theory for Multivariate Problems

Pointwise asymptotic results for estimators in basic shape constrained models are derived in Chapter 3. For estimators in more complicated models and estimators satisfying certain smoothess conditions, pointwise asymptotic results are derived in Chapter 11. In this chapter, additional results will be discussed for multivariate models, as introduced in Chapter 5.

In Section 12.1, the local asymptotic distribution theory for the maximum likelihood estimator (MLE) in the competing risk model with current status data will be given. In this model, several subdistribution functions are estimated simultaneously. Section 12.2 discusses the (multivariate) smoothed maximum likelihood estimator (SMLE) for this model and gives an asymptotic normality result for the SMLE in Theorem 12.6.

Finally, the bivariate current status model is considered in Section 12.3. A cube root n consistent estimator for the distribution function is defined, which is different from the MLE in this model. We also discuss the SMLE for this model, for which we give the conjectured (normal) limit distribution. The rate the MLE achieves is still unknown, but it is conjectured to be cube root n with possibly an extra logarithmic factor.

12.1 The ML Estimator in the Competing Risk Model with Current Status Data

It is considerably more difficult to derive the local asymptotics for the MLE for competing risk models with current status data than to do this for the MLE based on one-dimensional current status data as considered in Section 3.8. The reason for the increased difficulty lies in the fact that, in contrast with the situation for ordinary current status data, the relevant cumulative sum diagram does in fact depend on the solution itself: the cumulative sum diagram is self-induced.

This cumulative sum diagram is given by (5.10) in Section 5.1. Note that the second coordinate

$$\int_{t_k \le t} \delta_k \, d\mathbb{P}_n(u,\delta)$$

is just as it is in the ordinary current status (and, in particular, is not self-induced), but that the first coordinate

$$\int_{u \le t} \frac{\delta_{K+1}}{1-\hat{F}_{n+}(u)} \, d\mathbb{P}_n(u,\delta) + \lambda_n 1_{\{u=T_{(p)}\}}$$

involves $\hat{F}_{n+}$, which is part of the solution of the maximization problem. In fact, defining $\delta_+ = \sum_{k=1}^K \delta_k$, this expression can be rewritten as

$$\int_{u \le t} \frac{\delta_{K+1}}{1-\hat{F}_{n+}(u)}\, d\mathbb{P}_n(u,\delta) + \lambda_n 1_{\{u=T_{(p)}\}}$$

$$= \mathbb{G}_n(t) + \int_{u \le t} \frac{\hat{F}_{n+}(u) - \delta_+}{1-\hat{F}_{n+}(u)}\, d\mathbb{P}_n(u,\delta) + \lambda_n 1_{\{u=T_{(p)}\}},$$

where $\mathbb{G}_n$ is the empirical distribution function of the T_i. This shows that, up to the perturbation

$$\int_{u \le t} \frac{\hat{F}_{n+}(u) - \delta_+}{1-\hat{F}_{n+}(u)}\, d\mathbb{P}_n(u,\delta) + \lambda_n 1_{\{u=T_{(p)}\}},$$

the first coordinate is, just as in the ordinary current status case, the empirical distribution function $\mathbb{G}_n$. Also note that the population equivalent of the last perturbation term satisfies

$$\int_{u \le t} \frac{F_{0+}(u) - \delta_+}{1-F_{0+}(u)}\, d\mathbb{P}_0(u,\delta) + 1 - \int \frac{1-\delta_+}{1-F_{0+}(u)}\, d\mathbb{P}_0(u,\delta) = 0.$$

Therefore, the perturbation can be expected to go to zero. However, this does not mean that the asymptotics of the MLE can be reduced to the asymptotics of the MLE for ordinary current status.

To develop the asymptotics it is important to get hold of the asymptotic behavior of $\hat{F}_{n+}$, and this not only on neighborhoods of order $n^{-1/3}$ of a fixed point t, but also outside these neighborhoods. It is to be shown that no strange things will happen outside the $n^{-1/3}$ neighborhood, which could spoil the cube root asymptotics. The crucial result for this is the following theorem.

Theorem 12.1 (Theorem 4.10 in Groeneboom et al., 2008a) *Let $t_0 > 0$ be such that for all $k = 1, \ldots, K$, $0 < F_{0k}(t_0) < F_{0k}(\infty)$. Also, let F_{0k} and G be continuously differentiable at t_0 with strictly positive derivatives $f_{0k}(t_0)$ and $g(t_0)$. For $\beta \in (0,1)$, define*

$$v_n(t) = \begin{cases} n^{-1/3} & \text{if } |t| \le n^{-1/3}, \\ n^{-(1-\beta)/3}\,|t|^\beta & \text{if } |t| > n^{-1/3}. \end{cases} \tag{12.1}$$

Then there exists a constant $r > 0$ so that

$$\sup_{t \in [t_0 - r, t_0 + r]} \frac{|\hat{F}_{n+}(t) - F_{0+}(t)|}{v_n(t - t_0)} = O_p(1). \tag{12.2}$$

The method of proof of Theorem 12.1 is very similar to the methods of proof of Theorem 11.3 in Section 11.2. We can, for example, take $\beta = 1/2$. Then $v_n(t)$ becomes

$$v_n(t) = n^{-1/6}\left(|t|^{1/2} \vee n^{-1/6}\right).$$

Using $\beta = 1/2$, we get the following corollary.

Corollary 12.1 (Corollary 4.16 in Groeneboom et al., 2008a) *Let the conditions of Theorem 12.1 be satisfied. Then there exists an $r > 0$ such that for every $\epsilon > 0$ and $M_1 > 0$ there* *exist* $C > 0$ *and* $n_1 \in \mathbb{N}$ *such that*

$$\mathbb{P}\left\{ \sup_{t\in[t_0-r,s]} \frac{\left|\int_t^s \{\hat{F}_{n+}(u) - F_{0+}(u)\}\, d\mathbb{G}_n(u)\right|}{n^{-2/3} \vee n^{-1/6}(s-t)^{3/2}} > C \right\} < \epsilon, \text{ for } s \in [t_0 - r, t_0 + r],\ n > n_1.$$

The main point of this corollary is that it shows that $\int_t^s \{\hat{F}_{n+}(u) - F_{0+}(u)\} d\mathbb{G}_n(u)$ increases slower than $(s-t)^2$ if $t - s$ is of larger order than $n^{-1/3}$ and that therefore the quadratic drift in the second coordinate of the cusum diagram will cause the asymptotics to be localized to an $n^{-1/3}$ neigborhood around t_0. As a consequence we have the following result.

Theorem 12.2 (Theorem 4.17 in Groeneboom et al., 2008a) *Let the conditions of Theorem 12.1 be satisfied. Then there exists an* $r > 0$ *such that for every* $\epsilon > 0$ *and* $M_1 > 0$ *there exist* $M > 0$ *and* $n_1 \in \mathbb{N}$ *such that*

$$\mathbb{P}\left\{ \sup_{t\in[-M_1,M_1]} n^{1/3} \left|\hat{F}_{nk}(s + n^{-1/3}t) - F_{0k}(s)\right| > M \right\} < \epsilon, \qquad k = 1, \ldots, K$$

for all $n > n_1$ *and* $s \in [t_0 - r, t_0 + r]$.

This shows that we can indeed localize and that, for the asymptotic distribution, we only have to consider $n^{-1/3}$ neighborhoods of t_0. We now define the drifting Brownian motion processes that will be needed in the limit in Definition 12.1, and their convex minorants in Definition 12.2.

Definition 12.1 Let $W = (W_1, \ldots, W_K)$ be a K-tuple of two-sided Brownian motion processes originating from zero, with mean zero and covariances

$$E\{W_j(t)W_k(s)\} = (|s| \wedge |t|)1\{st > 0\}\Sigma_{jk}, \qquad s, t \in \mathbb{R},$$

where $\Sigma_{jk} = g(t_0)^{-1}\left\{1_{\{j=k\}}F_{0k}(t_0) - F_{0j}(t_0)F_{0k}(t_0)\right\}$, for $j, k \in \{1, \ldots, K\}$. Furthermore, let

$$V_k(t) = W_k(t) + \tfrac{1}{2} f_{0k}(t_0)t^2, \qquad k = 1, \ldots, K,\ t \in \mathbb{R}.$$

Remark The processes V_k are the limits of the rescaled and recentered processes $V_{nk}(t) = \int_{u\le t} \delta_k d\mathbb{P}_n(u, \delta)$. The correspondence is:

$$\frac{n^{2/3}}{g(t_0)} \int_{(t_0, t_0+n^{-1/3}t]} \{\delta_k - F_{0k}(t_0)\} d\mathbb{P}_n(u, \delta) \xrightarrow{\mathcal{D}} V_k(t)$$

in the Skorohod topology on $D(\mathbb{R})$.

Definition 12.2 For each $k \in \{1, \ldots, K\}$, let $\tilde{H}_k$ be the greatest convex minorant of V_k, i.e., $\tilde{H}_k$ is the unique convex function satisfying the following conditions:

$$\tilde{H}_k(t) \le V_k(t), \qquad t \in \mathbb{R},$$
$$\int \{\tilde{H}_k(t) - V_k(t)\}\, d\tilde{H}_k'(t) = 0.$$

Let $\tilde{H} = (\tilde{H}_1, \ldots, \tilde{H}_K)$, and let $\tilde{F} = (\tilde{F}_1, \ldots, \tilde{F}_K)$ be the vector of right derivatives of $\tilde{H}$.

The limiting distribution of the naive estimator $\tilde{F}_n$, where the component subdistribution functions $\tilde{F}_{nk}$ are estimated as in ordinary current status, is given in Theorem 12.3.

Theorem 12.3 *Fix $t_0 > 0$. For each $k = 1, \ldots, K$, let F_{0k} be continuously differentiable at t_0 with strictly positive derivative $f_{0k}(t_0)$. Furthermore, let G be continuously differentiable at t_0 with strictly positive derivative $g(t_0)$. Then*

$$n^{1/3}\{\tilde{F}_n(t_0 + n^{-1/3}t) - F_0(t_0)\} \xrightarrow{\mathcal{D}} \tilde{F}(t)$$

in the Skorohod topology on $D(\mathbb{R})$, where $\tilde{F}$ is as defined in Definition 12.2.

We let $a_k = (F_{0k}(t_0))^{-1}$ for $k = 1, \ldots, K+1$, where $a_{K+1} = (F_{0+}(t_0))^{-1}$. The limiting process $\hat{H}$ for the MLE is a self-induced process. The difference with $\tilde{H}$ is caused by extra terms involving the sum of the drifting Brownian motions and the sum of the components $\hat{H}_k$. Theorem 12.4 characterizes $\hat{H}$, and establishes its existence and uniqueness. The limiting distribution of the MLE is given in Theorem 12.5.

Theorem 12.4 *There exists an almost surely unique K-tuple $\hat{H} = (\hat{H}_1, \ldots, \hat{H}_K)$ of convex functions with right continuous derivatives $\hat{F} = (\hat{F}_1, \ldots, \hat{F}_K)$, satisfying the following conditions:*

(i) $a_k\hat{H}_k(t) + a_{K+1}\hat{H}_+(t) \le a_k V_k(t) + a_{K+1}V_+(t)$, *for* $k = 1, \ldots, K$, $t \in \mathbb{R}$.
(ii) $\int\{a_k\hat{H}_k(t) + a_{K+1}\hat{H}_+(t) - a_k V_k(t) - a_{K+1}V_+(t)\}d\hat{F}_k(t) = 0$, $k = 1, \ldots, K$.
(iii) For each $M > 0$ and each $k = 1, \ldots, K$, there exist points $\tau_{1k} < -M$ and $\tau_{2k} > M$ so that

$$a_k\hat{H}_k(t) + a_{K+1}\hat{H}_+(t) = a_k V_k(t) + a_{K+1}V_+(t) \quad \text{for} \quad t = \tau_{1k} \text{ and } t = \tau_{2k}.$$

Theorem 12.5 *Let the conditions of Theorem 12.3 be satisfied. Then*

$$n^{1/3}\{\hat{F}_n(t_0 + n^{-1/3}t) - F_0(t_0)\} \xrightarrow{\mathcal{D}} \hat{F}(0)$$

in the Skorohod topology on $D(\mathbb{R})$.

Proofs of these results are given in Groeneboom et al., 2008b.

12.2 The SMLE in the Current Status Competing Risk Model

In Li and Fine, 2013, smoothed nonparametric estimation for current status competing risk data is studied. It can be expected that the treatment will proceed along similar lines as the treatment of the SMLE for ordinary current status model, but there are additional difficulties, since the tools we were using for ordinary current status have to be extended to this situation. Li and Fine, 2013, note that the proofs for the naive MLE estimators, studied in Jewell et al., 2003, where the individual component distribution functions are estimated as if they were generated by ordinary current status data, can be transformed to SMLEs just as in Groeneboom et al., 2010. They refer for this to Lemma 5.9 in Part II of Groeneboom and Wellner, 1992, where it is proved that for the MLE $\hat{F}_n$ for ordinary current status data

$$\sup_{x\in[0,M]}\left|\hat{F}_n(x) - F_0(x)\right| = O_p\left(n^{-1/3}\log n\right),$$

and where it is also proved that the distance between the locations of successive jumps is of the same order.

In our present treatment of the SMLE for current status data in Section 11.3, however, we do not need these supremum bounds any more, but use instead L_2 methods, just as for the global smooth functionals theory, applying Theorem 11.3 in Section 11.2. So the theory for the naive estimators, studied in Li and Fine, 2013, which treats the component distribution functions as if they were just ordinary current status data, proceeds just as in Section 11.3, using the L_2 bounds instead of the supremum bounds.

For the real SMLE for the competing risk model, however, we need more, because the SMLE is based on a self-induced cusum diagram, depending on $\hat{F}_{n+}$, given by (5.10) in Section 5.1. Nevertheless, the argument is analogous to the argument in Section 11.3.

Fix the point $t > 0$ at which we want to study the asymptotic behavior of the SMLE, and define the (canonical) score function or efficient influence function in the hidden space:

$$\mathbb{K}_h(t-x) + \mathbb{K}_h(t+x-2a) - \mathbb{K}_h(2b-t-x), \qquad \mathbb{K}_h(x) = \int_{-\infty}^{x/h} K(u)\,du,$$

for a symmetric positive kernel K with support $[-1, 1]$, where $[a, b]$ is the observation interval. Letting $\delta = (\delta_1, \dots, \delta_{K+1})$, the images of the score operator in the observation space are given by:

$$\theta_{t,h,F}(u,\delta) = E_F\left\{\alpha(X,Y)|(T,\Delta) = (u,\delta)\right\}, \tag{12.3}$$

where Y is the failure cause, and α is a score in the hidden space. By p. 102 in Groeneboom, 1996, we get

$$\begin{aligned}
E_F\{a(X,Y)|(T,\Delta) = (u,\delta)\} &= \sum_{k=1}^{K}\left\{\frac{\delta_k \int_{[0,t]} \alpha(x,k)\,dF_k(x)}{F_k(t)} + \frac{\delta_{K+1}\int_{(t,\infty)} \alpha(x,k)\,dF_k(x)}{1-F_+(t)}\right\}\\
&= \sum_{k=1}^{K}\left\{\frac{\delta_k \int_{[0,t]} \alpha(x,k)\,dF_k(x)}{F_k(t)} - \frac{\delta_{K+1}\int_{[0,t]} \alpha(x,k)\,dF_k(x)}{1-F_+(t)}\right\}\\
&= \sum_{k=1}^{K}\left\{\frac{\delta_k}{F_k(t)} - \frac{\delta_{K+1}}{1-F_+(t)}\right\}\int_{[0,t]} \alpha(x,k)\,dF_k(x),
\end{aligned}$$

where we use $\alpha \in L_2^0(Q)$ in the second equality, and where Q is the probability distribution of (X, Y).

Analogously to the approach followed in the ordinary current status problem, the following equation has to be solved

$$\begin{aligned}
&E\{\theta_{t,h,F}(T,\Delta)\mid X=x,\ Y=k\} = \int_{u\ge x}\theta_{t,h,F}(u,e_k)g(u)\,du + \int_{u<x}\theta_{t,h,F}(u,e_{K+1})g(u)\,du\\
&= \mathbb{K}_h(t-x) + \mathbb{K}_h(t+x-2a) - \mathbb{K}_h(2b-t-x)\\
&\quad - \int \{\mathbb{K}_h(t-u) + \mathbb{K}_h(t+u-2a) - \mathbb{K}_h(2b-t-u)\}\,dF_k(u).
\end{aligned} \tag{12.4}$$

Here g is the density of the observation time distribution and e_k denotes a unit vector with 1 as the kth component and zeroes otherwise. The solution to this equation is given in the following lemma.

Lemma 12.1 *Equation (12.4) is solved by*

$$\theta^{(k)}_{t,h,F}(u,\delta) = \{\delta_k - F_k(u)\}\,\frac{K_h(t-u) - K_h(t+u-2a) - K_h(2b-t-u)}{g(u)},\ k = 1,\dots,K.$$

Proof We have, for $1 \le k \le K$:

$$\begin{aligned}
&\int_{u\ge x} \theta^{(k)}_{t,h,F}(u,e_k)g(u)\,du + \int_{u<x} \theta^{(k)}_{t,h,F}(u,e_{K+1})g(u)\,du\\
&= \int_{u\ge x} \{1 - F_k(u)\}\,\frac{K_h(t-u) - K_h(t+u-2a) - K_h(2b-t-u)}{g(u)}g(u)\,du\\
&\quad - \int_{u<x} F_k(u)\frac{K_h(t-u) - K_h(t+u-2a) - K_h(2b-t-u)}{g(u)}g(u)\,du\\
&= \int_{u\ge x} \{K_h(t-u) - K_h(t+u-2a) - K_h(2b-t-u)\}\,du\\
&\quad - \int F_k(u)\,\{K_h(t-u) - K_h(t+u-2a) - K_h(2b-t-u)\}\,du\\
&= \mathbb{K}_h(t-x) + \mathbb{K}_h(t+x-2a) - \mathbb{K}_h(2b-x-u)\\
&\quad - \int \{\mathbb{K}_h(t-x) + \mathbb{K}_h(t+x-2a) - \mathbb{K}_h(2b-x-u)\}\,dF_k(u).
\end{aligned}$$

□

We now have, just as in the ordinary current status model:

$$\begin{aligned}
&\int \{\mathbb{K}_h(t-x) + \mathbb{K}_h(t+x-2a) - \mathbb{K}_h(2b-x-u)\}\,d\big(\hat F_{nk} - F_{0k}\big)(x)\\
&= -\int \theta^{(k)}_{t,h,\hat F_n}(u,\delta)\,dP_0(u,\delta).
\end{aligned} \tag{12.5}$$

We next define, as in Section 11.3,

$$\psi_{t,h}(u) = \frac{K_h(t-u) - K_h(t+u-2a) - K_h(2b-t-u)}{g(u)}, \tag{12.6}$$

and

$$\bar\psi^{(k)}_{t,h}(u) = \begin{cases} \psi_{t,h}(\tau_i), & \text{, if } F_{0k}(u) > \hat F_{nk}(\tau_i),\ u \in [\tau_i,\tau_{i+1}),\\ \psi_{t,h}(s), & \text{, if } F_{0k}(s) = \hat F_{nk}(s),\ \text{for some } s \in [\tau_i,\tau_{i+1}),\\ \psi_{t,h}(\tau_{i+1}), & \text{, if } F_{0k}(u) < \hat F_{nk}(\tau_i),\ u \in [\tau_i,\tau_{i+1}), \end{cases}$$

where the τ_i are successive points of jump of $\hat F_{nk}$. We are now going to use another representation of $\theta^{(k)}_{t,h,\hat F_n}$, given in the following lemma (see Exercise 12.4).

Lemma 12.2 *The function $\theta^{(k)}_{t,h,F}$ has the alternative representation:*

$$\theta^{(k)}_{t,h,F}(u,\delta) = \sum_{j=1}^{K}\left\{\frac{\delta_j}{F_j(u)} - \frac{1-\delta_+}{1-F_+(u)}\right\}\{1_{\{j=k\}} - F_j(u)\}\,\psi_{t,h}(u),$$

where $\psi_{t,h}$ is defined by (12.6).

We assume that t is an interior point of the support of the distributions F_k. Hence, for shrinking h the maximum point $T_{(p)}$ of Lemma 5.2 will not belong to $[t-h, t+h]$. Therefore, defining

$$\bar{\theta}_{t,h,F}^{(k)}(u,\delta) = \sum_{j=1}^{K} \left\{ \frac{\delta_j}{F_j(u)} - \frac{1-\delta_+}{1-F_+(u)} \right\} \{1_{\{j=k\}} - F_j(u)\}\, \bar{\psi}_{t,h}^{(k)}(u),$$

we obtain from Lemma 5.2 that

$$\int \bar{\theta}_{t,h,F}^{(k)}(u,\delta)\, d\mathbb{P}_n(u,\delta) = 0$$

for large n. Thus:

$$\int \mathbb{K}_h(t-u)\, d\left(\hat{F}_{nk} - F_{0k}\right)(u) = \int \{\delta_k - \hat{F}_{nk}(u)\}\bar{\psi}_{t,h}^{(k)}(u)\, d\left(\mathbb{P}_n - P_0\right)(u,\delta)$$
$$+ \int \{F_0(u) - \hat{F}_n(u)\} \left\{\bar{\psi}_{t,h}^{(k)}(u) - \psi_{t,h}(u)\right\} dG(u);$$

see (11.47).

Hence, assuming that, similarly as in Section 11.3,

$$\left|\bar{\psi}_{t,h}^{(k)}(x) - \psi_{t,h}(x)\right| \le \frac{K}{h^2}\left|\hat{F}_{nk}(x) - F_{0k}(x)\right|,$$

and

$$\left\|\bar{\psi}_{t,h}^{(k)} - \psi_{t,h}\right\|_{2,G} = O_p\left(h^{-3/2}\|\hat{F}_{nk} - F_{0k}\|_{2,G}\right),$$

we obtain the following result, similar to Theorem 1 in Li and Fine, 2013, p. 177.

Theorem 12.6 *Assume that, for $1 \le k \le K$, F_{0k} has a positive subdensity, strictly staying away from zero on $[0, M]$. Moreover, suppose the density g also stays away from zero on $[0, M]$, with a bounded derivative g'. Finally, let t be an interior point of $[0, M]$ such that f_{0k} has a continuous derivative f_{0k}' at t. Then, if $h \sim cn^{-1/5}$ and the SMLE $\hat{F}_{nk}^{(SML)}$ is defined by*

$$\hat{F}_{nk,h}^{(SML)}(x) = \int \mathbb{K}_h(x-y)\, d\hat{F}_{nk}(y),$$

where $\hat{F}_n = (\hat{F}_{n1}, \ldots, \hat{F}_{nK})$ is the MLE, we have:

$$n^{2/5}\left\{\hat{F}_{nk,h}^{(SML)}(t) - F_{0k}(t)\right\} \xrightarrow{\mathcal{D}} N\left(\mu_k, \sigma_k^2\right),$$

where

$$\mu_k = \tfrac{1}{2}c^2 f_{0k}'(t) \int u^2 K(u)\, du$$

and

$$\sigma_k^2 = \frac{F_{0k}(t)\{1 - F_{0k}(t)\}}{cg(t)} \int K(u)^2\, du.$$

(A) The Bangkok Cohort Study

In Section 1.4 we discussed the Bangkok cohort study, which was interpreted in the context of a competing risk model in Maathuis and Hudgens, 2011, and Li and Fine, 2013. We now show how to construct confidence intervals for the SMLE and the hazard estimates. For this, we use exactly the same methods as were used in Section 9.5.

The data consist of 1,365 subjects, of whom 973 subjects were not infected, 114 subjects were infected with subtype B, 237 subjects were infected with subtype E, and 41 subjects were infected with another subtype. The confidence intervals were computed in the following way. We took 1,000 bootstrap samples of size 1,365 with replacement from the pairs (T_i, Δ^i), where T_i runs through the ages of the subjects and the first component of Δ^i is equal to 1 if the subject is not infected, the second component is 1 when the subject is infected with subtype B, the third component equals 1 when the subject is infected with subtype E, and the fourth component equals 1 otherwise. For each of these bootstrap samples the MLE and SMLE were computed. Denoting the bootstrap SMLE of the subdistribution function F_k by $\tilde F_{nk}^*$, we obtained in this way, analogously to (9.76), 1,000 values of

$$Z_{nk,h}^*(t) = \frac{\tilde F_{nk,h}^*(t) - \tilde F_{nk,h}(t)}{\sqrt{n^{-2}\sum_{i=1}^n \left\{K_h(t - T_i^*) - K_h(t + T_i^* - 2a) - K_h(2b - t - T_i^*)\right\}^2 \left(\Delta_{ik}^* - \hat F_{nk}^*(T_i^*)\right)^2}},$$

where $\hat F_{nk}^*$ is the kth component of the ordinary MLE (not the SMLE) of the bootstrap sample $(T_1^*, (\Delta^1)^*), \dots, (T_n^*, (\Delta^n)^*)$ and where we use the notation Δ_{ik}^* for the kth component of the ith value of $(\Delta^i)^*$ in the bootstrap sample (see Section 5.1). We let t run through the values on an equidistant grid of 100 points in the interval $[15, 35]$ (the age interval for the subjects), and $\tilde F_{nk,h}$ is the SMLE in the original sample.

The MLEs in the original sample and in the bootstrap samples were computed by the iterative convex minorant algorithm, as described in Section 7.5 (for the more general interval censoring model). This algorithm turned out to be about 100 times faster than the support reduction algorithm, used in the `R`-package `MLEcens`, which is important if one starts bootstrapping or performing simulations. The SMLE subdistribution functions were computed from the MLE subdistribution functions, using definition (9.75), where $a = 15$ and $b = 35$, so:

$$\tilde F_{nk,h}(t) = \int \left\{\mathbb{K}\left(\frac{t-x}{h}\right) + \mathbb{K}\left(\frac{t+x-2a}{h}\right) - \mathbb{K}\left(\frac{2b-t-x}{h}\right)\right\} d\hat F_{nk}(x),$$

and similarly

$$\tilde F_{nk}^*(t) = \int \left\{\mathbb{K}\left(\frac{t-x}{h}\right) + \mathbb{K}\left(\frac{t+x-2a}{h}\right) - \mathbb{K}\left(\frac{2b-t-x}{h}\right)\right\} d\hat F_{nk}^*(x),$$

for the bootstrap samples. Let $U_{\alpha,k}^*(t)$ be the αth percentile of the 1,000 values $Z_{nk,h}^*(t)$. Then the $1-\alpha$ bootstrap confidence interval for $F_{0k}(t)$ is given by:

$$\left[\tilde F_{nk}(t) - U_{1-\alpha/2,k}^*(t) S_{nk,h}(t),\ \tilde F_{nk}(t) - U_{\alpha/2,k}^*(t) S_{nk,h}(t)\right], \tag{12.7}$$

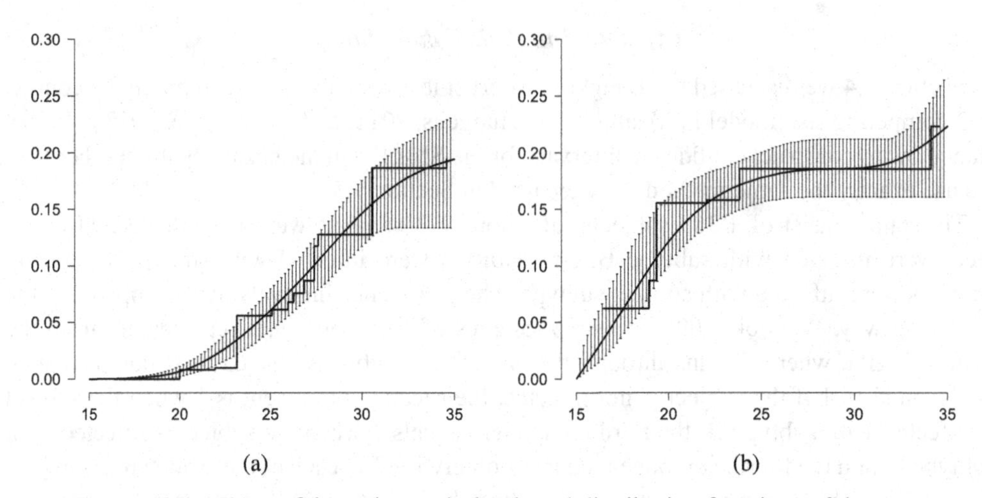

Figure 12.1 95% confidence intervals for the subdistribution functions of the ages for the group with type B (a) and type E (b) in the Bangkok cohort data.

where

$$S_{nk,h}(t)^2 = n^{-2}\sum_{i=1}^{n}\{K_h(t-T_i)-K_h(t+T_i-2a)-K_h(2b-t-T_i)\}^2\left(\Delta_{ik}-\hat{F}_{nk}(T_i)\right)^2.$$

In the present case we took as bandwidth $h = 35n^{-1/4} \approx 5.75818$ (undersmoothing as in Section 9.5), and for $\mathbb{K}$ we took the integrated triweight kernel. The resulting confidence intervals for the subdistribution functions for the subjects with type B and type E infections are shown in Figure 12.1. The confidence upper and lower bounds were made monotone from the middle, by taking iteratively the maximum of two neighboring values, going to the right of 25.

In constructing the confidence intervals for the hazard, we estimated the hazard of the kth component by:

$$\frac{\tilde{f}_{nk,h_1}(t)}{1-\tilde{F}_{nk,h_2}(t)}$$

where $h_1 = Bn^{-1/6}$ and $h_2 = Bn^{-1/5}$. The variance of this estimator is dominated by the variance of $\tilde{f}_{nk,h_1}(t)$ and estimated by

$$S_{nk}(t)^2 = \frac{1}{n^2\left(1-\tilde{F}_{nk,h_2}(t)\right)^2}\sum_{i=1}^{n}\left(\left[K_{h_1}^t\right]'(t-T_i)\right)^2\left(\Delta_{ik}-\hat{F}_{nk}(T_i)\right)^2, \tag{12.8}$$

apart from a factor depending on the density of the observation times T_i, where $\hat{F}_{nk}$ is the kth component of the MLE, and where $\tilde{f}_{nk,h_1}(t)$ is given by

$$\tilde{f}_{nk,h_1}(t) = \int K_{h_1}^t(t-x)\,d\hat{F}_{nk}(x).$$

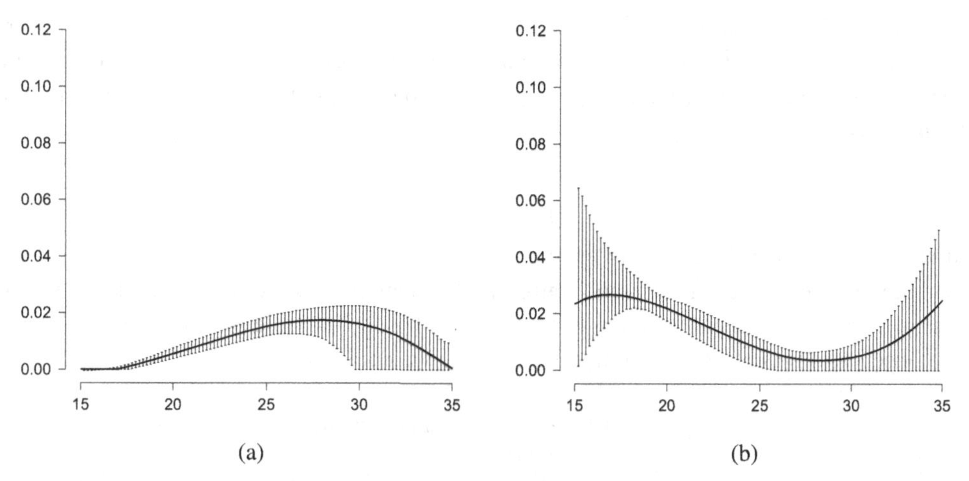

Figure 12.2 95% confidence intervals for the hazards for the group with type B (a) and type E (b) in the Bangkok cohort data.

We now define

$$Z_{nk}^*(t) = \left\{ \frac{\tilde{f}_{nk}^*(t)}{1 - \tilde{F}_{nk}^*(t)} - \frac{\tilde{f}_{nk}(t)}{1 - \tilde{F}_{nk}(t)} \right\} \Big/ S_{nk}^*(t), \tag{12.9}$$

where $S_{nk}^*(t)^2$ is the variance in the bootstrap sample, defined as (12.8), with $\tilde{F}_{nk,h_2}$, Δ_{ik} and $\hat{F}_{nk}$ replaced by $\tilde{F}_{nk,h_2}^*$, Δ_{ik}^* and $\hat{F}_{nk}^*$, respectively.

Let $U_{\alpha,k}^*(t)$ be the αth percentile of the 1,000 values $Z_{nk}^*(t)$. Then the $1-\alpha$ confidence interval is defined by:

$$\left[\frac{\tilde{f}_{nk}(t)}{1 - \tilde{F}_{nk}(t)} - U_{1-\alpha/2,k}^*(t) S_{nk}(t), \frac{\tilde{f}_{nk}(t)}{1 - \tilde{F}_{nk}(t)} - U_{\alpha/2,k}^*(t) S_{nk}(t) \right].$$

The result is shown in Figure 12.2. It is seen that the results are similar to the results in Li and Fine, 2013, but that the confidence bounds are somewhat narrower for the hazard estimates. We kept the same scaling as in Li and Fine, 2013, for an easy comparison.

It turns out that in the present case, confidence intervals directly based on

$$Z_{nk}^*(t) = \frac{\tilde{f}_{nk}^*(t)}{1 - \tilde{F}_{nk}^*(t)} - \frac{\tilde{f}_{nk}(t)}{1 - \tilde{F}_{nk}(t)}$$

instead of (12.9) lead to almost exactly the same confidence intervals for the hazard, indicating that scaling by S_{nk}^* and S_{nk} is not really necessary here.

12.3 The Bivariate Current Status Model

Basically, the MLE for the one-dimensional current status model as introduced in Section 2.3 is the monotone derivative of the cusum diagram

$$(\mathbb{G}_n(t), V_n(t)), \ t \in I, \qquad V_n(t) = \int_{u \le t} \delta \, d\mathbb{P}_n(u, \delta).$$

Here I is the observation interval and $\mathbb{G}_n$ the empirical distribution function of the observations $T_1, \ldots, T_n$. So it can be considered to be a monotone version of the derivative $dV_n(t)/d\mathbb{G}_n(t)$. Note that if we replace V_n and $\mathbb{G}_n$ by their deterministic equivalents, the derivative becomes

$$\frac{F_0(t)g(t)}{g(t)} = F_0(t),$$

so it is indeed the object we want to estimate.

For the simplest bivariate current status model, which is sometimes called the in-out model, we only have the information of whether the hidden variable is below and to the left of the observation point (T_i, U_i) or not. In this case we could also define

$$V_n(t,u) = \int_{v \le t,\, w \le u} \delta \, d\mathbb{P}_n(v, w, \delta),$$

where $\delta = 1$ represents the situation that the hidden variable is below and to the left of (v, w). If the empirical observation distribution is again denoted by $\mathbb{G}_n$, we this time want to estimate the "derivative" $dV_n(t,u)/d\mathbb{G}_n(t,u)$, since, replacing V_n and $\mathbb{G}_n$ by their deterministic equivalents, the derivative becomes

$$\frac{F_0(t,u)g(t,u)}{g(t,u)} = F_0(t,u).$$

So we want to find a version of the derivative $dV_n(t,u)/d\mathbb{G}_n(t,u)$, under the (shape) restriction that it is a bivariate distribution function.

However, a natural cusum diagram for this situation does not seem to exist. But we can define a two-dimensional Fenchel process, incorporating the duality conditions for a solution of the optimization problem. Analogously to the one-dimensional current status model, the Fenchel duality conditions for the isotonic least squares (LS) estimate, minimizing

$$\sum_{i=1}^{n} \{\Delta_i - F(T_i, U_i)\}^2, \qquad \Delta_i = 1_{\{X_i \le T_i,\, Y_i \le U_i\}}$$

over all bivariate distribution functions F, where the (X_i, Y_i) are the hidden variables, are:

$$\int_{v \ge t,\, w \ge u} \{\delta - F(v,w)\} \, d\mathbb{P}_n(v, w, \delta) \le 0, \tag{12.10}$$

with equality if (t, u) is a point of mass of the solution. So we have to deal with a process

$$(t,u) \mapsto \int_{v \ge t,\, w \ge u} F(v,w) \, d\mathbb{G}_n(v,w) \tag{12.11}$$

that has to lie above the process

$$(t,u) \mapsto \int_{v \ge t,\, w \ge u} \delta \, d\mathbb{P}_n(v, w, \delta),$$

with points of touch at points of mass of F. Denoting temporarily the process (12.11) by Q_n, we get that the isotonic least squares estimator can (formally) be denoted by $dQ_n(t,u)/d\mathbb{G}_n(t,u)$ (which no longer necessarily coincides with the MLE!). Note, however, that the function Q_n is not necessarily close to a convex or concave function, so here the analogy with 1-dimensional current status breaks down. But it must have the property that its derivative with respect to $d\mathbb{G}_n$ must be a distribution function, which is analogous to the fact that the derivative of the convex minorant of the cusum diagram must be a distribution function in the one-dimensional case.

For the full bivariate current status model the situation is more complicated, since we then have to deal with four regions instead of two per observation time; see also Section 5.2. From (5.16) we get:

$$\int_{[\mathbf{t},\infty)} \frac{\delta_1\delta_2}{\hat{F}_n(u,v)}\, d\,\mathbb{P}_n + \int_{[t_1,\infty)\times[0,t_2)} \frac{\delta_1(1-\delta_2)}{\hat{F}_{n1}(u)-\hat{F}_n(u,v)}\, d\,\mathbb{P}_n$$
$$+ \int_{[0,t_1)\times[t_2,\infty)} \frac{(1-\delta_1)\delta_2}{\hat{F}_{n2}(v)-\hat{F}_n(u,v)}\, d\,\mathbb{P}_n$$
$$+ \int_{[0,t_1)\times[0,t_2)} \frac{(1-\delta_1)(1-\delta_2)}{1-\hat{F}_{n1}(u)-\hat{F}_{n2}(v)+\hat{F}_n(u,v)}\, d\,\mathbb{P}_n$$
$$\le 1,$$

where $\mathbf{t} = (t_1, t_2)$, with equality if (t_1, t_2) is a point of mass of $\hat{F}_n$.

It has been conjectured that the MLE in the bivariate current status model converges locally at rate $n^{1/3}$, just as in the one-dimensional current status model (with smooth underlying distribution functions). Song, 2001, proves a minimax lower bound of order $n^{-1/3}$. It would be somewhat surprising if the one-dimensional rate would be preserved in dimension two, since in general one gets lower rates for density estimators if the dimension gets up, and the estimation of the distribution function in the current status model is similar to density estimation problems, as argued earlier.

To show that it is in principle possible to attain the local rate $n^{1/3}$, we construct a purely discrete estimator, converging locally at rate $n^{1/3}$. We restrict ourselves for simplicity to distributions with support $[0, 1]^2$ in the remainder of this section, but the generalization to more general rectangles is obvious. We have the following result, which is proved in Groeneboom, 2013a.

Theorem 12.7 *Consider an interior point (t, u), and define the square A_n, with midpoint (t, u), by:*

$$A_n = [t - h_n, t + h_n] \times [u - h_n, u + h_n].$$

Moreover, suppose that the observation distribution G is twice continuously differentiable at (t, u) with a strictly positive density $g(t, u)$ at (t, u), and that F_0 is twice continuously differentiable at (t, u). Moreover, suppose

$$\lim_{n\to\infty} h_n^2 n^{1/3} = c > 0. \tag{12.12}$$

Then the estimator

$$\tilde{F}_n(t,u) \stackrel{def}{=} \frac{\int_{A_n} \delta_1\delta_2\, d\mathbb{P}_n(v, w, \delta_1, \delta_2)}{\int_{A_n} d\mathbb{G}_n(v, w)}, \tag{12.13}$$

where $\mathbb{G}_n$ is the empirical distribution function of the observations (T_i, U_i) and $\mathbb{P}_n$ is the empirical distribution function of the observations

$$(T_i, U_i, \Delta_{i1}, \Delta_{i2}), \; i = 1, \ldots, n,$$

satisfies:

$$n^{1/3}\left\{\tilde{F}_n(t,u) - F_0(t,u)\right\} \xrightarrow{\mathcal{D}} N\left(\beta, \sigma^2\right),$$

where $N(\beta, \sigma^2)$ is a normal distribution with first moment

$$\beta = c\left\{\tfrac{1}{6}\left\{\partial_1^2 F_0(t,u) + \partial_2^2 F_0(t,u)\right\} + \frac{\partial_1 F_0(t,u)\partial_1 g(t,u) + \partial_2 F_0(t,u)\partial_2 g(t,u)}{3g(t,u)}\right\},$$

and variance

$$\sigma^2 = \frac{F_0(t,u)\left\{1 - F_0(t,u)\right\}}{4cg(t,u)}.$$

We now allow as possible points of mass the points $(t_{n1}, u_{n1}), \ldots, (t_{n,m_n}, u_{n,m_n})$, running through a rectangular grid, where the distances between the points on the x and y axis are of order $n^{-1/3}$, and define the estimate $\tilde{F}_n$ at each point (t_{ni}, u_{ni}) as in Theorem 12.7. Next we define the masses p_{ni} at the points (t_{ni}, u_{ni}) by the equations

$$\sum_{j: t_{nj} \le t_{ni},\, u_{nj} \le u_{ni}} p_{nj} = \tilde{F}_n(t_{ni}, u_{ni}), \; i = 1, \ldots, m_n.$$

Note that the estimate $\tilde{F}_n$ we obtain in this way is not necessarily a distribution function and that the masses p_{nj} can have negative values.

Also note that we get roughly order $n^{1/3} \times n^{1/3}$ equations in this way, which turns out to be solvable, although it is not clear beforehand that the system is nonsingular. Nevertheless, one can build the system from left and below to the right and above, where one gets more and more values in the corresponding matrix, so it seems likely that in general the solution exists. This is a point for further research. A picture of $\tilde{F}_n$, together with the MLE, computed on the sieve of points of mass of the plug-in estimator, is shown in Figure 12.3. The sieved MLE is a proper (discrete) distribution function, so all its masses are nonnegative.

Let K be a symmetric nonnegative kernel. Moreover, let the integrated kernel $\mathbb{K}$ be defined by

$$\mathbb{K}(x) = \int_{-\infty}^{x} K(y)\, dy.$$

We follow the approach for the one-dimensional case, discussed in Section 11.3 and the references Geskus and Groeneboom, 1996, and Groeneboom et al., 2010.

At an interior point (t, u), not too close to the boundary, the smoothed maximum likelihood estimator (SMLE) is just defined by

$$\hat{F}_{n,h}^{(SML)}(t,u) = \int \mathbb{K}_h(t - v)\mathbb{K}_h(u - w)\, d\hat{F}_n(v,w), \qquad \mathbb{K}_h(x) = \mathbb{K}(x/h). \qquad (12.14)$$

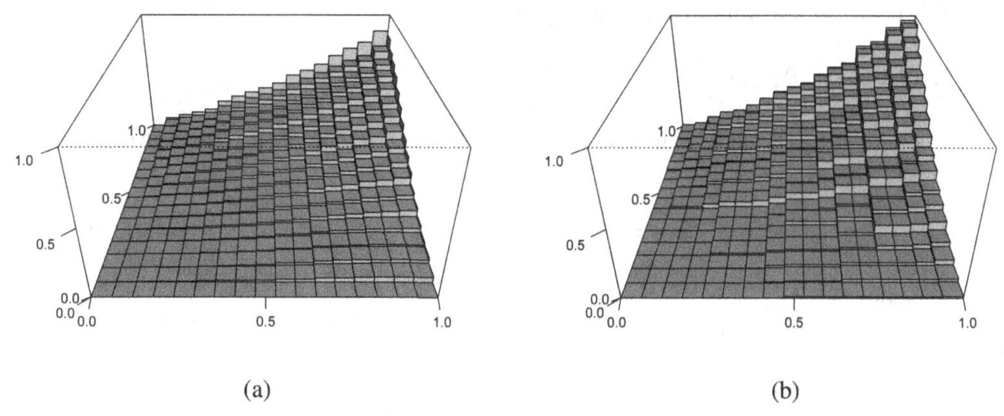

(a) (b)

Figure 12.3 The plug-in estimator $\tilde{F}_n$ (a) and the MLE on points of mass of the plug-in estimator (b), for a sample of size $n = 5000$ of bivariate current status data, where the hidden variables have a distribution with density $f_0(x, y) = x + y$, and the observation distribution is uniform on $[0, 1]^2$.

To prevent the negative bias at the right and upper boundary of the support, we now generally define

$$\hat{F}_{n,h}^{(SML)}(t, u) = \int \{\mathbb{K}_h(t - v) + \mathbb{K}_h(t + v) - \mathbb{K}_h(2 - t - v)\} \\ \times \{\mathbb{K}_h(u - w) + \mathbb{K}_h(u + w) - \mathbb{K}_h(2 - u - w)\}\, d\hat{F}_n(v, w). \quad (12.15)$$

This definition of the (integrated) boundary kernel is based on the reflection boundary used in dimension 1 in (11.37). Note that the definitions (12.14) and (12.15) coincide if $h \le v, w \le 1 - h$.

We next define the score function in the hidden space:

$$\kappa_{(t,u)}(x, y) \\ = \{\mathbb{K}_h(t - v) + \mathbb{K}_h(t + v) - \mathbb{K}_h(2 - t - v)\} \{\mathbb{K}_h(u - w) + \mathbb{K}_h(u + w) - \mathbb{K}_h(2 - u - w)\}.$$

Scores in the observation space are given by

$$\theta_{F_0}(v, w, \delta_1, \delta_2) = E\left\{a(X, Y) \,\middle|\, (T, U, \Delta_1, \Delta_2) = (v, w, \delta_1, \delta_2)\right\}, \quad (12.16)$$

where a is a score in the hidden space. We have, for example,

$$E\left\{a(X, Y) \,\middle|\, (T, U, \Delta_1, \Delta_2) = (v, w, 1, 1)\right\} = \frac{\int_{x \le v,\, y \le w} a(x, y)\, dF_0(x, y)}{F_0(v, w)}.$$

With this notation, we want to solve the equation

$$\begin{aligned}
&E\left\{\theta_{F_0}(T, U, \Delta_1, \Delta_2) \,\middle|\, (X, Y) = (x, y)\right\}\\
&\quad= \int_{v \ge x,\, w \ge y} \theta_{F_0}(v, w, 1, 1)\, g(v, w)\, dv\, dw + \int_{v \ge x,\, w < y} \theta_{F_0}(v, w, 1, 0)\, g(v, w)\, dv\, dw\\
&\qquad+ \int_{v < x,\, w \ge y} \theta_{F_0}(v, w, 0, 1)\, g(v, w)\, dv\, dw + \int_{v < x,\, w < y} \theta_{F_0}(v, w, 0, 0)\, g(v, w)\, dv\, dw\\
&\quad= \kappa_{(t,u)}(x, y).
\end{aligned} \tag{12.17}$$

Defining

$$\phi_{F_0}(x, y) = \int_{v \le x,\, w \le y} a(v, w)\, dF_0(v, w), \tag{12.18}$$

where a is a score function in the hidden space, and differentiating (12.17) with respect to x and y, we now obtain the equation:

$$\begin{aligned}
&\frac{\phi_{F_0}(x, y)}{F_0(x, y)} - \frac{\phi_{F_0}(x, 1) - \phi_{F_0}(x, y)}{F_0(x, 1) - F_0(x, y)} - \frac{\phi_{F_0}(1, y) - \phi_{F_0}(x, y)}{F_0(1, y) - F_0(x, y)}\\
&\qquad + \frac{\phi_{F_0}(x, y) - \phi_{F_0}(x, 1) - \phi_{F_0}(1, y)}{1 - F_0(x, 1) - F_0(1, y) + F_0(x, y)}\\
&\quad= g(x, y)^{-1} \frac{\partial^2 \kappa_{(t,u)}(x, y)}{\partial x \partial y}\\
&\quad= \frac{\{K_h(t - x) - K_h(t + x) - K_h(2 - t - x)\}\{K_h(u - y) - K_h(u + y) - K_h(2 - u - y)\}}{g(x, y)}.
\end{aligned}$$

This equation has the solution

$$\begin{aligned}
&\phi_{F_0}(x, y)\\
&= \frac{\{K_h(t - x) - K_h(t + x) - K_h(2 - t - x)\}\{K_h(u - y) - K_h(u + y) - K_h(2 - u - y)\}}{g(x, y)}\\
&\quad \cdot \left\{\frac{1}{F_0(x, y)} + \frac{1}{F_0(x, 1) - F_0(x, y)} + \frac{1}{F_0(1, y) - F_0(x, y)}\right.\\
&\quad \left. + \frac{1}{1 - F_0(x, 1) - F_0(1, y) + F_0(x, y)}\right\}^{-1}.
\end{aligned} \tag{12.19}$$

Note that the solution satisfies:

$$\phi_{F_0}(1, y) = \phi_{F_0}(x, 1) = 0, \; x, y \in [0, 1].$$

This suggests that the asymptotic behavior of the SMLE is given by:

$$\int \theta_{F_0}(x, y, \delta_1, \delta_2)\, d\left(\mathbb{P}_n - P\right)(x, y, \delta_1, \delta_2), \tag{12.20}$$

where

$$\theta_{F_0}(v, w, \delta_1, \delta_2) = \frac{\delta_1\delta_2\phi_{F_0}(x, y)}{F_0(x, y)} - \frac{\delta_1(1-\delta_2)\phi_{F_0}(x, y)}{F_0(x, 1) - F_0(x, y)} - \frac{(1-\delta_1)\delta_2\phi_{F_0}(x, y)}{F_0(1, y) - F_0(x, y)} + \frac{(1-\delta_1)(1-\delta_2)\phi_{F_0}(x, y)}{1 - F_0(1, y) - F_0(x, 1) + F_0(x, y)},$$

leading at interior points (t, u) to an asymptotic variance, given by:

$$\frac{1}{n}\int_{(x,y)\in[0,1]^2}\left\{\frac{1}{F_0(x, y)} + \frac{1}{F_0(x, 1) - F_0(x, y)} + \frac{1}{F_0(1, y) - F_0(x, y)} + \frac{1}{1 - F_0(1, y) - F_0(x, 1) + F_0(x, y)}\right\}^{-1}$$

$$\cdot\frac{\{K_h(t-x) - K_h(t+x) - K_h(2-t-x)\}^2\{K_h(u-y) - K_h(u+y) - K_h(2-u-y)\}^2}{g(x, y)}\,dx\,dy$$

$$\sim \left(nh^2\right)^{-1}\left\{\frac{1}{F_0(t, u)} + \frac{1}{F_0(t, 1) - F_0(t, u)} + \frac{1}{F_0(1, t) - F_0(t, u)} + \frac{1}{1 - F_0(1, u) - F_0(t, 1) + F_0(t, u)}\right\}^{-1} g(t, u)^{-1}\left\{\int K(v)^2\,dv\right\}^2, \qquad h \downarrow 0.$$

Assume that $h \le t, u \le 1 - h$. Then the bias is given by:

$$\begin{aligned}
&\int \mathbb{K}_h(t-v)\mathbb{K}_h(u-w)f_0(v, w)\,dv\,dw - F_0(t, u)\\
&= \int\left\{\int K_h(t-v)\int_0^v f_0(x, w)\,dx\,dv\right\}\mathbb{K}_h(u-w)\,dw - F_0(t, u)\\
&= \int K_h(t-v)K_h(u-w)F_0(v, w)\,dv\,dw - F_0(t, u)\\
&= \int K(v)K(w)\{F_0(t-hv, u-hw) - F_0(t, u)\}\,dv\,dw\\
&= \tfrac{1}{2}\left\{\partial_1^2F_0(t, u) + \partial_2^2F_0(t, u)\right\}h^2\left\{\int x^2K(x)\,dx\right\}^2 + o\left(h^2\right).
\end{aligned}$$

The SMLE is compared with the MLE in Figure 12.4. Using (12.20), we get the following conjectured result.

Conjecture 12.1 *Under the conditions of Theorem 12.7 we have, for each point $(t, u) \in (0, 1)^2$ satisfying these conditions, if $c = \lim_{n\to\infty} n^{1/3}h_n^2$,*

$$n^{1/3}\left\{\hat F_{n,h_n}^{(SML)}(t, u) - F_0(t, u)\right\} \xrightarrow{\mathcal{D}} N\left(\beta, \sigma^2\right),$$

where $N(\beta, \sigma^2)$ is a normal distribution with first moment

$$\beta = \tfrac{1}{2}c\left\{\partial_1^2F_0(t, u) + \partial_2^2F_0(t, u)\right\}\left\{\int x^2K(x)\,dx\right\}^2,$$

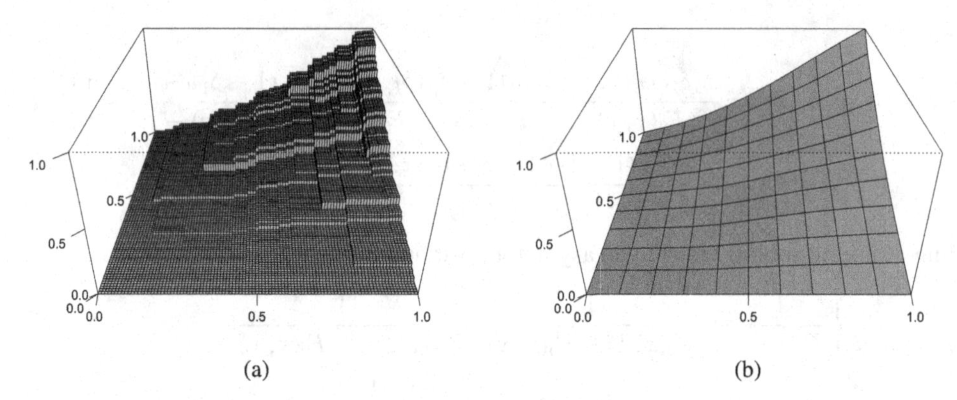

Figure 12.4 The MLE (a) and SMLE (b) for for a sample of size $n = 1000$ of bivariate current status data, where the hidden variables have a distribution with density $f_0(x, y) = x + y$, and the observation distribution is uniform on $[0, 1]^2$.

and variance

$$\sigma^2 = c^{-1}\left\{\frac{1}{F_0(t,u)} + \frac{1}{F_0(t,1) - F_0(t,u)} + \frac{1}{F_0(1,t) - F_0(t,u)} + \frac{1}{1 - F_0(1,u) - F_0(t,1) + F_0(t,u)}\right\}^{-1} g(t,u)^{-1}\left\{\int K(v)^2\,dv\right\}^2.$$

Remark Note that choosing $h_n \asymp n^{-1/6}$ is the asymptotically optimal choice (modulo constants) since the variance is of order $1/(nh_n^2)$ and the bias of order h_n^2, unless the bias is of order $o(h_n^2)$ (as happens when F_0 is the uniform distribution function on $[0, 1]^2$). Also note that the bias term, caused by the interaction of the observation distribution G and the distribution function F_0, which entered into the bias term of Theorem 12.7, plays no role here.

Simulation results, comparing the plug-in estimator, the SMLE and the MLE, are given in Groeneboom, 2013a.

Exercises

12.1 Let the conditions of Theorem 12.1 be satisfied and let, respectively, for $k = 1, \ldots, K$, $\tau_{nk}^-(s)$ and $\tau_{nk}^+(s)$ be the largest jump point $\le s$ and the smallest jump point $> s$ of $\hat F_{nk}$. Deduce from Theorem 12.2 that there exists an $r > 0$ such that for all $\epsilon > 0$ there exist $n_1, C > 0$ such that

$$\mathbb{P}\left\{\tau_{nk}^+(s) - \tau_{nk}^-(s) > Cn^{-1/3}\right\} < \epsilon, \qquad \forall n > n_1,\ s \in [t_0 - r, t_0 + r].$$

12.2 Suppose that all subdistribution functions in the current status competing risk model are defined on $[0, B]$ and that $F_{0+}(B) = 1$. The SMLE for a component subdistribution function is then generally defined by

$$\tilde F_{nk}(t) = \int \{\mathbb{K}_h(t - x) + \mathbb{K}_h(t + x) - \mathbb{K}_h(2B - t - x)\}\, d\hat F_{nk}(x), \qquad k = 1, \ldots, K.$$

However, we do not have the property $F_{0k}(B) = 1$, as in the ordinary current status model. What does this mean for the bias of $\tilde F_{nk}$ on $[B - h, B]$?

12.3 In Section 9.5 bootstrap confidence intervals were constructed for the distribution function in the current status model, using the SMLE. Describe an analogous procedure for confidence intervals for the subdistribution functions in the competing risk model, based on the SMLEs.

12.4 Show that the representation of $\theta^{(k)}_{t,h,\hat{F}_n}$ given in Lemma 12.2 holds.

12.5 In Section 9.5 likelihood ratio tests, based on the MLE, as discussed in Banerjee and Wellner, 2005, were introduced and compared with the confidence intervals, based on the SMLE. Is a construction of this type, using LR tests, also possible for the competing risk model?

12.6 Give a sketch of proof of Theorem 12.6, using methods analogous to the methods used in Section 11.3.

12.7 In the study of the SMLE for the one-dimensional current status model the toy estimator (11.53) was introduced, which is asymptotically equivalent with the SMLE for this model. The treatment in Section 12.3 suggests a similar toy estimator equivalent to the SMLE in the bivariate current status model. How is this estimator defined?

Bibliographic Remarks

This chapter treats material that is still in full development. The local asymptotic theory for the MLE in the competing risk model with current status data was developed in the papers Groeneboom et al., 2008a,b, but although the rate of convergence has been determined and the limit distribution has been characterized, an analytic characterization of this limit distribution is still not available.

The limit distribution of the SMLE, density and hazard in this model has been developed in Li and Fine, 2013. Their treatment relies in last instance on a conjectured relation (see (S18) and (S19) on p. 10 and 11 in the supplementary material to their article), assuming an analogue of Lemma 5.9 in Part II of Groeneboom and Wellner, 1992. We proposed in Section 12.2 to replace this by an L_2-type bound, as in our treatment of the SMLE in the current status model in Section 11.3.

If we take a continuum of competing risks, we get the so-called continuous mark model as introduced in Section 5.3. For this model the MLE is inconsistent, see Maathuis and Wellner, 2008, but the MSLE will be consistent if one chooses the right smoothing parameters, see Groeneboom, Jongbloed and Witte, 2012.

Still less is known about the distribution theory in the bivariate current status model. Theorem 12.7, which is proved in Groeneboom, 2013a, is the only distributional result we are aware of at present; the limit distribution and even the rate of convergence of the MLE is still unknown. Nevertheless it seems rather likely that the limit behavior for the SMLE, as stated in Conjecture 12.1 (and also in Groeneboom, 2013a), will hold. In fact, since the tools in proving limit results for the MLE and the SMLE are rather different, we probably do not need to know the limit distribution of the MLE for the derivation of the limit distribution of the SMLE. We still need certain bounds for the SMLE, though, analogous to the bounds we were using for the SMLE in the one-dimensional current status problem, to prove Conjecture 12.1.

13

Asymptotic Distribution of Global Deviations

Chapter 3 and Chapter 11 deal with pointwise asymptotic distribution theory for shape constrained estimators in a variety of models. In Chapter 9, several testing problems are discussed where test statistics are based on global deviations of the estimator from the hypothesized function. In this chapter asymptotic theory will be developed for global deviation measures in three particular models.

The first is the monotone density model introduced in Section 2.2. The asymptotic distribution of the L_1-distance between the Grenander estimator and the underlying (decreasing) density will be derived. The method used relies heavily on the switch relation introduced in Section 3.3. In Section 13.2 one of the global test statistics for testing the monotonicity hypothesis of a hazard function as discussed in Section 9.1 is studied. The method applied is based on first approximating the test statistic by a random variable that depends on Brownian motion rather than the empirical process. After that, this random variable (which is an integral over an interval $[0, a]$) is written as a sum of local integrals. Finally, this sum is dealt with using an appropriate central limit result.

In Section 13.3 the asymptotic distribution (under the null hypothesis) of the two sample quasi-LR test, introduced in Section 9.3 for current status data, based on maximum smoothed likelihood estimators (MSLEs), is studied. This time the treatment is based on the asymptotic theory of degenerate U-statistics, taking advantage of the asymptotic linearization performed by the MSLEs.

13.1 The L_1 Loss of the Grenander Estimator

The MLE (Grenander estimator) of a decreasing density is introduced in Section 2.2. Its pointwise asymptotic behavior is derived in Section 3.2 and, more formally, in Section 3.6. Under some smoothness assumption on the underlying density f, the pointwise rate of convergence of the Grenander estimator is $n^{1/3}$, the same as the lower bound on the minimax risk derived in Example 6.2. In this section, the asymptotic behavior of the L_1 loss of the Grenander estimator is studied. The following result is stated in Groeneboom, 1985, with a sketch of proof.

Theorem 13.1 *Define*

$$V(c) = \sup\{t : W(t) - (t - c)^2 \textit{ is maximal}\}, \tag{13.1}$$

where $\{W(t) : -\infty < t < \infty\}$ denotes standard two sided Brownian motion on $\mathbb{R}$ originating from zero. Let f be a twice differentiable decreasing density on [0,1], satisfying

(A1) $0 < f(1) \le f(t) \le f(s) \le f(0) < \infty$, *for* $0 \le s \le t \le 1$.

(A2) $0 < \inf_{t \in (0,1)} |f'(t)| \le \sup_{t \in (0,1)} |f'(t)| < \infty$.

(A3) $\sup_{t \in (0,1)} |f''(t)| < \infty$.

Then, with $\mu = 2E|V(0)| \int_0^1 |\frac{1}{2} f'(t) f(t)|^{1/3}\, dt$,

$$n^{1/6}\left\{n^{1/3}\int_0^1 |\hat{f}_n(t) - f(t)|\, dt - \mu\right\} \xrightarrow{\mathcal{D}} N(0, \sigma^2),$$

where $N(0, \sigma^2)$ *is a normal distribution with mean zero and variance*

$$\sigma^2 = 8\int_0^\infty \mathrm{covar}\,(|V(0)|, |V(c) - c|)\, dc.$$

The L_1 loss of the Grenander estimator, scaled at rate $n^{1/3}$, converges in probability to a constant μ. The difference between this rescaled loss and μ, blown up at rate $n^{1/6}$, converges in distribution to a normal limit distribution. The result shows that, approximately, the L_1 loss multiplied by $n^{1/2}$ is normally distributed with expectation of the order $\mu n^{1/6}$ and variance σ^2. The details of the proof are provided in Groeneboom et al., 1999. Just as in the analysis of the local behavior, the inverse process that can be obtained via the switch relation plays an important role in the proof, that is, the process $\{U_n(a) : a \in [f(1), f(0)]\}$, where $U_n(a)$ is defined as the last time that the process $\mathbb{F}_n(t) - at$ attains its maximum:

$$U_n(a) = \sup\{t \in [0, M] : \mathbb{F}_n(t) - at \text{ is maximal}\}, \tag{13.2}$$

where we assume that [0, 1] is the support of the density. The following result is a crucial tool for the proof of Theorem 13.1.

Theorem 13.2 *Let the conditions of Theorem 13.1 be satisfied and let* $U(a) = f^{-1}(a)$. *Then there exists a constant* $C > 0$, *only depending on* f, *such that for all* $n \ge 1$, $a \in [f(1), f(0)]$ *and* $x > 0$,

$$P\left\{n^{1/3}\,|U_n(a) - U(a)| > x\right\} \le 2e^{-Cx^3}.$$

Note that this result is similar to Theorem 11.3 in Section 11.2. In both cases the result is proved by martingale methods, but the martingales are rather different in the two cases. We will use the following notation to describe the relevant martingales. For $s \le t$, we write:

$$\mathbb{F}_n(s, t) = \mathbb{F}_n(t) - \mathbb{F}_n(s),$$

$$F(s, t) = F(t) - F(s),$$

where F is the underlying distribution function. In Groeneboom et al., 1999, the following lemma is proved.

Lemma 13.1 *Let* $a \in [f(1), f(0)]$ *and let* $t_0 = U(a)$. *Moreover, let* V_n *be defined by:*

$$V_n(a) = n^{1/3}(U_n(a) - U(a)). \tag{13.3}$$

Then

$$P\{V_n(a) > x\} \leq P\left\{\sup_{t\in[t_0+xn^{-1/3},1]} \frac{\mathbb{F}_n(t_0,t)}{F(t_0,t)} \geq \frac{f(t_0)xn^{-1/3}}{F(t_0,t_0+xn^{-1/3})}\right\},$$

for each x *such that* $t_0 < t_0 + xn^{-1/3} \leq 1$, *and*

$$P\{V_n(a) < -x\} \leq P\left\{\inf_{t\in[0,t_0-xn^{-1/3}]} \frac{\mathbb{F}_n(t,t_0)}{F(t,t_0)} \leq \frac{f(t_0)xn^{-1/3}}{F(t_0-xn^{-1/3},t_0)}\right\},$$

for each x *such that* $0 \leq t_0 - xn^{-1/3} < t_0$.

The following lemma introduces the martingales.

Lemma 13.2 *Let* $0 \leq t_0 \leq 1$. *Consider, for* n *fixed, the processes*

$$t \mapsto M_{1n}(t) = \frac{\mathbb{F}_n(t_0,t)}{F(t_0,t)}, \quad t \in (t_0, 1]$$

and

$$t \mapsto M_{2n}(t) = \frac{\mathbb{F}_n(t,t_0)}{F(t,t_0)}, \quad t \in [0, t_0).$$

Let $\mathcal{F}_s = \sigma\{\mathbb{F}_n(t) : t \in [s,1]\}$ *and* $\mathcal{G}_s = \sigma\{\mathbb{F}_n(t) : t \in [0,s]\}$. *Then, conditionally on* $\mathbb{F}_n(t_0)$, *the process* M_{1n} *is a reverse time martingale with respect to the filtration* $\{\mathcal{F}_s : s \in (t_0,1]\}$ *and* M_{2n} *is a forward time martingale with respect to the filtration* $\{\mathcal{G}_s : s \in [0,t_0)\}$.

Next, Doob's inequality is used to obtain the following lemma.

Lemma 13.3 *Let* $h(y) = 1 - y + y\log y$, $y > 0$; *see Figure 13.1. Then, for* $t_0 \in [0,1)$, $y \geq 1$ *and* $\delta > 0$ *such that* $t_0 + \delta < 1$:

$$P\left\{\sup_{t\in[t_0+\delta,1]} M_{1n}(t) \geq y\right\} \leq \exp\{-nF(t_0,t_0+\delta)h(y)\}$$

and for $t_0 \in (0,1]$, $0 < y \leq 1$ *and* $\delta > 0$ *such that* $t_0 - \delta > 0$:

$$P\left\{\inf_{t\in[0,t_0-\delta]} M_{2n}(t)) \leq y\right\} \leq \exp\{-nF(t_0-\delta,t_0)h(y)\}.$$

Proof We start with the proof of the first inequality. According to Lemma 13.2 we have that for each $r > 0$, conditionally on $\mathbb{F}_n(t_0)$, the process $\exp\{rM_{1n}(t)\}$ is a reverse time submartingale. Hence, by Doob's inequality,

$$\begin{aligned}
P\left\{\sup_{t\in[t_0+\delta,1]} M_{1n}(t) \geq y\right\} &= E\left[P\left\{\sup_{t\in[t_0+\delta,1]} M_{1n}(t) \geq y \,\middle|\, \mathbb{F}_n(t_0)\right\}\right] \\
&= E\left[P\left\{\sup_{t\in[t_0+\delta,1]} e^{rM_{1n}(t)} \geq e^{ry} \,\middle|\, \mathbb{F}_n(t_0)\right\}\right] \\
&\leq E\left[e^{-ry}E\left(e^{rM_{1n}(t_0+\delta)} \,\middle|\, \mathbb{F}_n(t_0)\right)\right] \\
&= e^{-ry}Ee^{rM_{1n}(t_0+\delta)}.
\end{aligned}$$

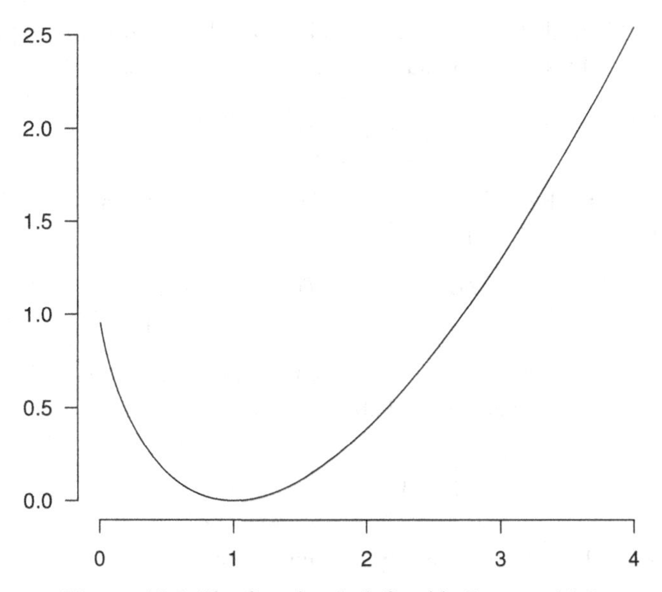

Figure 13.1 The function h defined in Lemma 13.3.

Since $n\mathbb{F}_n(t_0, t_0 + \delta)$ has a binomial distribution with parameters n and $p = F(t_0, t_0 + \delta)$, the last expression is equal to:

$$e^{-ry}\left(1 + p(e^{r/np} - 1)\right)^n \le e^{-ry} \exp\left(np(e^{r/np} - 1)\right) = e^{-nph(y)},$$

by putting $r = np \log y$ in the last equality. This proves the first exponential bound.

For the proof of the second inequality, note that, for $y \in (0, 1]$:

$$\begin{aligned} P\left\{\inf_{t\in[0,t_0-\delta]} M_{2n}(t) \le y\right\} &= E\left[P\left\{\sup_{t\in[0,t_0-\delta]} -M_{2n}(t) \ge -y \,\middle|\, \mathbb{F}_n(t_0)\right\}\right] \\ &\le E\left[e^{ry} E\left(e^{-rM_{2n}(t_0-\delta)} \,\middle|\, \mathbb{F}_n(t_0)\right)\right] \\ &= e^{ry} E e^{-rM_{2n}(t_0-\delta)}, \end{aligned}$$

where again Doob's inequality is used. Taking $p = F(t_0 - \delta, t_0)$ and $r = -np \log y$, we get

$$e^{ry} E e^{-rM_{2n}(t_0-\delta)} \le e^{-nph(y)}.$$

□

Remark The function $y \mapsto h(y)$, used in Lemma 13.3, is a well known function in large deviation theory. It is nonnegative and convex on $(0, \infty)$, and can be written as $h(y) = \int_1^y \log u \, du$, $y > 0$. Its minimum 0 is attained at $y = 1$; see Exercise 13.1.

We can now prove Theorem 13.2.

Proof of Theorem 13.2 We write $\delta_n = xn^{-1/3}$. First consider the probability

$$P\{V_n(a) > x\}. \tag{13.4}$$

If $U(a)+\delta_n \ge 1$, this probability is zero, in which case there is nothing to prove, so we can restrict ourselves to values of $x > 0$, such that $U(a)+\delta_n < 1$. Let

$$y_n = \frac{f(t_0)\delta_n}{F(t_0, t_0+\delta_n)},$$

where $t_0 = U(a)$. Note that $y_n > 1$, since f is strictly decreasing. We also have, using assumption (A1) of Theorem 13.1,

$$y_n = \frac{f(t_0)\delta_n}{F(t_0, t_0+\delta_n)} \le \frac{f(t_0)}{f(t_0+\delta_n)} \le \frac{f(0)}{f(1)} < \infty.$$

Hence $1 < y_n < c_1$, for a constant $c_1 > 0$, independent of x such that $t_0+\delta_n < 1$. By Lemma 13.1, the probability in (13.4) is bounded above by

$$P\left\{\sup_{t\in[t_0+\delta_n,1]} M_{1n}(t) \ge y_n\right\}.$$

According to Lemma 13.3 this probability is bounded by

$$\exp\{-nF(t_0, t_0+\delta_n)h(y_n)\}. \tag{13.5}$$

Using a Taylor expansion with a Lagrangian remainder term of the convex function $u \mapsto h(u)$ at $u = 1$, we get

$$h(y_n) = \frac{1}{2}h''(\xi_n)(y_n-1)^2 \ge \frac{1}{2}c_1^{-1}(y_n-1)^2, \tag{13.6}$$

where $1 \le \xi_n \le c_1$. But

$$|y_n - 1| \ge \frac{\delta_n \inf_{u\in(0,1)}|f'(u)|}{2f(0)},$$

and hence, by (13.6),

$$h(y_n) \ge c_2\delta_n^2,$$

for a constant $c_2 > 0$, independent of x such that $t_0+\delta_n < 1$. Since $F(t_0, t_0+\delta_n) \ge f(1)\delta_n$, it now follows that (13.5) is bounded above by $\exp(-Cx^3)$.

Now consider the probability

$$P\{V_n(a) < -x\}. \tag{13.7}$$

If $U(a) - xn^{-1/3} \le 0$, this probability is zero, so we can restrict ourselves to consider an $x > 0$ such that $U(a) - xn^{-1/3} > 0$. Define

$$y_n = \frac{f(t_0)\delta_n}{F(t_0-\delta_n, t_0)}.$$

The fact that f is strictly decreasing this time implies that $y_n < 1$. Using Lemma 13.1 it is seen that (13.7) is bounded above by

$$P\left\{\inf_{t\in[0,t_0-\delta_n]} M_{2n}(t) \le y_n\right\},$$

which, by Lemma 13.2, leads to the upper bound

$$\exp\{-nf(1)\delta_n h(y_n)\}.$$

We have, using $h''(x) \ge 1,\ x \in (0, 1]$:

$$h(y_n) = \frac{1}{2}h''(\xi_n)(y_n - 1)^2 \ge \frac{1}{2}(y_n - 1)^2,$$

where in this case $0 < \xi_n \le 1$. Following the same line of argument as was used earlier, we get the upper bound $\exp\{-Cx^3\}$. □

The preceding results show that the difference between the L_1 risk

$$\|\hat{f}_n - f\|_1 = \int_0^1 |\hat{f}_n(t) - f(t)|\, dt \tag{13.8}$$

and the integral

$$\int_{f(1)}^{f(0)} |U_n(a) - U(a)|\, da,$$

defined in terms of the inverse process, is of order $o_p(n^{-1/2})$. In fact, we have the following important corollary as a consequence of Theorem 13.2.

Corollary 13.1 *Let $\hat{f}_n$ be the Grenander estimator and let U_n be defined in (13.2). Then*

$$\|\hat{f}_n - f\|_1 - \int_{f(1)}^{f(0)} |U_n(a) - U(a)|\, da = O_p\left(n^{-2/3}\right). \tag{13.9}$$

Proof The difference on the left hand side of (13.9) can be written as

$$\int_0^1 [\hat{f}_n(t) - f(0)]^+\, dt + \int_0^1 [f(1) - \hat{f}_n(t)]^+\, dt,$$

where $x^+ = \max(0, x)$, $x \in \mathbb{R}$. We show that the first term is $O_p\left(n^{-2/3}\right)$. The second term can be treated similarly.

We have:

$$\int_0^1 \left[\hat{f}_n(t) - f(0)\right]^+ dt = \int_0^{U_n(f(0))} \left(\hat{f}_n(t) - f(0)\right) dt = F_n(U_n(f(0))) - f(0)U_n(f(0))$$
$$= F_n(U_n(f(0))) - F(U_n(f(0))) + F(U_n(f(0))) - f(0)U_n(f(0)).$$

According to Theorem 13.2 we have for the second difference on the right hand side:

$$|F(U_n(f(0))) - f(0)U_n(f(0))| \le \frac{1}{2} \sup_{x\in[0,1]} |f'(x)|\, U_n(f(0))^2 = O_p(n^{-2/3}). \tag{13.10}$$

Let $Z_n = F_n(U_n(f(0))) - F(U_n(f(0)))$ and $\delta_n = n^{-1/3}\log n$. Then write

$$Z_n = Z_n 1_{\{U_n(f(0))>\delta_n\}} + Z_n 1_{\{U_n(f(0))\le\delta_n\}}.$$

Then, according to Theorem 13.2,

$$E|Z_n| 1_{\{U_n(f(0))>\delta_n\}} \le 2P\{U_n(f(0)) > \delta_n\} \le 4e^{-C(\log n)^3}.$$

Hence by the Markov inequality we can conclude that

$$Z_n 1_{\{U_n(f(0))>\delta_n\}} = o_p(n^{-2/3}). \tag{13.11}$$

Let (B_n) be a sequence of Brownian bridges given by the Hungarian embedding approximating $n^{1/2}(F_n - F)$, see Komlós et al., 1975. Then

$$|Z_n| 1_{\{U_n(f(0))\le\delta_n\}} \le n^{-1/2} \sup_{t\in[0,F(\delta_n)]} |B_n(t)| + O_p(n^{-1}\log n).$$

Since $B_n(t) \stackrel{d}{=} W(t) + tW(1)$, where W denotes Brownian motion, the right hand side can be bounded by a random variable that has the same distribution as

$$n^{-1/2} \sup_{t\in[0,F(\delta_n)]} |W(t)| + n^{-1/2} F(\delta_n)|W(1)| + O_p(n^{-1}\log n).$$

Note that $F(\delta_n)|W(1)| = O_p(\delta_n)$. Furthermore, since for any $\epsilon > 0$,

$$P\left\{ \sup_{t\in[0,F(\delta_n)]} |W(t)| > \epsilon \right\} \le 4P\left\{ W(1) \ge \frac{\epsilon}{F(\delta_n)^{1/2}} \right\},$$

we have that

$$n^{-1/2} \sup_{t\in[0,F(\delta_n)]} |W(t)| = o_p(n^{-2/3}),$$

which implies that $Z_n 1_{\{U_n(f(0))\le\delta_n\}} = o_p(n^{-2/3})$. Together with (13.10) and (13.11) this proves that

$$\int_0^1 \left[\hat{f}_n(t) - f(0)\right]^+ dt = O_p(n^{-2/3}).$$

□

The preceding result shows that in Theorem 13.1 the proof of

$$n^{1/6}\left\{ n^{1/3} \int_0^1 |\hat{f}_n(t) - f(t)|\, dt - \mu \right\} \xrightarrow{\mathcal{D}} N(0, \sigma^2)$$

can be reduced to the proof of

$$n^{1/6}\left\{ n^{1/3} \int_{f(1)}^{f(0)} |U_n(a) - U(a)|\, da - \mu \right\} \xrightarrow{\mathcal{D}} N(0, \sigma^2).$$

Moreover, it is shown in Groeneboom et al., 1999, that we can replace the process U_n by the argmax process $U_n^W(a)$, where

$$U_n^W(a) = \mathrm{argmax}_{t\in[0,1]} \left\{ W_n(F(t)) + \sqrt{n}\,(F(y) - at) \right\},$$

where $\{W_n; n = 1, 2, \ldots$ is a sequence of Brownian motion processes on [0, 1], coupled to the sample processes. The process U_n^W is strong mixing, with mixing function

$$\alpha_n(d) = 12 \exp\left\{ -Cnd^3 \right\},$$

so the dependence dies out rather rapidly with d (see Theorem 3.3 of Groeneboom et al., 1999). This shows that, in fact, the central limit theorem can be reduced to a theorem on strongly mixing sequences of stationary random variables, given in Ibragimov and Linnik, 1971. For more details, see Groeneboom et al., 1999.

These results have been further developed by Durot et al., 2012; see also Section 9.2.

13.2 Empirical L_1-Test for a Monotone Hazard

In the context of testing monotonicity of a hazard function on the interval $[0, a]$, the following test statistic is introduced in Section 9.1:

$$S_n = \int_{[0,a]} \{\mathbb{H}_n(x-) - \hat{H}_n(x)\} \, d\mathbb{F}_n(x). \tag{13.12}$$

This is the empirical L_1-distance between the empirical cumulative hazard function and its greatest convex minorant on $[0, a]$; see also Section 2.6. In this section we point out the main steps to derive the asymptotic distribution of this test statistic. This distribution is derived by considering the quantity

$$\tilde{S}_n = \int_{[0,a]} \{\mathbb{H}_n(x-) - \hat{H}_n(x)\} \, dF_0(x) \tag{13.13}$$

that turns out to have the same asymptotic distribution as S_n. Using a convenient representation of $\tilde{S}_n$ in terms of integrals of functionals of Brownian motions, the integral is viewed as the sum of increasingly (as n grows) many local integrals. For this sum a central limit theorem can be used to derive its asymptotic normal distribution. To this end, the first two moments of the local integrals will be needed. These moments are related to the moments of a process associated with two sided Brownian motion. One could say that the asymptotic distribution of $\tilde{S}_n$ is derived using standard asymptotics for a sum of independent random variables (CLT) where the terms in the sum need to be handled using nonstandard asymptotics involving the convex minorant of Brownian motion with a parabolic drift.

We start by representing $\tilde{S}_n$ in terms of a Brownian motion process rather than the empirical process. To this end, approximate the process $\mathbb{H}_n(x) - \hat{H}_n(x)$ in $\tilde{S}_n$ by the process

$$x \mapsto H_0(x) + \frac{E_n(x)}{\sqrt{n}\{1 - F_0(x)\}} - \tilde{H}_n(x), \; x \in [0, a], \tag{13.14}$$

where E_n is the empirical process $\sqrt{n}\{\mathbb{F}_n - F_0\}$ and $\tilde{H}_n$ is the greatest convex minorant of the process

$$x \mapsto H_0(x) + \frac{E_n(x)}{\sqrt{n}(1 - F_0(x))}; \tag{13.15}$$

see Exercise 13.2. Then use the strong (Hungarian) approximation of the empirical process by a Brownian bridge $\mathbb{B}_n$. This approximation entails that Brownian bridges $\mathbb{B}_n$ on the same sample space as $\mathbb{F}_n$ can be constructed such that

$$Y_n = \sup_{x \in [0,a]} \frac{n^{1/2} \, |E_n(x) - \mathbb{B}_n(F_0(x))|}{2 \vee \log n}$$

is a random variable with with $EY_n \le C < \infty$ for all n. This yields approximation

$$x \mapsto H_0(x) + \frac{\mathbb{B}_n(F_0(x))}{\sqrt{n}(1 - F_0(x))}, \; x \in [0, a],$$

$\tilde{J}_{n,1}$ $I_{n,1}$ $\bar{J}_{n,2}$ $\tilde{J}_{n,2}$ $I_{n,2}$ $\bar{J}_{n,3}$ I_{n,m_n} $\bar{J}_{n,m_n+1}$

0 $L_{n,2}$ a

Figure 13.2 Partition of [0, a] in big blocks separated by small blocks.

to the process (13.15). See Theorem 3 in Komlós et al., 1975, for the approximation and Exercise 3.9 for the definition of Brownian bridge. This process is distributed as

$$x \mapsto V_n(x) := H_0(x) + n^{-1/2} W\left(\frac{F_0(x)}{1 - F_0(x)}\right), \ x \in [0, a], \tag{13.16}$$

where W is standard Brownian motion on $\mathbb{R}_+$; see Exercise 13.3. So, if C_n is the greatest convex minorant of the process V_n on $[0, a]$, we have (where we add an O_P-term we do not discuss in detail here):

$$\begin{aligned}&\int_{[0,a]} \left\{H_0(x) + \frac{E_n(x)}{\sqrt{n}(1 - F_0(x))} - \tilde{H}_n(x)\right\} dF_0(x)\\ &\stackrel{\mathcal{D}}{=} \int_{[0,a]} \left\{H_0(x) + n^{-1/2} W\left(\frac{F_0(x)}{1 - F_0(x)}\right) - C_n(x)\right\} dF_0(x) + O_P\left(\frac{\log n}{n}\right).\end{aligned} \tag{13.17}$$

The integral at the right hand side of (13.17) shows that for the asymptotic distribution of $\tilde{S}_n$, properties of the (well understood) Brownian motion process are needed.

Having established representation (13.17), we proceed by breaking up the interval $[0, a]$ into m_n intervals $I_{n,k}$ with (equal) length of order $n^{-1/3} \log n$ (big blocks), separated by intervals $J_{n,k}$ $(k = 2, 3, \dots, m_n)$ with length of order $2n^{-1/3}\sqrt{\log n}$ (small blocks). The idea is that the bigger intervals dominate the behavior of the whole integral, whereas the smaller intervals ensure that the integrals over the bigger intervals are approximately independent. The small interval $J_{n,1}$ to the left of $I_{n,1}$ has half the length of the other separating blocks as has the small interval J_{n,m_n+1} to the right of I_{n,m_n}. Hence,

$$[0, a] = J_{n,1} \cup I_{n,1} \cup J_{n,2} \cup I_{n,2} \cdots \cup J_{n,m_n} \cup I_{n,m_n} \cup J_{n,m_n+1}.$$

For $k = 2, 3, \dots, m_n$, let $\tilde{J}_{n,k}$ be the interval with the same right endpoint as $J_{n,k}$ with half the length of $J_{n,k}$ and take $\tilde{J}_{n,1} = J_{n,1}$. For $k = 1, 2, \dots, m_n - 1$ let $\bar{J}_{n,k+1}$ be the interval with the same left endpoint as $J_{n,k+1}$ with half the length of $J_{n,k+1}$ and $\bar{J}_{n,m_n+1} = J_{n,m_n+1}$. Then

$$[0, a] = \tilde{J}_{n,1} \cup I_{n,1} \cup \bar{J}_{n,2} \cup \tilde{J}_{n,2} \cup I_{n,2} \cdots \cup \tilde{J}_{n,m_n} \cup I_{n,m_n} \cup \bar{J}_{n,m_n+1}$$

where all I intervals have the same length of order $n^{-1/3} \log n$ and the J intervals have the same length of (smaller) order $n^{-1/3}\sqrt{\log n}$. Finally, let the interval $L_{n,k}$ be defined by

$$L_{n,k} = \tilde{J}_{n,k} \cup I_{n,k} \cup \bar{J}_{n,k+1} = [a_{nk}, a_{n,k+1}), \quad k = 1, 2, \dots, m_n, \ \text{yielding } [0, a) = \cup_{k=1}^{m_n} L_{n,k}. \tag{13.18}$$

Note that $m_n \sim an^{1/3}/\log n$ (see Exercise 13.4) and see Figure 13.2 for the structure of the partition.

Lemma 13.4 states that on intervals $I_{n,k}$ the global convex minorant of V_n (defined in (13.16)) over $[0, a]$ coincides with high probability with the restriction to I_{nk} of the local convex minorant of the process V_n on the interval $L_{n,k}$.

Lemma 13.4 *Let h_0 be strictly positive on $[0, a]$, with a strictly positive continuous derivative h_0' on $(0, a)$, which also has a strictly positive right limit at 0 and a strictly positive left limit at a. Then:*

(i) The probability that there exists a k, $1 \le k \le m_n$, such that the greatest convex minorant C_n of V_n is different on the interval I_{nk} from the restriction to I_{nk} of the (local) greatest convex minorant of V_n on L_{nk}, is bounded above by

$$c_1 \exp\left\{-c_2(\log n)^{3/2}\right\},$$

for constants $c_1, c_2 > 0$, uniformly in n.

(ii) The probability that there exists a k, $1 \le k \le m_n$, such that C_n has no change of slope in an interval $\bar{J}_{nk}$ or $\tilde{J}_{nk}$ is also bounded by

$$c_1 \exp\left\{-c_2(\log n)^{3/2}\right\},$$

for constants c_1, c_2, uniformly in n.

This is an important lemma, as it allows us to focus on the individual contributions to (13.17) of the intervals L_{nk}. For each $n \ge 1$ and $1 \le k \le m_n$ define independent standard Brownian motions $W_{n1}, \dots, W_{n,m_n}$ and consider the (local) processes

$$x \mapsto H_0(x) - H_0(a_{nk}) + n^{-1/2} W_{nk}\left(\frac{F_0(x)}{1 - F_0(x)} - \frac{F_0(a_{nk})}{1 - F_0(a_{nk})}\right), \; x \in L_{nk}.$$

Denote the greatest convex minorants of these processes (on L_{nk}) by C_{nk}. Furthermore, define the processes S_{nk} by

$$S_{nk}(x) = H_0(x) - H_0(a_{nk}) + n^{-1/2} W_{nk}\left(\frac{F_0(x)}{1 - F_0(x)} - \frac{F_0(a_{nk})}{1 - F_0(a_{nk})}\right) - C_{nk}(x), \; x \in L_{nk}. \tag{13.19}$$

A consequence of the “big blocks separated by small blocks” construction, with interval lengths as described, is that it can be shown that

$$\int_{L_{nk}} (S_{nk}(x) - ES_{nk}(x))\, dF_0(x) = (1 + o_p(1)) \int_{I_{nk}} (S_{nk}(x) - ES_{nk}(x))\, dF_0(x).$$

All this leads to the following approximation to $\tilde{S}_n$:

$$\tilde{S}_n = (1 + o_P(1)) \sum_{k=1}^{m_n} \int_{I_{nk}} (S_{nk}(x) - ES_{nk}(x))\, dF_0(x). \tag{13.20}$$

To derive the asymptotic distribution of this sum, we need to consider the asymptotic behavior of the individual terms. It will be seen soon that the following process (also encountered in Section 3.9) will play an important role in this behavior

$$x \mapsto V(x) = W(x) + x^2, \; x \in \mathbb{R} \tag{13.21}$$

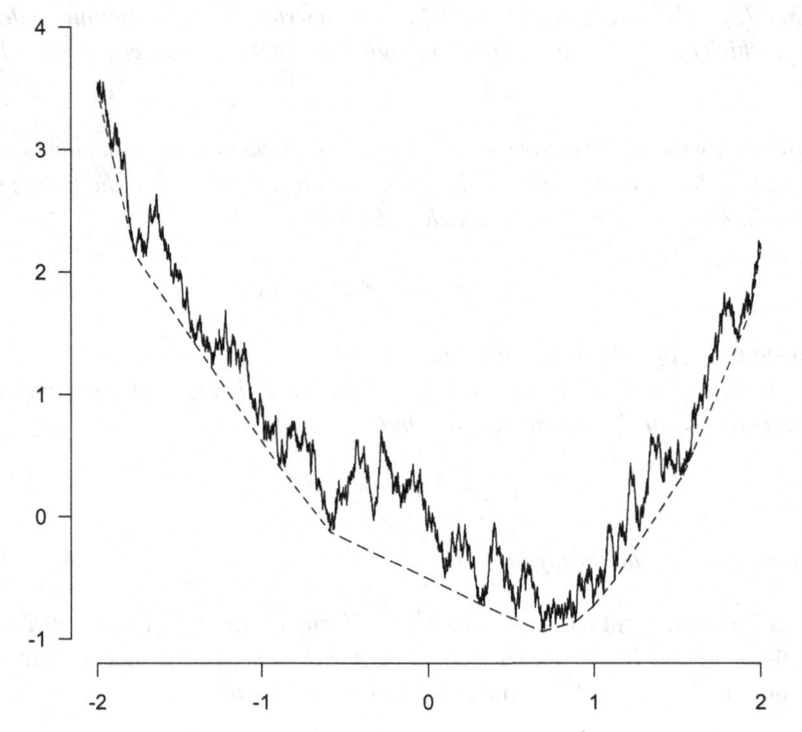

Figure 13.3 The greatest convex minorant of $W(x)+x^2$, restricted to $[-2,2]$.

with W standard two sided Brownian motion on $\mathbb{R}$ and C the greatest convex minorant of V on $\mathbb{R}$. In particular, for $c>0$, the integral Q_c with

$$Q_c = \int_0^c \{V(x) - C(x)\}\, dx \tag{13.22}$$

will be important. See Figure 13.3 for a picture of the process V and its greatest convex minorant, restricted to the interval $[-2,2]$. For Q_c, the following lemma can be proved.

Lemma 13.5

$$c^{-1/2}\{Q_c - cE|C(0)|\} \xrightarrow{\mathcal{D}} N(0,\sigma^2),\ c\to\infty,$$

where $C(0)$ is the value of the greatest convex minorant C of the process V at zero, and

$$\sigma^2 = 2\int_0^\infty \mathrm{covar}(-C(0), V(x) - C(x))\, dx.$$

All moments of $c^{-1/2}\{Q_c - cE|C(0)|\}$ exist and (in particular) the fourth moment is uniformly bounded in c and converges to the fourth moment of the normal $N(0,\sigma^2)$ distribution, as $c\to\infty$.

We now connect the individual terms in (13.20) to the quantity Q_c. Let $c_n = n^{1/3}|L_{nk}| \sim \log n$ and $I_{nk} = [a_{nk} + n^{-1/3}\sqrt{c_n}, a_{nk} + n^{-1/3}(c_n - \sqrt{c_n})]$. We then have

$$\begin{aligned}
&n \int_{I_{nk}} S_{nk}(x)\, dF_0(x) \\
&= \int_{\sqrt{c_n}}^{c_n - \sqrt{c_n}} \Bigg\{ n^{2/3} \left\{ H_0(a_{nk} + n^{-1/3}x) - H_0(a_{nk}) - n^{-1/3} x h_0(a_{nk}) \right\} \\
&\quad + n^{1/6} W_{nk} \left(\frac{F_0(a_{nk} + n^{-1/3}x)}{1 - F_0(a_{nk} + n^{-1/3}x)} - \frac{F_0(a_{nk})}{1 - F_0(a_{nk})} \right) - n^{2/3} C_{nk}(a_{nk} + n^{-1/3}x) \Bigg\} \\
&\quad \times f_0\left(a_{nk} + n^{-1/3}x\right)\, dx,
\end{aligned}$$

where C'_{nk} is the greatest convex minorant of the process

$$\begin{aligned}
x \mapsto\ & n^{2/3} \left\{ H_0(a_{nk} + n^{-1/3}x) - H_0(a_{nk}) - n^{-1/3} x h_0(a_{nk}) \right\} \\
&+ n^{1/6} W \left(\frac{F_0(a_{nk} + n^{-1/3}x)}{1 - F_0(a_{nk} + n^{-1/3}x)} - \frac{F_0(a_{nk})}{1 - F_0(a_{nk})} \right) \\
&\approx \frac{1}{2} h'_0(a_{nk}) x^2 + W \left(\frac{f_0(a_{nk})}{(1 - F_0(a_{nk}))^2} x \right)
\end{aligned} \tag{13.23}$$

on $[0, c_n]$. See Exercise 13.5 for the latter approximation. We also use that adding a linear function to a function does not change the difference between this function and its greatest convex minorant. Process (13.23) is two sided Brownian motion with a quadratic drift, but not yet precisely like (13.21). Using Brownian scaling (see Exercise 13.6) it can be seen that this process on $[0, c_n]$ has the same distribution as the process

$$x \mapsto k_2 \left((k_1 x)^2 + W(k_1 x) \right) \tag{13.24}$$

on the interval $[0, c_n]$, where

$$k_1 = \frac{(\frac{1}{2} h'_0(a_{nk})(1 - F_0(a_{nk})))^{2/3}}{f_0(a_{nk})^{1/3}} \text{ and } k_2 = \frac{f_0(a_{nk})^{2/3}}{(1 - F_0(a_{nk}))^{4/3}(\frac{1}{2} h'_0(a_{nk}))^{1/3}}. \tag{13.25}$$

Therefore, writing C''_{nk} for the greatest convex minorant of the process (13.24) on $[0, c_n]$, and noting that this convex minorant can be expressed as $k_2 C'''_{nk}(k_1 x)$ where C'''_{nk} is the convex minorant of the process defined in (13.21) on $[0, k_1 c_n]$, we can write

$$\begin{aligned}
&f_0(a_{nk}) \int_{\sqrt{c_n}}^{c_n - \sqrt{c_n}} \left(\frac{1}{2} h'_0(a_{nk}) x^2 + W \left(\frac{f_0(a_{nk})}{(1 - F_0(a_{nk}))^2} x \right) - C'_{nk}(x) \right) dx \\
&\approx^{\mathcal{D}} f_0(a_{nk}) \int_{\sqrt{c_n}}^{c_n - \sqrt{c_n}} \left(k_2 (k_1 x)^2 + k_2 W(k_1 x) - C''_{nk}(x) \right) dx \\
&= k_2 f_0(a_{nk}) \int_{\sqrt{c_n}}^{c_n - \sqrt{c_n}} \left((k_1 x)^2 + W(k_1 x) - C'''_{nk}(k_1 x) \right) dx \\
&= \frac{k_2 f_0(a_{nk})}{k_1} \int_{k_1 \sqrt{c_n}}^{k_1 (c_n - \sqrt{c_n})} \left(x^2 + W(x) - C'''_{nk}(x) \right) dx.
\end{aligned}$$

"Chopping off" the boundary parts in the integral was needed to decouple the global integral independent local integrals. This modification of the integral can be shown to be small enough to guarantee that the latter integral behaves like the integral over the whole interval,

$$\frac{k_2 f_0(a_{nk})}{k_1}\int_0^{k_1 c_n}\left(x^2+W(x)-C(x)\right)\,dx=\frac{k_2 f_0(a_{nk})}{k_1}Q_{k_1 c_n},$$

where C is the convex minorant of $x\mapsto x^2+W(x)$ taken over $\mathbb{R}$, rather than $[0,k_1c_n]$, and Q_c is as defined in (13.22). This means that

$$n\int_{I_{nk}} S_{nk}(x)\,dF_0(x)\sim\frac{k_2 f_0(a_{nk})}{k_1}Q_{k_1c_n}.$$

Therefore, using Lemma 13.5, it can be seen that

$$\mathrm{Var}\left(\frac{n}{\sqrt{c_n}}\int_{I_{nk}} S_{nk}(x)\,dF_0(x)\right)\sim\sigma_{nk}^2,\ n\to\infty,$$

uniformly in $k=1,\ldots,m_n$, where

$$\begin{aligned}\sigma_{nk}^2&=\frac{k_2^2 f_0(a_{nk})^2}{k_1^2}k_1\sigma^2=\frac{k_2^2 f_0(a_{nk})^2}{k_1}\sigma^2\\&=\frac{f_0(a_{nk})^{11/3}}{(1-F_0(a_{nk}))^{10/3}(\frac{1}{2}h_0'(a_{nk}))^{4/3}}\sigma^2=\frac{2^{4/3}h_0(a_{nk})^3\left\{h_0(a_{nk})f_0(a_{nk})\right\}^{1/3}}{h_0'(a_{nk})^{4/3}}\sigma^2,\qquad(13.26)\end{aligned}$$

and σ^2 is defined as in Theorem 13.5. Likewise, also with $C(0)$ as defined in Theorem 13.5,

$$\begin{aligned}\frac{n}{c_n}\int_{I_{nk}} ES_{nk}(x)\,dF_0(x)&\sim\frac{k_2 f_0(a_{nk})}{k_1}k_1E|C(0)|\\&=\frac{2^{1/3}h_0(a_{nk})\left\{h_0(a_{nk})f_0(a_{nk})\right\}^{1/3}}{h_0'(a_{nk})^{1/3}}E|C(0)|;\qquad(13.27)\end{aligned}$$

see Exercise 13.7.

Finally, all contributions to the full integral over $[0,a]$ are added. Since the fourth moments of

$$\frac{n}{\sqrt{c_n}}\int_{I_{nk}}\left\{S_{nk}(x)-ES_{nk}(x)\right\}\,dF_0(x)$$

can be shown to be uniformly bounded, Chebyshev's inequality yields that for each $\epsilon>0$,

$$\sum_{k=1}^{m_n}\mathbb{P}\left\{m_n^{-1/2}\left|\frac{n}{\sqrt{c_n}}\int_{I_{nk}}\left\{S_{nk}(x)-ES_{nk}(x)\right\}\,dF_0(x)\right|\ge\epsilon\right\}\to 0,\ n\to\infty.$$

Using that $m_n^{-1}\sim a^{-1}n^{-1/3}\log n$ and that the intervals I_{nk} have lengths of order $n^{-1/3}\log n$, we get:

$$\begin{aligned}m_n^{-1}\sum_{k=1}^{m_n}\sigma_{nk}^2&\sim m_n^{-1}\sum_{k=1}^{m_n}\frac{2^{4/3}h_0(a_{nk})^3\left\{h_0(a_{nk})f_0(a_{nk})\right\}^{1/3}}{h_0'(a_{nk})^{4/3}}\sigma^2\\&\longrightarrow\frac{2^{4/3}\sigma^2}{a}\int_0^a\frac{h_0(t)^3\{h_0(t)f_0(t)\}^{1/3}}{h_0'(t)^{4/3}}\,dt.\end{aligned}$$

Using that $m_n = an^{1/3}/c_n$, the normal convergence criterion on p. 316 of Loève, 1963, gives:

$$n^{5/6}\sum_{k=1}^{m_n}\int_{I_{nk}}\{S_{nk}(x)-ES_{nk}(x)\}\,dF_0(x)$$

$$= m_n^{-1/2}\sum_{k=1}^{m_n}\frac{n\sqrt{a}}{\sqrt{c_n}}\int_{I_{nk}}\{S_{nk}(x)-ES_{nk}(x)\}\,dF_0(x)\xrightarrow{\mathcal{D}} N\left(0,\sigma_{H_0}^2\right),$$

where

$$\sigma_{H_0}^2 = 2^{4/3}\sigma^2\int_0^a \frac{h_0(t)^2\{h_0(t)f_0(t)\}^{1/3}}{h_0'(t)^{4/3}}\,dH_0(t). \tag{13.28}$$

Also note that

$$m_n^{-1/2}\sum_{k=1}^{m_n}\frac{n}{c_n^{1/2}}\int_{I_{nk}}ES_{nk}(x)\,dF_0(x)\sim m_n^{-1/2}c_n^{1/2}\sum_{k=1}^{m_n}\frac{2^{1/3}h_0(a_{nk})\,\{h_0(a_{nk})f_0(a_{nk})\}^{1/3}}{h_0'(a_{nk})^{1/3}}E|C(0)|$$

$$\sim\sqrt{m_nc_n}\,E|C(0)|\int_0^a\frac{2^{1/3}h_0(t)\,\{h_0(t)f_0(t)\}^{1/3}}{h_0'(t)^{1/3}}\,dt$$

$$= n^{1/6}E|C(0)|\int_0^a\left(\frac{2h_0(t)f_0(t)}{h_0'(t)}\right)^{1/3}dH_0(t).$$

This reasoning leads to the following lemma.

Lemma 13.6 *Let h_0 be strictly positive on $[0,a]$, with a strictly positive continuous derivative h_0' on $(0,a)$, which also has a strictly positive right limit at 0 and a strictly positive left limit at a. Moreover, let $C(0)$ be defined as in Lemma 13.5 and $\sigma_{H_0}^2$ as in (13.28). Then:*

$$n^{5/6}\sum_{k=1}^{m_n}\int_{I_{nk}}\{S_{nk}(x)-ES_{nk}(x)\}\,dF_0(x)\xrightarrow{\mathcal{D}} N(0,\sigma_{H_0}^2),\ n\to\infty,$$

where

$$n^{2/3}\sum_{k=1}^{m_n}\int_{I_{nk}}ES_{nk}(x)\,dF_0(x)\to E|C(0)|\int_0^a\left(\frac{2h_0(t)f_0(t)}{h_0'(t)}\right)^{1/3}dH_0(t)\,,\ n\to\infty, \tag{13.29}$$

Finally, this leads to the following asymptotic result for test statistic S_n.

Theorem 13.3 *Let h_0 satisfy the conditions of Lemma 13.6. Moreover, let V_n be as defined in (13.16) and C_n be its convex minorant. Define*

$$\mu_n = E\int_0^a (V_n(x)-C_n(x))\,dF_0(x)\sim n^{-2/3}E|C(0)|\int_0^a\left(\frac{2h_0(t)f_0(t)}{h_0'(t)}\right)^{1/3}dH_0(t).$$

Then, for S_n defined in (13.12),

$$n^{5/6}\{S_n-\mu_n\}\xrightarrow{\mathcal{D}} N(0,\sigma_{H_0}^2),\ n\to\infty,$$

where $\sigma_{H_0}^2$ is as defined in (13.28).

13.3 Two-Sample Tests for Current Status Data

For the current status model introduced in Section 2.3, a two sample test is introduced in Section 9.3. It is based on the likelihood ratio statistic V_N, defined by (9.48), using MSLEs. In this section the asymptotic behavior of this global deviation measure is studied. Full proofs of Theorems 9.5 through 9.8 are given in the online supplement of Groeneboom, 2012b. Here we will give a sketch of the main line of the argument.

By Lemma 8.5 (also Corollary 3.4 in Groeneboom et al., 2010), using the two sample problem notation, we have, with probability tending to one,

$$\tilde F_{N1}(t) = \frac{\tilde h_{N1}(t)}{\tilde g_{N1}(t)},\ t \in [a,b]. \tag{13.30}$$

Here $\tilde h_{N1}$ and $\tilde g_{N1}$ are the kernel estimators defined in (9.52). Hence, the MSLE is just equal to the ratio of two kernel estimators for $t \in [a,b]$, with probability tending to one. Similarly, with probability tending to one,

$$\tilde F_{N2}(t) = \frac{\tilde h_{N2}(t)}{\tilde g_{N2}(t)},\ t \in [a,b], \tag{13.31}$$

again using the definitions given in (9.52) and

$$\tilde F_N(t) = \frac{\alpha_N \tilde h_{N1}(t) + \beta_N \tilde h_{N2}(t)}{\alpha_N \tilde g_{N1}(t) + \beta_N \tilde g_{N2}(t)},\ t \in [a,b], \qquad \alpha_n = m/N, \qquad \beta_N = 1 - \alpha_N. \tag{13.32}$$

Hence we assume in the following that $\tilde F_N$, $\tilde F_{N1}$ and $\tilde F_{N2}$ have the representations (13.30), (13.31) and (13.32), respectively.

Next, using Giné and Guillou, 2002, which in turn relies on Talagrand, 1994, and Talagrand, 1996, if the bandwidth b_N satisfies $b_N \asymp N^{-\alpha}$, it is proved that, under the conditions of any Theorems 9.5 to 9.7:

$$\sup_{t\in[a,b]} \left|\tilde g_{Nj}(t) - g_j(t)\right| = O_p\left(N^{-(1-\alpha)/2}\sqrt{\log N}\right) \tag{13.33}$$

and

$$\sup_{t\in[a,b]} \left|\tilde h_{Nj}(t) - F(t)g_j(t)\right| = O_p\left(N^{-(1-\alpha)/2}\sqrt{\log N}\right),\ j = 1,2, \tag{13.34}$$

implying that also:

$$\sup_{t\in[a,b]} \left|\tilde F_{Nj}(t) - F(t)\right| = O_p\left(n^{-(1-\alpha)/2}\sqrt{\log N}\right),\ j = 1,2.$$

This leads to the representation of the following lemma.

Lemma 13.7 *Let any of the conditions of Theorems 9.5 to 9.7 be satisfied and let the bandwidth satisfy $b_N \sim N^{\alpha}$, where b_N (and hence α) satisfies the conditions of the corresponding Theorem 9.5, 9.6 or 9.7. Then*

$$V_n = \alpha_N \beta_N \int_{t\in[a,b]} \frac{\{\tilde g_{N2}(t)\tilde h_{N1}(t) - \tilde g_{N1}(t)\tilde h_{N2}(t)\}^2}{F(t)\{1-F(t)\}\tilde g_N(t) g_1(t) g_2(t)}\,dt + O_p\left(N^{-3(1-\alpha)/2}(\log N)^{3/2}\right), \tag{13.35}$$

where

$$\bar{g}_N = \alpha_N g_1 + \beta_N g_2.$$

Moreover,

$$\alpha_N\beta_N \int_{t\in[a,b]} \frac{\{\tilde{g}_{N2}(t)\tilde{h}_{N1}(t) - \tilde{g}_{N1}(t)\tilde{h}_{N2}(t)\}^2}{F(t)\{1-F(t)\}\bar{g}_N(t)g_1(t)g_2(t)}\,dt - \frac{b-a}{Nb_N}\int K(u)^2\,du$$
$$= A_N + B_N - C_N + D_N + o_p\left(\frac{1}{N\sqrt{b_N}}\right), \tag{13.36}$$

where

$$A_N = \frac{2\alpha_N\beta_N}{m^2}\sum_{1<i<j\le m}\{\Delta_i - F(T_i)\}\{\Delta_j - F(T_j)\}\int_{t=a}^{b}\frac{g_2(t)K_{b_N}(t-T_i)K_{b_N}(t-T_j)}{g_1(t)\bar{g}_N(t)F(t)\{1-F(t)\}}\,dt,$$

$$B_N = \frac{2\alpha_N\beta_N}{n^2}\sum_{m<i<j\le N}\{\Delta_i - F(T_i)\}\{\Delta_j - F(T_j)\}\int_{t=a}^{b}\frac{g_1(t)K_{b_N}(t-T_i)K_{b_N}(t-T_j)}{g_2(t)\bar{g}_N(t)F(t)\{1-F(t)\}}\,dt,$$

$$C_N = \frac{2\alpha_N\beta_N}{mn}\sum_{i=1}^{m}\sum_{j=m+1}^{N}\{\Delta_i - F(T_i)\}\{\Delta_j - F(T_j)\}\int_{t=a}^{b}\frac{K_{b_N}(t-T_i)K_{b_N}(t-T_j)}{\bar{g}_N(t)F(t)\{1-F(t)\}}\,dt$$

and

$$D_N = \alpha_N\beta_N\int_{t=a}^{b}\frac{f(t)^2\{g_1'(t)g_2(t) - g_2'(t)g_1(t)\}^2}{F(t)\{1-F(t)\}\bar{g}_N(t)g_1(t)g_2(t)}\,dt\left\{\int u^2K(u)\,du\right\}^2 b_N^4.$$

Note that $D_N = 0$ *if* $g_1 = g_2$.

Lemma 13.7 gives an (asymptotic) representation of the statistic V_N in terms of the U-statistics with degenerate kernel A_N, B_N and C_N, together with an extra bias term D_N. Note that (13.35) of Lemma 13.7 gives that V_N has the asymptotic representation as a weighted integral of

$$\{\tilde{g}_{N2}(t)\tilde{h}_{N1}(t) - \tilde{g}_{N1}(t)\tilde{h}_{N2}(t)\}^2$$

which is an estimate of

$$\{g_1(t)g_2(t)\}^2\{F_1(t) - F_2(t)\}^2.$$

Under the null hypothesis $F_1 = F_2 = F$ we get, if $b_N \ll N^{-1/5}$,

$$\alpha_N\beta_N\int_{t\in[a,b]}\frac{Nb_N E\left(\{\tilde{g}_{N2}(t)\tilde{h}_{N1}(t) - \tilde{g}_{N1}(t)\tilde{h}_{N2}(t)\}^2 \mid T_1,\dots,T_N\right)}{F(t)\{1-F(t)\}\bar{g}_N(t)g_1(t)g_2(t)}\,dt \xrightarrow{p} (b-a)\int K(u)^2\,du.$$

If $b_N \asymp N^{-1/5}$ and $g_1 \ne g_2$, also the second bias term D_N comes into play.

This type of decomposition is somewhat typical for the representation of global deviation measures in density estimation. A_N, B_N and C_N are U-statistics with degenerate kernels, which are, for example, analyzed in Hall, 1984. The sum of the U-statistics A_N, B_N and $-C_N$ can be treated by using a central limit theorem for martingale difference arrays. This is slightly tricky, due to the conditioning on $T_1, \dots, T_N$; the unconditional version of the result is easier, but leads to a bootstrap procedure where one has to resample both the T_i and Δ_i instead of only the Δ_i, as is done in the conditional approach.

In Groeneboom, 2012b, the σ-algebras $\mathcal{F}_j$, $j = 0, \dots, N$, are introduced, where $\mathcal{F}_0$ is the trivial σ-algebra, and $\mathcal{F}_j$ is the σ-algebra, generated by $Y_1, \dots, Y_j$, where Y_j is of the form

$$Y_j = \sum_{1 \le i < j} c_{m,n}(T_i, T_j)\{\Delta_i - F(T_i)\}\{\Delta_j - F(T_j)\}.$$

As an example,

$$A_N = \sum_{j=2}^{m} Y_j,$$

where

$$Y_j = \sum_{1 \le i < j} c_{m,n}(T_i, T_j)\{\Delta_i - F(T_i)\}\{\Delta_j - F(T_j)\},$$

and

$$c_{m,n}(T_i, T_j) = \frac{2\alpha_N \beta_N}{m^2} \int_{t=a}^{b} \frac{g_2(t) K_{b_N}(t - T_i) K_{b_N}(t - T_j)}{g_1(t)\bar{g}_N(t) F(t)\{1 - F(t)\}}\, dt.$$

The sequence $Y_1, \dots, Y_m$ is a martingale, since

$$E\{Y_j \mid \mathcal{F}_{j-1}\} = 0, \; j = 1, \dots, m.$$

Furthermore we have, in probability,

$$E\left\{Y_j^2 \mid \mathcal{F}_{j-1}\right\} = F(T_j)\{1 - F(T_j)\} \left\{ \sum_{1 \le i < j} c_{m,n}(T_i, T_j)\{\Delta_i - F(T_i)\} \right\}^2$$

$$\sim \frac{4(j-1)\alpha_N^2 \beta_N^2}{m^4} \frac{g_2(T_j)^2 1_{[a-b_N, b+b_N]}(T_j)}{g_1(T_j)\bar{g}_N(T_j)^2 b_N} \int \left\{ \int K(v)K(v+x)\, dv \right\}^2 dx,$$

which gives for A_N, in probability,

$$\sum_{j=2}^{m} E\left\{Y_j^2 \mid \mathcal{F}_{j-1}\right\} \sim \sum_{j=1}^{m} \frac{4(j-1)\beta_N^2}{m^2 N^2 b_N} \frac{g_2(T_j)^2 1_{[a-b_N, b+b_N]}(T_j)}{\bar{g}_N(T_j)^2 g_1(T_j)} \int \left\{ \int K(v)K(v+x)\, dv \right\}^2 dx$$

$$\sim \frac{2m(m-1)\beta_N^2}{m^2 N^2 b_N} \int_a^b \frac{g_2(t)^2}{\bar{g}_N(t)^2}\, dt \int \left\{ \int K(v)K(v+x)\, dv \right\}^2 dx$$

$$\sim \frac{2\beta_N^2}{N^2 b_N} \int_a^b \frac{g_2(t)^2}{\bar{g}_N(t)^2}\, dt \int \left\{ \int K(v)K(v+x)\, dv \right\}^2 dx, \; m \to \infty,$$

see Groeneboom, 2012b. We next can apply the martingale convergence theorem on p. 171 for martingale difference arrays in Pollard, 1984, to prove that A_N is asymptotically normal.

The terms B_N and C_N can be treated in a similar way. In this way we obtain the asymptotic normality of the test statistic V_N, conditioning on $T_1, \dots, T_N$, where the asymptotic normality holds in probability, that is: for each x the probability that the rescaled V_N is smaller than x tends to $\Phi(x/\sigma_K)$, the standard normal distribution function, evaluated at x/σ_K.

Proving that bootstrapping from $\tilde{F}_{N,\tilde{b}_N}$ works is a bit subtle and not at all trivial. First of all, if we take bandwidth $\tilde{b}_N \sim N^{-1/5}$, the second derivative is not uniformly bounded, and hence $\tilde{F}_{N,\tilde{b}_N}$ does not satisfy the same differentiability conditions as the underlying distributions in Theorems 9.5 to 9.7. Second, we have to show that this way of bootstrapping does not change the bias, and that, for example, exactly the same bias as in Theorem 9.6 is reproduced. That this indeed happens is due to the fact that we keep the observation times T_i of the original sample fixed in the bootstrap procedure. We give here the main line of the proof of Theorem 9.8.

We may assume that, for large N, $\tilde{F}_{N,\tilde{b}_N}$ has the representation

$$\tilde{F}_{N,\tilde{b}_N}(t) = \frac{\int \delta K_{\tilde{b}_N}(t-u)\,d\mathbb{P}_N(u,\delta)}{\int K_{\tilde{b}_N}(t-u)\,d\mathbb{G}_N(u)}$$

for $t \in [a,b]$, where $\tilde{b}_N \asymp N^{-1/5}$. This gives

$$\tilde{f}_{N,\tilde{b}_N}(t) = \frac{\int \delta K'_{\tilde{b}_N}(t-u)\,d\mathbb{P}_N(u,\delta)}{\tilde{g}_{N,\tilde{b}_N}(t)} - \frac{\tilde{g}'_{N,\tilde{b}_N}(t)\int \delta K_{\tilde{b}_N}(t-u)\,d\mathbb{P}_N(u,\delta)}{\tilde{g}_{N,\tilde{b}_N}(t)^2},$$

where

$$\tilde{g}_{N,\tilde{b}_N}(t) = \int K_{\tilde{b}_N}(t-u)\,d\mathbb{G}_N(u), \qquad \tilde{g}'_{N,\tilde{b}_N}(t) = \int K'_{\tilde{b}_N}(t-u)\,d\mathbb{G}_N(u),$$

and $K'_{\tilde{b}_N}$ is defined by:

$$K'_{\tilde{b}_N}(t-u) = \frac{1}{\tilde{b}_N^2}K'\left(\frac{t-u}{\tilde{b}_N}\right).$$

By the assumptions on g, and using $\tilde{b}_N \asymp n^{-1/5}$, we have

$$\sup_{t\in[a,b]}\left|\tilde{g}_{N,\tilde{b}_N}(t) - \bar{g}_N(t)\right| = O_p\left(N^{-2/5}\sqrt{\log n}\right) \text{ and } \sup_{t\in[a,b]}\left|\tilde{g}'_{N,\tilde{b}_N}(t) - g'(t)\right|$$
$$= O_p\left(N^{-1/5}\sqrt{\log n}\right)$$

uniformly for $t \in [a,b]$ (note that the O_p-term is an O_p-term *conditionally* on $(T_1,\Delta_1),\ldots,(T_n,\Delta_n)$). Furthermore, since

$$\int \delta K'_{\tilde{b}_N}(t-u)\,d\mathbb{P}_N(u,\delta) = \frac{1}{N\tilde{b}_N^2}\sum_{i=1}^N K'\left(\frac{t-T_i}{\tilde{b}_N}\right)\Delta_i$$

we get:

$$\int \delta K'_{\tilde{b}_N}(t-u)\,d\mathbb{P}_N(u,\delta) - \int K'_{\tilde{b}_N}(t-u)F(u)\,dG(u)$$
$$= \int \{\delta - F(u)\}\,K'_{\tilde{b}_N}(t-u)\,d\mathbb{P}_N(u,\delta) + \int F(u)K'_{\tilde{b}_N}(t-u)\,d\left(\mathbb{G}_N - G\right)(u)$$

and hence

$$\sup_{t\in[a,b]}\left|\tilde f_{N,\tilde b_N}(t)-f(t)\right| = O_p\left(N^{-1/5}\sqrt{\log N}\right). \tag{13.37}$$

It can be proved in a similar way that

$$\sup_{x\in[a,b]}\left|\tilde F_{N,\tilde b_N}(t)-F(t)\right| = O_p\left(N^{-2/5}\sqrt{\log N}\right).$$

The bootstrap test statistic V_N^* now has the representation

$$V_N = \frac{2m}{N}\int_{t\in[a,b]}\left\{\tilde h_{N1}^*(t)\log\frac{\tilde F_{N1}^*(t)}{\tilde F_N^*(t)} + \{\tilde g_{N1}(t)-\tilde h_{N1}^*(t)\}\log\frac{1-\tilde F_{N1}^*(t)}{1-\tilde F_N^*(t)}\right\}dt$$
$$+\frac{2n}{N}\int_{t\in[a,b]}\left\{\tilde h_{N2}^*(t)\log\frac{\tilde F_{N2}^*(t)}{\tilde F_N^*(t)} + \{\tilde g_{N2}(t)-\tilde h_{N2}^*(t)\}\log\frac{1-\tilde F_{N2}^*(T_i)}{1-\tilde F_N^*(T_i)}\right\}dt,$$

where

$$\tilde h_{Nj}^*(t) = \int \delta^* K_{b_N}(t-u)\,d\mathbb{P}_{Nj}(u,\delta^*),\ j=1,2,$$

and the Δ_i^* are defined by

$$\Delta_i^* = 1_{\left[0,\tilde F_{N,\tilde b_N}(T_i)\right]}(U_i^*),$$

for independent random variables $U_1^*,\dots,U_N^*$, independent of the random variables (T_i,Δ_i), $i=1,\dots,N$, and where we may assume, as before, that

$$\tilde F_{Nj}^*(t) = \frac{\int \delta^* K_{b_N}(t-u)\,d\mathbb{P}_{Nj}(u,\delta^*)}{\tilde g_{Nj}(t)},\ j=1,2.$$

Note that the only extra randomness is introduced by the uniform random variables U_i^*, and that the bandwidth b_N, used here, may be smaller than the bandwidth $\tilde b_N$, used in the computation of $\tilde F_{N,\tilde b_N}$. In fact, b_N is the bandwidth that is used in the original sample and we have, by assumption,

$$b_N \asymp N^{-\alpha},$$

where $1/3<\alpha<1/5$, and where we allow $\alpha=1/5$ if it is assumed that $g_1=g_2$. The densities $\tilde g_{Nj}$ have been computed in the original sample, using this possibly smaller bandwidth b_N.

We now get:

$$\alpha_N\beta_N\int_{t\in[a,b]}\frac{\left\{\tilde g_{N2}(t)\tilde h_{N1}^*(t)-\tilde g_{N1}(t)\tilde h_{N2}^*(t)\right\}^2}{\tilde F_{N,\tilde b_N}(t)\{1-\tilde F_{N,\tilde b_N}(t)\}\tilde g_N(t)g_1(t)g_2(t)}\,dt - \frac{b-a}{Nb_N}\int K(u)^2\,du$$
$$= A_N^*+B_N^*-C_N^*+D_N+o_p\left(\frac{1}{N\sqrt{b_N}}\right), \tag{13.38}$$

where

$$A_N^* = \frac{2\alpha_N\beta_N}{m^2}\sum_{1<i<j\le m}\{\Delta_i^* - \tilde F_{N,\tilde b_N}(T_i)\}\{\Delta_j^* - \tilde F_{N,\tilde b_N}(T_j)\}$$

$$\cdot\int_{t=a}^{b}\frac{g_2(t)K_{b_N}(t-T_i)K_{b_N}(t-T_j)}{g_1(t)\bar g_N(t)\tilde F_{N,\tilde b_N}(t)\{1-\tilde F_{N,\tilde b_N}(t)\}}\,dt,$$

$$B_N^* = \frac{2\alpha_N\beta_N}{n^2}\sum_{m<i<j\le N}\{\Delta_i^* - \tilde F_{N,\tilde b_N}(T_i)\}\{\Delta_j^* - F(T_j)\}$$

$$\cdot\int_{t=a}^{b}\frac{g_1(t)K_{b_N}(t-T_i)K_{b_N}(t-T_j)}{g_2(t)\bar g_N(t)\tilde F_{N,\tilde b_N}(t)\{1-\tilde F_{N,\tilde b_N}(t)\}}\,dt,$$

$$C_N^* = \frac{2\alpha_N\beta_N}{mn}\sum_{i=1}^{m}\sum_{j=m+1}^{N}\{\Delta_i^* - \tilde F_{N,\tilde b_N}(T_i)\}\{\Delta_j^* - \tilde F_{N,\tilde b_N}(T_j)\}$$

$$\cdot\int_{t=a}^{b}\frac{K_{b_N}(t-T_i)K_{b_N}(t-T_j)}{\bar g_N(t)\tilde F_{N,\tilde b_N}(t)\{1-\tilde F_{N,\tilde b_N}(t)\}}\,dt,$$

and the bias term D_N is given by:

$$D_N = \alpha_N\beta_N\int_{t=a}^{b}\frac{f(t)^2\{g_1'(t)g_2(t)-g_2'(t)g_1(t)\}^2}{F(t)\{1-F(t)\}\bar g_N(t)g_1(t)g_2(t)}\,dt\left\{\int u^2K(u)\,du\right\}^2 b_N^4.$$

Note (again) that $D_N = 0$ if $g_1 = g_2$ and that the bias term D_N is equal to the bias term of the original statistic.

However, the distribution function $\tilde F_{N,\tilde b_N}$ does not satisfy the condition that the second derivative is uniformly bounded on an interval (a', b'), containing $[a, b]$, which is a condition on F in Theorems 9.5 to 9.7. But this condition was only needed to take care of bias terms of the form

$$n^{-1}\sum_{i=m+1}^{N}K_{b_N}(t-T_i)\,m^{-1}\sum_{i=1}^{m}\{F(T_i)-F(t)\}K_{b_N}(t-T_i)$$

$$-\,m^{-1}\sum_{i=1}^{m}K_{b_N}(t-T_i)\,n^{-1}\sum_{i=m+1}^{n}\{F(T_i)-F(t)\}K_{b_N}(t-T_i), \tag{13.39}$$

in the term

$$\tilde g_{N2}(t)\tilde h_{N1}(t) - \tilde g_{N1}(t)\tilde h_{N2}(t),$$

which appears in the numerator of the integrand in the first term of the representation of V_N:

$$V_n = \alpha_N\beta_N\int_{t\in[a,b]}\frac{\{\tilde g_{N2}(t)\tilde h_{N1}(t)-\tilde g_{N1}(t)\tilde h_{N2}(t)\}^2}{F(t)\{1-F(t)\}\bar g_N(t)g_1(t)g_2(t)}\,dt + O_p\left(N^{-3(1-\alpha)/2}(\log N)^{3/2}\right), \tag{13.40}$$

see (13.35).

But (13.39) transforms in the bootstrap representation into

$$n^{-1}\sum_{i=m+1}^{N} K_{b_N}(t-T_i)\,m^{-1}\sum_{i=1}^{m}\{\tilde F_{N,\tilde b_N}(T_i)-\tilde F_{N,\tilde b_N}(t)\}K_{b_N}(t-T_i)$$
$$-\,m^{-1}\sum_{i=1}^{m} K_{b_N}(t-T_i)\,n^{-1}\sum_{i=m+1}^{n}\{\tilde F_{N,\tilde b_N}(T_i)-\tilde F_{N,\tilde b_N}(t)\}K_{b_N}(t-T_i).$$

Since $\tilde F_{N,\tilde b_N}(t)$ is represented by

$$\frac{\int \delta K_{\tilde b_N}(t-u)\,d\mathbb{P}_N(u,\delta)}{\int K_{\tilde b_N}(t-u)\,d\mathbb{G}_N(u)},$$

and since

$$\int \delta K''_{\tilde b_N}(t-u)\,d\mathbb{P}_N(u,\delta)=\frac{1}{N\tilde b_N^3}\sum_{i=1}^{N}K''\left(\frac{t-T_i}{\tilde b_N}\right)\Delta_i=O_p\left(\sqrt{\log N}\right),$$

uniformly in $t\in[a,b]$, we get:

$$n^{-1}\sum_{i=m+1}^{N} K_{b_N}(t-T_i)\,m^{-1}\sum_{i=1}^{m}\{\tilde F_{N,\tilde b_N}(T_i)-\tilde F_{N,\tilde b_N}(t)\}K_{b_N}(t-T_i)$$
$$-\,m^{-1}\sum_{i=1}^{m} K_{b_N}(t-T_i)\,n^{-1}\sum_{i=m+1}^{n}\{\tilde F_{N,\tilde b_N}(T_i)-\tilde F_{N,\tilde b_N}(t)\}K_{b_N}(t-T_i)$$
$$=\tilde b_N^2 f(t)\left\{g_1(t)g_2'(t)-g_1'(t)g_2(t)\right\}\int u^2K(u)\,du+O_p\left(\sqrt{\frac{\tilde b_N\log N}{n}}\right)+O\left(b_N^4\sqrt{\log N}\right).$$

So the only thing that has changed with respect to the representation of the original statistic is that the remainder term $O\left(b_N^4\right)$, coming from the bias, has been changed to $O\left(b_N^4\sqrt{\log N}\right)$, and the dominating term for the bias is again

$$\tilde b_N^2 f(t)\left\{g_1(t)g_2'(t)-g_1'(t)g_2(t)\right\}\int u^2K(u)\,du,$$

which appears squared in the numerator of the integrand in the first term of the representation (13.40) of V_N. Theorem 9.8 now follows.

Exercises

13.1 Consider the function h defined in Lemma 13.3 and shown in Figure 13.1,

$$h(y)=1-y+y\log y.$$

a) Show that h can be represented as

$$h(y)=\int_1^y \log u\,du.$$

b) Show that h is nonnegative and convex on $(0,\infty)$ and attains its minimum value at $y=1$.

13.2 Observe that the integrand in (13.13) can be rewritten as

$$H_0(x) - \log\left(1 - \frac{E_n(x-)}{\sqrt{n}(1 - F_0(x))}\right) - \hat{H}_n(x)$$

where $E_n = \sqrt{n}(\mathbb{F}_n - F_0)$ is the empirical process. Verify that (13.14) indeed approximates this integrand on $[0, a]$ up to an $O_P(n^{-1})$ remainder term.

13.3 Let W be the standard Wiener process on $[0, \infty)$ and $\mathbb{B}$ Brownian bridge on $[0, 1]$. Show that the processes

$$x \mapsto \frac{\mathbb{B}(F_0(x))}{1 - F_0(x)},\ x \in [0, a] \text{ and } x \mapsto W\left(\frac{F_0(x)}{1 - F_0(x)}\right),\ x \in [0, a]$$

have the same distributions. Use that both processes are Gaussian and compute their mean and covariance functions. For the covariance functions, see Exercise 3.9.

13.4 Show that the number of $I_{n,k}$-intervals m_n is indeed of the order $n^{1/3}/\log n$ if the lengths of the $I_{n,k}$- and $J_{n,k}$ intervals is of order $n^{-1/3}\log n$ and $2n^{-1/3}\sqrt{\log n}$ respectively, as described in Section 13.2.

13.5 Let (a_n) be a sequence in $(0, a)$ and let the distibution function F_0 on $[0, \infty)$ be twice continuously differentiable on $[0, a]$. As usual, denote by f_0, h_0 and H_0 the density, hazard rate and cumulative hazard rate associated with F_0. Show that for $n \to \infty$

$$n^{1/3}\left\{F_0(a_n + n^{-1/3}x) - F_0(a_n)\right\} = f_0(a_n)x(1 + o(1)),$$

$$n^{2/3}\left\{H_0(a_n + n^{-1/3}x) - H_0(a_n) - n^{-1/3}xh_0(a_n)\right\} = \frac{1}{2}h_0'(a_n)x^2(1 + o(1))$$

$$\text{and } n^{1/3}\left\{\frac{F_0(a_n + n^{-1/3}x)}{1 - F_0(a_n + n^{-1/3}x)} - \frac{F_0(a_n)}{1 - F_0(a_n)}\right\} = \frac{f_0(a_n)x}{(1 - F_0(a_n))^2}(1 + o(1)).$$

13.6 Brownian scaling says that for any $a > 0$, the Wiener process $t \mapsto W(t)$ on $[0, \infty)$ has the same distribution as the process $t \mapsto a^{-1/2}W(at)$; see also (3.41). In the context of Section 13.2, write

$$\alpha = \frac{1}{2}h_0'(x_0) \text{ and } \beta = \frac{f_0(x_0)}{(1 - F_0(x_0))^2}.$$

a) Use Brownian scaling to show that the processes

$$x \mapsto \alpha x^2 + W(\beta x) \text{ and } x \mapsto k_2(k_1x)^2 + k_2W(x)$$

have the same distributions whenever $k_1 = \alpha^{2/3}\beta^{-1/3}$ and $k_2 = \alpha^{-1/3}\beta^{2/3}$.

b) Verify (13.25) by substituting α and β in the equations derived in (a).

13.7 Verify (13.26) and (13.27), using (13.25).

Bibliographic Remarks

In Groeneboom, 1985, the asymptotic distribution of the L_1-loss was stated and a sketch of the proof given. The result was proved rigorously in Groeneboom et al., 1999. Results on other global deviation measures for the Grenander estimator, such as the supremum distance, can be found in Durot et al., 2012. The asymptotic distribution of the empirical L_1 test statistic for testing the hypothesis of monotonicity of a hazard rate is studied in Groeneboom and Jongbloed, 2013c. The "big blocks separated by small blocks" approach to deriving the distribution of a sum of dependent random variables was introduced in Rosenblatt, 1956a. Asymptotic results for two sample tests in the current status model are derived in Groeneboom, 2012b.

References

Albers, M.G. 2012. *Boundary Estimation of Densities with Bounded Support.* Master's thesis. ETH Zürich.

Andersen, P.K., and Rønn, B.B. 1995. A nonparametric test for comparing two samples where all observations are either left- or right-censored. *Biometrics*, **51**, 323–329.

Andersen, P.K., Borgan, O., Gill, R.D., and Keiding, N. 1993. *Statistical models based on counting processes.* New York: Springer.

Anevski, D. 2003. Estimating the derivative of a convex density. *Statistica neerlandica*, **57**(2), 245–257.

Anevski, D. 2007. Interarrival times in a counting process and bird watching. *Statistica Neerlandica*, **61**, 198–208.

Ayer, M., Brunk, H.D., Ewing, G.M., Reid, W.T., and Silverman, E. 1955. An empirical distribution function for sampling with incomplete information. *Ann. Math. Statist.*, **26**, 641–647.

Balabdaoui, F., and Wellner, J.A. 2007. Estimation of a k-monotone density: limit distribution theory and the spline connection. *Ann. Statist.*, **35**, 2536–2564.

Balabdaoui, F., Rufibach, K., and Wellner, J.A. 2009. Limit distribution theory for maximum likelihood estimation of a log-concave density. *Ann. Statist.*, **37**, 1299–1331.

Ball, K., and Pajor, A. 1990. The entropy of convex bodies with "few" extreme points. Pages 25–32 of: *Geometry of Banach spaces (Strobl, 1989).* London Math. Soc. Lecture Note Ser., vol. 158. Cambridge: Cambridge Univ. Press.

Banerjee, M., and Wellner, J.A. 2001. Likelihood ratio tests for monotone functions. *Ann. Statist.*, **29**, 1699–1731.

Banerjee, M., and Wellner, J.A. 2005. Confidence intervals for current status data. *Scand. J. Statist.*, **32**, 405–424.

Barlow, R.E., Bartholomew, D.J., Bremner, J.M., and Brunk, H.D. 1972. *Statistical inference under order restrictions. The theory and application of isotonic regression.* John Wiley & Sons, London–New York–Sydney. Wiley Series in Probability and Mathematical Statistics.

Bazaraa, M.S., Sherali, H.D., and Shetty, C.M. 2006. *Nonlinear programming.* Third ed. Hoboken, NJ: Wiley-Interscience [John Wiley & Sons]. Theory and algorithms.

Bennett, C., and Sharpley, R.C. 1988. *Interpolation of operators.* Vol. 129. Access Online via Elsevier.

Betensky, R.A., and Finkelstein, D.M. 1999. A non-parametric maximum likelihood estimator for bivariate interval censored data. *Statist. Med.*, **18**, 3089–3100.

Bickel, P.J., Klaassen, C.A.J., Ritov, Y., and Wellner, J.A. 1998. *Efficient and adaptive estimation for semiparametric models.* New York: Springer-Verlag. Reprint of the 1993 original.

Billingsley, P. 1995. *Probability and measure.* Third ed. Wiley Series in Probability and Mathematical Statistics. New York: John Wiley & Sons. A Wiley-Interscience Publication.

Birgé, L. 1999. Interval censoring: a nonasymptotic point of view. *Math. Methods Statist.*, **8**, 285–298.

Birke, M., and Dette, H. 2007. Testing strict monotonicity in nonparametric regression. *Math. Methods Statist.*, **16**, 110–123.

Birman, M.Š., and Solomjak, M.Z. 1967. Piecewise polynomial approximations of functions of classes W_p^{α}. *Mat. Sb. (N.S.)*, **73 (115)**, 331–355.

Bogaerts, K., and Lesaffre, E. 2004. A new, fast algorithm to find the regions of possible support for bivariate interval-censored data. *J. Comput. Graph. Statist.*, **13**, 330–340.

Böhning, D. 1982. Convergence of Simar's algorithm for finding the maximum likelihood estimate of a compound Poisson process. *Ann. Statist.*, **10**, 1006–1008.

Böhning, D. 1986. A vertex-exchange-method in D-optimal design theory. *Metrika*, **33**, 337–347.

Carolan, C.A., and Dykstra, R.L. 1999. Asymptotic behavior of the grenander estimator at density flat regions. *Canad. J. Statist.*, **27**, pp. 557–566.

Chernoff, H. 1964. Estimation of the mode. *Ann. Inst. Statist. Math.*, **16**, 31–41.

Cule, M., Gramacy, R., and Samworth, R.J. 2009. LogConcDEAD: an R package for maximum likelihood estimation of a multivariate log-concave density. *Journal of Statistical Software*, **29**(2).

Cule, M., and Samworth, R.J. 2010. Theoretical properties of the log-concave maximum likelihood estimator of a multidimensional density. *Electron. J. Stat.*, **4**, 254–270.

Cule, M., Samworth, R.J., and Stewart, M.I. 2010. Maximum likelihood estimation of a multi-dimensional log-concave density. *J. Roy. Statist. Soc. Ser. B (Statistical Methodology)*, **72**(5), 545–607.

Dabrowska, D.M. 1988. Kaplan-Meier estimate on the plane. *Ann. Statist.*, **16**, 1475–1489.

Daniels, H.E., and Skyrme, T.H.R. 1985. The maximum of a random walk whose mean path has a maximum. *Adv. in Appl. Probab.*, **17**, 85–99.

Dempster, A.P., Laird, N.M., and Rubin, D.B. 1977. Maximum likelihood from incomplete data via the EM algorithm. *J. Roy. Statist. Soc. Ser. B*, **39**, 1–38. With discussion.

Dette, H., Neumeyer, N., and Pilz, K.F. 2006. A simple nonparametric estimator of a strictly monotone regression function. *Bernoulli*, **12**, 469–490.

Dietz, K., and Schenzle, D. 1985. Proportionate mixing models for age-dependent infection transmission. *J. Math. Biol.*, **22**, 117–120.

Donoho, D.L., and Liu, R.C. 1991. Geometrizing rates of convergence. II, III. *Ann. Statist.*, **19**, 633–667, 668–701.

Dümbgen, L., and Rufibach, K. 2009. Maximum likelihood estimation of a log-concave density and its distribution function: basic properties and uniform consistency. *Bernoulli*, **15**, 40–68.

Dümbgen, L., and Rufibach, K. 2011. logcondens: computations related to univariate log-concave density estimation. *Journal of Statistical Software*, **39**, 1–28.

Durot, C. 2007. On the $\mathbb{L}_p$-error of monotonicity constrained estimators. *Ann. Statist.*, **35**, 1080–1104.

Durot, C. 2008. Testing convexity or concavity of a cumulated hazard rate. *IEEE Trans. on Rel.*, **57**, 465–473.

Durot, C., Kulikov, V.N., and Lopuhaä, H.P. 2012. The limit distribution of the L_∞-error of Grenander-type estimators. *Ann. Statist.*, **40**(3), 1578–1608.

Durot, C., Groeneboom, P., and Lopuhaä, H.P. 2013. Testing equality of functions under monotonicity constraints. *J. Nonparametr. Stat.*, **25**, 939–970.

Eggermont, P.P.B., and LaRiccia, V.N. 2000. Maximum likelihood estimation of smooth monotone and unimodal densities. *Annals of statistics*, 922–947.

Eggermont, P.P.B., and LaRiccia, V.N. 2001a. *Maximum penalized likelihood estimation: density estimation*. Vol. 1. New York: Springer.

Eggermont, P.P.B., and LaRiccia, V.N. 2001b. *Maximum penalized likelihood estimation: regression*. Vol. 2. New York: Springer.

Fan, J. 1993. Local linear regression smoothers and their minimax efficiencies. *Ann. Statist.*, **21**, 196–216.

Fedorov, V.V. 1971. Experimental design under linear optimality criteria. *Teor. Verojatnost. i Primenen.*, **16**, 189–195.

Feuerverger, A., and Hall, P. 2000. Methods for density estimation in thick-slice versions of Wicksell's problem. *J. Amer. Statist. Assoc.*, **95**(450), 535–546.

Feuerverger, A., Kim, P.T., and Sun, J. 2008. On optimal uniform deconvolution. *J. Stat. Theory Pract.*, **2**, 433–451.

Fleming, T.R., and Harrington, D.P. 2011. *Counting processes and survival analysis*. Wiley.

Fougères, A.-L. 1997. Estimation de densités unimodales. *Canadian Journal of Statistics*, **25**(3), 375–387.

Gentleman, R., and Vandal, A.C. 2002. Nonparametric estimation of the bivariate CDF for arbitrarily censored data. *Canad. J. Statist.*, **30**, 557–571.

Geskus, R.B. 1997. *Estimation of Smooth Functionals with Interval Censored Data*. Ph.D. thesis. Delft University of Technology.

Geskus, R.B., and Groeneboom, P. 1996. Asymptotically optimal estimation of smooth functionals for interval censoring. I. *Statist. Neerlandica*, **50**, 69–88.

Geskus, R.B., and Groeneboom, P. 1997. Asymptotically optimal estimation of smooth functionals for interval censoring. II. *Statist. Neerlandica*, **51**, 201–219.

Geskus, R.B., and Groeneboom, P. 1999. Asymptotically optimal estimation of smooth functionals for interval censoring, case 2. *Ann. Statist.*, **27**, 627–674.

Gijbels, I., and Heckman, N.E. 2004. Nonparametric testing for a monotone hazard function via normalized spacings. *J. Nonparametr. Stat.*, **16**, 463–477.

Gill, R.D., and Levit, B.Y. 1995. Applications of the Van Trees inequality: a Bayesian Cramér-Rao bound. *Bernoulli*, **1**, 59–79.

Giné, E., and Guillou, A. 2002. Rates of strong uniform consistency for multivariate kernel density estimators. *Ann. Inst. H. Poincaré Probab. Statist.*, **38**, 907–921. En l'honneur de J. Bretagnolle, D. Dacunha-Castelle, I. Ibragimov.

Good, I.J., and Gaskins, R.A. 1971. Nonparametric roughness penalties for probability densities. *Biometrika*, **58**, 255–277.

Grenander, U. 1956. On the theory of mortality measurement. II. *Skand. Aktuarietidskr.*, **39**, 125–153 (1957).

Groeneboom, P. 1983. The concave majorant of Brownian motion. *Ann. Probab.*, **11**, 1016–1027.

Groeneboom, P. 1985. Estimating a monotone density. Pages 539–555 of: *Proceedings of the Berkeley conference in honor of Jerzy Neyman and Jack Kiefer, Vol. II (Berkeley, Calif., 1983)*. Wadsworth Statist./Probab. Ser. Belmont, CA: Wadsworth.

Groeneboom, P. 1989. Brownian motion with a parabolic drift and Airy functions. *Probab. Theory Related Fields*, **81**, 79–109.

Groeneboom, P. 1996. Lectures on inverse problems. Pages 67–164 of: *Lectures on probability theory and statistics (Saint-Flour, 1994)*. Lecture Notes in Math., vol. 1648. Berlin: Springer.

Groeneboom, P. 2010. The maximum of Brownian motion minus a parabola. *Electron. J. Probab.*, **15**, no. 62, 1930–1937.

Groeneboom, P. 2011. Vertices of the least concave majorant of Brownian motion with parabolic drift. *Electron. J. Probab.*, **16**, no. 84, 2234–2258.

Groeneboom, P. 2012a. Convex hulls of uniform samples from a convex polygon. *Adv. in Appl. Probab.*, **44**, 330–342.

Groeneboom, P. 2012b. Likelihood ratio type two-sample tests for current status data. *Scand. J. Statist.*, **39**, 645–662.

Groeneboom, P. 2013a. The bivariate current status model. *Electron. J. Stat.*, **7**, 1797–1845.

Groeneboom, P. 2013b. Nonparametric (smoothed) likelihood and integral equations. *J. Statist. Plann. Inference*, **143**, 2039–2065.

Groeneboom, P. 2014. Maximum smoothed likelihood estimators for the interval censoring model. *Ann. Statist.*, **42**, 2092–2137.

Groeneboom, P., and Jongbloed, G. 2003. Density estimation in the uniform deconvolution model. *Statist. Neerlandica*, **57**, 136–157.

Groeneboom, P., and Jongbloed, G. 2012. Isotonic L_2-projection test for local monotonicity of a hazard. *J. Statist. Plann. Inference*, **142**, 1644–1658.

Groeneboom, P., and Jongbloed, G. 2013a. Smooth and non-smooth estimates of a monotone hazard. Volume in the IMS Lecture Notes Monograph Series in honor of the 65th birthday of Jon Wellner. IMS.

Groeneboom, P., and Jongbloed, G. 2013b. Smooth and non-smooth estimates of a monotone hazard. Pages 174–196 of: *From Probability to Statistics and Back: High-Dimensional Models and Processes–A Festschrift in Honor of Jon A. Wellner*. Institute of Mathematical Statistics.

Groeneboom, P., and Jongbloed, G. 2013c. Testing monotonicity of a hazard: asymptotic distribution theory. *Bernoulli*, **19**, 1965–1999.

Groeneboom, P., and Jongbloed, G. 2014. Nonparametric confidence intervals for monotone functions. http://arxiv.org/abs/1407.3491.

Groeneboom, P., and Ketelaars, T. 2011. Estimators for the interval censoring problem. *Electron. J. Stat.*, **5**, 1797–1845.

Groeneboom, P., and Lopuhaä, H.P. 1993. Isotonic estimators of monotone densities and distribution functions: basic facts. *Statist. Neerlandica*, **47**, 175–183.

Groeneboom, P., and Pyke, R. 1983. Asymptotic normality of statistics based on the convex minorants of empirical distribution functions. *Ann. Probab.*, **11**, 328–345.

Groeneboom, P., and Temme, N.M. 2011. The tail of the maximum of Brownian motion minus a parabola. *Electron. Commun. Probab.*, **16**, 458–466.

Groeneboom, P., and Wellner, J.A. 1992. *Information bounds and nonparametric maximum likelihood estimation*. DMV Seminar, vol. 19. Basel: Birkhäuser Verlag.

Groeneboom, P., and Wellner, J.A. 2001. Computing Chernoff's distribution. *J. Comput. Graph. Statist.*, **10**, 388–400.

Groeneboom, P., Hooghiemstra, G., and Lopuhaä, H.P. 1999. Asymptotic normality of the L_1 error of the Grenander estimator. *Ann. Statist.*, **27**, 1316–1347.

Groeneboom, P., Jongbloed, G., and Wellner, J.A. 2001a. Estimation of a convex function: characterizations and asymptotic theory. *Ann. Statist.*, **29**, 1653–1698.

Groeneboom, P., Jongbloed, G., and Wellner, J.A. 2001b. A canonical process for estimation of convex functions: the "invelope" of integrated Brownian motion $+t^4$. *Ann. Statist.*, **29**, 1620–1652.

Groeneboom, P., Jongbloed, G., and Wellner, J.A. 2008. The support reduction algorithm for computing non-parametric function estimates in mixture models. *Scand. J. Statist.*, **35**, 385–399.

Groeneboom, P., Maathuis, M.H., and Wellner, J.A. 2008a. Current status data with competing risks: consistency and rates of convergence of the MLE. *Ann. Statist.*, **36**, 1031–1063.

Groeneboom, P., Maathuis, M.H., and Wellner, J.A. 2008b. Current status data with competing risks: limiting distribution of the MLE. *Ann. Statist.*, **36**, 1064–1089.

Groeneboom, P., Jongbloed, G., and Witte, B.I. 2010. Maximum smoothed likelihood estimation and smoothed maximum likelihood estimation in the current status model. *Ann. Statist.*, **38**, 352–387.

Groeneboom, P., Jongbloed, G., and Michael, S. 2012. Consistency of maximum likelihood estimators in a large class of deconvolution models. *Canad. J. Statist.*

Groeneboom, P., Jongbloed, G., and Witte, B.I. 2012. A maximum smoothed likelihood estimator in the current status continuous mark model. *J. Nonparametr. Stat.*, **24**, 85–101.

Groeneboom, P., Lalley, S.P., and Temme, N.M. 2013. *Chernoff's distribution and differential equations of parabolic and Airy type*. Submitted.

Hall, P. 1984. Central limit theorem for integrated square error of multivariate nonparametric density estimators. *J. Multivariate Anal.*, **14**, 1–16.

Hall, P. 1992. Effect of bias estimation on coverage accuracy of bootstrap confidence intervals for a probability density. *Ann. Statist.*, **20**, 675–694.

Hall, P., and Horowitz, J.L. 2013. A simple bootstrap method for constructing nonparametric confidence bands for functions. *Ann. Statist.*, **41**, 1892–1921.

Hall, P., and Smith, R.L. 1988. The kernel method for unfolding sphere size distributions. *J. Comput. Phys.*, **74**, 409–421.

Hall, P., and Van Keilegom, I. 2005. Testing for monotone increasing hazard rate. *Ann. Statist.*, **33**, 1109–1137.

Hampel, F.R. 1987. Design, modelling, and analysis of some biological datasets. Pages 111–115 of: *Design, data and analysis, by some friends of Cuthbert Daniel*. New York: Wiley.

Hansen, B.E. 1991. *Nonparametric estimation of functionals for interval censored observations*. Master's thesis. Delft University of Technology.

Hanson, D.L., and Pledger, G. 1976. Consistency in Concave Regression. *Ann. Statist.*, **4**, 1038–1050.

Hoel, D.G., and Walburg, H.E. 1972. Statistical analysis of survival experiments. *Journal of the National Cancer Institute*, **49**, 361–372.

Huang, J., and Wellner, J.A. 1995. Asymptotic normality of the NPMLE of linear functionals for interval censored data, case 1. *Statist. Neerlandica*, **49**, 153–163.

Huang, Y., and Louis, T.A. 1998. Nonparametric estimation of the joint distribution of survival time and mark variables. *Biometrika*, **85**, 7856–7984.

Hudgens, M.G., Satten, G.A., and Longini, Jr., I.M. 2001. Nonparametric maximum likelihood estimation for competing risks survival data subject to interval censoring and truncation. *Biometrics*, **57**, 74–80.

Hudgens, M.G., Maathuis, M.H., and Gilbert, P.B. 2007. Nonparametric estimation of the joint distribution of a survival time subject to interval censoring and a continuous mark variable. *Biometrics*, **63**, 372–380.

Ibragimov, I.A., and Linnik, Yu.V. 1971. *Independent and stationary sequences of random variables*. Wolters-Noordhoff Publishing, Groningen. With a supplementary chapter by I. A. Ibragimov and V. V. Petrov, Translation from the Russian edited by J. F. C. Kingman.

Janson, S. 2013. Moments of the location of the maximum of Brownian motion with parabolic drift. *Electron. Commun. Probab.*, **18**, no. 15, 1–8.

Janson, S., Louchard, G., and Martin-Löf, A. 2010. The maximum of Brownian motion with parabolic drift. *Electron. J. Probab.*, **15**, no. 61, 1893–1929.

Jewell, N.P. 1982. Mixtures of exponential distributions. *Ann. Statist*, **10**, 479–484.

Jewell, N.P., and Kalbfleisch, J.D. 2004. Maximum likelihood estimation of ordered multinomial parameters. *Biostatistics*, **5**, 291–306.

Jewell, N.P., van der Laan, M.J., and Henneman, T. 2003. Nonparametric estimation from current status data with competing risks. *Biometrika*, **90**, 183–197.

Johnstone, I.M., and Raimondo, M. 2004. Periodic boxcar deconvolution and Diophantine approximation. *Ann. Statist.*, **32**, 1781–1804.

Jongbloed, G. 1998a. Exponential deconvolution: two asymptotically equivalent estimators. *Statist. Neerlandica*, **52**, 6–17.

Jongbloed, G. 1998b. The iterative convex minorant algorithm for nonparametric estimation. *J. Comput. Graph. Statist.*, **7**, 310–321.

Jongbloed, G. 2001. Sieved maximum likelihood estimation in Wicksell's problem and related deconvolution problems. *Scand. J. Statist.*, **28**, 161–183.

Jongbloed, G. 2009. Consistent likelihood-based estimation of a star-shaped distribution. *Metrika*, **69**, 265–282.

Jongbloed, G., and van der Meulen, F.H. 2009. Estimating a concave distribution function from data corrupted with additive noise. *Ann. Statist.*, **37**, 782–815.

Kaipio, J.P., and Somersalo, E. 2005. *Statistical and computational inverse problems*. Vol. 160. New York: Springer.

Keiding, N. 1991. Age-specific incidence and prevalence: a statistical perspective. *J. Roy. Statist. Soc. Ser. A*, **154**(3), 371–412. With discussion.

Keiding, N., Begtrup, K., Scheike, T.H., and Hasibeder, G. 1996. Estimation from Current Status Data in Continuous Time. *Lifetime Data Anal.*, **2**, 119–129.

Keiding, N., Højbjerg Hansen, O.K., Sørensen, D.N., and Slama, R. 2012. The current duration approach to estimating time to pregnancy. *Scand. J. Statist.*, **39**, 185–204.

Kim, J.K., and Pollard, D. 1990. Cube root asymptotics. *Ann. Statist.*, **18**, 191–219.

Kitayaporn, D., Vanichseni, S., Mastro, T.D., Raktham, S., Vaniyapongs, T., Des Jarlais, D.C., Wasi, C., Young, N.L., Sujarita, S., Heyward, W.L., and Esparza, J. 1998. Infection with HIV-1 subtypes B and E in injecting drug users screened for enrollment into a prospective cohort in Bangkok, Thailand. *J. Acquir. Immune Defic. Syndr. Hum. Retrovirol.*, **19**, 289–295.

Klein, J.P., and Moeschberger, M.L. 2003. *Survival Analysis: Techniques for Censored and Truncated Data*. Statistics for Biology and Health. New York: Springer.

Komlós, J., Major, P., and Tusnády, G. 1975. An approximation of partial sums of independent RV's and the sample DF. I. *Z. Wahrscheinlichkeitstheorie und Verw. Gebiete*, **32**, 111–131.

Kosorok, M.R. 2008a. Bootstrapping the Grenander estimator. Pages 282–292 of: *Beyond parametrics in interdisciplinary research: Festschrift in honor of Professor Pranab K. Sen*. Inst. Math. Stat. Collect., vol. 1. Beachwood, OH: Inst. Math. Statist.

Kosorok, M.R. 2008b. *Introduction to empirical processes and semiparametric inference*. New York: Springer.

Kress, R. 1989. *Linear integral equations*. Applied Mathematical Sciences, vol. 82. Berlin: Springer-Verlag.

Kulikov, V.N. 2003. *Direct and Indirect Use of Maximum Likelihood*. Ph.D. thesis. Delft University of Technology.

Lesperance, M.L., and Kalbfleisch, J.D. 1992. An algorithm for computing the nonparametric MLE of a mixing distribution. *J. Amer. Statist. Assoc.*, **87**, 120–126.

Li, C., and Fine, J.P. 2013. Smoothed nonparametric estimation for current status competing risks data. *Biometrika*, **100**, 173–187.

Lindsay, B.G. 1995. Mixture models: theory, geometry and applications. Pages i–163 of: *NSF-CBMS regional conference series in probability and statistics*. JSTOR.

Loève, M. 1963. *Probability theory*. Third ed. Princeton, NJ–Toronto–London: D. Van Nostrand Co.

Maathuis, M.H. 2005. Reduction algorithm for the NPMLE for the distribution function of bivariate interval-censored data. *J. Comput. Graph. Statist.*, **14**, 352–362.

Maathuis, M.H., and Hudgens, M.G. 2011. Nonparametric inference for competing risks current status data with continuous, discrete or grouped observation times. *Biometrika*, **98**, 325–340.

Maathuis, M.H., and Wellner, J.A. 2008. Inconsistency of the MLE for the joint distribution of interval censored survival times and continuous marks. *Scand. J. Statist.*, **35**, 83–103.

Mackowiak, P.A., Wasserman, S.S., and Levine, M.M. 1992. A critical appraisal of 98.6 degrees F, the upper limit of the normal body temperature, and other legacies of Carl Reinhold August Wunderlich. *Journal of the American Medical Association*, **268**, 1578–1580.

Mammen, E. 1991. Nonparametric regression under qualitative smoothness assumptions. *Ann. Statist.*, **19**, 741–759.

Marshall, A.W. 1969. Discussion on Barlow and van Zwets paper. Pages 174–176 of: *Nonparametric Techniques in Statistical Inference. Proceedings of the First International Symposium on Nonparametric Techniques held at Indiana University, June*.

Marshall, A.W., and Proschan, F. 1965. Maximum likelihood estimation for distributions with monotone failure rate. *Ann. Math. Statist*, **36**, 69–77.

McGarrity, K.S., Sietsma, J., and Jongbloed, G. 2014. *Nonparametric inference in a stereological model with oriented cylinders applied to dual phase steel.* Submitted for publication.

McLachlan, G.J., and Krishnan, T. 2007. *The EM algorithm and extensions*. Vol. 382. Hoboken, NJ: John Wiley & Sons.

Meister, A. 2009. *Deconvolution problems in nonparametric statistics*. Lecture Notes in Statistics, vol. 193. Berlin: Springer-Verlag.

Meyer, M.C. 2008. Inference using shape-restricted regression splines. *Ann. Appl. Stat.*, **2**, 1013–1033.

Nane, G.F. 2013. *Shape Constrained Nonparametric Estimation in the Cox Model*. Ph.D. thesis. Delft University of Technology.

Neuhaus, G. 1993. Conditional rank tests for the two-sample problem under random censorship. *Ann. Statist.*, **21**, 1760–1779.

Newcomb, S. 1886. A generalized theory of the combination of observations so as to obtain the best result. *American Journal of Mathematics*, 343–366.

Ohser, J., and Mücklich, F. 2000. *Statistical analysis of microstructures in materials science*. New York: John Wiley.

Pal, J.K. 2008. Spiking problem in monotone regression: Penalized residual sum of squares. *Statist. Probab. Lett.*, **78**, 1548–1556.

Patil, G.P., and Rao, C.R. 1978. Weighted distributions and size-biased sampling with applications to wildlife populations and human families. *Biometrics*, **34**, 179–189.

Peto, R., and Peto, J. 1972. Asymptotically efficient rank invariant test procedures. *J.R. Statist. Soc. Series A*, **135**, 184–207.

Pimentel, L.P.R. 2014. *On the location of the maximum of a continuous stochastic process*. *J. Appl. Prob.*, **51**, 152–161.

Pitman, J.W. 1983. Remarks on the convex minorant of Brownian motion. Pages 219–227 of: *Seminar on stochastic processes, 1982 (Evanston, Ill., 1982)*. Progr. Probab. Statist., vol. 5. Boston, MA: Birkhäuser Boston.

Politis, D.N., Romano, J.P., and Wolf, M. 1999. *Subsampling*. Springer Series in Statistics. New York: Springer-Verlag.

Pollard, D. 1984. *Convergence of stochastic processes*. Springer Series in Statistics. New York: Springer-Verlag.

Prakasa Rao, B.L.S. 1969. Estimation of a unimodal density. *Sankhyā Ser. A*, **31**, 23–36.

Preusser, F., Degering, D., Fuchs, M., Hilgers, A., Kadereit, A., Klasen, N., Krbetschek, M., Richter, D., and Spencer, J.Q.G. 2008. Luminescence dating: basics, methods and applications. *Quaternary Science Journal*, **57**, 95–149.

Proschan, F., and Pyke, R. 1967. Tests for monotone failure rate. Pages 293–312 of: *Proc. Fifth Berkeley Sympos. Mathematical Statistics and Probability (Berkeley, Calif., 1965/66), Vol. III: Physical Sciences*. Berkeley, CA: University of California Press.

R Development Core Team. 2011. *R: A Language and Environment for Statistical Computing*. R Foundation for Statistical Computing, Vienna, Austria.

Ramsay, J.O. 1998. Estimating smooth monotone functions. *J. R. Statist. Soc.: Series B (Statistical Methodology)*, **60**, 365–375.

Rebolledo, R. 1980. Central limit theorems for local martingales. *Z. Wahrsch. Verw. Gebiete*, **51**, 269–286.

Robertson, T., Wright, F.T., and Dykstra, R.L. 1988. *Order restricted statistical inference*. Wiley Series in Probability and Mathematical Statistics: Probability and Mathematical Statistics. Chichester: John Wiley & Sons.

Rosenblatt, M. 1956a. A central limit theorem and a strong mixing condition. *Proc. Nat. Acad. Sci. U.S.A.*, **42**, 43–47.

Rosenblatt, M. 1956b. Remarks on some nonparametric estimates of a density function. *Ann. Math. Statist.*, **27**, 832–837.

Ross, S.M. 2010. *Introduction to probability models*. Tenth ed. Burlington, MA: Harcourt/Academic Press.

Ruppert, D., Wand, M.P., and Carroll, R.J. 2003. *Semiparametric regression*. Cambridge Series in Statistical and Probabilistic Mathematics, vol. 12. Cambridge: Cambridge University Press.

Schoemaker, A.L. 1996. What's normal? Temperature, gender, and heart rate. *Journal of Statistics Education*, **4**.

Schuhmacher, D., Hüsler, A., and Dümbgen, L. 2011. Multivariate log-concave distributions as a nearly parametric model. *Statistics & Risk Modeling with Applications in Finance and Insurance*, **28**(3), 277–295.

Schuster, E.F. 1985. Incorporating support constraints into nonparametric estimators of densities. *Comm. Statist. A—Theory Methods*, **14**, 1123–1136.

Sen, B., Banerjee, M., and Woodroofe, M.B. 2010. Inconsistency of bootstrap: the Grenander estimator. *Ann. Statist.*, **38**, 1953–1977.

Silvapulle, M.J., and Sen, P.K. 2005. *Constrained statistical inference: inequality, order and shape constraints*. Hoboken, NJ: Wiley.

Silverman, B.W. 1978. Weak and strong uniform consistency of the kernel estimate of a density and its derivatives. *Ann. Statist.*, **6**, 177–184.

Silverman, B.W. 1986. *Density estimation for statistics and data analysis*. Vol. 26. Boca Raton, FL: CRC press.

Simar, L. 1976. Maximum likelihood estimation of a compound Poisson process. *Ann. Statist.*, **4**, 1200–1209.

Slama, R., Højbjerg Hansen, O.K., Ducot, B., Bohet, A., Sørensen, D., Allemand, L., Eijkemans, M.J., Rosetta, L., Thalabard, J.C., Keiding, N., et al. 2012. Estimation of the frequency of involuntary infertility on a nation-wide basis. *Human reproduction*, **27**, 1489–1498.

Song, S. 2001. *Estimation with Bivariate Interval Censored data*. Ph.D. diss. University of Washington.

Sparre Andersen, E. 1954. On the fluctuations of sums of random variables. II. *Math. Scand.*, **2**, 195–223.

Steinsaltz, D., and Orzack, S.H. 2011. Statistical methods for paleodemography on fossil assemblages having small numbers of specimens: an investigation of dinosaur survival rates. *Paleobiology*, **37**, 113–125.

Sun, J. 2006. *The statistical analysis of interval-censored failure time data*. Statistics for Biology and Health. New York: Springer.

Talagrand, M. 1994. Sharper bounds for Gaussian and empirical processes. *Ann. Probab.*, **22**, 28–76.

Talagrand, M. 1996. New concentration inequalities in product spaces. *Invent. Math.*, **126**, 505–563.

Temme, N.M. 1985. A convolution integral equation solved by Laplace transformations. Pages 609–613 of: *Proceedings of the international conference on computational and applied mathematics (Leuven, 1984)*, vol. 12/13.

Tsai, W.-Y., Leurgans, S.E., and Crowley, J.J. 1986. Nonparametric estimation of a bivariate survival function in the presence of censoring. *Ann. Statist.*, **14**, 1351–1365.

Tsybakov, A.B., and Zaiats, V. 2009. *Introduction to nonparametric estimation*. Vol. 11. New York: Springer.

Van de Geer, S.A. 1996. Rates of convergence for the maximum likelihood estimator in mixture models. *J. Nonparametr. Statist.*, **6**, 293–310.

Van de Geer, S.A. 2000. *Applications of empirical process theory*. Cambridge Series in Statistical and Probabilistic Mathematics, vol. 6. Cambridge: Cambridge University Press.

Van der Laan, M.J. 1996. Efficient estimation in the bivariate censoring model and repairing NPMLE. *Ann. Statist.*, **24**, 596–627.

Van der Vaart, A.W. 1991. On differentiable functionals. *Ann. Statist.*, **19**, 178–204.

Van der Vaart, A.W., and Van der Laan, M.J. 2003. Smooth estimation of a monotone density. *Statistics: A Journal of Theoretical and Applied Statistics*, **37**, 189–203.

Van der Vaart, A.W., and Wellner, J.A. 1996. *Weak convergence and empirical processes*. Springer Series in Statistics. New York: Springer-Verlag.

Van Eeden, C. 1956. Maximum likelihood estimation of ordered probabilities. *Nederl. Akad. Wetensch. Proc. Ser. A.* **59** =*Indag. Math.*, **18**, 444–455.

Van Es, A.J., and Hoogendoorn, A.W. 1990. Kernel estimation in Wicksell's corpuscle problem. *Biometrika*, **77**, 139–145.

Van Es, A.J., and van Zuijlen, M.C.A. 1996. Convex minorant estimators of distributions in non-parametric deconvolution problems. *Scand. J. Statist.*, **23**, 85–104.

Van Es, A.J., Jongbloed, G., and van Zuijlen, M.C.A. 1998. Isotonic inverse estimators for nonparametric deconvolution. *Ann. Statist.*, **26**, 2395–2406.

Van Trees, H.L. 1968. *Detection, estimation, and modulation theory*. New York: Wiley.

Vanichseni, S., Kitayaporn, D., Mastro, T.D., Mock, P.A., Raktham, S., D.C., Des Jarlais, Sujarita, S., Srisuwanvilai, L.O., Young, N.L., Wasi, C., Subbarao, S., Heyward, W.L., Esparza, J., and Choopanya, K. 2001. Continued high HIV-1 incidence in a vaccine trial preparatory cohort of injection drug users in Bangkok, Thailand. *AIDS*, **15**, 397–405.

Vardi, Y. 1982. Nonparametric estimation in the presence of length bias. *Ann. Statist.*, **10**, 616–620.

Walther, G. 2001. Multiscale maximum likelihood analysis of a semiparametric model, with applications. *Ann. Statist.*, **29**, 1297–1319.

Wand, M.P., and Jones, M.C. 1995. *Kernel smoothing*. Vol. 60. Boca Raton, FL: Crc Press.

Watson, G.S. 1971. Estimating functionals of particle size distributions. *Biometrika*, **58**, 483–490.

Wellner, J.A. 1995. Interval censoring, case 2: alternative hypotheses. Pages 271–291 of: *Analysis of censored data (Pune, 1994/1995)*. IMS Lecture Notes Monogr. Ser., vol. 27. Hayward, CA: Inst. Math. Statist.

Wellner, J.A., and Zhan, Y. 1997. A hybrid algorithm for computation of the nonparametric maximum likelihood estimator from censored data. *J. Amer. Statist. Assoc.*, **92**, 945–959.

Wicksell, S.D. 1925. The corpuscle problem. *Biometrika*, **17**, 84–99.

Wicksell, S.D. 1926. The corpuscle problem: second memoir: case of ellipsoidal corpuscles. *Biometrika*, **18**, 151–172.

Woodroofe, M.B., and Sun, J. 1993. A penalized maximum likelihood estimate of $f(0+)$ when f is nonincreasing. *Statist. Sinica*, **3**, 501–515.

Wright, S.J. 1997. *Primal-dual interior-point methods*. Vol. 54. Philadelphia, PA: Siam.

Wu, C.-F.J. 1983. On the convergence properties of the EM algorithm. *Ann. Statist.*, **11**, 95–103.

Wynn, H.P. 1970. The sequential generation of D-optimum experimental designs. *Ann. Math. Statist.*, **41**, 1655–1664.

Yu, B. 1997. Assouad, Fano, and Le Cam. Pages 423–435 of: *Festschrift for Lucien Le Cam*. Springer.

Zeidler, E. 1985. *Nonlinear functional analysis and its applications. III*. Variational methods and optimization. Translated from the German by Leo F. Boron. New York: Springer-Verlag.

Zhang, S., Karunamuni, R.J., and Jones, M.C. 1999. An improved estimator of the density function at the boundary. *J. Amer. Statist. Assoc.*, **94**, 1231–1240.

Zhang, Y. 2006. Nonparametric k-sample tests with panel count data. *Biometrika*, **93**, 777–790.

Zhang, Y., Liu, W., and Zhan, Y. 2001. A nonparametric two-sample test of the failure function with interval censoring case 2. *Biometrika*, **88**, 677–686.

Author Index

Subject Index

Printed in the United States
by Baker & Taylor Publisher Services